Comprehensive Virology 10

Comprehensive Virology

Edited by Heinz Fraenkel-Conrat
University of California at Berkeley

and Robert R. Wagner
University of Virginia

Comprehensive

Edited by

Heinz Fraenkel-Conrat
Department of Molecular Biology and Virus Laboratory
University of California, Berkeley, California

and

Robert R. Wagner
Department of Microbiology
University of Virginia, Charlottesville, Virginia

Virology

10

Regulation and Genetics

Viral Gene Expression and Integration

PLENUM PRESS · NEW YORK AND LONDON

Library of Congress Cataloging in Publication Data

Fraenkel-Conrat, Heinz, 1910-
  Regulation and genetics.

  (*Their* Comprehensive virology; v. 10)
  Includes bibliographies and index.
  1. Viral genetics. 2. Gene expression. 3. Genetic regulation. 4. Host-virus relation-
ships. I. Wagner, Robert R., 1923-   joint author. II. Title. III. Series.
QR357.F72 vol. 10 [QH434] 576'.64'08s
ISBN-13: 978-1-4684-0834-8   e-ISBN-13: 978-1-4684-0832-4
DOI: 10.1007/978-1-4684-0832-4   [576'.64]   77-7199

© 1977 Plenum Press, New York
Softcover reprint of the hardcover 1st edition 1977
A Division of Plenum Publishing Corporation
227 West 17th Street, New York, N.Y. 10011

Foreword

The time seems ripe for a critical compendium of that segment of the biological universe we call viruses. Virology, as a science, having passed only recently through its descriptive phase of naming and numbering, has probably reached that stage at which relatively few new—truly new—viruses will be discovered. Triggered by the intellectual probes and techniques of molecular biology, genetics, biochemical cytology, and high-resolution microscopy and spectroscopy, the field has experienced a genuine information explosion.

Few serious attempts have been made to chronicle these events. This comprehensive series, which will comprise some 6000 pages in a total of about 22 volumes, represents a commitment by a large group of active investigators to analyze, digest, and expostulate on the great mass of data relating to viruses, much of which is now amorphous and disjointed, and scattered throughout a wide literature. In this way, we hope to place the entire field in perspective, and to develop an invaluable reference and sourcebook for researchers and students at all levels.

This series is designed as a continuum that can be entered anywhere, but which also provides a logical progression of developing facts and integrated concepts.

Volume 1 contains an alphabetical catalogue of almost all viruses of vertebrates, insects, plants, and protists, describing them in general terms. Volumes 2–4 deal primarily, but not exclusively, with the processes of infection and reproduction of the major groups of viruses in their hosts. Volume 2 deals with the simple RNA viruses of bacteria, plants, and animals; the togaviruses (formerly called arboviruses), which share with these only the feature that the virion's RNA is able to act as messenger RNA in the host cell; and the reoviruses of animals and plants, which all share several structurally singular features, the most important being the double-strandedness of their multiple RNA molecules.

Volume 3 addresses itself to the reproduction of all DNA-containing viruses of vertebrates, encompassing the smallest and the largest viruses known. The reproduction of the larger and more complex RNA viruses is the subject matter of Volume 4. These viruses share the property of being enclosed in lipoprotein membranes, as do the togaviruses included in Volume 2. They share as a group, along with the reoviruses, the presence of polymerase enzymes in their virions to satisfy the need for their RNA to become transcribed before it can serve messenger functions.

Volumes 5 and 6 represent the first in a series that focuses primarily on the structure and assembly of virus particles. Volume 5 is devoted to general structural principles involving the relationship and specificity of interaction of viral capsid proteins and their nucleic acids, or host nucleic acids. It deals primarily with helical and the simpler isometric viruses, as well as with the relationship of nucleic acid to protein shell in the T-even phages. Volume 6 is concerned with the structure of the picornaviruses, and with the reconstitution of plant and bacterial RNA viruses.

Volumes 7 and 8 deal with the DNA bacteriophages. Volume 7 concludes the series of volumes on the reproduction of viruses (Volumes 2–4 and Volume 7) and deals particularly with the single- and double-stranded virulent bacteriophages.

Volumes 8, the first of the series on regulation and genetics of viruses, covers the biological properties of the lysogenic and defective phages, the phage-satellite system P 2–P 4, and in-depth discussion of the regulatory principles governing the development of selected lytic phages.

Volume 9 provides a truly comprehensive analysis of the genetics of all animal viruses that have been extensively studied to date. Described in ten detailed chapters are genotypes and phenotypic expression of conditional, host range, and deletion mutants of three major classes of animal DNA viruses followed by seven genera of RNA viruses. Principles and methodology are presented and compared to provide insight into mechanisms of mutagenesis, selection of mutants, complementation analysis, and gene mapping with restriction endonucleases and other methods. Whenever appropriate, the genetic properties of viruses are related to nucleic acid structure and function as well as recombination, integration of viral with host genome, malignant transformation, and alteration of host cell functions.

The present volume deals with transcriptional and translational regulation of viral gene expression, defective virions, and integration of tumor virus genomes into host cell chromosomes. Later volumes will be concerned with regulation of plant virus development, covirus systems, satellitism, and viroids. Two or three additional volumes will be devoted largely to structural aspects and the assembly of bacteriophages and animal viruses, as well as to special groups of newer viruses.

The complete series will endeavor to encompass all aspects of the molecular biology and the behavior of viruses. We hope to keep this series up to date at all times by prompt and rapid publication of all contributions, and by encouraging the authors to update their chapters by additions or corrections whenever a volume is reprinted.

Contents

Chapter 2

Defective Interfering Animal Viruses

Alice S. Huang and David Baltimore

Chapter 3

Virion Polymerases

David H. L. Bishop

Chapter 4

Animal Virus–Host Genome Interactions

Walter Doerfler

Chapter 5

Cell Transformation by RNA Tumor Viruses

Hidesaburo Hanafusa

CHAPTER 1

Translation of Animal Virus mRNAs *in Vitro*

A. J. Shatkin, A. K. Banerjee, and G. W. Both

Roche Institute of Molecular Biology
Nutley, New Jersey 07110

1. INTRODUCTION

Studies of the translation of bacteriophage RNAs in cell-free systems have contributed much to our understanding of many important aspects of protein synthesis, e.g., the elucidation of RNA virus gene order and the mechanism of suppression of nonsense mutations. Cell-free systems have also been useful for studying the regulation of cellular polypeptide formation, notably the role of cAMP and its binding protein in the expression of the *gal* operon. Recently, eukaryotic cellular and viral mRNAs that contain 3′-terminal poly(A) have been purified by selective binding to oligo(dT)-cellulose or poly(U)-sepharose. In addition, many animal virus mRNAs can be prepared in large quantities *in vitro* by taking advantage of the respective virion-associated transcriptases, and several heterologous cell-free systems synthesize authentic viral and cellular proteins in response to these purified mRNAs. The potential of *in vitro* systems for studying eukaryotic gene expression is likely to continue to attract the interest and attention of increasing numbers of investigators. It therefore seems appropriate to consider some of the basic characteristics of the cell-free systems that are available as of June 1975, when this chapter was written, to summarize the results of current studies, and to discuss how future work on the *in vitro* translation of animal virus mRNAs may increase our knowledge of the biochemistry of animal virus multiplication and eukaryotic cell growth.

2. PREPARATION AND COMPARATIVE PROPERTIES OF *IN VITRO* PROTEIN-SYNTHESIZING SYSTEMS

The basic mechanisms of prokaryotic protein synthesis are now reasonably well understood (Lucas-Lenard and Lipmann, 1971), and in many ways protein synthesis in eukaryotes is similar (Haselkorn and Rothman-Denes, 1973). With a general understanding of this process and with hindsight gained from work with prokaryotes, it has been possible to prepare *in vitro* protein-synthesizing systems from a variety of eukaryotic cells and organisms. As discussed in this chapter, many viral mRNAs have recently been isolated and can be faithfully translated in these systems. As such, *in vitro* translation is a particularly useful method for identifying an unknown mRNA on the basis of the protein for which it codes. Moreover, the primary translation product of a purified viral mRNA can often be identified in a cell-free system whereas it may be lost by rapid proteolytic cleavage *in vivo*. This information is helpful in elucidating the events which take place in virus-infected cells.

In this section, the general properties of cell-free protein-synthesizing systems derived from frog eggs and oocytes, tissue culture and ascites cells, reticulocytes, and wheat germ will be compared with emphasis on the basic methods for preparing active extracts, their efficiencies of translation, and the relative advantages and disadvantages of each system. In this chapter, an *in vitro* protein-synthesizing system (*in vitro* or cell-free system for short) is defined as one in which amino acid incorporation into authentic polypeptides is directed by exogenous messenger RNA. Although eggs and oocytes are not, strictly speaking, *in vitro* systems and certainly not cell free, they have been used for translating viral mRNAs and are therefore included in this discussion.

2.1. Frog Eggs and Oocytes

The preparation of frog oocytes for studies on protein synthesis was first described with the South African clawed toad, *Xenopus laevis,* by Gurdon (1968), but oocytes have also been obtained from *Pleurodeles waltlii* (Brachet *et al.,* 1973) and the Queensland cane toad,

Bufo marinus (May and Glenn, 1974). Oocytes from the former are larger than those of *Xenopus* and are better able to survive microinjection (Brachet *et al.,* 1973). The *Bufo* species is widely distributed and, unlike *Xenopus,* is found on all continents of the world (May and Glenn, 1974). Active oocytes from ovarian tissue and unfertilized eggs (Gurdon, 1967) are obtained from sacrificed female frogs which had been induced to ovulate by hormonal injection between 2 and 4 weeks previously (Gurdon *et al.,* 1971). A radioactive amino acid or mRNA is introduced by injection with a micropipette into actively growing oocytes (May and Glenn, 1974; Gurdon *et al.,* 1971). The oocytes are then incubated in culture medium at 19°C, where they remain synthetically active for up to 3 days (Gurdon, 1968). Radioactive amino acids can also be introduced into the cells by addition to the culture medium. The choice of labeling procedure is determined mainly by (1) the rate at which the label in the culture medium penetrates the cells, (2) the rate at which injected label leaks out, and (3) the duration of the labeling period. Radioactive amino acid leaks out of injected oocytes more quickly than from injected eggs. However, the label (at least for [^{3}H]histidine) penetrates oocytes much more quickly than eggs. In general, when short labeling periods (up to 1 h) or unfertilized eggs are used, the highest amount of labeled protein is synthesized when the radioactive amino acid is injected. When oocytes are used, especially for labeling periods of more than 2 h, it is best to introduce label by incubation.

2.2. Reticulocytes

Circulating reticulocytes are collected by bleeding rabbits made anemic with daily subcutaneous injections of acetylphenylhydrazine (Adamson *et al.,* 1968; Housman *et al.,* 1970; Villa-Komaroff *et al.,* 1974*b*; Gilbert and Anderson, 1970). The blood is filtered through cheesecloth into chilled saline (Villa-Komaroff *et al.,* 1974*b*), and the cells are washed, packed by centrifugation, and lysed at 0°C by the addition of an equal volume of water (Adamson *et al.,* 1968; Housman *et al.,* 1970; Villa-Komaroff *et al.,* 1974*b*) or hypotonic buffer (Gilbert and Anderson, 1970; Schreier and Staehelin, 1973). After 60 s, the lysate is centrifuged at 30,000*g* for 15 min (Villa-Komaroff *et al.,* 1974*b*; Schreier and Staehelin, 1973) and the supernatant is frozen in aliquots at −80°C, at which temperature activity remains stable for several months.

The reticulocyte protein-synthesizing system has been extensively fractionated, and many of the factors involved in protein synthesis have been purified (see Vol. 30 of *Methods in Enzymology*, Academic Press, New York). One of the simplest fractionated systems derived from reticulocyte lysates and used to translate eukaryotic viral mRNA consists of a high-speed supernatant fraction (S100) derived by centrifugation of the reticulocyte lysate, ribosomes washed with 0.5 M KCl, and the ribosomal wash fraction (Cancedda and Schlesinger, 1974; Woodward *et al.*, 1974). This system preferentially translates exogenous viral mRNA (Cancedda and Schlesinger, 1974) and is more active than most of the more highly fractionated systems (Woodward *et al.*, 1974).

The most efficient mammalian cell-free protein-synthesizing system, described by Schreier and Staehelin (1973), was originally developed for the *in vitro* translation of exogenous rabbit globin mRNA. It is prepared by a procedure which maintains the structural and functional integrity of the ribosomes. The basic system consists of purified ribosomal subunits from mouse liver, rabbit reticulocytes, or guinea pig brain, partially purified initiation factors from rabbit reticulocytes, and elongation factors, termination factors, aminoacyl tRNA synthetases, and tRNA from rat liver in the form of pH 5 enzymes (Schreier and Staehelin, 1973). The system has been adapted for the translation of adenovirus-2-specific mRNA by the preparation of ribosomal subunits and the pH 5 enzyme fraction from ascites cells (Anderson *et al.*, 1974).

2.3. Ascites and Tissue Culture Cells

2.3.1. Propagation of Ascites Cells

Ascites tumor cells can be propagated in various strains of mice by intraperitoneal injection of 0.1–0.2 ml of ascitic fluid containing $5-10 \times 10^7$ cells/ml. Bright yellow fluid, which is probably contaminated with bacteria, or very bloody fluid should not be used for passaging (Mathews and Korner, 1970; Aviv *et al.*, 1971; Villa-Komaroff *et al.*, 1974b). The cells are harvested and propagated every 7–10 days (Martin *et al.*, 1961; Mathews and Korner, 1970; Aviv *et al.*, 1971; Jacobs-Lorena and Baglioni, 1972; McDowell *et al.*, 1972; Villa-Komaroff *et al.*, 1974b; Samuel and Joklik, 1974). One mouse provides 2–10 ml of fluid containing about 10^8 cells/ml (Martin *et al.*, 1961; Villa-Komaroff *et al.*, 1974b), and stocks can be frozen for recourse, should the cells in passage become unsuitable for use due to bloody or clotted tumors (Mathews and Korner, 1970; Aviv *et al.*, 1971). Tumor cells can be passaged for up to 30–40 generations by this procedure without any obvious change in the relevant properties of the tumor (Mathews and Korner, 1970). In addition, ascites cells maintained in tissue culture in Eagle's medium retain the ability to cause tumors in animals (Van Venrooij *et al.*, 1970).

2.3.2. Growth of Tissue Culture Cells

Cell-free extracts used for translation of animal virus mRNAs have been prepared from HeLa cells (McDowell *et al.*, 1972; Villa-Komaroff *et al.*, 1974*b*; Eggen and Shatkin, 1972), Chinese hamster ovary (CHO) cells (McDowell *et al.*, 1972; Villa-Komaroff *et al.*, 1974*b*), mouse L-cell fibroblasts (McDowell *et al.*, 1972; Villa-Komaroff *et al.*, 1974*b*; Eggen and Shatkin, 1972; Friedman *et al.*, 1972*a*; Graziadei and Lengyel, 1972), and MOPC 460 tumor cells (Lawrence and Thach, 1974). Most of the cell lines can be grown in Eagle's minimum essential medium supplemented with 7–10% calf or bovine serum (McDowell *et al.*, 1972; Villa-Komaroff *et al.*, 1974*b*; Eggen and Shatkin, 1972; Friedman *et al.*, 1972*a*; Graziadei and Lengyel, 1972; Samuel and Joklik, 1974), but MOPC 460 tumor cells are grown in Liebowitz L15 medium (Lawrence and Thach, 1974). CHO cells should be further supplemented with nonessential amino acids (McDowell *et al.*, 1972). For the preparation of cell extracts, cultures are generally grown to densities of $2–10 \times 10^5$ cells/ml for L, HeLa, and CHO cells (McDowell *et al.*, 1972; Villa-Komaroff *et al.*, 1974*b*; Friedman *et al.*, 1972*a*; Graziadei and Lengyel, 1972) and $4–5 \times 10^6$ cells/ml for MOPC 460 cells (Lawrence and Thach, 1974).

2.3.3. Preparation of Cell Extracts

The basic method used for the preparation of cell extracts is that described by Mathews and Korner (1970); a similar procedure may be used for all cell types. Tissue culture cells ($1–2 \times 10^9$ cells) are harvested by centrifugation and resuspended in cold isotonic buffer. Ascites cells from five mice, harvested by draining the ascitic fluid from the opened peritoneal cavity into a sterile, precooled beaker, are diluted with cold isotonic buffer (Martin *et al.*, 1961; Mathews and Korner, 1970; McDowell *et al.*, 1972). The ascites cells may be filtered through two layers of cheesecloth as they are collected (McDowell *et al.,* 1972); Samuel and Joklik, 1974). All subsequent steps, unless otherwise noted, are carried out at 0–4°C. Cells are washed three times by centrifugation and resuspension, and after the final wash are resuspended in a small volume of the isotonic washing buffer in a graduated conical centrifuge tube. After being firmly packed by centrifugation, the cell pellet is drained to remove excess buffer, resuspended in 2–3 packed-cell volumes of hypotonic buffer, and allowed to swell for 5–10 min at 0°C. The cells are lysed with about 30 strokes of a tight-fitting plunger in a glass Dounce homogenizer, and the tonicity of the lysate is restored immediately by the addition of one-tenth volume of $10\times$ concentrated incubation buffer. The concentration of this buffer depends on the ionic conditions required in the cell extract. Generally, 20–30 mM tris-HCl (pH 7.5) or 20 mM Hepes (pH 7.5) buffer is used with KCl and Mg acetate concentrations of 80–120 mM and 3.5–5 mM, respectively; β-mercaptoethanol (4–7 mM) or dithiothreitol (1 mM) is also included. The homogenate is centrifuged at 10,000g.(Eggen and Shatkin, 1972; Friedman *et al.,* 1972a; Samuel and Joklik, 1974) (S10 extract) or 30,000g for 10–20 min (Mathews and Korner, 1970; Aviv *et al.,* 1971; Jacobs-Lorena and Baglioni, 1972; McDowell *et al.,* 1972; Villa-Komaroff *et al.,* 1974b; Graziadei and Lengyel, 1972; Lawrence and Thach, 1974) (S30 extract). The supernatant is carefully removed, passed through Sephadex G25 (see below), and used directly (Mathews and Korner, 1970) or preincubated as follows. ATP, GTP, creatine phosphate, creatine phosphokinase, and in some cases CTP and all 20 amino acids (Aviv *et al.,* 1971; Jacobs-Lorena and Baglioni, 1972; McDowell *et al.,* 1972; Villa-Komaroff *et al.,* 1974b; Eggen and Shatkin, 1972; Graziadei and Lengyel, 1972; Friedman *et al.,* 1972a) are added. The cell extract is incubated at 37°C for 30–45 min in order to reduce endogenous protein synthesis observed as background in reactions stimulated by exogenous mRNA. If necessary, the extract may be centrifuged at 10,000g (McDowell *et al.,* 1972; Eggen and Shatkin, 1972) or 30,000g for 5–10 min (Aviv *et al.,* 1971; Jacobs-Lorena and Baglioni, 1972; McDowell *et al.,* 1972; Eggen and Shatkin, 1972) to remove any flocculent material: The extract may be desalted by dialysis for 6 h against incubation buffer (Aviv *et al.,* 1971) or by chromatography on a column of Sephadex G25 (medium) (2.5–3 by 30 cm) equilibrated with incubation buffer (McDowell *et al.,* 1972). The lysate is eluted from the column with this buffer, and the opalescent fractions are pooled, quickly frozen in small aliquots, and stored in liquid nitrogen. The protein concentration of the extracts should be in the range of 8–14 mg/ml.

2.4. Wheat Germ

2.4.1. Preparation of Wheat Embryos

The wheat germ cell-free protein-synthesizing system is prepared from wheat embryos which are either obtained commercially (Both *et al.*, 1975*b*; Roberts and Paterson, 1973; Davies and Kaesberg, 1973; Marcus and Dudock, 1974; Weeks and Marcus, 1971) or prepared from wheat seeds (Johnston and Stern, 1957; Shih and Kaesberg, 1973; Davies and Kaesberg, 1973). Basically, their preparation involves (1) the removal of the loosely attached embryos from the seeds by mechanical agitation for 5–6 s in a Waring blender and (2) the separation of embryos from endosperm fragments by sieving and blowing and a selective flotation procedure based on the differences in buoyant density between embryos and endosperm (Johnston and Stern, 1957; Marcus *et al.*, 1974*a*; Shih and Kaesberg, 1973). Embryos from the wheat variety Fortuna and Kenosha winter wheat give preparations with consistently high activities (Marcus *et al.*, 1974*a*; Shih and Kaesberg, 1973).

The commercial availability of raw wheat germ now makes it almost unnecessary to prepare embryos from wheat seed. Raw wheat germ can be obtained from a variety of sources (Both *et al.*, 1975*b*; Roberts and Paterson, 1973; Davies and Kaesberg, 1973; Marcu and Dudock, 1974). However, some varieties yield *in vitro* protein-synthesizing systems which have higher activities than others (Marcu and Dudock, 1974). In general, wheat germ preparations from Niblacks, Inc., Rochester, N.Y., General Mills, Inc., Vallejo, Calif., and the "Bar-Rav" Mill, Tel Aviv, Israel, have yielded consistently good protein-synthesizing extracts. No doubt some local varieties of wheat germ will also provide suitably active extracts.

2.4.2. Preparation of Wheat Germ Extracts

Wheat germ protein-synthesizing extracts dependent on exogenous mRNA for translation are prepared as follows. Wheat embryos are ground for 30–60 s (Both *et al.*, 1975*b*; Marcu and Dudock, 1974) in a mortar at 4°C, using broken glass or sand as an abrasive, in a small amount of solution containing 50–100 mM KCl, 1 mM magnesium acetate, and 2 mM $CaCl_2$; 6 mM β-mercaptoethanol may also be present. The pH of the grinding solution varies depending on the method of preparation (Marcus *et al.*, 1974*a*; Shih and Kaesberg, 1973; Both *et al.*, 1975*b*; Roberts and Paterson, 1973). A pH greater than 6.8 apparently results in the release of endogenous mRNA (Weeks and Marcus, 1971); nevertheless, the extract (see below) is almost totally dependent on the addition of exogenous mRNA (Both *et al.*, 1975*b*; Roberts and Paterson, 1973). The ground wheat embryos are centrifuged at 23,000*g* (Marcus *et al.*, 1974*a*; Shih and Kaesberg, 1973; Both *et al.*, 1975*b*; Marcu and Dudock, 1974) or 30,000*g* (Roberts and Paterson, 1973) for 10 min at 4°C and the supernatant is removed; the surface layer of lipid is avoided. This supernatant, called the S23 or S30, may be stored undialyzed at −20°C, with only moderate loss in activity (Marcus *et al.*, 1974*a*) but must be dialyzed before use. Alternatively, the extracts can be dialyzed prior to storage (Shih and Kaesberg, 1973). In addition, the S23 or S30 may be preincubated for 10–15 min at 30°C with the components required for protein synthesis (Both *et al.*, 1975*b*; Roberts and Paterson, 1973), but a low background is also obtained if this step is omitted. The S23 or S30 preincubated extract is desalted by passing it through a column of Sephadex G25 (coarse or medium) equilibrated with the required incubation buffer (Roberts and Paterson, 1973; Both *et al.*, 1975*b*; Marcu and Dudock, 1974). The turbid eluate fractions are collected, pooled, and stored frozen in aliquots in liquid nitrogen (Both *et al.*, 1975*b*; Roberts and Paterson, 1973). The crude S23 extracts can also be used as a basic starting material for further fractionation of the wheat germ system (Allende and Bravo, 1966; Marcus *et al.*, 1974*b*). However, to date there are no reports on the translation of animal virus mRNAs in a fractionated wheat germ system.

2.5. Comparison of Properties

2.5.1. Availability and Maintenance of Tissue

Of the *in vitro* systems described above, the one prepared from wheat germ is perhaps best in terms of availability and cost of the starting material; e.g., a jar of vacuum packed raw (not toasted) wheat germ (350 g) can be purchased from the supermarket for about a dollar. A routine procedure for the preparation of the extract requires only 6 g of wheat germ (Both *et al.*, 1975*b*; Roberts and Paterson, 1973); hence the economy of the system. Surplus wheat germ may be stored for long periods under reduced pressure at 0–4°C (Roberts and Paterson, 1973) without significant loss in activity of the protein-synthesizing systems prepared from it (Marcus *et al.*, 1974*a*).

The South African clawed toad, *Xenopus laevis,* is conveniently reared in the laboratory in an aquatic environment (Brown and Littna, 1964), while the species *Bufo* can be kept on damp sand (May and Glenn, 1974). However, the use of oocytes for protein synthesis requires a ready supply of micropipettes (10–15 μm diameter) calibrated to inject a volume of 50–70 nl (May and Glenn, 1974; Gurdon *et al.,* 1971).

Since most biochemical research institutions have facilities for housing animals, the availability of rabbit reticulocytes and mouse ascites cells generally presents no problem. Similarly, those cell lines maintained in tissue culture by standard procedures can be grown whenever they are required.

In general, therefore, except perhaps for the special requirements of micropipettes and the injection technique for the oocyte system, any of these *in vitro* systems can be quickly established as a routine procedure in a laboratory.

2.5.2. Requirements for Protein Synthesis

With regard to the reagents required for protein synthesis, injected oocytes clearly have the least requirements. Amphibian embryogenesis occurs within a "closed system" where all the organic nutrients are derived from substances present in the unfertilized egg (Brown and Littna, 1964). Thus oocytes are simply incubated at 19–22°C in a modified Barth's medium (Barth and Barth, 1959), which is essentially a buffered salts medium containing penicillin and streptomycin to inhibit bacterial growth (Gurdon, 1968). The desired radioactive amino acid(s) is added to this medium or injected directly into the oocyte (Gurdon *et al.,* 1971).

The reagents necessary for the study of protein synthesis in the various cell-free extracts discussed in this chapter are virtually identical. Each extract requires ATP, GTP, creatine phosphate and creatine phosphokinase, dithiothreitol, Hepes or tris-HCl buffer, KCl, magnesium acetate, and amino acids. Phosphoenolpyruvate and pyruvate kinase can be used as an alternative energy-regenerating system in wheat germ (Shih and Kaesberg, 1973). The concentrations of individual components vary depending on the system and are detailed in the literature (Villa-Komaroff *et al.,* 1974*b*; Schreier and Staehelin, 1973; Cancedda and Schlesinger, 1974; Shih and Kaesberg, 1973; Both *et al.,* 1975*b*; Roberts and Paterson, 1973; Eggen and Shatkin, 1972). In addition, hemin is added to reticulocyte lysates in order to prolong the period of initiation of protein synthesis (Villa-Komaroff *et al.*, 1974*b*; Hunt *et al.*, 1971). In general, these unfractionated cell-free systems do not require exogenous tRNA; however, the efficiency of translation in the ascites *in vitro* system may be considerably improved by the addition of the homologous tRNA (Aviv *et al.*, 1971). The conditions for protein synthesis in these cell-free systems should be optimized for each mRNA translated by systematically varying the K^+, Mg^{2+}, and mRNA concentrations. The optimum incubation temperature (generally 25–30°C) should also be determined. Incorporation of added radioactive amino acid into material precipitable with 10% trichloroacetic acid is a suitable parameter for optimizing the system (Villa-Komaroff *et al.*, 1974*b*; Both *et al.*, 1975*b*). However, it may be necessary to vary the concentration of components in the system for efficient synthesis of apparently authentic polypeptide products in response to an exogenous mRNA (Both *et al.*, 1975*b*) (see below).

2.5.3. Analysis and Identification of Viral Polypeptides

There is little difference among any of these *in vitro* systems in the ease with which radiolabeled polypeptides can be analyzed, except that oocytes must first be homogenized (May and Glenn, 1974; Gurdon *et al.*, 1971; Laskey *et al.*, 1972). A variety of analytical procedures can be used, the most common being the specific immunoprecipitation of a polypeptide with its homologous antiserum (Eron *et al.*, 1974*a*; Öberg *et al.*, 1975), tryptic peptide mapping, and SDS-polyacrylamide gel electrophoresis. Both cylindrical gels and slab gels may be used, but the latter generally provide better resolution. Resolution of radiolabeled viral polypeptides on SDS-polyacrylamide gels is often complicated by the presence of proteins synthesized in these *in vitro* systems in the absence of added exogenous viral mRNA, i.e., background or endogenous protein synthesis. In reticulocyte lysates and oocytes, which are not preincubated systems, this background is very high (Laskey *et al.*, 1972; McDowell *et al.*, 1972). In the former, the major polypeptide product, globin, has a molecular weight of only 16,000 and therefore does not obfuscate high molecular weight products in the gel. However, the other significant endogenous polypeptides (McDowell *et al.*, 1972) sometimes obscure the viral polypeptide pattern (Morrison *et al.*, 1974). *In vitro* systems derived from ascites, CHO, and L cells also have considerable background levels of protein synthesis (Graziadei and Lengyel, 1972; Eron *et al.*, 1974*a*; Öberg *et al.*, 1975; Lodish *et al.*, 1974) which often make it necessary to immunoprecipitate specific polypeptide products in order to separate them from background proteins (Eron *et al.*, 1974*a*; Öberg *et al.*, 1975). The wheat germ system is particularly useful in this respect because it has very low levels of background protein synthesis and consequently polypeptide products coded for by exogenous viral mRNA can be readily identified (Both *et al.*, 1975*b,c,d*; Roberts and Paterson, 1973; Prives *et al.*, 1974*a*; Roberts *et al.*, 1975).

When the *in vitro* synthesized presumptive viral polypeptide and the corresponding authentic viral protein can be purified by polyacrylamide gel electrophoresis, or some other method, their degree of relatedness can be conveniently compared by tryptic peptide mapping. Two methods are commonly used. The purified proteins, differentially labeled in the same amino acid residue, e.g., [^{3}H]- and [^{35}S]methionine, are mixed together and digested with trypsin. The peptides are resolved by chromatography on an ion exchange resin (Both *et al.*, 1975*c,d*; Eron *et al.*, 1974*b*), where identical peptides comigrate. Alternatively, the two-dimensional tryptic peptide fingerprints of the *in vitro* synthesized and the authentic viral protein can be compared (Anderson *et al.*, 1974; Eron *et al.*, 1974*a*). Both procedures allow the degree of similarity between two proteins to be compared.

2.5.4. Identification of Primary Viral Polypeptides

In many infections, viral structural proteins are derived from higher molecular weight precursor polypeptides by proteolytic cleavage which may occur before translation of the precursor is completed (see later). Therefore, in infected cells it is sometimes difficult to determine which viral proteins are primary translation products of specific viral mRNAs and which arise by cleavage. With the availability of many extensively purified viral mRNA species, this problem can be approached by the use of *in vitro* systems which lack proteolytic activity. In principle, in the absence of proteolytic cleavage, the primary *in vitro* translation product of an mRNA is likely to be the largest viral mRNA-directed polypeptide with a molecular weight less than the coding potential of the mRNA. However, the currently used *in vitro* systems also synthesize many short polypeptides ("early quitters") which presumably result from premature termination events during translation (see below). Therefore, in practice this approach can be applied only for the largest mRNA added to a translation system, if more than one mRNA is added. Since wheat germ extracts have little or no proteolytic activity (Roberts *et al.*, 1974), it has been possible to determine in them the primary gene products of reovirus and VSV (Both *et al.*, 1975*b,c,d*). Similarly, the Schreier–Staehelin *in vitro* system and extracts prepared from uninfected ascites cells also apparently lack specific proteolytic activities since several adenovirus-2 proteins, known to be precursors to smaller viral polypeptides, were synthesized but not cleaved in these *in vitro* systems (Eron *et al.*, 1974*a*; Öberg *et al.*, 1975; Anderson *et al.*, 1974). In addition, some of the primary gene products of VSV have been translated in ascites extracts, and results similar to those obtained with the wheat germ *in vitro* system were obtained (Knipe *et al.*, 1975). There appears to be no evidence for cleavage of high molecular weight polypeptides coded for by Sindbis and encephalomyocarditis (EMC) virus RNA in extracts prepared from uninfected ascites cells (Öberg and Shatkin, 1972; Boime and Leder, 1972; Kerr *et al.*, 1972; Öberg and Shatkin, 1974; Smith, 1973; Mathews and Osborn, 1974; Cancedda and Schlesinger, 1974), nor apparently for cleavage of polio RNA-directed proteins synthesized in uninfected ascites and HeLa cell extracts (Villa-Komaroff *et al.*, 1974*a*). However, a protease activity may exist in a partially fractionated system prepared from reticulocytes (Cancedda *et al.*, 1974*b*). Proteolytic activity is also very low or absent in extracts prepared from uninfected L cells, although in extracts prepared from EMC virus-infected L cells the initial cleavage of the nascent viral polyprotein chain synthesized in response to exogenous EMC RNA occurred, suggesting that the activity of the enzyme(s) responsible for this cleavage increased after infection of the cells (Esteban ·and Kerr, 1974). Consistent with this observation is the finding that EMC virus-infected ascites cells appear to contain a proteolytic activity which is not found in uninfected ascites cell extracts (Lawrence and Thach, 1975). Moreover, EMC RNA-directed polypeptides synthesized in injected oocytes appear to be processed by proteolytic cleavage in a manner similar to that which occurs in EMC-infected ascites cells (Laskey *et al.*, 1972). This result may be explained by the observation that an EMC polypeptide, which is one of the major EMC-directed products in oocytes (Laskey *et al.*, 1972), may itself be a proteolytic enzyme involved in the maturation of other EMC proteins (Lawrence and Thach, 1975).

Another difficulty in identifying the primary gene product of a viral mRNA occurs when the virion protein is a glycoprotein. No cell-free system has yet been shown to glycosylate a polypeptide coded for by exogenous mRNA. The *in vitro* synthesized polypeptide usually has a slightly lower apparent molecular weight, i.e., migrates more rapidly in polyacrylamide gels than the authentic viral glycoprotein, presumably because of the absence of the carbohydrate moieties (Both *et al.*, 1975*d*; Knipe *et al.*, 1975; Clegg and Kennedy, 1975).

Although the biochemical basis for premature termination of protein synthesis is unclear, it is a feature common to all *in vitro* systems, with the possible exception of oocytes. In general, all the systems translate small mRNAs with considerable efficiency, and the extent of incomplete translation increases with increasing molecular weight of the mRNA (Eron *et al.*, 1974*b*), suggesting that random scission of the mRNA by nuclease(s) in the extract may account at least in part for "premature termination." In ascites cell extracts, the rate of translation of EMC RNA is dependent on the K^+ concentration and is slower than the rate of translation observed *in vivo*. A reduced rate of amino acid polymerization *in vitro* presumably reduces the chances for complete translation of very long mRNAs (Mathews and Osborn, 1974). However, by adjusting the K^+ and Mg^{2+} concentrations it is possible to synthesize polypeptides of molecular weight 150,000 in extracts of wheat germ (Both *et al.*, 1975*b*). In addition, the inclusion of polyamines in some translation systems seems to reduce the number of "early quitters" and preferentially increases the amount of large proteins that are synthesized, possibly by inhibiting the activity of ribonuclease(s) or stabilizing polysomes (Roberts *et al.*, 1975; Atkins *et al.*, 1975).

2.5.5. Relative Efficiency of Translation

The overall efficiency of synthesis of authentic polypeptides in an *in vitro* system is affected by many factors, among them the efficiency of initiation of protein synthesis, the fidelity of elongation and termination, and the ability to reinitiate protein synthesis. Moreover, the efficiency of *in vitro* translation may depend to some extent on the mRNA used to direct protein synthesis since it has been suggested that some mRNA molecules may possess an intrinsic ability to initiate protein synthesis more efficiently than others (Lodish, 1974; Both *et al.*, 1975*b*). Most of the studies on the efficiency of translation *in vitro* have been done with eukaryotic cellular mRNAs, e.g., globin and ovalbumin, but these results are probably applicable to the translation of animal virus mRNAs.

The oocytes and reticulocyte lysate systems are undoubtedly the best in terms of overall efficiency of translation. Injected rabbit globin

mRNA is translated in oocytes at the same relative rate as endogenous
mRNA for at least 24 h and as many as 24 molecules of globin can be
synthesized per molecule of exogenous globin mRNA. In effect, this
represents 100–1000 times more efficient synthesis than that observed
in the best cell-free systems. In fact, the efficiency of translation of
globin mRNA in frog oocytes compares favorably with that in intact
reticulocytes (Gurdon *et al.*, 1971). Since oocytes are not totally
disrupted by the injection of mRNA, translation in this system is
probably closest to the *in vivo* situation. In addition, crude reticulocyte
lysates are capable of translating ovalbumin mRNA with nearly the
same rate of chain initiation and elongation as observed in the intact
hen oviduct (Palmiter, 1973). Moreover, both frog oocytes and the
reticulocyte lysate system are capable of utilizing minute amounts of
exogenous mRNA, presumably because they are virtually free of
ribonuclease activity (Gurdon *et al.*, 1971; Palmiter, 1973). The
Schreier–Staehelin mixed cell-free system also translates exogenous
globin mRNA efficiently. It has been estimated that the rate of globin
synthesis in this system is 4–5 times slower than that of the best crude
reticulocyte lysate systems, which approach the efficiency of *in vivo*
synthesis. Reinitiation is also efficient in the Schreier–Staehelin system,
and as many as three globin polypeptide chains per ribosome can be
synthesized (Schreier and Staehelin, 1973). There are considerably
fewer data concerning the translation efficiency of cell-free systems
prepared from ascites cells, tissue culture cells, and wheat germ. The
efficiency of these systems appears to be generally inferior to that of
the systems discussed above. In the ascites system, initiation of protein
synthesis directed by exogenous EMC RNA is complete within 15 min
(Eggen and Shatkin, 1972; Öberg and Shatkin, 1972; Smith and Wigle,
1973), and the rate of elongation of protein synthesis is considerably
slower than that observed during virus replication in infected cells
(Mathews and Osborn, 1974). The inefficiency may be caused by
damage to, or loss of, vital components during the manipulations
involved in preparation of the cell extracts (Mathews and Osborn,
1974), and it has been suggested that factors stimulatory to protein
synthesis may be artifacts of such damage (Schreier and Staehelin,
1973; Lodish, 1974). In any event, despite the differences among the
various *in vitro* systems, it seems that they have a common feature, i.e.,
the apparent lack of species specificity as to the type of mRNA that
can be translated (Gurdon *et al.*, 1971; Davies and Kaesberg, 1973;
Villa-Komaroff *et al.*, 1974*b*). Hence, in each of these *in vitro* systems,
a putative mRNA from almost any source can be identified by the

polypeptide for which it codes, and in this sense each of these systems can be a very useful tool. However, a major limitation, which applies to all but the oocyte system and possibly the reticulocyte lysate system, is that they do not faithfully represent the *in vivo* situation. Therefore, interpretation of data concerning regulation of protein synthesis must be considered with this important reservation (Lodish, 1974).

3. TRANSLATION OF DNA VIRUS mRNAs

3.1. Adenoviruses

3.1.1. Virion Proteins

The double-stranded DNA genome of adenovirus is a linear structure of molecular weight 20–25 $\times$ 10^6, i.e., sufficient to code for about 25–50 polypeptides. Of this number, about ten are present in purified virions of human adenovirus type 2, the serotype used for the majority of the *in vitro* translation studies. A detailed review of the morphological and antigenic features of adenoviruses has appeared (Philipson and Lindberg, 1974), and only a brief description of type 2 adenovirus proteins follows. Two internal basic proteins (V and VII, molecular weights 48.5K* and 18.5K, respectively) are associated with the DNA in a core structure. Enclosing the core in an icosahedral arrangement are 240 morphological subunits, the hexons, which each consist of three copies of polypeptide II (molecular weight 120K), the biochemical subunit. Capsids also contain 12 penton subunits, one at each vertex of the icosahedron; the penton subunit includes a fiber antigen (IV, 62K) projecting from a base (III, 70K). When hexons are isolated in groups of nine subunits (the triangular faces of the icosahedron), they have associated an additional protein (IX, 12K) which may serve as an interhexon "cement." Similarly, polypeptide IIIa (66K) appears to be part of the vertex grouping of five hexons surrounding the penton base.

3.1.2. Proteins in Infected Cells

Adenovirus type 2 infection of human KB cells results in a marked inhibition of cellular protein synthesis. Consequently, viral proteins can

---

* K will be used to denote 1000 hereafter.

be specifically labeled with [³⁵S]methionine at late times in the infectious cycle and analyzed by SDS-polyacrylamide gel electrophoresis in comparison with virion structural polypeptides. In one study (Anderson *et al.*, 1973), about 20 virus-induced proteins were resolved including hexon, penton base, penton fiber, V, IIIa, IX, and nonstructural polypeptides, as well as other structural components. Notably absent from pulse-labeled infected cell lysates was core protein VII, which was replaced by a band of slightly higher molecular weight (20K). Pulse-chase experiments established that the 20K protein (PVII) was a precursor of the major core protein, VII. Other adenovirus structural proteins, including hexon-associated polypeptide VI, are also derived by posttranslational cleavage of larger precursors (Anderson *et al.*, 1973; Lewis *et al.*, 1975).

3.1.3. Cell-Free Synthesis

Authentic adenovirus polypeptides have been synthesized in cell-free extracts of wheat embryo cells (Öberg *et al.*, 1975) and mouse ascites cells (Eron *et al.*, 1974a, 1974b; Eron and Westphal, 1974; Öberg *et al.*, 1975) and in a reconstituted system from purified components of mammalian cell fractions (Anderson *et al.*, 1974). The viral mRNA was isolated from polyribosomes of infected KB cells at late times after infection. Authentic polypeptide products made in response to unfractionated viral mRNA included the four major virion proteins (hexon, penton fiber, base, and IIIa), the minor core protein V, and PVII, which is the precursor of the major core protein. The formation of PVII without the appearance of VII indicates that the posttranslational processing of this and probably other adenovirus proteins is mediated by virus-induced proteolytic enzyme(s). Several criteria were used to identify the virus-specific products including migration in SDS-polyacrylamide gels, immunoprecipitation with specific antisera, chromatography on SDS-hydroxylapatite columns, and tryptic peptide mapping. The relative amounts of the various *in vitro* products differed from those observed in infected cells or in purified virions. For example, the major viral structural protein II (120K) was synthesized in low amounts *in vitro*, presumably because of premature termination, whereas the minor structural component, IX (12K), was the predominant cell-free product. The low yield of high molecular weight products may also result from the increased nuclease susceptibility of larger mRNAs or the absence from heterologous cell extracts of virus-induced specific factor(s). Nevertheless, the results demonstrate that the synthesis of authentic adenovirus polypeptides can be programmed by viral mRNA in extracts of uninfected animal or plant cells.

Although temperature-sensitive, conditional lethal mutants of adenovirus have been isolated and studied, deletion and nonsense mutants are not generally available for standard genetic mapping of the adenovirus genome. A powerful alternative approach to this problem involves the use of cell-free extracts for translating selected viral mRNAs that are transcribed from different regions of the viral DNA. The mRNAs can be separated simply on the basis of molecular weight (Eron and Westphal, 1974; Eron *et al.*, 1974*a*; Öberg *et al.*, 1975; Anderson *et al.*, 1974) or more selectively by hybridization to defined fragments of the genome that have been derived by cleavage with restriction enzymes (Lewis *et al.*, 1975). A striking result obtained with separated size classes of adenovirus mRNAs is the finding that many of the high molecular weight mRNAs, i.e., those with large coding capacities, direct the synthesis of low molecular weight viral polypeptides. For example, PVII, the 20K precursor of the major core protein, is coded for by mRNA that has a sedimentation constant of 21 S. This value is consistent with a molecular weight of about 1×10^6 and a polypeptide coding capacity of about 100K daltons, i.e., a five-fold greater capacity than the size of the observed *in vitro* product. Similarly, 27 S mRNA directs the formation of two polypeptides of molecular weights 27K and 100K, while the 19 S mRNA codes for the fiber polypeptide (62K) and a 26K protein. In both cases, the coding capacity of a single mRNA molecule would be sufficient to code for the two polypeptides synthesized *in vitro*. Because the mRNA was treated with 90% formamide and sedimented under denaturing conditions, the results probably cannot be explained by mRNA aggregation. However, it remains to be determined whether the larger mRNAs (1) are processed during protein synthesis, (2) contain long untranslated regions, or (3) are polycistronic. At least for the two cases cited (27 S and 19 S mRNAs), results obtained with mRNAs selected by hybridization to specific fragments of viral DNA make it unlikely that two polypeptide products are coded for by a single, polycistronic adenovirus mRNA. *Escherichia coli* restriction endonuclease *Eco*RI was used to prepare six specific adenovirus genome fragments (Lewis *et al.*, 1975). Viral mRNA isolated at late times in the infectious cycle from polysomes of adenovirus type 2 infected KB cells was hybridized to the separated genome fragments, recovered from the DNA, and translated *in vitro*. The 27 S mRNAs which coded for 100K and 27K

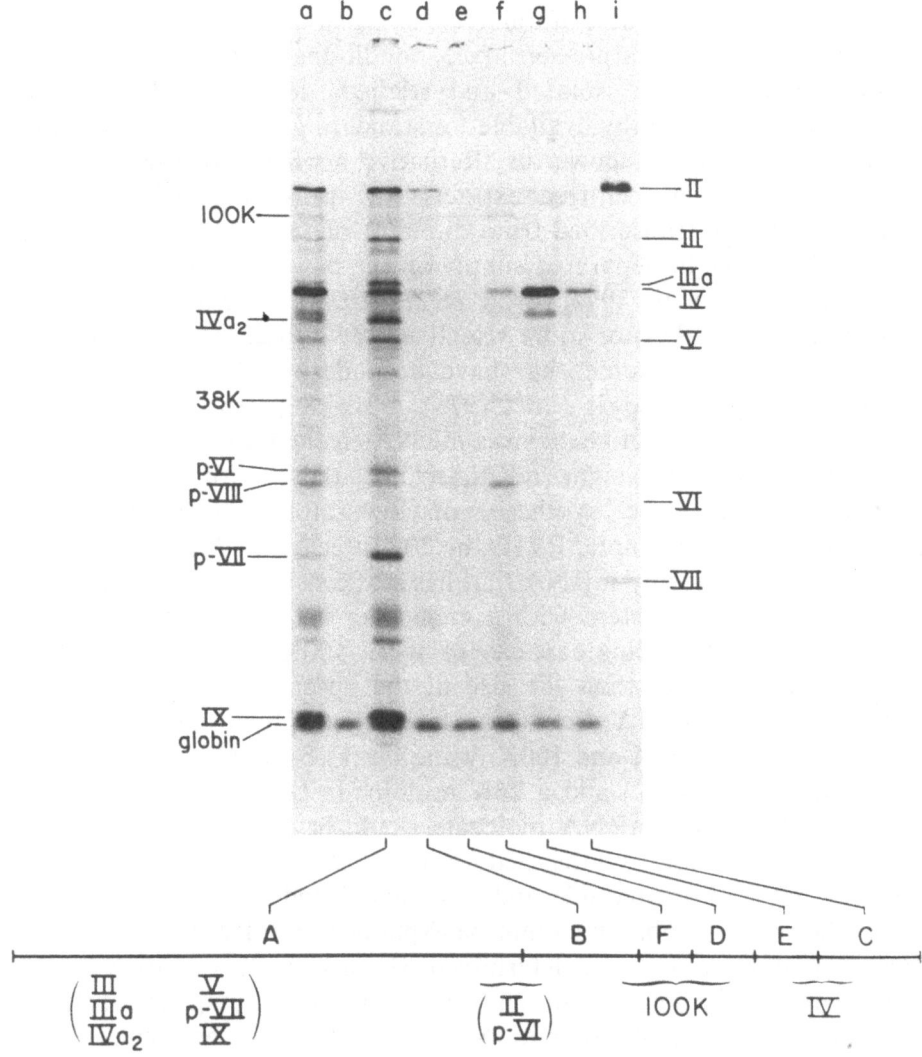

Fig. 1. Biochemical mapping of late adenovirus genes. Autoradiogram of a 17.5% sodium dodecylsulfate/polyacrylamide gel to display the products of cell-free syntheses. Protein synthesis was directed by the fraction of Ad2 RNA selected by hybridization to (a) Ad2 DNA, (b) λplac DNA, (c) R·EcoRI A fragment of Ad2 DNA, (d) B fragment, (e) F fragment, (f) D fragment, (g) E fragment, and (h) C fragment. A comparison with viral polypeptides is provided by disrupted virions (i). The positions of the genes for ten viral polypeptides are shown along the R·EcoRI map of Ad2 DNA at the bottom of the figure. From Lewis *et al.* (1975).

polypeptides hybridized to different genome fragments as did the 19 S mRNAs that directed synthesis of the 26K and fiber (62K) proteins.

A partial genetic map of adenovirus type 2 has been obtained by translating classes of mRNAs selected by hybridization to individual *Eco*RI fragments and analyzing the resulting subsets of polypeptide products (Fig. 1) (Lewis *et al.*, 1975). Polypeptides identical in electrophoretic mobility to penton base (III), IIIa, IVa2, minor core (V), major core precursor (PVII), and IX were coded for only by mRNAs that hybridized to the A fragment, i.e., the left half of the genome. The mRNAs for hexon (II) and protein PVI were homologous to both the A region and the contiguous B fragment. The mRNA for 100K protein hybridized to three adjacent DNA segments: B, F, and D. Fiber (IV) mRNA was transcribed from a region divided between the adjacent fragments E and C. It is encouraging to note that the partial map deduced from *in vitro* translation of specific mRNAs is consistent with the results of complementation studies of adenovirus type 5 temperature-sensitive mutants (Williams *et al.*, 1974) and analyses of the type 2 and type 5 adenovirus recombinants (Mautner *et al.*, 1975). It is reasonable to expect that a complete genetic map of adenovirus type 2 and other viruses will soon be established by this biochemical approach. Further experiments with early, late, and transformed cell mRNA selected on DNA fragments obtained with a variety of specific endonucleases should also yield new information on the regulation of transcription and translation in both permissive and nonpermissive infections.

3.2. Papovaviruses

3.2.1. Virion Proteins

The papova group includes simian virus 40 (SV40) and polyoma virus, two small viruses that contain closed, circular double-stranded DNA genomes of molecular weight about 3×10^6 (Salzman and Khoury, 1974). This is equivalent to a coding potential of only seven polypeptides, a number considerably lower than the multitude of new proteins that appear in cells following virus infection. The virion structural proteins of SV40 and polyoma viruses are similar in size and number but are unrelated immunologically. They include a major capsid polypeptide (VP1) of molecular weight about 45K and two minor proteins (VP2, VP3) of about 35K and 20K daltons. In addition, four low molecular weight basic proteins are present in association with the viral DNA. They apparently correspond to cellular histones F3, F2b, F2a2, and F2a1.

3.2.2. Early Antigens

Both SV40 and polyoma virus mediate a lytic, productive infection in permissive cells and the oncogenic transformation of cells that are nonpermissive for virus replication. Consequently, they have been intensively studied in an effort to determine which protein(s) are necessary for the initiation and maintenance of the transformed state. Among the virus-induced nonstructural proteins synthesized at early times in productively infected cells, i.e., before viral DNA replication, and in transformed cells is the immunologically identifiable, intranuclear T antigen. Virus mutants have been isolated that are temperature sensitive for the synthesis of T antigen, indicating that it is probably virus coded. Other early antigens detected in virus-infected and in transformed cells include the heat-stable, perinuclear U antigen and the tumor-specific transplantation antigen located at the cell surface.

3.2.3. Polyoma Virus

Because host protein synthesis continues in papovavirus-infected cells, it is difficult to identify nonstructural proteins and to differentiate between virus- and cell-coded proteins. One fruitful approach to the latter problem clearly is the use of temperature-sensitive mutants. Another is the translation *in vitro* of virus-specific mRNAs which should permit the identification of all the viral proteins that are coded for by the relatively small genomes of SV40 and polyoma virus. In an early study (Crawford and Gesteland, 1973), a coupled *E. coli* cell-free system was used for transcription and translation of polyoma virus DNA. The addition of viral DNA stimulated by sevenfold the incorporation of [^{3}H]methionine into polypeptide products. Most of the radioactivity did not comigrate in SDS-polyacrylamide gels with authentic virion proteins. However, the highest molecular weight product, a minor fraction of the total radioactivity, did correspond in electrophoretic mobility to VP1. Similarities in the methionine-containing tryptic peptides of the *in vitro* products and structural proteins indicated that at least some regions of the polyoma proteins were synthesized in phase, although correct initiation and termination of polypeptides probably did not occur in *E. coli* extracts. More recently, polyoma-specific mRNA isolated from infected mouse cells late in infection has been translated with fidelity in wheat germ extracts (A. E. Smith, S. T. Bayley, T. Wheeler, and W. F. Mangel, personal communication). Capsid protein VP1 was identified as the major *in vitro* product by immunoprecipitation and tryptic peptide analysis. mRNA for VP1 hybridized only to viral DNA fragments corresponding to the late region of the polyoma genome. Similarly, complementary RNA (cRNA) transcribed from polyoma closed superhelical DNA (form I) by *E. coli* RNA polymerase directed authentic VP1 synthesis *in vitro*. In contrast to SV40 cRNA, polyoma cRNA corresponds to late mRNA. Thus the results indicate that the major virion capsid protein is coded for entirely by the late strand of the polyoma genome.

3.2.4. SV40

Poly(A)-containing mRNA isolated from uninfected or SV40-infected monkey cells stimulated protein synthesis to a similar extent in cell-free extracts of rabbit reticulocytes, Chinese hamster ovary cells (CHO) (Lodish *et al.,* 1974), and wheat germ (Prives *et al.,* 1974*a*). Viral products were synthesized only in extracts that were programmed with mRNA from infected cells. The infected cell mRNA was isolated at late times after infection when virion structural proteins were being synthesized, and it directed the formation *in vitro* of the major capsid protein, VP1. The protein was identified as VP1 by SDS-polyacrylamide gel electrophoresis, immunoprecipitation, and fingerprint analysis. VP1 product synthesized by the mammalian cell extracts differed slightly from the virion-associated polypeptide in two ways. It eluted from SDS-hydroxylapatite after the authentic protein and contained a reduced amount of one of the six methionine-containing tryptic peptides that were resolved by paper electrophoresis at pH 3.5.

In wheat germ extracts stimulated by infected cell mRNA, 5–10% of the polypeptide products corresponded in mobility to VP1 (Prives *et al.,* 1974*a*). Trypsin digestion of the *in vivo* and *in vitro* synthesized VP1 yielded three major methionine-containing peptides that migrated identically by two-dimensional fingerprint analysis. A second virus-specific minor protein of molecular weight 35K was also synthesized in wheat germ extracts. mRNA extracted from infected cells that were treated with cytosine arabinoside to block DNA replication and virus structural protein formation was also tested for messenger activity; synthesis of both VP1 and the 35K protein was abolished or markedly reduced. As further evidence that viral mRNA was directing the synthesis of the VP1 and 35K polypeptides *in vitro,* the same two virus-specific proteins were made in response to mRNA selected by hybridization to viral DNA. The virus-specific mRNAs isolated at late times in the SV40 infectious cycle included 19 S and 16 S classes of mRNA; they have been separately translated in wheat germ extracts (Prives *et al.,* 1974*b*). The 16 S mRNA stimulated synthesis of VP1 as the major product. In contrast, the 19 S mRNA-directed products were a nonstructural 39K polypeptide, i.e., somewhat smaller than VP1, and three polypeptides of molecular weight greater than VP1. It was suggested that 19 S mRNA is partially translated to form the 39K protein, then cleaved to 16 S mRNA, which directs VP1 synthesis subsequently by initiation of protein synthesis at a new site that becomes available only after conversion of 19 S to 16 S RNA (Prives *et al.,* 1975*b*).

Recently, an efficient coupled system has been developed for the cell-free transcription and translation of SV40 (Roberts *et al.,* 1975). By incubating the viral DNA with *E. coli* RNA polymerase followed by the addition of wheat germ protein-synthesizing extract, an eightyfold stimulation in [^{35}S]methionine incorporation into acid-precipitable products was obtained. Protein synthesis was dependent on the formation, during the transcription step, of mRNA which was heterogeneous in size (4–26 S). The polypeptide products directed by the mRNA included a protein of molecular weight 48K that was specifically precipitated with antiserum against disrupted virions and yielded a tryptic peptide map almost identical to that of SV40 VP1. A single extra methionine-containing peptide was present in the *in vitro* product. It is of some interest that a late protein, VP1, was coded for by mRNA that was synthesized by *E. coli* polymerase because the bacterial enzyme transcribes the early strand of the viral genome. However, on the basis of self-annealing experiments, 10–20% of the mRNA apparently was transcribed asymmetrically, a finding which would account for VP1 synthesis. Of the several other products, three proteins (molecular weights 60K, 50K, and 25K) were precipitable by hamster antiserum to SV40 T antigen but not by normal hamster serum. A. E. Smith, S. T. Bayley, T. Wheeler, and W. F. Mangel (personal communication), using cRNA-stimulated wheat germ extract, have also detected the synthesis of a 59K polypeptide that reacted specifically with hamster anti-T serum. This finding suggests the exciting possibility that T antigen can be synthesized *in vitro* in a coupled system.

Preliminary results indicate that mapping studies similar to those initiated with adenovirus should also be possible with SV40 (Roberts *et al.*, 1975). Cleavage of supercoiled viral DNA form I to the linear duplex form III by restriction nucleases *Eco*RI and *Bam*I (from *Bacillus amyloliquefaciens* H) did not inactivate its template activity for VP1 synthesis in the coupled *in vitro* system. This observation localizes the information for VP1 synthesis in the late mRNA at 0.65–0 map unit in a clockwise direction (on the basis of cleavage at position zero by *Eco*RI). Treatment with enzyme *Hae*III (from *Haemophilus aegypticus*) which degrades the late region of the viral genome abolished VP1 synthesis, but it should be noted that the formation of the 60K protein which appears to be related to T antigen, an early protein, was also abolished. However, in another study SV40-specific mRNA selected by hybridization to a restriction endonuclease fragment containing the early region of the viral DNA (*Hae*III fragment A) directed the formation in wheat germ extract of a polypeptide that reacted with anti-T serum (A. E. Smith, S. T. Bayley, T. Wheeler, and W. F. Mangel, personal communication).

3.3. Vaccinia Virus

Vaccinia, a member of the poxvirus group, is among the largest of animal viruses and contains a DNA genome of molecular weight 150–200 $\times$ 10^6 (Moss, 1974). One-fifth of its coding capacity is utilized for structural protein formation, and about 30 polypeptides including one glycoprotein and one phosphoprotein are present in purified virions. They range in molecular weight from 8K to more than 200K. At least three major virion proteins are derived by posttranslational cleavage. As in many other animal viruses, purified vaccinia virus contains several enzymatic activities including DNA-dependent RNA polymerase, nucleoside triphosphate phosphohydrolase, RNA methylase(s), poly(A) polymerase, DNase, and protein kinase.

In virus-infected cells, host protein synthesis is rapidly inhibited, and virus-induced proteins in lysates of pulse-labeled cells can be readily resolved by SDS-polyacrylamide gel analysis. Early and late classes of viral proteins have been described. The early proteins are, at least initially, coded for by mRNA synthesized by the RNA polymerase in parental virions. mRNA produced by virions *in vitro* corresponds to the early mRNA of infected cells and represents about 7% of the genome, i.e., equivalent to 10–25 polypeptides. Cell-free extracts of mouse L or ascites cells or rabbit reticulocyte lysates synthesize a variety of polypeptides in response to vaccinia *in vitro* mRNA (Beaud *et al.*, 1972; Fournier *et al.*, 1973; Jaureguiberry *et al.*, 1975). Viral mRNA that sedimented broadly with a peak at 12–14 S in sucrose gradients stimulated synthesis of a similar array of proteins of molecular weight up to 40K in extracts of both cell types. Some of the *in vitro* products corresponded in electrophoretic mobility to the early virus-specific polypeptides present in lysates of infected cells, but they did not react with virus-specific antiserum. The four major methionine-containing tryptic peptides of one of the polypeptides (14K) made *in vitro* were similar to those of an *in vivo* synthesized viral protein of the same molecular weight when analyzed by one-dimensional chromatography (Jaureguiberry *et al.*, 1975). Further identification of the *in vitro* products as vaccinia virus coded will require additional studies.

4. TRANSLATION OF RNA VIRUS mRNAs

Animal RNA viruses have been classified into four groups according to the structure and mechanism of expression of their genomes (Baltimore, 1971; Shatkin, 1974*a*). The following discussion of viral mRNA translation is based on this classification.

4.1. Class 1 Viruses (mRNA = Genome)

4.1.1. Picornaviruses

4.1.1a. Virion Proteins

The picornaviruses include the structurally similar enteroviruses (poliovirus, coxsackieviruses), cardioviruses (encephalomyocarditis, EMC; mouse Elberfeld, mE; mengovirus), rhinoviruses, and foot-and-mouth disease viruses (Levintow, 1974). Each consists of four structural proteins of molecular weights about 35K, 30K, 27K, and 7K and a single-stranded RNA chain of molecular weight about 2.6×10^6. The genome RNA is infectious and apparently functions directly as the viral mRNA without an intervening virion-associated polymerase. Virion RNA, like many eukaryotic mRNAs (Brawerman, 1974) contains poly(A) at the 3′ terminus. The structure of the 5′ end, reported to be pAp for poliovirus (Wimmer, 1972), has not been determined for other picornaviruses.

In picornavirus-infected cells, host protein synthesis is markedly inhibited and, in the presence of actinomycin, can be almost completely abolished without affecting virus replication (Shatkin, 1968). Consequently, a spectrum of newly formed virus-specific polypeptides can be readily detected by SDS-polyacrylamide gel analysis of radioisotope-labeled infected cells. Both noncapsid viral proteins (NCVP) and virion structural proteins (VP) ranging in size from about 7K to as high as 200K daltons have been identified. Their combined molecular weight far exceeds the expected coding capacity of the viral genome, a paradox resolved by the suggestion that the multitude of picornavirus polypeptides are generated from a larger "polyprotein" by posttranslational cleavage (Summers and Maizel, 1968; Jacobson and Baltimore, 1968; Holland and Kiehn, 1968). This has since proven to be an important general mechanism for the formation of proteins of DNA viruses, other RNA viruses (Korant, 1975), and animal cells (Korant, 1975).

A model for picornavirus protein formation is shown in Fig. 2. The polypeptides are labeled according to the nomenclature used to identify poliovirus-specific products, but a similar cleavage pattern has been established for EMC and rhinoviruses (Butterworth, 1973; see also chapter by Rueckert in Vol. 6 of this series). The general features of the model have found experimental support from *in vivo* and *in vitro* studies, but many of the details including the exact number and size of the cleavage products and the origin of the required protease(s) remain

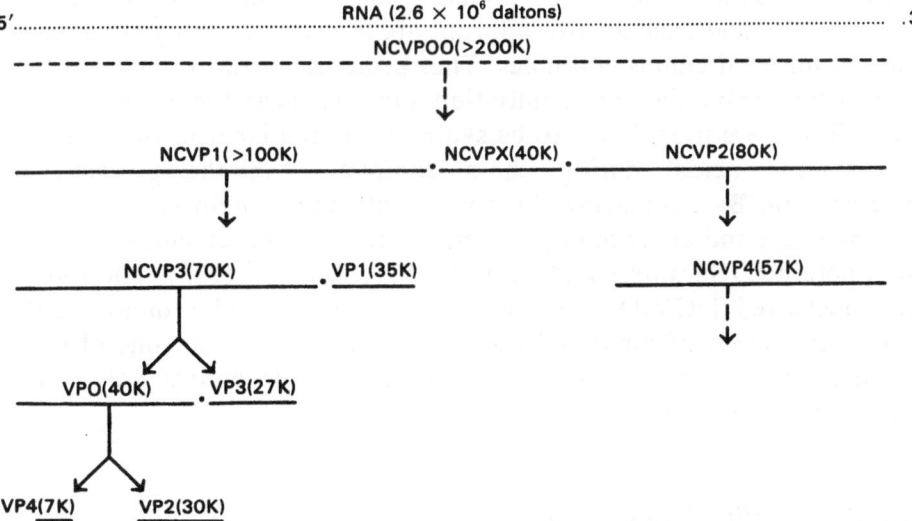

Fig. 2. A model for picornavirus protein cleavage based on studies of poliovirus polypeptide maturation.

to be established. Initiation of viral protein synthesis in infected cells apparently begins at a highly preferred site near the 5′ end of the RNA. A large "polyprotein," NCVP00, corresponding to most of the coding capacity of the RNA is synthesized. It is cleaved into three primary products, noncapsid viral proteins 1, X, and 2, during or soon after completion on polyribosomes. Cleavage can be inhibited by protease inhibitors (Korant, 1972; Summers *et al.,* 1972), by incorporation of amino acid analogues into nascent polyprotein (Jacobson and Baltimore, 1968, Kiehn and Holland, 1970), or by increase in the incubation temperature of virus-infected cells (Jacobson *et al.,* 1970). Under these conditions, larger virus-specific proteins accumulate and in coxsackievirus-infected, pulse-labeled HeLa cells a polyprotein of molecular weight 200–230K has been reported (Kiehn and Holland, 1970). Host proteases, either preexisting or activated upon infection, probably catalyze the primary cleavages and the subsequent processing of NCVP1 that results in the formation of the four capsid proteins. The final proteolytic step in capsid protein synthesis, i.e., conversion of VP0 to VP4 and VP2, may occur in assembled empty particles called procapsids, during RNA packaging (Levintow, 1974).

The genetic information for viral capsid polypeptides is clustered at the 5′ end of the RNA in 5′→3′ order VP4, VP2, VP3, VP1. The gene order of picornaviruses was established by the ingenious use of pactamycin at concentrations which block specifically the initiation of viral polypeptide chains. In infected cells labeled with radioactive amino acids and treated with appropriate levels of pactamycin, elongation of initiated chains continues. Thus proteins coded for by genes that are distant from the single initiation site, i.e., near the 3′ end of the viral RNA, are most likely to be synthesized (and labeled with radioactive amino acids) during chain completion in the presence of pactamycin. By comparing the amount of each protein synthesized in the presence and absence of pactamycin, the gene order shown in Fig. 2 was obtained (Summers and Maizel, 1971; Rekosh, 1972). The role of the uncleaved NCVPX and the processed noncapsid proteins in the replicative cycle of picornaviruses is not established, but one of these proteins is presumably the virus-induced RNA-dependent polymerase responsible for viral replication.

4.1.1b. *In Vitro* Translation

Deproteinized RNA isolated from purified picornaviruses is capable of initiating a productive infection and, in infected cells, is present in polyribosomes that synthesize virus-specific proteins (Levintow, 1974). These observations strongly indicate that the virion RNA functions both as genome and as viral messenger. The dual role was confirmed when it was shown that RNA from EMC virus directs the synthesis of virus-specific polypeptides *in vitro* (Smith *et al.*, 1970; Dobos *et al.*, 1971; Aviv *et al.*, 1971; Eggen and Shatkin, 1972; Kerr and Martin, 1971). Much of the pioneering work in the development of a dependable *in vitro* translating system from animal cells was done with EMC RNA as the messenger (Kerr *et al.*, 1962, 1966; Mathews and Korner, 1970). In extracts of mouse ascites cells or L cells, exogenous EMC RNA stimulates [^{35}S]methionine incorporation by more than fiftyfold; similar results were obtained with mE and mengovirus RNA. Both capsid and noncapsid polypeptides were synthesized as shown by comparative fingerprint analysis of the [^{35}S]methionine-labeled tryptic peptides from *in vitro* protein products and virus-specific polypeptides present in infected cells and purified virions (Dobos *et al.*, 1971; Aviv *et al.*, 1971; Eggen and Shatkin, 1972; Kerr and Martin, 1971; Kerr *et al.*, 1962, 1966; Mathews and Korner, 1970; Boime *et al.*, 1971).

Synthesis of EMC, mE, mengovirus, and poliovirus polypeptides in extracts of uninfected ascites cells is initiated predominantly at one site near the 5′ end of the viral mRNA (Öberg and Shatkin, 1972, 1974; Smith, 1973; Boime and Leder, 1972; Villa-Komaroff *et al.*, 1974a). ^{35}S-labeled yeast initiator Met tRNA$_F^{Met}$, formylated to prevent its cleavage from nascent protein products, was used as a precursor of virus-specific polypeptides (Öberg and Shatkin, 1972, 1974; Smith, 1973). Although protein chain elongation continued for 20 min or longer, initiation of new chains was completed during the first few minutes of incubation. The [^{35}S]methionine was retained in the high molecular weight products, indicating the absence of cleavage in the *in vitro* system. Most (~90%) of the [^{35}S]methionine was obtained in a single EMC-specific, *N*-terminal tryptic peptide, although a second minor *N*-terminal tryptic peptide was also found (Öberg and Shatkin, 1972). Recent experiments with poliovirus indicate that the relative amounts of each of the two *N*-terminal peptides synthesized *in vitro* varied as a function of the Mg^{2+} concentration (M. Celma and E. Ehrenfeld, personal communication). The origin of the two poliovirus-specific *N*-terminal peptides with respect to their corresponding positions in the RNA will be of great interest. The major EMC *N*-terminal peptide differed from the *N*-terminal peptide products directed by mE and mengovirus RNAs (Öberg and Shatkin, 1972, 1974). Moreover, the EMC peptide was absent from the virion capsid proteins, suggesting that a lead-in sequence at the 5′ end of the viral RNA precedes the information for coat protein synthesis. The amino acid composition of the lead-in, N-terminal sequence in EMC-directed polypeptide products was identical whether they were initiated with fMet tRNA$_F$^{Met} and obtained by digestion with pronase, chymotrypsin, or trypsin or derived by the use of unformylated Met tRNA$_F$^{Met} under conditions where chain elongation was inhibited with sparsomycin (Smith, 1973). A similar tryptic peptide, not found in uninfected cells, was detected in EMC virus-infected ascites cells pulse-labeled with [^{35}S]methionine (Smith et al., 1970; Smith, 1973). After longer labeling periods, the peptide was absent, consistent with its cleavage as an N-terminal lead-in sequence. Thus the in vitro lead-in sequence is probably initiated at the physiological site in the viral RNA.

The molecular weights of the in vitro synthesized products and virus-specific polypeptides from picornavirus-infected cells were similar, as determined by SDS-polyacrylamide gel electrophoresis. However, whereas the in vivo polypeptides are formed by a series of specific proteolytic cleavages, the in vitro products result largely from premature termination of protein synthesis. The biochemical basis for premature termination is unclear, but polypeptide chain elongation can be promoted by adjusting the K$^+$ concentration in the cell-free system (Mathews and Osborn, 1974). The similarity between the molecular weights of the in vivo and in vitro virus-specific polypeptides suggests that premature termination occurs at discrete sites located near the positions of proteolytic cleavage (Eggen and Shatkin, 1972). In contrast to the products made by extracts of virus-infected or uninfected ascites cells, a small fraction of the EMC-specific products synthesized by extracts of uninfected L cells had a molecular weight of ~230K (Friedman et al., 1972a). This value is consistent with the translation of the complete viral RNA chain, but the predominant products were smaller polypeptides.

Somewhat similar results were obtained with extracts prepared from EMC virus-infected cells (Esteban and Kerr, 1974). A striking finding was the increased amount of the three primary cleavage products among the polypeptides synthesized in vitro by extracts of infected L cells, although further cleavage of the capsid precursor polypeptide to the four structural proteins did not occur in vitro. In extracts prepared from cells at 2 h after infection, the predominant polypeptide synthesized had an apparent molecular weight of 130K and presumably corresponded to NCVP1 or its uncleaved precursor. The in vitro products synthesized by extracts prepared 4 h after virus infection contained a major polypeptide of 105K which was probably an intermediate between the primary cleavage product NCVP1 and NCVP3 + VP1 (model in Fig. 2). It may result from removal of a 25K, *N*-terminal lead-in fragment from NCVP1.

Findings related to those with L-cell extracts have been obtained with the EMC RNA-directed Krebs ascites cell-free extracts (Lawrence and Thach, 1975). The ascites extracts were not preincubated to reduce endogenous activity or treated by dialysis or gel filtration to remove low molecular weight constituents. A major protein product synthesized in the uninfected cell extracts had an apparent molecular weight of 112K. In infected extracts, the largest product was slightly smaller (100K) and was similar to NCVP1 by the criteria of gel electrophoresis and peptide analysis. Another major polypeptide synthesized in infected cell extracts had a molecular weight of 12.5K and was absent from the EMC RNA-directed products of uninfected extracts. The infected ascites cell lysates contained a proteolytic activity that yielded a 12K fragment from the products synthesized in uninfected cell extracts, apparently by cleaving the *N*-terminal region of the 112K protein. The protease activity copurified with VP3, suggesting that the viral capsid protein may participate in the cleavages necessary for maturation of picornaviruses. Similarly, foot-and-mouth disease virus RNA is translated in ascites cell extracts, and the polypeptide products include immunoprecipitable VP3, other capsid polypeptides, and high molecular weight viral precursor proteins (N. K. Chatterjee, personal communication).

4.1.2. Togaviruses

4.1.2a. Virion Proteins

The togaviruses, formerly grouped as arboviruses according to their mechanism of transmission, have been reclassified on the basis of their structure and subdivided into two antigenically distinct groups: A or alphaviruses and B or flaviviruses. Virions of both groups share many basic features including the presence of a lipid envelope enclosing the nucleocapsid that consists of the single-stranded RNA genome and multiple copies of the capsid protein (Pfefferkorn and Shapiro, 1974). Most of the *in vitro* translation studies designed to elucidate the mechanism of togavirus gene expression have been done with the group A togaviruses, Sindbis and Semliki Forest viruses. Although they may differ in some properties from other togaviruses, their general features are probably representative of the group. These viral genomes have a molecular weight of about 4×10^6 and a sedimentation constant of 42–49 S. In contrast to the large coding capacity of the genome RNA, only three structural proteins are present in purified virions. The envelope contains two glycoproteins, E1 and E2, of similar molecular weight (~ 50K); the nucleocapsids have a single protein, C, which is of molecular weight 30–36K.

The 49 S Sindbis genome RNA is infectious and is also present in polysomes of infected cells. In these respects, it resembles the genome RNA of picornaviruses. However, whereas the virus-specific polysomes of poliovirus-infected cells contain exclusively 35 S viral mRNA (Rich *et al.*, 1963), Sindbis virus-specific polysomes contain predominantly 26 S viral RNA and a considerably smaller fraction ($\sim 10\%$) of 49 S viral RNA (Kennedy, 1972; Mowshowitz, 1973; Rosemond and Sreevalsan, 1973; Simmons and Strauss, 1974*a*). The 26 S RNA has a molecular weight of about 1.6×10^6 and consists of one-third of the viral genome, i.e., sufficient to code for all three structural proteins (Simmons and Strauss, 1972, 1974*b*). The presence of 26 S RNA in infected cell polysomes that are synthesizing mostly virion proteins indicates that the strategy of the togaviruses is a variation on the picornavirus theme. Sindbis protein formation in infected cells appears not to occur by translation of the complete genome RNA from a single initiation site, followed by proteolytic cleavages. Instead, that portion of the genome which codes for structural proteins is amplified by synthesis of 26 S mRNA (Schlesinger and Schlesinger, 1972). If this RNA contains a single initiation site for protein synthesis, the translational product of 26 S RNA would be a "polyprotein" of molecular weight about 150K containing the amino acid sequences of all three structural proteins. Cells infected with temperature-sensitive mutants of Sindbis virus at nonpermissive temperatures accumulated a high molecular weight polypeptide that apparently corresponds to the precursor of the viral structural proteins (Scheele and Pfefferkorn, 1970; Schlesinger and Schlesinger, 1973). Similarly, in pulse-labeled Sindbis virus-infected BHK cells a large protein (>100K) was detected (Scheele and Pfefferkorn 1970). It contained many of the [^{14}C]arginine-containing tryptic peptides of the three viral structural proteins, and during a chase period a portion of the radioactivity in the presumptive precursor appeared in lower molecular weight polypeptides including the viral envelope proteins. From the results of studies on Sindbis virus-infected cells, it appears that the infectious 49 S genome RNA codes predominantly for nonstructural protein(s), e.g., the polymerase(s) necessary for viral RNA synthesis, and the 26 S RNA for a large precursor protein that is cleaved to the three structural proteins.

In Semliki Forest virus-infected BHK cells, the initiation of viral structural protein synthesis was synchronized by a procedure also used for studying picornavirus protein synthesis (Saborio *et al.*, 1974). It involves (1) raising the NaCl concentration to inhibit the start of new polypeptide chains, (2) restoring isotonicity to reverse the inhibitory effect, and (3) pulse-labeling the cells with radioactive amino acids during the initiation period (Clegg, 1975). As observed for poliovirus-infected HeLa cells by the same procedure (Saborio *et al.*, 1974), viral proteins were formed sequentially after the reversal of inhibition. Capsid protein labeled with [^{35}S]methionine was first apparent after 2 min of synthesis. The polypeptide precursors of envelope proteins were detected after 5 min. In pulse-chase experiments, a clearer picture of the order of synthesis of virion structural proteins emerged. A 1-min pulse of [^{35}S]methionine followed by a 15-min chase in the presence of nonradioactive methionine resulted in the appearance of capsid protein (36K) only. When the pulse time was increased to 2 min, envelope precursor polypeptides (molecular weights 97K and 63K) were observed and envelope protein E1 ($\sim$50K) was found when the pulse time was extended to 4 min. High molecular weight proteins ($>$100K) were also resolved in lysates prepared after a 3-min pulse. The results are consistent with a single initiation site, presumably in 26 S RNA, for the sequential synthesis of viral structural proteins. The first structural protein synthesized is capsid which is cleaved from the growing polypeptide chain before the envelope sequences are completed. The envelope proteins are formed from a nonglycosylated precursor of molecular weight 97K, which, after its synthesis is completed, is cleaved to yield the glycosylated envelope protein E1 and another partially glycosylated intermediate (63K). The latter is converted to envelope protein E2 by cleavage and further glycosylation.

4.1.2b. *In Vitro* Translation

Cell-free translating systems have been used for testing the coding functions of the 49 S and 26 S togavirus RNAs. Of particular interest is the question of the number of initiation sites in each type of RNA. Sindbis virion RNA functions as a messenger for virus-specific protein synthesis in rabbit reticulocyte and mouse ascites cell extracts. In one study (Cancedda *et al.*, 1975*a*; Cancedda and Schlesinger, 1974), capsid protein and high molecular weight products, including nonstructural proteins, were synthesized in both the rabbit and mouse cell extracts programmed with Sindbis virion RNA. In another report (Simmons and Strauss, 1974*b*), reticulocyte lysates synthesized eight or nine Sindbis-specific products ranging in molecular weight from 60K to 180K. However, none corresponded in molecular weight to the virion structural proteins, to their large precursors, or to any of the newly formed virus-specific polypeptides in infected cells. Their total molecular weight exceeded the theoretical coding capacity of 49 S RNA, suggesting that they share common amino acid sequences and may have been produced by proteolysis or premature termination. It should be noted that at least two laboratories have reported that different batches of cell-free extracts prepared from reticulocytes or ascites cells synthesized variable amounts of discrete polypeptides larger than capsid protein when programmed with Sindbis virus (Cancedda and Schlesinger, 1974) or Semliki Forest virus (Smith *et al.*, 1974) RNA.

Semliki Forest virus genome RNA also directed the formation of authentic capsid protein in an ascites cell-free system (Smith *et al.*, 1974). Another major *in vitro* virus-specific product had a molecular weight of only 15–20K but apparently contained most of the methionine tryptic peptides of the two envelope structural proteins. Its low molecular weight, compared to that of the virion proteins, may be due to degradation of the nonglycosylated *in vitro* products and premature termination. The results indicate that genome RNA, as well as the 26 S mRNA from togavirus-infected cells, can, under some conditions, code for viral structural proteins, suggesting that the larger RNA may have more than one initiation site for polypeptide synthesis.

The 26 S RNA isolated from Sindbis and Semliki Forest virus infected cells apparently codes for the synthesis of only the virion structural protein *in vitro*. In L-cell or ascites cell-free systems programmed with Sindbis (Cancedda *et al.*, 1975*a*; Cancedda and Schlesinger, 1974) or Semliki Forest virus 26 S mRNA (Wengler *et al.*, 1974; Clegg and Kennedy, 1975), the major product was capsid protein as identified by gel analysis and tryptic peptide mapping. In L-cell extracts programmed with Semliki Forest virus 26 S RNA, both of the envelope proteins were also synthesized in the same relative amounts as found in virions. The envelope polypeptides made *in vitro* were apparently nonglycosylated (Clegg and Kennedy, 1975). Other discrete, larger and smaller products were found in the Sindbis-specific products of reticulocyte and ascites extracts (Cancedda and Schlesinger, 1974), and they also contained capsid tryptic peptides. The time course of appearance of the capsid and envelope protein products supports the model of a single initiation site in the 26 S RNA. However, specific proteolytic cleavage has not been observed in most cell-free systems, and it is therefore of some interest that capsid and envelope proteins of the correct size are major products of the *in vitro* reactions.

In addition to capsid protein, 26 S Sindbis RNA coded for a polypeptide product of 100K in reticulocyte lysates (Simmons and Strauss, 1974*b*). It comigrated in SDS-polyacrylamide gels with the precursor polypeptide of the viral envelope proteins and was precipitated with antibodies against Sindbis virus. Ascites and L-cell extracts also synthesized immunologically detectable envelope proteins in response to 26 S RNA (Cancedda *et al.*, 1974*a*; Clegg and Kennedy, 1975). It has not been established if the envelope precursor is derived by cleavage of a polyprotein-containing capsid sequence or if initiation can also occur at a second site in the 26 S RNA. Experiments with viral mRNA from cells infected with temperature-sensitive mutants of Sindbis virus are consistent with a single initiation site. Cell-free extracts were directed by 26 S RNA that was isolated from cells infected with *ts*2, a temperature-sensitive mutant of Sindbis virus. At nonpermissive temperatures, *ts*2-infected cells accumulate a 130K precursor of the three viral structural proteins. Similarly, in reticulocyte extracts incubated at the restrictive temperature of 37°C with *ts*2 26 S RNA, a 130K protein was the only product synthesized. It comigrated in SDS-polyacrylamide gels with the viral structural protein precursor from infected cells (Simmons and Strauss, 1974*b*). At the permissive temperature of 27°C, *ts*2 26 S RNA directed the synthesis of capsid protein and a 100K product, the presumptive precursor of envelope proteins (Simmons and Strauss, 1974*b*). From the results of the *in vitro* studies, it appears that 26 S RNA codes for structural proteins, possibly in a manner analogous to that of picornavirus genome RNA.

The infectious togavirus genome RNA can also code for virion proteins under some *in vitro* conditions but in infected cells may function primarily as messenger for the virion RNA polymerase(s) and other nonstructural viral proteins. Further *in vitro* studies with initiator tRNA acylated with [^{35}S]methionine should help to establish whether structural and nonstructural proteins are synthesized directly from different initiation sites in the 49 S RNA or if a single site, presumably near the 5′ end of the genome RNA, is used to make a polyprotein that is subsequently cleaved. Similar studies with 26 S RNA may distinguish between *in vitro* translational models that involve (1) a single 5'-terminal initiation site with readthrough into envelope protein sequences and cleavage of the polyprotein products, (2) multiple initiation sites for the virion structural proteins, or (3) a combination of the two models, i.e., initiation sites for capsid protein and envelope precursor polypeptide with cleavage of the latter to form E1 and E2.

4.2. Class 2 Viruses (Genome Complementary to mRNA)

4.2.1. Rhabdoviruses

4.2.1a. Proteins

Among the various rhabdoviruses including those of mammals, plants, fish, and insects, vesicular stomatitis virus (VSV) has been most extensively studied (Wagner, 1975). Purified VSV contains five distinct structural proteins, which are the glycoprotein (G, molecular weight 70K), located as surface projections on the outer membrane of the virion; the matrix polypeptide (M, molecular weight 29K), which is nonglycosylated and is the constituent protein of the viral membrane; two other minor proteins (L, molecular weight 170K, and phosphoprotein NS, molecular weight 40K); and the nucleocapsid protein (N, molecular weight 50K), which is associated with the single-stranded genome RNA (molecular weight 3.6×10^6) to form the ribonucleoprotein (RNP) core. These five virus-coded proteins are all present in infected cells. Nonstructural viral proteins have not been detected in VSV-infected cells. In contrast to class 1 RNA viruses, the virus-specific proteins of VSV are formed in infected cells from monocistronic viral mRNA templates rather than by cleavage of a giant precursor polypeptide molecule. The total molecular weight of the individual VSV proteins corresponds closely to the predicted coding potential of the complete viral genome RNA.

4.2.1b. mRNA

The mRNAs released from polysomes of infected cells and analyzed by velocity sedimentation fall into two size groups: a relatively homogeneous 28–31 S group and a more heterogeneous ~15 S class (Huang *et al.*, 1970; Mudd and Summers, 1970). The ~15 S RNA can be resolved further into three distinct peaks at 17 S, 14.5 S, and 12 S by velocity sedimentation (Moyer *et al.*, 1975*a*) or formamide-polyacry-lamide gel electrophoresis (Knipe *et al.*, 1975; Moyer *et al.*, 1975*a*) of radioactive RNA. These mRNA species are complementary to the genome RNA and have been shown to contain poly(A) stretches ranging from 50 to 250 bases (Ehrenfeld and Summers, 1972; Soria and Huang, 1973). They also possess a high degree of secondary structure (Moyer *et al.*,1975*a*). the approximate molecular weights of the mRNAs based on their migration in polyacrylamide gels are as follows: 28–31 S, 1.7×10^6; 17 S, 0.8×10^6; 14.5 S, 0.6×10^6; 12 S, 0.33×10^6 (average values from the data of Moyer *et al.*, 1975*a*; Knipe *et al.*, 1975; Schincariol and Howatson, 1970). It has been shown that the different species of VSV mRNA in infected cells are compartmentalized (Grubman *et al.*, 1975); i.e., the 17 S mRNA is attached to membrane-bound polysomes while the other species are present in the free polysomes.

Purified VSV contains an RNA-dependent RNA polymerase (Baltimore *et al.*, 1970). Upon disruption of virions *in vitro* with a nonionic detergent, the polymerase synthesizes RNA that is complementary to the genome RNA (Baltimore *et al.*, 1970; Bishop and Roy, 1971). It has been shown that the RNA products synthesized *in vitro* by purified VSV consist of several distinct species with molecular weights similar to those of the mRNA found in VSV-infected BHK cells (Moyer and Banerjee, 1975). The *in vitro* mRNA molecules also contain 3′-terminal poly(A) (Banerjee and Rhodes, 1973; Banerjee *et al.*, 1974) and are similar in nucleotide sequence to *in vivo* mRNAs of the same size (Moyer *et al.*, 1975*a*). Thus, by several criteria, the VSV mRNAs from infected cells are very similar if not identical to those synthesized *in vitro* by the virion-associated polymerase. A typical sedimentation profile of the *in vitro* and *in vivo* VSV mRNAs is shown in Fig. 3.

4.2.1c. *In Vitro* Protein Synthesis

From a comparison of the molecular weights of the VSV mRNA species and the structural proteins, it seems likely that each species of RNA is monocistronic and codes for an individual polypeptide. Utilizing cytoplasmic fractions of VSV-infected cells, Grubman and Summers (1973) and Ghosh *et al.* (1973) demonstrated the synthesis *in vitro* of polypeptides similar in electrophoretic mobilities to the five VSV structural proteins. Furthermore, it was shown that only the

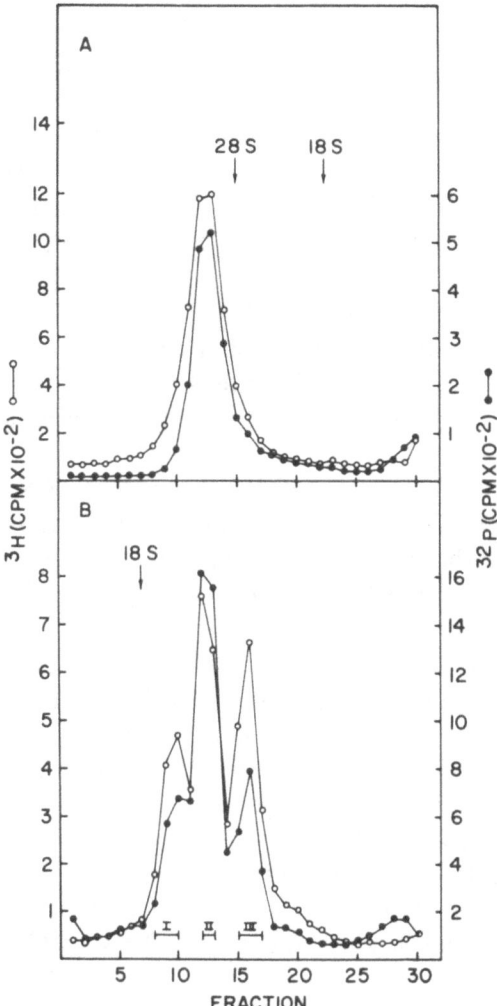

Fig. 3.   Velocity sedimentation analysis of the *in vitro* and *in vivo* VSV RNA species.
VSV product RNA was synthesized *in vitro* with purified VSV using [α-^{32}P]UTP as the
labeled precursor. *In vivo* VSV mRNAs labeled with [^{3}H]uridine were isolated from
infected cells. Each class of poly(A)-containing RNA was isolated and purified (Moyer
et al., 1975a). The corresponding species were mixed, and the larger class of RNA (A)
was analyzed on SDS-sucrose gradients in an SW41 rotor at 23,000 rpm for 18 h at
23°C, while the smaller class (B) was analyzed at 33,000 rpm for 17 h at 23°C. The
BHK cell ribosomal RNA markers, 28 S and 18 S, were analyzed in parallel gradients,
and their sedimentation positions are indicated by the arrows. O, ^{3}H-labeled *in vivo*
RNA; ●, ^{32}P-labeled *in vitro* RNA. From S. A. Moyer (unpublished data).

membrane-bound polysomes directed the synthesis of apparently nonglycosylated G protein *in vitro* (Grubman *et al.*, 1974; Toneguzzo and Ghosh, 1975*b*). Grubman and Summers (1973) found that a soluble (S10) fraction from uninfected cells stimulated protein synthesis by membrane-bound polysomes, and also was essential for synthesis of G protein *in vitro*.

More direct evidence concerning the coding capacity of the different VSV mRNA species has been obtained from translation studies of individual mRNA species in a variety of cell-free systems. Morrison *et al.* (1974) showed that mRNA species isolated from polyribosomes of VSV-infected cells can direct the synthesis of virus-specific proteins by cell-free extracts of rabbit reticulocytes and wheat germ. Utilizing the rabbit reticulocyte system, the VSV 28 S RNA apparently directed the synthesis of structural protein L, and the 13–15 S RNA species coded predominantly for the synthesis of polypeptides N, NS, and M. On the other hand, the wheat germ cell-free system synthesized N, NS, M, and also a G-like protein in response to the 13–15 S RNA. Analysis of VSV *in vitro* products labeled with [^{35}S]formylmethionine tRNA$_F{}^{Met}$ indicated that the 28 S RNA directed the initiation of a single protein, whereas the 13–15 S RNA directed the initiation of at least four proteins. These results suggested that VSV mRNA includes five monocistronic messages. More recently, the individual VSV mRNA species synthesized *in vitro* by the virion-associated polymerase were identified on the basis of their translation into authentic polypeptides in wheat germ extract (Both *et al.*, 1975*c*). The 17 S RNA directed the synthesis of two proteins of molecular weights 63K and 60K (P63 and P60). Since there is no evidence for *in vitro* glycosylation by wheat germ extracts, the P63 protein was assumed to be nonglycosylated G protein; tryptic digests of [^{35}S]methionine-labeled P63 were identical to those of [^{3}H]methionine-labeled authentic G protein. The P60 protein was probably prematurely terminated P63 protein. The 14.5 S RNA coded exclusively for VSV N protein, and the 12 S RNA coded for both M and NS proteins. The tryptic digests of these proteins were identical to those of the corresponding authentic viral proteins. It is interesting to note that the *in vitro* synthesized NS phosphoprotein, like the viral NS protein, had different relative electrophoretic mobilities in two gel systems of different pH (Both *et al.*, 1975*c,d*; Moyer and Summers, 1974; Imblum and Wagner, 1974). Toneguzzo and Ghosh (1975*a*) have also shown that mRNAs synthesized *in vitro* by viral ribonucleoprotein cores from infected cells can direct the synthesis of the VSV-specific polypeptides *in vitro*.

Similar *in vitro* translation studies identified the individual viral mRNAs from VSV-infected cells (Both *et al.*, 1975c; Knipe *et al.*, 1975). The 17 S RNA, which is present only in the membrane-bound polysomes (Grubman *et al.*, 1975), was shown to direct the synthesis of the P63 protein. Again, the 14.5 S RNA isolated from the cytoplasmic unbound polysomes directed the synthesis of the N protein, and the 12 S RNA coded for both the M and NS proteins (Fig. 4). Similar results were also obtained with L-cell (Toneguzzo and Ghosh, 1975b) and CHO-cell extracts (Morrison and Lodish, 1975). Knipe *et al.* (1975) using mRNA species separated by polyacrylamide gel electrophoresis in formamide obtained the same results. The consistent synthesis of both M and NS proteins in response to 12 S mRNA indicated that it consisted of two mRNA species (Both *et al.*, 1975c,d; Knipe *et al.*, 1975). This was confirmed recently by two-dimensional oligonucleotide fingerprints of the 12 S RNA. It was shown that the oligonucleotide pattern of digests by RNase T_1 plus RNase A was consistent with the presence of at least two different RNA molecules (Rose and Knipe, 1975). Table 1 summarizes the molecular weights of the VSV mRNA species, their translation products, and their coding potentials.

Breindl and Holland (1975) have developed a coupled transcription–translation system. VSV ribonucleoprotein cores were added directly to protein-synthesizing systems derived from Krebs II ascites cells or wheat embryos. Polypeptide synthesis commenced after an initial lag of 15–30 min and continued for 2–3 h. The newly synthesized RNAs formed during the first part of the incubation subsequently were translated into VSV-specific polypeptides, including N, NS, M, and a small amount of G precursor but no detectable L protein.

The important findings obtained by translating VSV mRNAs *in vitro* can be summarized as follows: (1) the *in vitro* synthesized mRNAs are biochemically active products; (2) the protein moiety of the G protein has an apparent molecular weight of 63K rather than 66K observed for fully glycosylated G protein; (3) the membrane-bound 17 S mRNA functions as messenger for the G protein; (4) although M is a membrane protein, its mRNA is present in the cytoplasm on free polysomes; (5) the 12 S RNA contains two distinct mRNA species coding for NS and N proteins; (6) the molecular weight of the mRNA coding for NS protein is approximately 0.3×10^6, i.e., considerably less than that expected for the NS protein which migrates as a 50K protein in polyacrylamide gels at pH 8.7; (7) the *in vitro* translation system is a potentially useful one for studying RNA–protein interactions and possibly for reconstituting subviral particles from purified genome

RNA and *in vitro* synthesized viral structural polypeptides. Dissociation of purified VSV virions and reconstitution of the RNP and its protein components with restoration of transcriptase activity and infectivity have already been achieved (Emerson and Wagner, 1973; Bishop *et al.*, 1974).

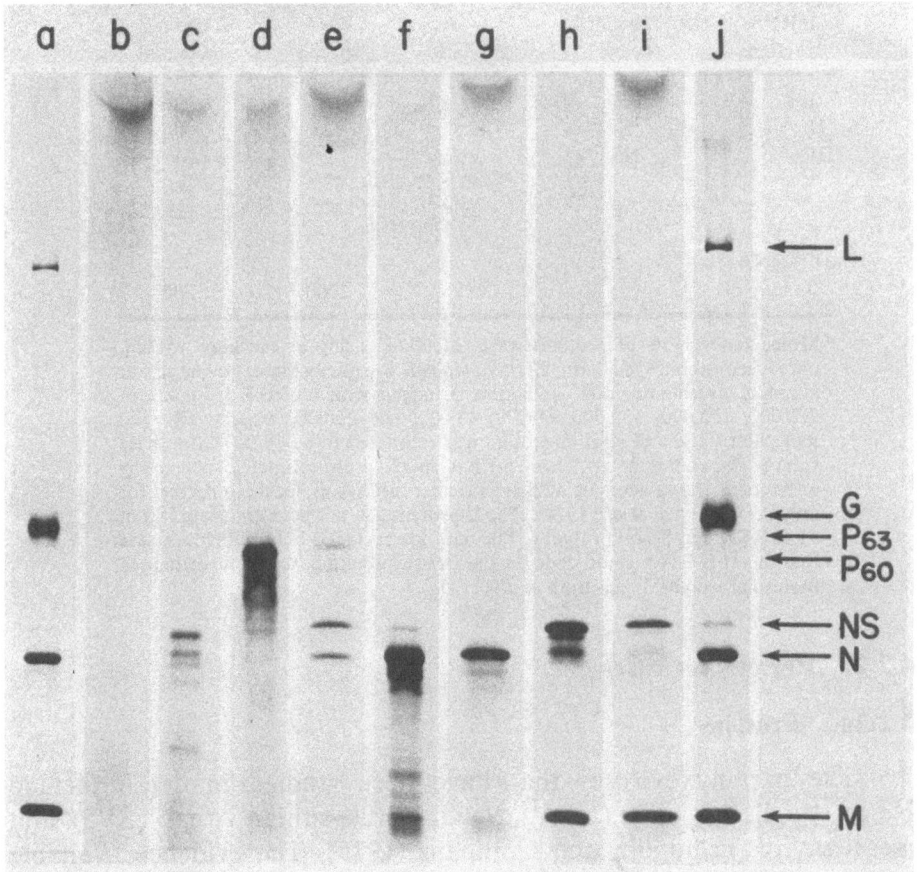

Fig. 4.   Autoradiogram of polypeptides synthesized by wheat germ extracts in response to VSV RNA synthesized *in vivo* or *in vitro*. VSV mRNA species were isolated, and the purified classes I, II, and III were obtained as described in Fig. 3. Migration is from top to bottom. (a) Proteins of the virion. Proteins synthesized in wheat germ extracts (b) in the absence of RNA and in response to (c) RNA from mock-infected cells. (d) Fraction I RNA from membrane-bound polysomes. (e) Fraction I RNA synthesized *in vitro*. (f) Fraction II RNA from infected cell cytoplasmic supernatant. (g) Fraction II RNA synthesized *in vitro*. (h) Fraction III RNA from infected cell cytoplasmic supernatants. (i) Fraction III RNA synthesized *in vitro*. (j) Marker proteins of purified virus. From Both *et al.* (1975c).

TABLE 1

Coding Potential of the VSV mRNA Species Synthesized *in Vitro*[a]

| mRNA fractions | Proteins coded | Molecular weights (in thousands) | | |
		Proteins	RNA species	Estimated coding potential
I	P63	63	700	66
II	N	47.5	525	48
III	NS	52		
		40	372	33
	M	29	372	33
31 S RNA *in vitro*	(L)	195	2000	206

[a] Molecular weights of proteins were calculated using as markers myosin, unreduced rabbit γ-globulin, *Escherichia coli* β-galactosidase, bovine serum albumin, ovalbumin, and α-chymotrypsinogen with molecular weights of 200,000, 150,000, 130,000, 68,000, 45,000, and 25,000, respectively. The molecular weight of viral G protein was estimated to be 66,000 from these values. The coding potentials of mRNA species were estimated subtracting an average of 140 adenylic acid residues per mRNA molecule to correct for poly(A) (Banerjee *et al.*, 1974). The L protein has not been translated from 31 S RNA synthesized *in vitro*. The data are included only to demonstrate that the 31 S RNA could code for the viral L protein based on its estimated molecular weight. From Both *et al.* (1975c).

4.2.2. Paramyxoviruses

4.2.2a. Proteins

The paramyxoviruses that have been studied in detail include Newcastle disease virus (NDV), Sendai virus, simian virus 5 (SV5), and measles virus (Choppin and Compans, 1975). The evidence available indicates that paramyxoviruses contain four major structural proteins and a few additional minor proteins that vary in different strains. Two glycoproteins (HN, 65–74K; F, 56K) are located at the virion's outer surface. Neuraminidase and hemagglutinin activities are both associated with the HN glycoprotein (Scheid and Choppin, 1973, 1974; Tozawa *et al.*, 1973). The F glycoprotein is involved in virus-induced cell fusion and hemolysis (Scheid and Choppin, 1974; Homma and Ohuchi, 1973). It is interesting to note that an additional glycoprotein (designated FO, molecular weight 65K) was found in Sendai virus grown in Madin–Darby bovine kidney (MDBK) cells; it was subsequently shown to be a precursor of glycoprotein F (Scheid and Chop-pin, 1974; Homma and Ohuchi, 1973). The nucleocapsid protein (N) appears to be similar in molecular weight (60K) in all paramyxoviruses studied. The other major structural protein is a nonglycosylated membrane protein (M, 38–41K). Several minor proteins have been described in paramyxoviruses, including (1) a protein in NDV similar to L protein of VSV, (2) a 69K protein in Sendai, and (3) a 50K protein in SV5. In paramyxovirus-infected cells, all virus-specific proteins apparently are independently synthesized and are found in the same proportions. The glycoprotein (F) appears to be the only paramyxo-virus protein which may arise by posttranslational cleavage.

4.2.2b. mRNA

The paramyxoviruses, like the rhabdoviruses, direct in infected cells the synthesis of multiple species of mRNA that are comple-mentary to the viral genome. Collins and Bratt (1973) have shown that the major single-stranded, complementary RNA in NDV-infected cells sediments at 18–22 S and contains poly(A) (Weiss and Bratt, 1974). The heterogeneous mixture of 18–22 S RNAs can be resolved by polyacrylamide gel electrophoresis into six or seven species with molecular weights ranging from 0.55 to 1.53×10^6. Although it was not clear from these results whether each RNA fraction contained a unique mRNA species, the molecular weight of each was consistent with its functioning as a monocistronic mRNA for the synthesis of a protein found in purified virions or infected cells. Assuming that each RNA species is unique, it is important to note that the sum of their molecular weights is approximately 6.8×10^6, in reasonable agreement with the coding potential of the NDV genome RNA (molecular weight 6×10^6). However, the 18–22 S RNA class, on the basis of hybridization results, appears to contain only 50% of the genome RNA sequences (Bratt and Robinson, 1967), a discrepancy that remains to be clarified by addi-tional experiments.

The paramyxoviruses which have been demonstrated to contain a virion-associated RNA polymerase include NDV (Huang *et al.*, 1971), Sendai (Robinson, 1971; Stone *et al.*, 1971), SV5 (Choppin, personal communication), and mumps virus (Bernard and Northrop, 1974). In each case, the RNA products synthesized *in vitro* which are comple-mentary to, but smaller than, the genome RNA sediment as a heterogeneous mixture at about 16 S and contain poly(A) (Weiss and Bratt, 1974).

4.2.2c. *In Vitro* Protein Synthesis

Neither the *in vitro* nor the *in vivo* mRNAs of paramyxoviruses have been thoroughly characterized, and only one study of their translation in cell-free systems has been reported (Kingsbury, 1973). A rabbit reticulocyte system programmed with 18 S complementary RNA isolated from Sendai-infected cells was shown to synthesize three predominant proteins *in vitro*. The proteins were precipitated with antisera prepared against the solubilized virions. The major polypeptide product was characterized as similar to the nucleocapsid protein by its tryptic peptide pattern and its mobility in SDS-polyacrylamide gels. Sendai virion RNA did not stimulate polypeptide synthesis *in vitro*, suggesting that it does not function as viral messenger RNA. The purified RNA from Newcastle disease virions also did not stimulate incorporation of radioactive amino acids into acid-precipitable polypeptides in a cell-free system obtained from *E. coli* (Siegert *et al.*, 1973).

4.3. Class 3 Viruses (Segmented Genomes)

4.3.1. Influenza Virus

Influenza virus contains a single-stranded genome of total molecular weight 4–5×10^6 that consists of multiple RNA segments ranging in weight from 3.4 to 9.8×10^5 (Compans and Choppin, 1975). The genome RNA, like class 2 virion RNAs, presumably is a "negative" strand, i.e., does not function as viral messenger RNA. It hybridizes well with the RNA products synthesized *in vitro* by the virion-associated RNA polymerase (Chow and Simpson, 1971) and with the RNA isolated from polysomes of infected cells (Pons, 1973). The influenza structural proteins consist of about seven polypeptides with an apparent molecular weight range of 25–94K. They include two outer surface glycoproteins: a hemagglutinin activity (HA, 75K) and a neuraminidase activity (NA, 55K). The HA protein undergoes cleavage during purification of virions, resulting in the appearance of two additional glycoproteins, HA1 (50K) and HA2 (25K). The nonglycosylated membrane protein (M, 26K) is located at the inner viral membrane. The segments of RNA form nucleocapsids (RNP) by association with a major protein (NP, 55K) and one or possibly two other proteins (P1, 94K; P2, 81K). Thus there appears to be a good correlation between the molecular weights of the structural proteins and the coding potentials of the genome segments of influenza virus. All of the major polypeptides of the virion as well as one or two virus-specific nonstructural polypeptides have been identified in virus-infected cells.

Influenza virus-specific mRNAs have been shown to be present in the polysomes of infected cells (Pons, 1973), but details of their size and sequence are not known. Moreover, the isolation of the virus-specific mRNAs from polysomes was complicated by the presence of viral genome RNA sedimenting in the position of polysomes. The report by Etkind and Krug (1974) that influenza virus-specific mRNAs contain poly(A), in contrast to the report of Avery (1974), may aid in the purification of viral mRNA from infected cells. The mRNAs can be separated from genome RNA that contains no poly(A) (Etkind and Krug, 1974; Avery, 1974) by binding to and elution from poly(U)-sepharose or oligo(dT)-cellulose.

mRNAs synthesized *in vitro* by the virion-associated RNA polymerase are heterogeneous and of low molecular weight (Chow and Simpson, 1971; Penhoet *et al.*, 1971; Skehel *et al.*, 1971) and are probably unsuitable for cell-free translation studies. In spite of these difficulties, Kingsbury and Webster (1973) were able to obtain virus-specific products by utilizing total RNA from influenza virus-infected chick embryo cells as messenger in a cell-free rabbit reticulocyte system. Rabbit antiserum prepared against a recombinant influenza virus (which yielded higher titers of antibody to internal virion components) specifically precipitated a viral RNA-directed *in vitro* product that migrated with M protein during polyacrylamide gel electrophoresis. Several other minor proteins were also resolved. Antiserum made against purified M protein also specifically precipitated the newly formed material that comigrated with M protein. From these results, it appears that the mRNA for the M protein was present in large amounts in the RNA preparation; alternatively, optimum conditions for translation of other mRNA species were not achieved. Influenza virion RNA failed to stimulate the incorporation of amino acids into products corresponding to viral polypeptides, and total acid-precipitable radioactivity as compared to endogenous reactions was depressed about threefold by the addition of virion RNA. In contrast to these findings with an eukaryotic cell-free system, Siegert *et al.* (1973) reported that in an *E. coli* system influenza virus genome RNA was translated into polypeptides that were antigenically indistinguishable from influenza structural protein N. If confirmed, these results may indicate that one (or more) of the genome segments of influenza virus has the polarity of messenger RNA.

4.3.2. Reoviridae

Human reoviruses can be considered the prototype of a wide variety of viruses that contain segmented double-stranded RNA genomes. These include viruses of vertebrates, insects, plants, molds, and bacteria. Of these, the reoviruses, and especially serotype 3, have been studied most intensively (Joklik, 1974).

4.3.2a. Virion Proteins

Reovirions contain double-stranded RNA of total molecular weight 15×10^6 that is reproducibly isolated as ten discrete segments. The double-stranded RNA mixture consists of molecules in three size categories: three species of large RNA (L1–L3, $2.3–2.7 \times 10^6$ daltons), three medium (M1–M3, $1.3–1.6 \times 10^6$ daltons), and four small (S1–S4, $0.6–0.9 \times 10^6$ daltons). These subgenomic segments are contained within a characteristic double capsid-shell structure. The proteins of purified reovirus also fall into three size classes designated λ, μ, and σ. The nomenclature for the reovirus proteins used in this chapter extends that of Zweerink *et al.* (1971). The polypeptides designated as primary gene products have been numbered according to their decreasing molecular weights within the λ, μ, and σ size classes. Other virus-specific proteins have been named according to their estimated molecular weights (Both *et al.*, 1975*b*). Four proteins in the λ size class ($\lambda 1$, $\lambda 2$, $\lambda 3$, and P135 of 153K, 148K, 143K, and 135K daltons, respectively), three in the μ size class ($\mu 1$, P73, and $\mu 3$ of 79K, 73K, and 72K daltons, respectively), and three in the σ size class ($\sigma 1$, $\sigma 2$, and $\sigma 4$ of 54K, 52K, and 43K daltons, respectively) have now been resolved in purified virions (Both *et al.*, 1975*b*); polypeptides P135, $\mu 3$, and $\lambda 3$ had not previously been detected. All of these polypeptides, except P135 and P73, are apparently primary translation products of the reovirus genome (Both *et al.*, 1975*b*). Presumably P135 is derived from one of the λ proteins by proteolytic cleavage; P73, the major virion protein in the μ size class, is probably also derived by proteolytic cleavage from the virion polypeptide $\mu 1$ during morphogenesis of the virus (Zweerink *et al.*, 1971).

The molecular weight relationship between the mRNAs transcribed from the individual genome duplex segments and the virion structural proteins suggests that they are coded for by monocistronic mRNAs. Although ten genome segments and their single-stranded transcripts are found in infected cells (Joklik, 1974; Ward *et al.*, 1972), only eight primary gene products are present in virions as structural polypeptides. Moreover, in infected cells analyzed 17.5–19.5 h after infection at 31°C (Zweerink *et al.*, 1971) only eight primary viral gene products were previously resolved. These included two λ, one μ, and three σ capsid proteins and one μ and one σ noncapsid polypeptide (Zweerink and Joklik, 1970). Thus proteins corresponding to the mRNA species *l*2 (or *l*3 and m3 were not detected. Greater resolution of the viral proteins has been obtained by SDS-polyacrylamide slab gel electrophoresis, and ten reovirus primary gene products have been detected in the solubilized infected cell lysates (Both *et al.*, 1975*b*). These included a minor gene product, μ2 (77K daltons), which, in addition to λ3, had not been previously described. In addition, several other virus-specific polypeptides were detected (see Table 2); presumably these arise by proteolytic cleavage during viral infection.

4.3.2b. mRNAs

Reovirus contains an RNA-dependent RNA polymerase (Shatkin and Sipe, 1968; Borsa and Graham, 1968) which transcribes *in vitro* one strand of each of the duplex segments into the corresponding single-stranded RNA molecules (Skehel and Joklik, 1969; Banerjee and Shatkin, 1970; Levin *et al.*, 1970). The same strand of the genome segments is transcribed into viral mRNA in the infected cells and can be detected in polysomes (Ward *et al.*, 1972; Hay and Joklik, 1971). Since purified reovirus cores continue to synthesize *in vitro* correctly initiated and terminated viral mRNA for many hours or even days, they provide an excellent source of large amounts of purified viral mRNA. The large (*l*), medium (m), and small (s) classes of *in vitro* mRNA can be separated and purified by density gradient centrifugation (Both *et al.*, 1975*b*), or the ten individual species can be resolved by polyacrylamide gel electrophoresis (Floyd *et al.*, 1974).

4.3.2c. *In Vitro* Protein Synthesis

Cytoplasmic fractions prepared from reovirus-infected cells containing virus-specific polysomes effectively incorporate radioactive amino acids into viral polypeptides *in vitro* (McDowell and Joklik, 1971). The polypeptide products included eight reovirus proteins (or

TABLE 2
Estimated Molecular Weights of Reovirus Proteins

Virus-specific proteins in infected cells[a]	Apparent molecular weight[b]	Virus-specific proteins synthesized *in vitro*	Virion structural proteins	Previous nomenclature[c]
λ1	153,000	+	+	λ1
λ2	148,000	+	+	λ2
λ3	143,000	+	+	
P135	135,000	−	+	
μ1	79,000	+	+	μ1
μ2	77,000	+	−	
μ3	72,000	+	+	μ0
P73	73,000	−	+	μ2
P69	69,000	−	−	
P66	66,000	−	−	
σ1	54,000	+	+	σ2
σ2	52,000	+	+	σ1
σ3	49,000	+	−	σ2A
σ4	43,000	+	+	σ3
P40	40,000	−	−	
P36	36,000	−	−	

[a] Those proteins designated as primary gene products are numbered according to their decreasing molecular weight within a size class. Other virus-specific proteins are labeled according to their molecular weight.

[b] The molecular weights of the λ and μ polypeptides were determined using the SDS-gel system described by Laemmli (1970) and Anderson *et al.* (1973) with myosin (200,000), unreduced rabbit γ-globulin (150,000), β-galatosidase (130,000), *E. coli* elongation factor G (80,000), BSA (68,000), and ovalbumin (45,000) as molecular weight markers. The σ polypeptides were compared to BSA and ovalbumin in the gel system described by Maizel (1971).

[c] Tentative assignment of previously described proteins (Zweerink and Joklik, 1970; Zweerink *et al.*, 1971) to those in the leftmost column is based on their relative amounts in purified virions and reovirus-infected cells. More definitive assignments await peptide analyses.

primary gene products), six of which comigrated with reovirus capsid proteins. Two additional noncapsid polypeptides in the μ and σ classes were also synthesized. In contrast to the other viral capsid proteins, the major μ polypeptide was not detected *in vitro,* but a large amount of the minor capsid μ polypeptide was present, suggesting that it is a precursor of the major μ protein in virions.

The first studies with a cell-free system directed by exogenous reovirus mRNA were reported by Levin *et al.* (1971), who observed that pretreatment with formaldehyde of reovirus mRNA synthesized *in vitro* resulted in an increased stimulation of amino acid incorporation into acid-precipitable products by L-cell extracts. The products obtained were of low molecular weight (10–15K). Subsequently Graziadei and Lengyel (1972), utilizing a supplemented S30 fraction from L cells, were able to translate the m class of reovirus mRNA into polypeptides which comigrated in polyacrylamide gels with μ polypeptides. Similarly, the small class of mRNA was translated into σ polypeptides. Later, the same group (Graziadei *et al.*, 1973) employed antisera against the three size classes of reovirus structural proteins to show that each serum immunoprecipitated only the polypeptide products that corresponded to the size class of virion proteins used to prepare the specific antiserum. Although the exact number of reovirus-specific polypeptides synthesized *in vitro* was not determined, the results indicated that the *in vitro* synthesized RNAs were active as messengers and were translated with considerable fidelity *in vitro*.

Translation of reovirus mRNAs was also achieved in cell-free systems derived from other eukaryotic cells. McDowell *et al.* (1972) utilized preincubated S30 extracts prepared from Krebs II mouse ascites cells, mouse L cells, Chinese hamster ovary cells, HeLa cells, and non-preincubated rabbit reticulocyte lysates. In each of these extracts (with the exception of rabbit reticulocytes), the addition of reovirus mRNA stimulated protein synthesis five- to tenfold, with maximal stimulation occurring between RNA concentrations of 50 and 120 μg/ml. In ascites cell-free extracts, reovirus mRNA-directed protein synthesis was abolished by greater than 90% upon additon of cycloheximide, emetine, or anisomysin, but was unaffected by chloramphenicol. The major viral polypeptides synthesized were two μ and σ products, as determined by their comigration in gels with authentic virion capsid proteins or with proteins present in infected cells. However, in non-preincubated rabbit reticulocyte lysates, reovirus mRNA inhibited the endogenous synthesis of globin but stimulated the synthesis of higher molecular weight polypeptides that coelectrophoresed with reovirus polypeptides; however, only eight primary gene products were detected.

In cell-free extracts from wheat germ, at the optimal concentrations of 90 mM K^+ and 3 mM Mg^+, Both *et al.* (1975*b*) detected the synthesis *in vitro* of all ten presumptive primary gene products of reovirus. SDS-polyacrylamide slab gels, rather than disc gels, were employed to achieve higher resolution of the products by electrophoresis. In these studies, a primary gene product was defined as a protein that was (1) present in infected cells pulse-labeled with radioactive amino acids for a short time, (2) comigrated in an SDS-polyacrylamide gel with a protein synthesized *in vitro* in response to purified viral mRNA, and (3) had a molecular weight consistent with the coding potential of the viral mRNA. By these criteria, 10 of the 16 virus-specific polypeptides in infected L cells were identified as primary gene products. Table 2 summarizes the various reovirus-specific polypeptides synthesized *in vitro* and *in vivo*. The nomenclature has been modified from that originally suggested by Zweerink *et al.* (1971), which is also included in the table. It can be seen that as many as ten polypeptides can be detected in virions. Among them $\lambda 3$, P135, and $\mu 3$ ($\mu 0$, Zweerink *et al.*, 1971) had not previously been detected probably because of their low level in the virion. Eight of the structural polypeptides are primary gene products. In reovirus-infected cells, polypeptides P135, P73, P69, P66, P40, and P36 presumably are cleavage products of other viral proteins. The major polypeptide of the μ class in infected cells, P73, is a cleavage product of $\mu 1$ and thus is not found among the *in vitro* products coded for by reovirus mRNA. This finding is consistent with the inability of most cell-free systems to process polypeptide products by specific proteolytic cleavage.

Reovirus mRNAs isolated from polysomes or synthesized *in vitro* do not contain detectable poly(A) sequences (Stoltzfus *et al.*, 1973; Both *et al.*, 1975*b*). The *in vitro* translation studies demonstrate that the presence of poly(A) sequences in mRNA is not needed for their effective translation. Moreover, the relative amounts and efficiency of synthesis of particular classes of polypeptides *in vitro* and *in vivo* are quite similar (Both *et al.*, 1975*b*). For example, polypeptide $\mu 3$ is the major protein synthesized *in vitro*, even when the three species of m class mRNAs are added in equal amounts to cell-free extracts. Thus the relative efficiency of translation of a particular mRNA may depend on an intrinsic property, such as its secondary structure. Levin *et al.* (1972) have shown that the three size classes of reovirus mRNA synthesized *in vitro* can form protein synthesis initiation complexes with rat liver [^{35}S]Met tRNA$_F^{Met}$ incubated with salt-washed 40 S and 60 S ribosomal subunits. This process required mild treatment of the mRNA with formaldehyde. The initiation complex reacted with puromycin, and the resultant puromycin-peptide was identical to the methionyl-puromycin formed in response to poly(A,G,U). These results indicate that reovirus mRNAs and possibly all eukaryotic mRNAs contain the AUG initiator codon which binds to the puromycin-reactive site on ribosomes in the presence of Met tRNA$_F^{Met}$.

4.4   Class 4 Viruses (RNA Tumor Viruses)

4.4.1.   Virion Proteins

Studies of the structure and function of RNA tumor viruses have increased in number and intensity during the past few years. Consequently, considerable new information is becoming available on the mechanism of viral protein biogenesis. Purified avian (e.g., avian myeloblastosis virus, AMV; Rous sarcoma virus, RSV) and mammalian (e.g., Rauscher murine leukemia virus, RLV) RNA tumor viruses consist of a ribonucleoprotein core surrounded by a lipoprotein envelope which has surface projections of glycoproteins (Bader, 1975). The core includes the ~70 S genome RNA, which consists of two or three ~35 S subunits, 4 S transfer RNA, and other low molecular weight RNAs. On the basis of oligonucleotide fingerprint patterns (Duesberg *et al.*, 1974; Quade *et al.*, 1974; Weissman *et al.*, 1974) and hybridization studies (Baluda *et al.*, 1974), it appears that the two or three genome subunits are very similar, if not the same. The estimated polypeptide coding capacity of the genome subunit RNA, like that of picornavirus genome RNA, is about 250–300K.

Avian virus structural proteins include four internal, group-specific (*gs*) antigens that are immunologically distinct but found in all avian leukemia and sarcoma virus isolates. Two of these four nonglycosylated internal proteins are in the virus core. The apparent molecular weights of the avian *gs* antigens 1–4 as determined by gel filtration are 27K, 19K, 15K, and 12K, respectively (Fleissner, 1971; August *et al.*, 1974). A fifth, minor internal protein with a molecular weight of about 10K but no known antigenic function is also part of the virion. The avian oncornavirus serotype specificity, which varies among different virus isolates and subgroups, resides in the two large glycoproteins that comprise the surface spike structures. The molecular weights of these two polypeptides are difficult to establish accurately because the constituent carbohydrate moieties are large, and the degree of glycosylation varies in different virus strains. Thus C-type viruses of the avian group contain a total of seven distinct structural proteins, each with a molecular weight considerably below the theoretical informational content of the 35 S RNA subunit.

The murine oncornaviruses also contain several structural polypeptides (Bader, 1975). They include three internal proteins of molecular weights 30K (the predominant core protein), 16K, and 14K that together comprise 50–75% of the total virion proteins. A fourth minor internal component has a molecular weight of 10K. Three large type-specific glycoproteins of apparent molecular weights about 180K, 80K, and 60K are also present in RLV and other murine C-type virion particles.

4.4.2. Virus-Specific Proteins in Productively Infected Cells

In infected chick cells producing AMV or RSV, cellular proteins continue to be synthesized, and only a small proportion (~1%) of the total newly synthesized intracellular proteins are virus specific as determined by immunoprecipitation with antiviral serum. Infected cells pulse-labeled for 10 min with [^{35}S]methionine contained no immunoprecipitable intracellular proteins that corresponded in size to virion structural components (Vogt and Eisenman, 1973). The largest and predominant polypeptide in the immune precipitate had an apparent molecular weight of 76K, i.e., large enough to include the amino acid sequences of all four internal proteins. After a 1-h "chase" period, the 76K polypeptide was absent and replaced by smaller products that comigrated in SDS-polyacrylamide gels with the ^{3}H-labeled proteins of purified virions. Further evidence that the 76K polypeptide is a precursor of the virion structural proteins was obtained by comparing its methionine-containing tryptic peptide pattern with that of total AMV; 10 of the 12 methionine peptides in virions were also present in the 76K protein. More recent studies (Eisenman et al., 1975) have confirmed that the 76K polypeptide contains the amino acid sequences of the four major internal gs antigens but not the minor internal protein. The methionine-containing tryptic peptides are unique to each of the four gs antigens, suggesting that they are produced by protein. By using pactamycin to inhibit initiation of polypeptide synthesis preferentially, an approach first employed successfully with the picornaviruses (see above), the gene order of the gs antigens in the 76K precursor was tentatively established as N-terminus, gs2, gs1, (gs3), gs4, C-terminus. A cleavage scheme for processing of the 76K precursor polypeptide similar to that for generating picornavirus structural proteins have been proposed (Vogt et al., 1975). The sequence of cleavages begins with the removal of a 12K fragment from the C-terminus, suggesting that, in contrast to the picornavirus polyprotein, cleavage begins after the 76K precursor is (almost) completed. The 12K polypeptide apparently corresponds to gs4. The resulting large

N-terminal fragment (~66K) is converted to a 60K intermediate that is subsequently processed to yield *gs*2, *gs*3, and *gs*1. In RSV-transformed hamster cells that do not produce virions, the 76K protein is present but is not cleaved (Eisenman *et al.*, 1975), consistent with a precursor role of the 76K polypeptide for virion structural proteins.

The mechanism of formation of the minor internal protein, the two glycoproteins, and the core-associated reverse transcriptase of oncornaviruses is unknown, but clearly the 76K *gs* precursor polypeptide corresponds to only a fraction of the coding capacity of a 35 S genome subunit. Attempts to find larger, virus-specific protein precursors in infected cells have been unsuccessful (Vogt *et al.*, 1975). The presence of ~20 S RNA, i.e., smaller than the genome subunits, in virus-specific polysomes of murine virus-infected cells (Shanmugam *et al.*, 1974) suggests that the mechanism of RNA tumor virus protein formation may resemble that of the togaviruses. It will be of interest to determine whether the 35 S genome subunit has one or multiple sites for initiation of polypeptide synthesis.

Evidence has also been obtained for the presence of high molecular weight virion polypeptide precursors in murine oncornavirus-infected cells (Naso *et al.*, 1975). In Rauscher leukemia virus-infected mouse cells pulse-labeled with [^{35}S]methionine for 10–20 min and analyzed by specific immunoprecipitation and SDS-polyacrylamide gel electrophoresis, the major virus-specific intracellular polypeptides migrated with apparent molecular weights of 180K, 140K, 110K, 80K, 65K, 60K, 50K, 40K, 30K, and 14–16K. The 180K, 80K, and 60K, polypeptides correspond to the virion structural glycoproteins; the 110K polypeptide is also glycosylated and may be a precursor of virion protein(s). On the basis of pulse-chase experiments, the 140K, 65K, and 50K proteins appear to be precursors of the major internal polypeptides. Since tryptic peptide analyses have not yet been done, it is not entirely clear if the 65K and 50K are intermediates formed by cleavage of the largest presumptive precursor (140K) or if each large polypeptide represents a separate precursor of one of the three major internal proteins of RLV. However, antiserum specific for the major core polypeptide, VP30, precipitated the 140K, 65K, and 50K polypeptides, suggesting that the 140K material may be sequentially cleaved to form VP30. Thus RLV-infected cells contain two distinct high molecular weight polypeptides (140K and 110K). with the characteristics of structural protein precursors. Together they constitute   250,000 daltons, i.e., close to the theoretical coding capacity of the ~35 S viral genome subunit RNA.

4.4.3. *In Vitro* Studies

Polysomes of oncornavirus-infected cells contain virus-specific RNA that corresponds in sequence to the virion genome RNA (Schincariol and Joklik, 1973). Consequently, the virion genome subunits rather than complementary RNA strands presumably function as the viral messenger RNA. However, in contrast to results obtained with many other virion RNAs that also function as mRNA (e.g., EMC, Sindbis), efforts to translate oncornavirus RNA in eukaryotic cell-free systems have met with only limited success. In a detailed study designed to achieve optimal conditions for translation of murine RLV RNA, ribosomal subunits isolated from RLV-infected mouse JLS-V5 cells bound 65 S RLV RNA and formed dimeric through tetrameric polyribosomes (Naso *et al.*, 1973). The ~35 S genome subunit RNA also bound to ribosomal subunits but formed predominantly monomers. The results suggest that there is only one ribosomal binding site per 35 S RNA subunit. Addition of 65 S viral RNA to preincubated cytoplasmic S30 extracts prepared from JLS-V5 cells resulted in a twofold stimulation of amino acid incorporation, mainly into two polypeptide products of molecular weights 185K and 125K. The 185K product was not synthesized in the absence of added viral RNA, and further studies may establish that it is a virus-specific polypeptide precursor of the virion proteins.

RSV 70 S or 35 S RNA added to preincubated S30 extracts of Krebs ascites cells also stimulated [^{35}S]methionine incorporation into acid-precipitable products (Von der Helm and Duesberg, 1975). Although the increase above the background level of protein synthesis was only twofold with 35 S RNA (and less with 70 S RNA), the products included a polypeptide in the size range of 75–80K. It was absent from the unstimulated reaction products. The 75–80K polypeptide product was immunoprecipitated by rat antiserum against RSV *gs* antigens. Furthermore, 11 of the 13 resolved peaks of methionine-containing tryptic peptides of the 75–80K product cochromatographed with peaks derived by digestion of a mixture of the virion *gs* proteins. Thus the 75–80K *in vitro* product may correspond to the 76K precursor protein synthesized in RSV-infected chick cells (Eisenman *et al.*, 1975). Tumor virus RNA has also apparently been translated into viral polypeptide products, including the AMV 30K core protein, in cell-free systems prepared from *E. coli* (Siegert *et al.*, 1972; Gielkens *et al.*, 1972; Twardzik *et al.*, 1973).

5. FACTORS AFFECTING EFFICIENCY OF TRANSLATION OF VIRAL mRNAs *IN VITRO*

5.1. Methylation-Dependent Translation of Viral mRNAs

It was recently reported that the RNA-synthesizing activity of the cytoplasmic polyhedrosis virus (CPV) associated transcriptase was stimulated by the addition of *S*-adenosylmethionine (SAM) to the *in vitro* reaction mixtures; furthermore, the viral mRNA products were methylated (Furuichi, 1974). Similarly, methylated mRNA was synthesized in the presence of SAM by purified reovirus (Shatkin, 1974*b*), vaccinia virus (Wei and Moss, 1974; Urushibara *et al.*, 1975), VSV (Rhodes *et al.*, 1974), and Newcastle disease virus (NDV) (Colonno and Stone, 1975). Subsequent work (summarized in Table 3) has shown that the methylation occurs exclusively at the 5′ termini of the viral mRNAs, which have the general structure $m^7G(5')ppp(5')N^m$; the 5′-terminal 7-methylguanosine is linked by an inverted, 5′-5′ triphosphate bridge to the penultimate nucleotide, which is usually methylated in the 2′-ribose position. In addition, as shown in Table 3, these 5′-terminal structures exist in the viral mRNAs in cells infected with reovirus, VSV, adenovirus, and SV40 and in the virion RNA of avian sarcoma viruses. They also occur in the "positive" strand of the double-stranded genome RNA of reovirus (Furuichi *et al.*, 1975*c*) and presumably CPV (Furuichi and Miura, 1975; Miura *et al.*, 1974). Similar 5′-terminal structures have also been found in HeLa cell poly(A)-containing mRNAs (Furuichi *et al.*, 1975*b*; Wei *et al.*, 1975), mouse myeloma mRNA (Adams and Cory, 1975), BHK-cell mRNAs (Moyer and Banerjee, unpublished results), mouse L-cell mRNA (Furuichi, LaFiandra, and Shatkin, unpublished results; Perry and Kelley, 1975), rabbit globin mRNA (Muthukrishnan *et al.*, 1975), *Artemia salina* mRNA (Muthukrishnan, Filipowicz, Sierra, Both, Shatkin, and Ochoa, unpublished results), and BSC-1 monkey cell poly(A)-containing RNA (Lavi and Shatkin, 1975). However, the pattern of methylation in cellular mRNAs is more complex because of the presence of N^6-methyladenosine within RNA chains and additional methylations of the cap structures, e.g., at the 2′-ribose position on the third base from the 5′ terminus, to give the structure $m^7GpppN^mpN^mp$ . . . .

The widespread occurrence of these unusual 5′-terminal structures implies that they have important biological function(s), and, indeed, the availability of both methylated and unmethylated VSV and reovirus

TABLE 3

The 5'-Terminal Structures of Viral mRNAs

Viral mRNA	Synthesized in vivo	Synthesized in vitro +SAM[a]	Synthesized in vitro −SAM or +SAH[a]	Reference
Cytoplasmic polyhedrosis virus (CPV)		m^7GpppA^m		Furuichi and Miura (1975)
Reovirus	m^7GpppG^m	m^7GpppG^m 75% ppG 25%	m^7GpppG^m (<2%) GpppG (27%) ppG (71%)	Furuichi (unpublished results), Furuichi et al. (1975a), Both et al. (1975e)
Vaccinia virus		m^7GpppA^m m^7GpppG^m	GpppA GpppG	Urushibara et al. (1975), Wei and Moss (1975), Moss (unpublished results)
Vesicular stomatitis virus (VSV)	$m^7GpppA^{m6}_{m2}$	m^7GpppA^m	GpppA	Abraham et al. (1975a,b), Moyer et al. (1975b), Moyer (unpublished results)
Simian virus 40 (SV40)	m^7GpppA^m m^7GpppG^m			Lavi and Shatkin (1975)
Avian sarcoma virus B77	m^7GpppG^m			Furuichi, Stavnezer, Bishop, and Shatkin (unpublished results), Keith and Fraenkel-Conrat (unpublished results)
Rous sarcoma virus	m^7GpppG^m			
Adenovirus	$m^7GpppA^{m6}_{m2}$			Koczat and Moss (unpublished results), Sommer et al. (unpublished results)

[a] SAM, S-adenosylmethionine; SAH, S-adenosylhomocysteine.

mRNAs has made it possible to determine one of these functions. In cell-free extracts of wheat germ, unmethylated VSV and reovirus mRNAs are translated into authentic viral polypeptides (Both *et al.* 1975*b,c*). However, in the presence of *S*-adenosylhomocysteine (SAH), an inhibitor of methylation, the mRNAs are translated poorly, if at all (Both *et al.*, 1975*a*). Addition of the methyl donor SAM to the translation reaction mixtures results in methylation of the exogenous, unmethylated viral mRNAs by the wheat germ extract and stimulation of authentic viral polypeptide synthesis (Both *et al.*, 1975*a*). The methylation is highly specific; the 5′-terminal structure GpppN . . . in unmethylated reovirus and VSV mRNAs is converted to m⁷GpppN . . . , and no methylation occurs in any other nucleotides in the mRNA (Muthukrishnan *et al.*, 1975). In addition, reovirus grows in mouse L-cell fibroblasts, and L-cell S10 extracts specifically methylate the reovirus mRNAs at the 5′ termini. However, in L-cell extracts the penultimate base is also methylated in the 2′-ribose position, yielding m⁷GpppGᵐ . . . . When the m⁷G is removed from the 5′ termini of reovirus mRNAs by the β-elimination reaction (Fraenkel-Conrat and Steinschneider, 1968), the ability of reovirus mRNAs to direct *in vitro* protein synthesis in wheat germ extract is reduced. The results indicate that m⁷G at the 5′ terminus of viral mRNAs is fundamentally involved in translation.

Some insight has recently been gained into the function of m⁷G in translation. There is a requirement for methylation of reovirus mRNA at the level of mRNA–80 S ribosome complex formation, and only methylated mRNA forms stable complexes with 40 S ribosomal subunits in wheat germ extract (Both *et al.*, 1975*e*). In addition, the 5′ termini of methylated reovirus mRNAs bound in mRNA–80 S ribosome complexes in wheat germ extract are partially resistant (15% and 50% respectively) to digestion with pancreatic and T_1 RNases; a fragment(s) of the type $m^7GpppG^mpCpUp(Np)_3Gp(Np)_{25-30}Gp$ is protected by ribosomes against RNase digestion. The results suggest that the extremity of the 80 S ribosome in initiation complexes lies within the 5′-terminal mRNA sequence, $m^7GpppG^mpCpUp(Np)_3Gp$, contained within the ten reovirus mRNAs and protects it from attack by RNases. Moreover, this 5′-terminal fragment(s), which lacks most of the ribosome-protected mRNA sequences, will not bind to wheat germ ribosomal subunits; however, some of the longer, ribosome-protected, 5′-terminal fragment(s), $m^7GpppG^mpCpUp(Np)_3^3Gp(Np)_{25-30}Gp$, will rebind to ribosomes in an RNase-resistant complex. The close proximity in at least 50% of the reovirus mRNAs of the 5′-terminal m⁷GpppGᵐ structure to the ribosome-protected sequence(s) of the mRNAs suggests that both structural features may be required simultaneously for correct ribosome binding during the formation of protein synthesis initiation complexes (Both *et al.*, 1975*e*). The position in reovirus mRNAs of the initiator codon AUG relative to the 5′-terminal m⁷GpppGᵐ is not known. However, in the case of brome mosaic virus, which grows in wheat, the mRNA for the viral coat protein has the 5′-terminal structure m⁷GpppG . . ., and the AUG sequence lies only ten bases from the 5′ end of the mRNA molecule within a 23-nucleotide fragment which binds to ribosomes (Dasgupta *et al.*, 1975). Thus it may be that in monocistronic, eukaryotic viral and cellular mRNAs, the close proximity to the 5′ end of the molecule of the initiator codon and the ribosome binding sequence imposes a requirement for m⁷G, possibly because structures such as hairpins that may be recognized by ribosomes cannot be formed. The presence of this unique 5′-terminal structure m⁷GpppNᵐ in a variety of eukaryotic mRNAs is consistent with its fundamental importance in protein synthesis as first demonstrated in a heterologous cell-free system.

However, it appears that picornavirus RNAs including EMC and poliovirus do not have blocked, methylated 5′ termini (unpublished observations of S. Muthukrishnan and A. Shatkin; H. Oppermann and Y. Furuichi; E. Wimmer *et al.*; D. Baltimore *et al.*; J. Darnell *et al.*; P. Fellner *et al.*; S. Mandeles *et al.*). EMC RNA is not translated or methylated in wheat germ extract under conditions that result in the methylation and translation of VSV and reovirus unmethylated RNAs (G. Both, unpublished observations). Furthermore, in ascites cell extracts EMC RNA is translated into virus-specific polypeptides in the absence of exogenous SAM or in the presence of SAH (M. Morgan and A. Shatkin, unpublished results). Similarly, poliovirus RNA can be translated into authentic viral products in HeLa cell extract and reticulocyte lysate without being affected by the addition of SAM or SAH (L. Villa-Komaroff and D. Baltimore, unpublished results). Thus it seems clear that in the case of the picornavirus RNAs that code for "polyproteins" the initiation of translation occurs by a 7-methylguanosine-independent mechanism. In picornavirus RNAs, there may be extensive base complementarity between the 3′-terminal portion of 18 S ribosomal RNA and a sequence on the 5′ side of the initiator AUG codon. This could result in stable mRNA–ribsomal subunit interactions without a requirement for 5′-terminal 7-methylguanosine (Shine and Dalgarno, 1974; Steitz and Jakes, 1975; Dasgupta *et al.*, 1975; Both *et al.*, 1975*e*). Alternatively, in mRNAs where the initiator codon is distant from the 5′ end, ribosome binding may depend on recognition of an intramolecular secondary structure in the mRNA.

5.2. Effects of Interferon on Cell-Free Protein Synthesis

It is well known that multiplication of RNA and DNA viruses is markedly inhibited in vertebrate cells that have been treated with homologous interferon (*Interferon and Interferon Inducers,* 1973). Although this phenomenon has been known for many years, the mechanism of interferon action remains unknown. Evidence has been presented that the primary targets of interferon action in viral gene expression are at the levels of both transcription and translation.

With the development of mammalian cell-free protein-synthesizing systems, it became possible to investigate the effects of interferon on translation of viral and cellular mRNAs *in vitro.* The first striking result was that interferon added directly to the *in vitro* system failed to inhibit translation of EMC or mengovirus RNA (Kerr, 1971; Falcoff *et al.,* 1972). However, when cell-free extracts were prepared from interferon-treated cells a differential stimulation in protein synthesis was observed with viral vs. endogenous mRNAs. Friedman *et al.* (1972*a,b*) were unable to detect an inhibition of EMC RNA-stimulated incorporation of labeled amino acids into polypeptides in extracts derived from L cells treated with interferon. In contrast, in nonpreincubated extracts prepared from cells that had been treated with highly purified interferon and also infected with vaccinia or EMC virus, there was a marked decrease in the translation of added EMC RNA. In contrast to viral mRNA, translation of poly(U) or endogenous mRNA was not decreased. Thus both interferon treatment and virus infection appeared to be required in order to obtain inhibition of viral mRNA translation *in vitro.* On the other hand, Falcoff *et al.* (1972) showed that L cells exposed to homologous interferon preparations yielded cell-free extracts which were totally inactive for the translation of added purified mRNAs, including mengovirus RNA and rabbit globin mRNA. Translation of poly(U) or endogenous mRNA was not impaired. Control experiments demonstrated that this differential block in protein synthesis was apparent only under conditions in which the antiviral activity of interferon was expressed.

Fractionation of inhibited cell-free extracts indicated that the interferon inhibitory effect was localized in the ribosome fraction and could be removed by washing the ribosomes in high-salt buffer (Falcoff *et al.*, 1973). A similar inhibitory activity in the ribosomes of extracts prepared from interferon-treated Ehrlich ascites tumor cells was detected by Gupta *et al.* (1973). In these extracts, protein synthesis directed by exogenous EMC RNA or L-cell mRNA was impaired, but endogenous protein synthesis was unaffected. The inhibitory effect was eliminated by the addition of purified mammalian transfer RNA to the *in vitro* system (Gupta *et al.*, 1974; Content *et al.*, 1974). Moreover, the tRNA species which restored the translation of mengovirus RNA and globin mRNA in extracts from interferon-treated cells were not identical, suggesting a possible basis for the selective effects of interferon in virus and host protein synthesis (Content *et al.*, 1974). Using pactamycin or aurintricarboxylic acid (ATA), it was shown that the addition of tRNA enhanced elongation of polypeptide chains as well as the initiation of new molecules (Gupta *et al.*, 1974; Content *et al.*, 1974). The need for added tRNA may be due to an impairment of amino acid acceptance by some of the endogenous tRNA species in the S30 fraction. Indeed, this impairment was pronounced for leucine (Gupta *et al.*, 1974), and only leucine showed significant charging in the mixture of tRNAs which counteracted the inhibitory effect on mengovirus RNA translation (Content *et al.*, 1974). Content *et al.* (1975) have shown that in cell-free extracts prepared from interferon-treated L cells an inhibition of peptide chain elongation occurs before the block in initiation. These effects also appeared to be due to a deficiency in certain tRNA species.

The effect of interferon treatment of Krebs II ascites tumor cells on the ability of cell-free extracts prepared from them to translate various mRNAs was also studied by Samuel and Joklik (1974). Although the treated cells were not tested for resistance to virus infection, extracts prepared from them efficiently translated exogenous cellular mRNA and poly(U), but viral mRNAs including reovirus and vaccinia mRNA were poorly translated. The inability of the extracts to translate viral mRNAs was due to the presence of some inhibitory protein(s) associated with ribosomes. A presumptive inhibitory polypeptide of molecular weight 48K in the 0.3–0.6 M KCl wash of the ribosomes was absent in the corresponding fraction from untreated cell ribosomes. This fraction inhibited the translation of viral mRNAs in cell-free extracts of untreated cells more than other salt wash fractions, suggesting that the antiviral activity of interferon is mediated by a ribosome-associated polypeptide that permits discrimination between cellular and viral mRNAs. A better understanding of the mechanism of action of interferon, in spite of its apparent complexity, may be achieved by further studies of cell-free systems.

6. CONCLUSIONS

Our understanding of animal virus genetic expression is likely to be extended by the increased use of cell-free protein-synthesizing systems. Genetic mapping of DNA viruses *in vitro* has recently had an auspicious start with the translation of specific mRNAs selected by hybridization to separated endonuclease cleavage fragments of viral genomes. Among the several classes of RNA viruses, progress has already been made in defining various schemes for viral protein formation, and further studies should provide information on specific points such as the number of polypeptide initiation sites in togavirus mRNAs and in the genome subunits of RNA tumor viruses. Important control processes are also being elucidated with the aid of *in vitro* systems, e.g., the role of mRNA methylation in protein synthesis and the mechanism of action of interferon. These and other aspects of the regulation of eukaryotic protein synthesis provide challenging problems for future studies.

ACKNOWLEDGMENT

We thank Christa Nuss for invaluable assistance.

7. REFERENCES

Abraham, G., Rhodes, D. P., and Banerjee, A. K., 1975a, The 5'-terminal structure of the methylated mRNA synthesized *in vitro* by vesicular stomatitis virus, *Cell* **5**:51.

Abraham, G., Rhodes, D. P., and Banerjee, A. K., 1975b, Novel initiation of RNA synthesis *in vitro* by vesicular stomatitis virus, *Nature (London)* **255**:37.

Adams, J. M., and Cory, S., 1975, Modified nucleosides and bizarre 5'-termini in mouse myeloma mRNA, *Nature (London)* **255**:28.

Adamson, S. D., Herbert, E., and Godchaux, W., III, 1968, Factors affecting the rate of protein synthesis in lysate systems from reticulocytes, *Arch. Biochem. Biophys.* **125**:671.

Allende, J. E., and Bravo, M., 1966, Amino acid incorporation and aminoacyl transfer in a wheat embryo system, *J. Biol. Chem.* **241**:2756.

Anderson, C. W., Baum, P. R., and Gesteland, R. F., 1973, Processing of adenovirus 2-induced proteins, *J. Virol.* **12**:241.

Anderson, C. W., Lewis, J. B., Atkins, J. F., and Gesteland, R. F., 1974, Cell-free synthesis of adenovirus 2 proteins programmed by fractionated messenger RNA: A comparison of polypeptide products and messenger RNA lengths, *Proc. Natl. Acad. Sci. USA* **71**:2756.

Atkins, J. F., Lewis, J. B., Anderson, C. W., and Gesteland, R. F., 1975, Enhanced differential synthesis of proteins in a mammalian cell-free system by addition of polyamines, *J. Biol. Chem.* **250**:5688.

August, J. T., Bolognesi, D. P., Fleissner, E., Gilden, R. V., and Nowinski, R. C., 1974, A proposed nomenclature for the virion proteins of oncogenic RNA viruses, *Virology* **60**:595.

Avery, R. J., 1974, The sub-cellular localization of virus-specific RNA in influenza virus-infected cells, *J. Gen. Virol.* **24**:77.

Aviv, H., Boime, I., and Leder, P., 1971, Protein synthesis directed by encephalomyocarditis virus RNA: Properties of a transfer RNA-dependent system, *Proc. Natl. Acad. Sci. USA* **68**:2303.

Bader, J. P., 1975, Reproduction of RNA tumor viruses, in: *Comprehensive Virology,* Vol. 4 (H. Fraenkel-Conrat and R. R. Wagner, eds.), pp. 253–315, Plenum Press, New York.

Baltimore, D., 1971, Expression of animal virus genomes, *Bacteriol. Rev.* **35**:235.

Baltimore, D., Huang, A. S., and Stampfer, M., 1970, Ribonucleic acid synthesis of vesicular stomatitis virus. II. An RNA polymerase in the virion, *Proc. Natl. Acad. Sci, USA* **66**:572.

Baluda, M. A., Shoyab, M., Markham, P. D., Evans, R. M., and Drohan, W. N., 1974, Base sequence complexity of 35 S avian myeloblastosis virus DNA determined by molecular hybridization kinetics, *Cold Spring Harbor Symp. Quant. Biol.* **39**:859.

Banerjee, A. K., and Rhodes, D. P., 1973, *In vitro* synthesis of RNA that contains polyadenylate by virion-associated RNA polymerase of vesicular stomatitis virus, *Proc. Natl. Acad. Sci. USA* **70**:3566.

Banerjee, A. K., and Shatkin, A. J., 1970, Transcription *in vitro* by reovirus-associated ribonucleic acid-dependent polymerase, *J. Virol.* **6**:1.

Banerjee, A. K., Moyer, S. A., and Rhodes, D. P., 1974, Studies on the *in vitro* adenylation of RNA by vesicular stomatitis virus, *Virology* **61**:547.

Barth, L. G., and Barth, L. J., 1959, Differentiation of cells of the *Rana pipiens* gastrula in unconditioned medium, *J. Embryol. Exp. Morphol.* **7**:210.

Beaud, G., Kirn, A., and Gros, F., 1972, *In vitro* protein synthesis directed by RNA transcribed from vaccinia DNA, *Biochem. Biophys, Res. Commun.* **49**:1459.

Bernard, J. P., and Northrop, R. L., 1974, RNA polymerase in mumps virion, *J. Virol.* **14**:183.

Bishop, D. H. L., and Roy, P., 1971, Properties of the product synthesized by vesicular stomatitis virus particles, *J. Mol. Biol.* **58**:799.

Bishop, D. H. L., Emerson, S. U., and Flamand, A., 1974, Reconstitution of infectivity and transcriptase activity of homologous and heterologous viruses: Vesicular stomatitis (Indiana serotype), and Chandipura, vesicular stomatitis (New Jersey serotype) and Cocal viruses, *J. Virol.* **14**:139.

Boime, I., and Leder, P., 1972, Protein synthesis directed by encephalomyocarditis virus mRNA. III. Discrete polypeptides translated from a monocistronic messenger *in vitro*, *Arch. Biochem. Biophys.* **153**:706.

Boime, I., Aviv, H. and Leder, P., 1971, Protein synthesis directed by encephalomyocarditis virus RNA. II. The *in vitro* synthesis of high molecular weight proteins and elements of the viral capsid, *Biochem. Biophys. Res. Commun.* **45**:788.

Borsa, J., and Graham, A. F., 1968, Reovirus RNA polymerase activity in purified virions, *Biochem. Biophys. Res. Commun.* **33**:895.

Both, G. W., Banerjee, A. K., and Shatkin, A. J., 1975a, Methylation-dependent translation of viral messenger RNAs *in vitro*, *Proc. Natl. Acad. Sci. USA* **72**:1189.

Both, G. W., Lavi, S., and Shatkin, A. J., 1975b, Synthesis of all the gene products of the reovirus genome *in vivo* and *in vitro*, *Cell* **4**:173.

Both, G. W., Moyer, S. A., and Banerjee, A. K., 1975c, Translation and identification of the mRNA species synthesized *in vitro* by the virion-associated RNA polymerase of vesicular stomatitis virus, *Proc. Natl. Acad. Sci. USA* **72**:274.

Both, G. W., Moyer, S. A., and Banerjee, A. K., 1975d, Translation and identification of the viral mRNA species isolated from subcellular fractions of vesicular stomatitis virus-infected cells, *J. Virol.* **15**:1012.

Both, G. W., Furuichi, Y., Muthukrishnan, S., and Shatkin, A. J., 1975e, Ribosome binding to reovirus mRNA in protein synthesis requires 5′-terminal 7-methyl-guanosine, *Cell* **6**:185.

Brachet, J., Huez, G., and Hubert, E., 1973, Microinjection of rabbit hemoglobin messenger RNA into amphibian oocytes and embryos, *Proc. Natl. Acad. Sci. USA* **70**:543.

Bratt, M. A., and Robinson, W. S., 1967, Ribonucleic acid synthesis in cells infected with Newcastle disease virus, *J. Mol. Biol.* **23**:1.

Brawerman, G., 1974, Eukaryotic messenger RNA, *Annu. Rev. Biochem.* **43**:621.

Breindl, M., and Holland, J. J., 1975, Coupled *in vitro* transcription and translation of vesicular stomatitis virus messenger RNA, *Proc. Natl. Acad. Sci. USA* **72**:2545.

Brown, D. D., and Littna, E., 1964, RNA synthesis during the development of *Xenopus laevis*, the South African clawed toad, *J. Mol. Biol.* **8**:669.

Butterworth, B. E., 1973, A comparison of the virus-specific polypeptides of encephalomyocarditis virus, human rhinovirus-1A, and poliovirus, *Virology* **56**:439.

Cancedda, R., and Schlesinger, M. J., 1974, Formation of Sindbis virus capsid protein in mammalian cell-free extracts programmed with viral messenger RNA, *Proc. Natl. Acad. Sci. USA* **71**:1843.

Cancedda, R., Swanson, R., and Schlesinger, M. J., 1974a, Effects of different RNAs and components of the cell-free system on *in vitro* synthesis of Sindbis viral proteins, *J. Virol.* **14**:652.

Cancedda, R., Swanson, R., and Schlesinger, M. J., 1974b, Viral proteins formed in a cell-free rabbit reticulocyte system programmed with RNA from a temperature-sensitive mutant of Sindbis virus, *J. Virol.* **14**:664.

Choppin, P. W., and Compans, R. W., 1975, Reproduction of paramyxoviruses, in: Comprehensive Virology, Vol. 4 (H. Fraenkel-Conrat and R. R. Wagner, eds.), pp. 95–178, Plenum Press, New York.

Chow, N., and Simpson, R. W., 1971, RNA dependent RNA polymerase activity associated with virions and subviral components of myxoviruses, *Proc. Natl. Acad. Sci. USA* **68**:752.

Clegg, C., and Kennedy, I., 1975, Translation of Semliki-Forest virus intracellular 26 S RNA: Characterisation of the products synthesized *in vitro*, *Eur. J. Biochem.* **53**:175.

Clegg, J. C. S., 1975, Sequential translation of capsid and membrane protein genes of alphaviruses, *Nature (London)* **254**:454.

Collins, B. S., and Bratt, M. A., 1973, Separation of the messenger RNAs of Newcastle disease virus by gel electrophoresis, *Proc. Natl. Acad. Sci. USA* **70**:2544.

Colonno, R. J., and Stone, H. O., 1975, *In vitro* methylation of Newcastle disease virus messenger RNA by a virion-associated enzyme, *Proc. Natl. Acad. Sci. USA* **72**:2611.

Compans, R. W., and Choppin, P. W., 1975, Reproduction of myxoviruses, in: *Comprehensive Virology*, Vol. 4 (H. Fraenkel-Conrat and R. R. Wagner, eds.), pp. 179–239, Plenum Press, New York.

Content, J., Lebleu, B., Zilberstein, A., Berissi, H., and Revel, M., 1974, Mechanism of the interferon-induced block of mRNA translation in mouse L cells: Reversal of the block by transfer RNA, *FEBS Lett.* **41**:125.

Content, J., Lebleu, B., Nudel, U., Zilberstein, A., Berissi, H., and Revel, M., 1975, Blocks in elongation and initiation of protein synthesis induced by interferon treatment in mouse L cells, *Eur. J. Biochem.* **54**:1.

Crawford, L. V., and Gesteland, R. F., 1973, Synthesis of polyoma proteins *in vitro, J. Mol. Biol.* **74**:627.

Dasgupta, R., Shih, D. S., Saris, C., and Kaesberg, P., 1975, Nucleotide sequence of a viral RNA fragment that binds to eukaryotic ribosomes *Nature (London)* **256**:624.

Davies, J. W., and Kaesberg, P., 1973, Translation of virus mRNA: Synthesis of bacteriophage Qβ proteins in a cell-free extract from wheat embryo, *J. Virol.* **12**:1434.

Dobos, P., Kerr, I. M., and Martin, E. M., 1971, Synthesis of capsid and noncapsid viral proteins in response to encephalomyocarditis virus ribonucleic acid in animal cell-free systems, *J. Virol.* **8**:491.

Duesberg, P., Vogt, P. K., Beemon, K., and Lai, M., 1974, Avian RNA tumor viruses; mechanism of recombination and complexity of the genome, *Cold Spring Harbor Symp. Quant. Biol.* **39**:847.

Eggen, K. L., and Shatkin, A. J., 1972, *In vitro* translation of cardiovirus ribonucleic acid by mammalian cell-free extracts, *J. Virol.* **9**:636.

Ehrenfeld, E., and Summers, D. F., 1972, Adenylate-rich sequences in vesicular stomatitis virus messenger ribonucleic acid, *J. Virol.* **10**:683.

Eisenman, R., Vogt, V. M., and Diggelmann, H., 1975, The synthesis of avian RNA tumor virus structural proteins, *Cold Spring Harbor Symp. Quant. Biol.* **39**:1067.

Emerson, S. U., and Wagner, R. R., 1973, L protein requirement for *in vitro* RNA synthesis by vesicular stomatitis virus, *J. Virol.* **12**:1325.

Eron, L., and Westphal, H., 1974, Cell-free translation of highly purified adenovirus messenger RNA, *Proc. Natl. Acad. Sci. USA* **71**:3385.

Eron, L., Callahan, R., and Westphal, H., 1974a, Cell-free synthesis of adenovirus coat proteins, *J. Biol. Chem.* **249**:6331.

Eron, L., Westphal, H., and Callahan, R., 1974b, *In vitro* synthesis of adenovirus core proteins, *J. Virol.* **14**:375.

Esteban, M., and Kerr, I. M., 1974, The synthesis of encephalomyocarditis virus polypeptides in infected L-cells and cell-free systems, *Eur. J. Biochem.* **45**:567.

Etkind, P. R., and Krug, R. M., 1974, Influenza viral messenger RNA, *Virology* **62**:38.

Falcoff, E., Falcoff, R., Lebleu, B., and Revel, M., 1972, Interferon treatment inhibits Mengo RNA and haemoglobin mRNA translation in cell-free extracts of L cells, *Nature (London) New Biol.* **240**:145.

Falcoff, E., Falcoff, R., Lebleu, B., and Revel, M., 1973, Correlation between the antiviral effect of interferon treatment and the inhibition of *in vitro* mRNA translation in noninfected L cells, *J. Virol.* **12**:421.

Fleissner, E., 1971, Chromatographic separation and antigenic analysis of protein of the oncornaviruses. I. Avian leukemia-sarcoma viruses, *J. Virol.* **8**:778.

Floyd, R. W., Stone, M. P., and Joklik, W. K., 1974, Separation of single-stranded ribonucleic acids by acrylamide-agrose-urea gel electrophoresis, *Anal. Biochem.* **59**:599.

Fournier, F., Tovell, D. R., Esteban, M., Metz, D. H., Ball, L. A., and Kerr, I. A., 1973, The translation of vaccinia virus messenger RNA in animal cell-free systems, *FEBS Lett.* **30**:268 ·

Fraenkel-Conrat, H., and Steinschneider, A., 1968, Stepwise degradation of RNA: Periodate followed by aniline cleavage, in: *Methods in Enzymology,* Vol. XII (L. Grossman and K. Moldave, eds.), pp. 243–246. Academic Press, New York.

Friedman, R. M., Metz, D. H., Esteban, R. M., Tovell, D. R., Ball, L. A., and Kerr, I. M., 1972a, Mechanism of interferon action: Inhibition of viral messenger ribonucleic acid translation in L-cell extracts, *J. Virol.* **10**:1184.

Friedman, R. M., Esteban, R. M., Metz, D. H., Tovell, D. R., Kerr, I. M., and Williamson, R., 1972b, Translation of RNA by L cell extracts; effect of interferon, *FEBS Lett.* **24**:273.

Furuichi, Y., 1974, Methylation-coupled transcription by virus-associated transcriptase of cytoplasmic polyhedrosis virus containing double-stranded RNA, *Nucleic Acids Res.* **1**:809.

Furuichi, Y., and Miura, K.-I., 1975, A blocked structure at the 5′ terminus of mRNA from cytoplasmic polyhedrosis virus, *Nature (London)* **253**:374.

Furuichi, Y., Morgan, M., Muthukrishnan, S., and Shatkin, A. J., 1975a, Reovirus messenger RNA contains a methylated, blocked 5′-terminal structure: $m^7G(5′)ppp(5′)G^mpCp$-, *Proc. Natl. Acad. Sci. USA* **72**:362.

Furuichi, Y., Morgan, M., Shatkin, A. J., Jelinek, W., Salditt-Georgieff, M., and Darnell, J. E., 1975b, Methylated, blocked 5′-termini in HeLa cell mRNA, *Proc. Natl. Acad. Sci. USA* **72**:1904.

Furuichi, Y., Muthukrishnan, S., and Shatkin, A. J., 1975c, 5′-Terminal $m^7G(5′)ppp(5′)G^mp$ *in vivo*: Identification in reovirus genome RNA, *Proc. Natl. Acad. Sci. USA* **72**:742.

Ghosh, H. P., Toneguzzo, F., and Wells, S., 1973, Synthesis *in vitro* of vesicular stomatitis virus proteins in cytoplasmic extracts of L cells, *Biochem. Biophys. Res. Commun.* **54**:228.

Gielkens, A. L. T., Salden, M. H. L., Bloemendal, H., and Konings, R. N. H., 1972, Translation of oncogenic viral RNA and eukaryotic messenger RNA in the *E. coli* cell-free system, *FEBS Lett.* **28**:348.

Gilbert, J. M., and Anderson, W. F., 1970, Cell-free hemoglobin synthesis. II. Characteristics of the transfer ribonucleic acid-dependent assay system. *J. Biol. Chem.* **245**:2342.

Graziadei, W. D., III, and Lengyel, P., 1972, Translation of *in vitro* synthesized reovirus messenger RNAs into proteins of the size of reovirus capsid proteins in a mouse L cell extract, *Biochem. Biophys. Res. Commun.* **46**:1816.

Graziadei, W. D., Roy, D., Konisberg, W., and Lengyel, P., 1973, Translation of reovirus messenger ribonucleic acids synthesized *in vitro* into reovirus proteins in a mouse L cell extract, *Arch. Biochem. Biophys.* **158**:266.

Grubman, M. J., and Summers, D. F., 1973, *In vitro* protein-synthesizing activity of vesicular stomatitis virus-infected cell extracts, *J. Virol.* **12**:265.

Grubman, M. J., Ehrenfeld, E., and Summers, D. F., 1974, *In vitro* synthesis of proteins by membrane-bound polyribosomes from vesicular stomatitis virus-infected HeLa cells, *J. Virol.* **14**:560.

Grubman, M. J., Moyer, S. A., Banerjee, A. K., and Ehrenfeld, E., 1975, Subcellular localization of vesicular stomatitis virus messenger RNAs, Biochem. Biophys. Res. Commun. **62**:531.

Gupta, S. L., Sopori, M. L., and Lengyel, P., 1973, Inhibition of protein synthesis directed by added viral and cellular messenger RNAs in extracts of interferon-treated Ehrlich ascites tumor cells: Location and dominance of the inhibitor(s), *Biochem. Biophys. Res. Commun.* **54**:777.

Gupta, S. L., Sopori, M. L., and Lengyel, P., 1974, Release of the inhibition of messenger RNA translation in extracts of interferon-treated Ehrlich ascites tumor cells by added transfer RNA, *Biochem. Biophys. Res. Commun.* **57**:763.

Gurdon, J. B., 1967, On the origin and persistence of a cytoplasmic state inducing nuclear DNA synthesis in frogs' eggs, *Proc. Natl. Acad. Sci. USA* **58**:545.

Gurdon, J. B., 1968, Changes in somatic cell nuclei inserted into growing and maturing amphibian oocytes, *J. Embryol. Exp. Morphol.* **20**:401.

Gurdon, J. B., Lane, C. D., Woodland, H. R., and Marbaix, G., 1971, Use of frog eggs and oocytes for the study of messenger RNA and its translation in living cells, *Nature (London)* **233**:177.

Haselkorn, R., and Rothman-Denes, L. B., 1973, Protein synthesis, *Annu. Rev. Biochem.* **42**:397.

Hay, A. J., and Joklik, W. K., 1971, Demonstration that the same strand of reovirus genome RNA is transcribed *in vitro* and *in vivo*, *Virology* **44**:450.

Holland, J. J., and Kiehn, D. E., 1968, Specific cleavage of viral proteins as steps in the synthesis and maturation of enteroviruses, *Proc. Natl. Acad. Sci. USA* **60**:1015.

Homma, M., and Ohuchi, M., 1973, Trypsin action on growth of Sendai virus in tissue culture cells. III. Structural differences of Sendai viruses grown in eggs and tissue culture cells, *J. Virol.* **12**:1457.

Housman, D., Jacobs-Lorena, M., Rajbhandary, U. L., and Lodish, H. F., 1970, Initiation of haemoglobin synthesis by methionyl-tRNA, *Nature (London)* **227**:913.

Huang, A. S., Baltimore, D., and Stampfer, M., 1970, RNA synthesis of vesicular stomatitis virus. III. Multiple complementary messenger RNA molecules, *Virology* **42**:946.

Huang, A. S., Baltimore, D., and Bratt, M. A., 1971, Ribonucleic acid polymerase in virions of Newcastle disease virus: Comparison with the vesicular stomatitis virus polymerase, *J. Virol.* **7**:389.

Hunt, T., Vanderhoff, G., and London, I. M., 1971, Control of globin synthesis: The role of heme, *J. Mol. Biol.* **66**:471.

Imblum, R. L., and Wagner, R. R., 1974, Protein kinase and phosphoproteins of vesicular stomatitis virus, *J. Virol.* **13**:113.

Interferon and Interferon Inducers, 1973, (N. B. Finter, ed.), North-Holland, New York.

Jacobs-Lorena, M., and Baglioni, C., 1972, Characterization of a mouse ascites cell-free system, *Biochemistry* **11**:4970.

Jacobson, M. F., and Baltimore, D., 1968, Polypeptide cleavages in the formation of poliovirus proteins, *Proc. Natl. Acad. Sci. USA* **61**:77.

Jacobson, M. F., Asso, J., and Baltimore, D., 1970, Further evidence on the formation of poliovirus proteins, *J. Mol. Biol.* **49**:657.

Jaureguiberry, G., Ben-Hamida, G., Chapeville, G., and Beaud, G., 1975, Messenger activity of RNA transcribed *in vitro* by DNA-RNA polymerase associated to vaccinia virus cores, *J. Virol.* **15**:1467.

Johnston, F. B., and Stern, H., 1957, Mass isolation of viable wheat embryos, *Nature (London)* **179**:160.

Joklik, W. K., 1974, Reproduction of reoviridae, in: *Comprehensive Virology*, Vol. 2 (H. Fraenkel-Conrat and R. R. Wagner, eds.), pp. 231–320, Plenum Press, New York.

Kennedy, S. I. T., 1972, Isolation and characterization of the virus-specific RNA species found on membrane-bound polyribosomes of chick embryo cells infected with Semliki Forest virus, *Biochem. Biophys. Res. Commun.* **48**:1245.

Kerr, I. M., 1971, Protein synthesis in cell-free systems: An effect of interferon, *J. Virol.* **7**: 448.

Kerr, I. M., and Martin, E. M., 1971, Virus protein synthesis in animal cell-free systems: Nature of the products synthesized in response to ribonucleic acid of encephalomyocarditis virus, *J. Virol.* **7**:438.

Kerr, I. M., Martin, M., Hamilton, M. G., and Work, T. S., 1962, The initiation of virus protein synthesis in Krebs ascites-tumor cells infected with EMC virus, *Cold Spring Harbor Symp. Quant. Biol.* **27**: 259.

Kerr, I. M., Cohen, N., and Work, T. S., 1966, Factors controlling amino acid incorporation by ribosomes from Krebs 2 mouse ascites-tumor cells, *Biochem. J.* **98**:826.

Kerr, I. M., Brown, R. E., and Tovell, D. R., 1972, Characterization of the polypeptides formed in response to encephalomyocarditis virus ribonucleic acid in a cell-free system from mouse ascites tumor cells, *J. Virol.* **10**:73.

Kiehn, E. D., and Holland, J. J., 1970, Synthesis and cleavage of enterovirus polypeptides in mammalian cells, *J. Virol.* **5**:358.

Kingsbury, D. W., 1973, Cell-free translation of paramyxovirus messenger RNA, *J. Virol.* **12**:1020.

Kingsbury, D. W., and Webster, R. G., 1973, Cell-free translation of influenza virus messenger RNA, *Virology* **56**:654.

Knipe, D., Rose, J. K., and Lodish, H. F., 1975, Translation of individual species of vesicular stomatitis virus mRNA, *J. Virol.* **15**:1004.

Korant, B. D., 1972, Cleavage of viral precursor proteins *in vivo* and *in vitro, J. Virol.* **10**:751.

Korant, B. D., 1975, Regulation of animal virus replication by protein cleavage, in: *Proteases and Biological Control* (Symposium), Cold Spring Harbor Press, Cold Spring Harbor, NY.

Laemmli, U. K., 1970, Cleavage of structural proteins during the assembly of bacteriophage T$_4$, *Nature (London)* **227**:680.

Laskey, R. A., Gurdon, J. B., and Crawford, L. V., 1972, Translation of encephalomyocarditis viral RNA in oocytes of *Xenopus laevis, Proc. Natl. Acad. Sci. USA* **69**:3665.

Lavi, S., and Shatkin, A. J., 1975, Methylated SV$_{40}$-specific RNA from nuclei and cytoplasm of infected BSC-1 cells, *Proc. Natl. Acad. Sci. USA* **72**:2012.

Lawrence, C., and Thach, R. E., 1974, Encephalomyocarditis virus infection of mouse plasmacytoma cells. I. Inhibition of cellular protein synthesis, *J. Virol.* **14**:598.

Lawrence, C., and Thach, R. E., 1975, Identification of a viral protein involved in posttranslational maturation of the EMC virus capsid precursor, *J. Virol.* **15**:918.

Levin, D. H., Mendelsohn, N., Schonberg, M., Klett, H., Silberstein, S., Kapular, A. M., and Acs, G., 1970, Properties of RNA transcriptase in reovirus subviral particles, *Proc. Natl. Acad. Sci. USA* **66**:890.

Levin, D. H., Kyner, D., and Acs, G., 1971, Messenger activity in mammalian cell-free extracts of reovirus single-stranded RNA prepared *in vitro, Biochem. Biophys. Res. Commun.* **42**:454.

Levin, D. H., Kyner, D., and Acs, G., 1972, Formation of a mammalian initiation complex with reovirus messenger RNA, methionyl-tRNA$_F$, and ribosomal subunits, *Proc. Natl. Acad. Sci. USA* **69**:1234.

Levintow, L., 1974, The reproduction of picornaviruses, in: *Comprehensive Virology,* Vol. 3 (H. Fraenkel-Conrat and R. R. Wagner, eds.), pp. 109–169, Plenum Press, New York.

Lewis, J. B., Atkins, J. F., Anderson, C. W., Baum, P. R., and Gesteland, R. F., 1975, Mapping of late adenovirus genes by cell-free translation of RNA selected by hybridization to specific DNA fragments, *Proc. Natl. Acad. Sci. USA* **72**:1344.

Lodish, H. F., 1974, Model for the regulation of mRNA translation applied to haemoglobin synthesis, *Nature (London)* **251**:385.

Lodish, H. F., Weinberg, R., and Ozer, H. L., 1974, Translation of mRNA from simian virus 40-infected cells into simian virus 40 capsid protein by cell-free extracts, *J. Virol.* **13**:590.

Lucas-Lenard, J., and Lipmann, F., 1971, Protein synthesis, *Annu. Rev. Biochem.* **40**:409.

Maizel, J. V., 1971, Polyacrylamide gel electrophoresis of viral proteins, in: *Methods in Virology,* Vol. 5 (K. Maramorosch and H. Koprowski, eds.), pp. 179–246, Academic Press, New York.

Marcu, K. and Dudock, B., 1974, Characterization of a highly efficient protein synthesizing system derived from commercial wheat germ, *Nucleic Acid Res.* **1**:1385.

Marcus, A., Efron, D., and Weeks, D. P., 1974a, The wheat embryo cell-free system, in: *Methods in Enzymology,* Vol. 30 (K. Moldave and L. Grossman, eds.), pp. 749–754, Academic Press, New York.

Marcus, A., Seal, S. N., and Weeks, D. P., 1974b, Protein chain initiation in wheat embryo, in: *Methods in Enzymology,* Vol. 30 (K. Moldave and L. Grossman, eds.), pp. 94–101, Academic Press, New York.

Martin, E. M., Malec, J., Sved, S., and Work, T. S., 1961, Studies on protein and nucleic acid metabolism in virus-infected mammalian cells, *Biochem. J.* **80**:585.

Mathews, M. B., and Korner, A., 1970, Mammalian cell-free protein synthesis directed by viral ribonucleic acid, *Eur. J. Biochem.* **17**:328.

Mathews, M. B., and Osborn, M., 1974, The rate of polypeptide chain elongation in a cell-free system from Krebs II ascites cells, *Biochim. Biophys. Acta* **340**:147.

Mautner, V., Williams, J., Sambrook, J., Sharp, P. A., and Grodzicker, T., 1975, The location of the genes coding for hexon and fiber proteins in adenovirus DNA, *Cell* **5**:93.

May, B. K., and Glenn, A. R., 1974, Translation of rabbit haemoglobin mRNA in oocytes of the Queensland cane toad, *Bufo marinus, Aust. J. Biol. Sci.* **27**:585.

McDowell, M. J., and Joklik, W. K., 1971, An *in vitro* protein synthesizing system from mouse L fibroblasts infected with reovirus, *Virology* **45**:724.

McDowell, M., Joklik, W. K., Villa-Komaroff, L., and Lodish, H. F., 1972, Translation of reovirus messenger RNAs synthesized *in vitro* into reovirus polypeptides by several mammalian cell-free extracts, *Proc. Natl. Acad. Sci. USA* **69**:2649.

Miura, K.-I., Watanabe, K., and Sugiura, M., 1974, 5′-Terminal nucleotide sequences of the double-stranded RNA of silkworm cytoplasmic polyhedrosis virus, *J. Mol. Biol.* **86**:31.

Morrison, T. G., and Lodish, H. F., 1975, The site of synthesis of membrane and non-membrane proteins of vesicular stomatitis virus, *J. Biol. Chem.* **250**:6955.

Morrison, T., Stampfer, M., Baltimore, D., and Lodish, H. F., 1974, Translation of vesicular stomatitis messenger RNA by extracts from mammalian and plant cells, *J. Virol.* **13**:62.

Moss, B., 1974, Reproduction of poxviruses, in: *Comprehensive Virology,* Vol. 3 (H. Fraenkel-Conrat and R. R. Wagner, eds.), pp. 405–474, Plenum Press, New York.

Mowshowitz, D., 1973, Identification of polysomal RNA in BHK cells infected by Sindbis virus, *J. Virol.* **11**:535.

Moyer, S. A., and Banerjee, A. K., 1975, Messenger RNA species synthesized *in vitro* by the virion-associated RNA polymerase of vesicular stomatitis virus, *Cell* **4**:37.

Moyer, S. A., and Summers, D. F., 1974, Phosphorylation of vesicular stomatitis virus *in vivo* and *in vitro, J. Virol.* **13**:455.

Moyer, S. A., Grubman, M. J., Ehrenfeld, E., and Banerjee, A. K., 1975*a*, Studies on the *in vivo* and *in vitro* messenger RNA species of vesicular stomatitis virus, *Virology* **67**:463.

Moyer, S. A., Abraham, G., Adler, R., and Banerjee, A. K., 1975*b*, Methylated and blocked 5′ termini in vesicular stomatitis virus *in vivo* mRNAs, *Cell* **5**:59.

Mudd, J. A., and Summers, D. F., 1970, Polysomal ribonucleic acid of vesicular stomatitis virus-infected HeLa cells, *Virology* **42**:958.

Muthukrishnan, S., Both, G. W., Furuichi, Y., and Shatkin, A. J., 1975, 5′-Terminal 7-methylguanosine in eukaryotic mRNA is required for translation, *Nature (London)* **255**:33.

Naso, R. B., Wang, C. S., Tsai, S., and Arlinghaus, R. B., 1973, Ribosomes from Rauscher leukemia virus-infected cells and their response to Rauscher viral RNA and polyuridylic acid, *Biochim. Biophys. Acta* **324**:346.

Naso, R. B., Arcement, L. J., and Arlinghaus, R. B., 1975, Biosynthesis of Rauscher leukemia viral proteins, *Cell* **4**:31.

Öberg, B. F., and Shatkin, A. J., 1972, Initiation of picornavirus protein synthesis in ascites cell extracts, *Proc. Natl. Acad. Sci. USA* **69**:3589.

Öberg, B. F., and Shatkin, A. J., 1974, Translation of mengovirus RNA in Ehrlich ascites cell extracts, *Biochem. Biophys. Res. Commun.* **57**:1186.

Öberg, B. F., Saborio, J., Persson, T., Everitt, E., and Philipson, L., 1975, Identification of the *in vitro* translation products of adenovirus mRNA by immunoprecipitation, *J. Virol.* **15**:199.

Palmiter, R. D., 1973, Ovalbumin messenger ribonucleic acid translation: Comparable rates of polypeptide initiation and elongation on ovalbumin and globin messenger ribonucleic acid in a rabbit reticulocyte lysate, *J. Biol. Chem.* **248**:2095.

Penhoet, E., Miller, H., Doyle, M., and Blatti, S., 1971, RNA-dependent RNA polymerase activity in influenza virions, *Proc. Natl. Acad. Sci. USA* **68**:1369.

Perry, R. P., and Kelley, D. E., 1975, The methylated constituents of L cell messenger RNA: Evidence for an unusual cluster at the 5′ terminus, *Cell* **4**:387.

Pfefferkorn, E. R., and Shapiro, D., 1974, Reproduction of togaviruses, in: *Comprehensive Virology,* Vol. 2 (H. Fraenkel-Conrat and R. R. Wagner, eds.), pp. 171–230, Plenum Press, New York.

Philipson, L., and Lindberg, U., 1974, Reproduction of adenoviruses, in: *Comprehensive Virology,* Vol. 3 (H. Fraenkel-Conrat and R. R. Wagner, eds.), pp. 143–227, Plenum Press, New York.

Pons, M. W., 1973, The inhibition of influenza virus RNA synthesis by actinomycin D and cycloheximide, *Virology* **51**:120.

Prives, C. L., Aviv, H., Paterson, B. M., Roberts, B. E., Rozenblatt, S., Revel, M., and Winocour, E., 1974*a*, Cell-free translation of messenger RNA of simian virus 40: Synthesis of the major capsid protein, *Proc. Natl. Acad. Sci. USA* **71**:302.

Prives, C. L., Aviv, H., Gilboa, E., Revel, M., and Winocour, E., 1974*b*, The cell-free translation of SV40 messenger RNA, *Cold Spring Harbor Symp. Quant, Biol.* **39**:309.

Quade, K., Smith, R. E., and Nichols, J. L., 1974, Evidence for common nucleotide sequences in the RNA subunits comprising Rous sarcoma virus 70 S RNA, *Virology* **61**:287.

Rekosh, D. M., 1972, Gene order of the poliovirus capsid proteins, *J. Virol.* **9**:479.

Rhodes, D. P., Moyer, S. A., and Banerjee, A. K., 1974, *In vitro* synthesis of methylated messenger RNA by the virion-associated polymerase of vesicular stomatitis virus, *Cell* **3**:327.

Rich, A., Penman, S., Becker, Y., Darnell, J. E., and Hall, C., 1963, Polyribosomes: Size in normal and polio-infected HeLa cells, *Science* **142**:1658.

Roberts, B. E., and Paterson, B. M., 1973, Efficient translation of tobacco mosaic virus RNA and rabbit globin 9 S RNA in a cell-free system from commercial wheat germ, *Proc. Natl. Acad. Sci. USA* **70**:2330.

Roberts, B. E., Paterson, B. M., and Sperling, R., 1974, The cell-free synthesis and assembly of viral specific polypeptides into TMV particles, *Virology* **59**:307.

Roberts, B. E., Gorecki, M., Mulligan, R. C., Danna, K. J., Rozenblatt, S., and Rich, A., 1975, SV$_{40}$ DNA directs the synthesis of authentic viral polypeptides in a linked transcription-translation cell-free system, *Proc. Natl. Acad. Sci. USA* **72**:1922.

Robinson, W. S., 1971, Ribonucleic acid polymerase activity in Sendai virions and nucleocapsid, *J. Virol.* **8**:81.

Rose, J. K., and Knipe, D., 1975, Nucleotide complexities, molecular weights, and poly(A) content of the vesicular stomatitis virus messenger RNA species, *J. Virol.* **15**:994.

Rosemond, H., and Sreevalsan, T., 1973, Viral RNA's associated with ribosomes in Sindbis virus-infected HeLa cells, *J. Virol.* **11**:399.

Saborio, J. L., Pong, S.-S., and Koch, G., 1974, Selective and reversible inhibition of initiation of protein synthesis in mammalian cells, *J. Mol. Biol.* **85**:195.

Salzman, N. P., and Khoury, G., 1974, Reproduction of papovaviruses, in: *Comprehensive Virology,* Vol. 3 (H. Fraenkel-Conrat and R. R. Wagner, eds.), pp. 63–141, Plenum Press, New York.

Samuel, C. E., and Joklik, W. K., 1974, A protein synthesizing system from interferon-treated cells that discriminates between cellular and viral messenger RNAs, *Virology* **58**:476.

Scheele, C. M., and Pfefferkorn, E. R., 1970, Virus-specific proteins synthesized in cells infected with RNA$^+$ temperature-sensitive mutants of Sindbis virus, *J. Virol.* **5**:329.

Scheid, A., and Choppin, P. W., 1973, Isolation and purification of the envelope proteins of Newcastle disease virus, *J. Virol.* **11**:263.

Scheid, A., and Choppin, P. W., 1974, Identification of biological activities of paramyxovirus glycoproteins: Activation of cell fusion, hemolysis, and infectivity by proteolytic cleavage of an inactive precursor protein of Sendai virus, *Virology* **56**:475.

Schincariol, A. L., and Howatson, A. F., 1970, Replication of vesicular stomatitis virus. I. Viral specific RNA and nucleoproteins in infected L-cells, *Virology* **42**:732.

Schincariol, A. L., and Joklik, W. K., 1973, Early synthesis of virus-specific RNA and DNA in cells rapidly transformed with Rous sarcoma virus, *Virology* **56**:532.

Schlesinger, S., and Schlesinger, M. J., 1972, Formation of Sindbis virus proteins: Identification of a precursor for one of the envelope proteins, *J. Virol.* **10**:925.

Schlesinger, M. J., and Schlesinger, S., 1973, Large-molecular-weight precursors of Sindbis virus proteins, *J. Virol.* **11**:1013.

Schreier, M. H., and Staehelin, T., 1973, Initiation of mammalian protein synthesis: The importance of ribosome and initiation factor quality for the efficiency of *in vitro* systems, *J. Mol. Biol.* **73**:329.

Shanmugam, G., Bhaduri, S., and Green, M., 1974, The virus-specific RNA species in free and membrane-bound polyribosomes of transformed cells replicating murine sarcoma-leukemia viruses, *Biochem. Biophys. Res. Commun.* **56**:697.

Shatkin, A. J., 1968, Effects of actinomycin on virus replication, in: *Actinomycin* (S. A. Waksman, ed.), pp. 69–86, Interscience, New York.

Shatkin, A. J., 1974a, Animal RNA viruses: genome structure and function, *Annu. Rev. Biochem.* **43**:643.

Shatkin, A. J., 1974b, Methylated messenger RNA synthesis *in vitro* by purified reovirus, *Proc. Natl. Acad. Sci. USA* **71**:3204.

Shatkin, A. J., and Sipe, J. D., 1968, RNA polymerase activity in purified reoviruses, *Proc. Natl. Acad. Sci. USA* **61**:1462.

Shih, D. S., and Kaesberg, P., 1973, Translation of brome mosaic viral ribonucleic acid in a cell-free system derived from wheat embryo, *Proc. Natl. Acad. Sci. USA* **70**:1799.

Shine, J., and Dalgarno, L., 1974, Identical 3′-terminal octanucleotide sequence in 18 S ribosomal ribonucleic acid from different eukaryotes, *Biochem. J.* **141**:609.

Siegert, W., Konings, R. N. H., Bauer, H., and Hofschneider, P. H., 1972, Translation of avian myeloblastosis virus RNA in a cell-free lysate of *Escherichia coli*, *Proc. Natl. Acad. Sci. USA* **69**:888.

Siegert, W., Bauer, G., and Hofschneider, P. H., 1973, Direct evidence for messenger activity of influenza virion RNA, *Proc. Natl. Acad. Sci. USA* **70**:2960.

Simmons, D. T., and Strauss, J. H., 1972, Replication of Sindbis virus. I. Relative size and genetic content of 26 S and 49 S RNA, *J. Mol. Biol.* **71**:599.

Simmons, D. T., and Strauss, J. H., 1974a, Replication of Sindbis virus. V. Polyribosomes and mRNA in infected cells, *J. Virol.* **14**:552.

Simmons, D. T., and Strauss, J. H., 1974b, Translation of Sindbis virus 26 S and 49 S RNA in lysates of rabbit reticulocytes. *J. Mol. Biol.* **86**:397.

Skehel, J. J., 1971, RNA-dependent RNA polymerase activity of the influenza virus, *Virology* **45**:793.

Skehel, J. J., and Joklik, W. K., 1969, Studies on the *in vitro* transcription of reovirus RNA catalyzed by reovirus cores, *Virology* **39**:822.

Smith, A. E., 1973, The initiation of protein synthesis directed by the RNA from encephalomyocarditis virus, *Eur. J. Biochem.* **33**:301.

Smith, A. E., and Wigle, D. T., 1973, A rapid assay for the initiation of protein synthesis in extracts of animal cells, *Eur. J. Biochem.* **35**:566.

Smith, A. E., Marcker, K. A., and Mathews, M. B., 1970, Translation of RNA from encephalomyocarditis virus in a mammalian cell-free system, *Nature (London)* **225**:184.

Smith, A. E., Wheeler, T., Glanville, N., and Kääriäinen, L., 1974, Translation of Semliki-Forest virus 42 S RNA in a mouse cell-free system to give virus-coat proteins, *Eur. J. Biochem.* **49**:101.

Soria, M., and Huang, A. S., 1973, Association of polyadenylic acid with messenger RNA of vesicular stomatitis virus, *J. Mol. Biol.* **77**:449.

Steitz, J. A., and Jakes, K., 1975, How ribosomes select initiator regions in messenger RNA: Direct evidence for the formation of base pairs between the 3′ terminus of 16

S rRNA and the mRNA during initiation of protein synthesis in *E. coli, Proc. Natl. Acad. Sci. USA* (in press).

Stoltzfus, C. M., Shatkin, A. J., and Banerjee, A. K., 1973, Absence of polyadenylic acid from reovirus messenger ribonucleic acid, *J. Biol. Chem.* **248**:7993.

Stone, H. O., Portner, A., and Kingsbury, D. W., 1971, Ribonucleic acid transcriptases in Sendai virions and infected cells, *J. Virol.* **8**:174.

Summers, D. F., and Maizel, J. V., 1968, Evidence for large precursor proteins in poliovirus synthesis, *Proc. Natl. Acad. Sci. USA* **59**:966.

Summers, D. F., and Maizel, J. V., Jr., 1971, Determination of the gene sequence of poliovirus with pactamycin, *Proc. Natl. Acad. Sci. USA* **68**:2852.

Summers, D. F., Shaw, E. N., Stewart, M. L., and Maizel, J. V., Jr., 1972, Inhibition of cleavage of large poliovirus specific precursor proteins in infected HeLa cells by inhibitors of proteolytic enzymes, *J. Virol.* **10**:880.

Toneguzzo, F., and Ghosh, H. P., 1975a, Synthesis *in vitro* of vesicular stomatitis virus specific mRNA and its translation in viral proteins, *Fed. Proc.* **34**:675.

Toneguzzo, F., and Ghosh, H. P., 1975b, Cell-free synthesis of vesicular stomatitis virus proteins: Translation of membrane-bound polyribosomal mRNAs, *FEBS Lett.* **50**:369.

Tozawa, H., Watanabe, M., and Ishida, N., 1973, Structural components of Sendai virus, serological and physiochemical characterization of hemagglutinin subunit associated with neuraminidase activity, *Virology* **55**:242.

Twardzik, D., Simonds, J., Oskarsson, M., and Portugal, F., 1973, Translation of AKR-murine leukemia viral RNA in an *E. coli* cell-free system, *Biochem. Biophys. Res. Commun.* **52**:1108.

Urushibara, T., Furuichi, Y., Nishimura, C., and Miura, K.-I., 1975, A modified structure at the 5′-terminus of mRNA of vaccinia virus, *FEBS Lett.* **49**:385.

Van Venrooij, W. J. W., Henshaw, E. C., and Hirsch, C. A., 1970, Nutritional effects on the polyribosome distribution and rate of protein synthesis in Ehrlich ascites tumor cells in culture, *J. Biol. Chem.* **245**:5947.

Villa-Komaroff, L., Baltimore, D., and Lodish, H. F., 1974a, Translation of poliovirus mRNA in mammalian cell-free systems, *Fed. Proc.* **33**:531.

Villa-Komaroff, L., McDowell, M., Baltimore, D., and Lodish, H. F., 1974b, Translation of reovirus mRNA, poliovirus RNA, and bacteriophage Qβ RNA in cell-free extracts of mammalian cells, in: *Methods in Enzymology,* Vol. 30 (K. Moldave and L. Grossman, eds.), pp. 709–723, Academic Press, New York.

Vogt, V. M., and Eisenman, R., 1973, Identification of a large polypeptide precursor of avian oncornavirus proteins, *Proc. Natl. Acad. Sci. USA* **70**:1734.

Vogt, V. M., Eisenman, R., and Diggelmann, H., 1975, Generation of avian myeloblastosis virus structural proteins by proteolytic cleavage of a precursor polypeptide, *J. Mol. Biol.* **96**:471.

Von der Helm, K., and Duesberg, P. H., 1975, Translation of Rous sarcoma virus RNA in a cell-free system from ascites Krebs II cells, *Proc. Natl. Acad. Sci. USA* **72**:614.

Wagner, R. R., 1975, Reproduction of rhabdoviruses, in: *Comprehensive Virology,* Vol. 4 (H. Fraenkel-Conrat and R. R. Wagner, eds.), pp. 1–80, Plenum Press, New York.

Ward, R., Banerjee, A. K., LaFiandra, A., and Shatkin, A. J., 1972, Reovirus-specific ribonucleic acid from polysomes of infected L cells, *J. Virol.* **9**:61.

Weeks, D. P., and Marcus, A., 1971, Preformed messenger of quiescent wheat embryos, *Biochim. Biophys. Acta* **232**:671.

Wei, C. M., and Moss, B., 1974, Methylation of newly synthesized viral messenger RNA by an enzyme in vaccinia virus, *Proc. Natl. Acad. Sci. USA* **71**:3014.

Wei, C. M., and Moss, B., 1975, Methylated nucleotides block 5′-terminus of vaccinia virus messenger RNA, *Proc. Natl. Acad. Sci. USA* **72**:318.

Wei, C. M., Gershowitz, A., and Moss, B., 1975, Methylated nucleotides block 5′ terminus of HeLa cell messenger RNA, *Cell* **4**:379.

Weiss, S. R., and Bratt, M. A., 1974, Polyadenylate sequences on Newcastle disease virus messenger RNA synthesized *in vivo* and *in vitro, J. Virol.* **13**:1220.

Weissman, C., Parsons, J. T., Coffin, J. W., Rymo, L., Billeter, M. A., and Hofstetter, H., 1974, Studies on the structure and synthesis of Rous sarcoma virus RNA. *Cold Spring Harbor Symp. Quant. Biol.* **39**:1043.

Wengler, G., Beato, M., and Hackemack, B.-A., 1974, Translation of 26 S virus-specific RNA from Semliki-Forest virus-infected cells *in vitro. Virology* **61**:120.

Williams, J. F., Young, C. S. H., and Austin, P. E., 1974, Genetic analysis of human adenovirus type 5 in permissive and nonpermissive cells, *Cold Spring Harbor Symp. Quant. Biol.* **39**:427.

Wimmer, E., 1972, Sequence studies of poliovirus RNA. I. Characterization of the 5′ terminus, *J. Mol. Biol.* **68**:537.

Woodward, W. R., Ivey, J. L., and Herbert, E., 1974, Protein synthesis with rabbit reticulocyte preparations, in: *Methods in Enzymology,* Vol. 30 (K. Moldave and L. Grossman, eds.), pp. 724–731, Academic Press, New York.

Zweerink, H. J., and Joklik, W. K., 1970, Studies on the intracellular synthesis of reovirus-specified proteins, *Virology* **41**:501.

Zweerink, H. J., McDowell, M. J., and Joklik, W. K., 1971, Essential and nonessential noncapsid reovirus proteins, *Virology* **45**:716.

CHAPTER 2

Defective Interfering Animal Viruses

Alice S. Huang

Department of Microbiology and Molecular Genetics
Harvard Medical School
Boston, Massachusetts 02115

and

David Baltimore

Department of Biology and Center for Cancer Research
Massachusetts Institute of Technology
Cambridge, Massachusetts 02139

1. INTRODUCTION

Defective interfering (DI) particles were discovered three decades ago by von Magnus (1947) using the influenza virus system. He called them "incomplete" or "immature" particles. Even though they could not be isolated or characterized biochemically, a great deal was learned about them from their physiological interactions with the host and from their interference with the multiplication of "infectious" standard virus (Gard and von Magnus, 1947; Bernkopf, 1950; von Magnus, 1951). Reviews by Henle (1950), by von Magnus (1954), and by Schlesinger (1959) on viral interference discuss the earlier work on this particular homologous interference caused by incomplete virus.

A similar type of interference was soon discovered for other virus systems, most notably that of vesicular stomatitis virus (VSV) (Cooper and Bellett, 1959; Bellett and Cooper, 1959). This virus system has the added advantage that its DI particles are bullet-shaped particles shorter than standard virus (Hackett, 1964), and thus they could be purified by rate zonal centrifugation in sucrose (Huang *et al.*, 1966; Crick *et al.*, 1966; Hackett *et al.*, 1967). Because it was possible to separate them from standard virus, DI particles of VSV could be characterized biochemically and quantitative analysis of their effects on standard virus could be carried out (Huang and Wagner, 1966*a*). Studies on other virus systems show that the effect of DI particles on virus–host interactions is a general one. Although in some virus systems DI particles cannot be separated from standard virus, general criteria can be established for DI particles of animal viruses.

DI particles were named and their properties codified by Huang and Baltimore (1970), who postulated a significant role for DI particles in acute and persistent viral disease. The molecular biology of DI particles was last reviewed in 1973 along with a discussion of which animal viruses are known to produce DI particles (Huang, 1973).

In this chapter, the properties of DI particles will be presented first in general terms. Detailed discussions will follow on those prototype virus systems in which something is known about the mechanism of interference and about the characteristics of the DI particles. There will be only cursory coverage of the many very interesting biological studies on DI particles published before 1966. Only certain articles published between 1966 and 1973 will be mentioned because that period has already been reviewed (Huang, 1973). Lastly, studies on the presence of DI particles in laboratory infections of animals will be reviewed to see what evidence there is to support the hypothesis that DI particles play a role in determining the outcome of a viral infection.

Interference caused by DI-like particles has now been discovered for almost every animal virus group (Table 1). Although the DI particles have not been thoroughly characterized in each system, there is little doubt that DI particles are virtually universal and can play significant roles in virus multiplication. Moreover, DI particles are not limited to animal viruses alone: they have been reported for plant viruses (Huang, 1973) as well as for bacteriophage (Enea and Zinder, 1975).

2. GENERAL PROPERTIES

DI particles are a class of animal virus mutants that have a number of common properties. These properties are inability to propa-

TABLE 1

Catalogue of Viruses with DI Particles

Group name	Examples	References
Picorna	Polio	Cole *et al.* (1971)
Toga	Sindbis	Inglot and Chudzio (1972)[a]
		Schlesinger *et al.* (1972)[a]
	Rift Valley fever	Mims (1956)[a]
Orthomyxo	Influenza	von Magnus (1947)[a]
	Fowl plague	Rott and Scholtissek (1963)[a]
Paramyxo	Sendai	Sokol *et al.* (1964)[a]
	Mumps	Cantell (1961)[a]
		East and Kingsbury (1971)
	Measles	Parfanovich *et al.* (1971)[a]
		Schluederberg (1971)[a]
Rhabdo	Vesicular stomatitis	Bellett and Cooper (1959)[a]
	Rabies	Yoshino *et al.* (1966)[a]
		Sokol *et al.* (1968)[a]
	Infectious hematopoietic necrosis	McAllister and Pilcher (1974)
Arena	Lymphocytic choriomeningitis	Lehmann-Grube *et al.* (1969)[a]
		Welsh and Pfau (1972)
	Parana	Staneck and Pfau (1974)
Reo	reo	Nonoyama *et al.* (1970)
	Infectious pancreatic necrosis	Nicholson and Dunn (1972)[a]
Retro	Murine sarcoma	Bondurant *et al.* (1973)[a]
Papova	SV40	Uchida *et al.* (1966)
	Polyoma	Blackstein *et al.* (1969)
Adeno	Adeno	Mak (1971)[a]
Herpes	Herpes simplex	Bronson *et al.* (1973)
	Pseudorabies	Ben-Porat *et al.* (1974)

[a] Although DI particles were not characterized in these studies, enough evidence is presented regarding homologous interference during undiluted passages ("von Magnus-like" effect) to justify inclusion of these virus systems in this list.

gate themselves in the absence of a helper virus (defectiveness), ability to be complemented by helper virus and so to multiply in the presence of helper, ability to decrease the yield of the wild-type virus (interference), and ability to increase their proportion of the yield from cells coinfected with wild-type virus (enrichment). Because all of these are defining properties for DI particles, only a small class of deletion mutants are DI particles. For instance, viruses with lesions that confer temperature sensitivity can spawn DI particles; such particles can have two kinds of mutations. To distinguish the DI particle from the virus that gives rise to it, we have called the parental virus the *standard* virus (von Magnus, 1954; Huang and Baltimore, 1970) and we will continue that terminology here.

To provide a framework for considering the specific DI particles of different animal viruses, it is worth analyzing the general properties of DI particles in some detail.

2.1. Defectiveness

The defectiveness of DI particles implies that they have a lesion in some necessary gene. For all the DI particles studied, this lesion appears to be a deletion, but there could conceivably be DI particles that are point mutations or rearrangements. Some DI particles have heavily deleted genomes; in theory, the genome could be deleted of all complementable functions and could consist of the one critical region for a DI particle genome, the site for initiation of nucleic acid replication. Although many DI particles do carry information for functional products, some DI's of papovaviruses and VSV carry less than 30% of the total information found in standard genomes and do not code for any functional proteins (Leamnson and Reichmann, 1974; Lee *et al.*, 1975; Griffin and Fried, 1975).

The mechanism of deleting genetic material to form DI nucleic acid is not well understood, but appears to depend on the genetic organization of the specific virus system. There may be terminal as well as internal deletions. Segmented genomes may become deleted by losing specific pieces as well as by deletion of portions of a segment. Intramolecular and intermolecular recombinational events may delete specific regions in DNA molecules. Only further mapping and sequencing will demonstrate the exact deletion and, perhaps, give some idea of how these deletions arise. Sizing of DI genomes shows that, in general, they contain less nucleic acid than standard virus. In certain virus systems, tandem duplication of deleted genomes results in nucleic acid pieces of about the same size as the genome of standard virus, although such nucleic acids are still deletions genetically.

DI nucleic acids not only must be capable of being replicated, but also must have the property of being recognized by structural proteins for encapsidation. Most probably, although not necessarily, this involves specific sequences for the initial interaction between nucleic acids and structural proteins. The amount of nucleic acid in DI particles is determined by the interaction of the nucleic acid with structural proteins to form stable transmissible DI progeny. The lower limit of nucleic acid size varies considerably among the virus systems and depends on the specific morphological features of the virion. For example, VSV DI particles can still form very short bullet-shaped virions when only 25% of the standard genome is present (Reichmann *et al.*, 1971), whereas SV40 DI genomes appear to be constrained to sizes over about 70% of the size of standard supercoiled DNA in order to form a stable icosahedral particle (Brockman and Nathans, 1974).

When a great deal of the genetic information is deleted, as in the SV40 case, tandem repeats of the same small region of viral information and/or covalently incorporated host genetic information can build up the supercoiled molecule so that the prerequisite size for encapsidation is reached. Further studies on DI particles will show not only the minimal nucleic acid sequences necessary for encapsidation but also the role that nucleic acid may play in determining the size of virions during assembly.

2.2. Interference

Interference between a DI particle and the homologous standard virus is measured as a reduction of the total yield of standard virus caused by coinfection of cells with the DI particle. By inhibiting the production of its helper virus, the DI virus usually limits its own production indirectly. This inherent ability to inhibit virus multiplication makes DI particles particularly interesting in relation to disease.

The following properties distinguish this type of interference from other interference phenomena. Interference by DI particles is homologous; that is, interference occurs only against the standard virus from which a particular DI particle is derived or a serotypically closely related virus. Interferon plays no role in this interference, although DI particles may, by themselves, induce interferon synthesis. Prior ultraviolet irradiation of the DI particle destroys its ability to interfere. Lastly, the interference by DI particles occurs during an intracellular step of the viral multiplication cycle.

The ability of different host cells to support interference by DI particles varies considerably. The host functions that control these variations are, at present, unknown. In a particular case, one dominant gene has been shown to favor a high degree of interference by DI particles (Darnell *et al.*, 1974). Therefore, in any attempt to demonstrate interference, a minimum test would be to use cells from different genetic backgrounds or from different tissues of the same species.

The detailed mechanism of interference is likely to differ from virus system to virus system, depending on the particular strategy of nucleic acid replication and pattern of morphogenesis. Because the mechanism of interference is related to the ability of DI particles to enrich themselves, further details are discussed in Section 2.3, on enrichment.

2.3. Enrichment

Another of the necessary characteristics of the DI particle is its ability to enrich itself at the expense of standard virus. Operationally, this means that the proportion of DI particles in the progeny from a coinfection must be higher than the proportion in the input. This ability is necessary for the DI particle to be able to become an increasingly larger part of the progeny from multiple passages. Only if the DI particle can enrich itself this way will it ultimately dominate the yield and become detectable.

Most models for enrichment involve the preferential replication or the preferential encapsidation of DI nucleic acid. Whereas with poliovirus there is some evidence of preferential encapsidation (Cole and Baltimore, 1973c), such evidence in other systems is lacking and it seems likely that in general the mechanism of enrichment, as well as interference, involves preferential replication.

The mechanism for preferential replication of a DI nucleic acid will depend on the details of replication of the specific virus. There is, however, one principle which can be stated in general terms: DI nucleic acid must compete with standard nucleic acid for whatever factor is limiting in the rate of viral nucleic acid replication. If we assume that any replication system involves a single initiation point on the nucleic acid and propagation of the newly synthesized chain from that point to the end of the molecule, then the number of molecules synthesized will be determined entirely by the number of initiations which have occurred. In such a system, the length of the template to be copied is irrelevant. Thus, to the extent that enrichment involves only nucleic acid replication and nucleic acid replication is a linear process, for a DI nucleic acid to be replicated in preference to the standard nucleic acid either the DI nucleic acid must increase the affinity of its initiation site or it must increase the number of initiation sites. As we shall see, the mechanism of enrichment for viruses with supercoiled DNA generally involves an increased number of initiation sites. The mechanism of enrichment for at least some RNA viruses probably involves the production or unmasking of a higher affinity initiation site. In regard to viruses with segmented genomes, this higher affinity initiation site must be on each competing segment or, as is more readily envisaged, there is loss of an initiation site on one segment. These hypotheses for linear molecules remain to be tested experimentally.

2.4. Hunting for the DI Particle

Although DI particles are often easily demonstrated, in some viral systems they have either been found only after very deliberate searching or been present but difficult to find. As a guide to identification of DI particles in new systems, it is worth considering how DI particles present themselves.

2.4.1. The von Magnus Phenomenon

Initial indications that suggest the presence of DI particles in a virus system usually come from the biological observation that a high multiplicity of infection results in a lower yield of standard virus without a concomitant reduction in the total amount of viral hemagglutinin produced. Such were the results of successive undiluted passages first described by von Magnus (1947). In contrast, infection at lower multiplicities, or with diluted passages, will result in an increased production of standard virus. The effect of high multiplicities can sometimes be quite dramatic. In a plaque assay, sometimes no plaques will appear at low dilutions of the virus preparation, but they will be present at higher dilutions. In animal studies, inoculation of undiluted virus will result in healthy animals with no apparent sequelae, but diluted inocula will cause acute paralysis and death.

2.4.2. Proving That Interference Is by DI Particles

When such initial observations are made, the next experiment is to demonstrate that interference is caused by the presence of transmissible DI progeny from undiluted passages. To do so, the progeny from an undiluted passage are mixed with diluted virus. Interference is the inhibition of the yield of standard progeny in a mixed infection compared to cells infected by only diluted virus preparations. To demonstrate that interference is specific to DI particles, it is necessary to rule out other interference phenomena (see Section 2.2). Should DI particles induce interferon, then it is necessary to show that the amount of interferon produced and the degree of sensitivity of the cells to interferon action are not correlated with the amount of inhibition of standard virus growth. If the presence of actinomycin D does not prevent interference, then the interference is not mediated by interferon action. Also, interference by DI particles should be susceptible to neutralization of DI particles with antiserum made against the standard virus. In these experiments, it is well worth remembering that interferon action is species specific for the host cells but not virus specific.

With asynchronous infections and multiple cycles of virus growth, DI particles are generally detected late during the infection. To assure that DI particles are synthesized and are not breakdown products of standard virus, it is necessary to demonstrate that both DI and standard viruses are produced by cells in a single growth cycle. Also, the identification of DI-specific nucleic acids intracellularly demonstrates that DI-related events occur prior to viral maturation.

2.4.3. Obtaining Pure Populations of DI and Standard Virions

Biological experiments can establish the presence of DI particles in a virus system independently of whether or not the DI particles can be isolated in pure form. Further studies on the mechanism of interference and quantitative enrichment require relatively pure DI particles. Purification of DI particles has usually been achieved by physical means. DI particles may differ in size or buoyant density from standard virus, allowing them to be separated by rate zonal or equilibrium centrifugation. Another approach is to complement DI particles by the use of temperature-sensitive mutants under conditions where the progeny virions produced are only DI particles. Such a method has been successfully utilized with papovaviruses (Brockman and Nathans, 1974; Mertz and Berg, 1974) and reovirus (Spandidos and Graham, 1975a,b). Another method for the production of relatively pure DI particles is to establish carrier cultures where enrichment for DI particles is close to 100% (Staneck et al., 1972).

Purification of standard virus can, similarly, be accomplished by gradient centrifugation, especially if several cycles are used. Another method which is very effective in removing DI particles from a viral preparation is by cloning of standard virus with several successive plaque isolations. Early passages from such cloned viruses are likely to contain only standard virus. Serial undiluted passages of cloned virus have shown that DI particles can become a significant part of the viral preparation by as early as one passage or as late as 40 passages depending on the virus system. Strains of virus within a virus system may also vary considerably in their rate of production of DI particles.

2.4.4. Other Detection Methods for the DI Particles

When DI particles are not separable from standard particles by the usual methods, other approaches can be taken to characterize some of the constituents, especially the nucleic acids of the DI particles. Straightforward extraction of nucleic acids from a mixture of DI and standard particles can result in two distinct size classes or density classes of nucleic acids, with one of them representing the DI genome. Other DI particles may lack the largest segment of a segmented genome or lack the site for cleavage by a restriction endonuclease. Another method of increasing the differences between DI and standard particles is by partial digestion or removal of some structural proteins to utilize the differences in the density or the size of the cores from the two particles.

Occasionally, although DI particles have been characterized for a virus system, there has been need to detect small numbers of DI particles in a mixed population of virus particles. A most sensitive assay for DI particles is a biological one recently developed by Popescu (personal communication) for lymphocytic choriomeningitis (LCM) virus. This is a negative plaque assay based on zones of cells protected by DI particles against the cytopathic effect of standard LCM virus. Such a biological assay requires that cells remain viable after infection with DI particles and standard virus. This assay may prove useful for other virus systems as well. Another sensitive technique has been the amplification assay utilized by Holland and Villareal (1975). In order to detect the presence of VSV DI particles in carrier cultures, the virus preparations from the carrier cultures are passed once in a cell line which is known to favor the production of DI particles. From the number of radioactively labeled DI particles made on this amplifying passage, the amount of DI particles in the inoculum can be calculated by extrapolation. Such an amplification procedure can be used only if the number of successive undiluted passages is known for the production of detectable amounts of DI particles after cloning of the standard virus.

3. VIRAL SYSTEMS CONTAINING DI PARTICLES

3.1. Papovavirus

3.1.1. Discovery of DI Particles

Uchida *et al.* (1966) were the first to report that serial high-multiplicity passage of papovaviruses leads to the formation of DI particles. They found that after three consecutive high-multiplicity passages of a cloned stock, the ratio of plaque-forming units to physical particles falls dramatically. They were able to demonstrate that deleted SV40 molecules appear and that the particles containing deleted DNA interfere with the growth of standard virus.

Extensive investigations of both SV40 and polyoma DI particles since that time have shown that four effects occur during their generation: deletion of viral information, duplication of the origin of DNA replication in the remaining molecules, insertion of host cell DNA into the closed circular viral DNA, and polymerization of small monomer units into multimers about the same size as the standard genome. Papovavirus DI particles differ from pseudovirions, which contain only host cell DNA (Aposhian, Vol. 5 of this series).

3.1.2. Study of Mass Populations

Many investigators have reported that during serial passage of a cloned stock of either SV40 or polyoma virus the titer begins to decrease by the third passage and during later passages it decreases further (Uchida *et al.*, 1966; Yoshiike, 1968*a*; Blackstein *et al.*, 1969; Fried, 1974; Risser and Mulder, 1974). Brockman *et al.* (1973) have shown that after about the eighth passage the titer rises and then falls again. Such cyclic behavior is characteristic of systems which generate DI particles (Huang, 1973). Numerous recombinational events occur as the DI particles appear, leading to the deletion of viral information, duplication of specific regions, and acquisition of host cell information (Yoshiike, 1968*a,b*; Yoshiike and Furuno, 1969; Thorne, 1968; Thorne *et al.*, 1968; Lavi and Winocour, 1972; Tai *et al.*, 1972; Lavi *et al.*, 1973; Rozenblatt *et al.*, 1973; Brockman *et al.*, 1973; Folk and Wang, 1974; Frenkel *et al.*, 1975*b*). Extensive acquisition of host cell sequences to the point where they dominate the closed circular DNA molecules occurs after ten passages (Brockman *et al.*, 1973). During serial passage, no specific viral sequences are lost but there is a tendency for the sequences encoding the late proteins to be lost preferentially (Brockman *et al.*, 1973; Risser and Mulder, 1974).

3.1.3. Studies on Cloned Variants

Two different methods have been used to clone papovavirus DI particles. Mertz and Berg (1974) and Brockman and Nathans (1974) took advantage of the ability of some DI particles to complement a temperature-sensitive mutant. They isolated plaques formed by complementation between temperature-sensitive mutants and DI particles. From these plaques the DI particle was then purified and characterized. The second method of cloning DI particles, used by Lee *et al.* (1975) and by Fried (1974), involves the generation of virus stocks having a great excess of DI particles and then isolation of individual plaques from this stock. The plaques often contain both a wild-type virus and a DI particle; the wild type acts as helper for the growth of the DI particle. The first method, by its nature, involves the cloning only of DI particles that can still express the genetic information of one of the viral genes. The second method allows the cloning of DI particles that may not express any viral information but are able to replicate.

Mertz *et al.* (1975) and Brockman *et al.* (1975) have characterized the structure of certain cloned complementing variants. They found that these variants uniformly contain a single deletion as well as a single duplication. In at least one case, it is possible to say from the structure of the molecule that the deletion occurred before the duplication (Mertz *et al.*, 1975). In all cases, the duplicated segment involves a tandem duplication of the site of initiation of DNA synthesis. In studies of either mass DI populations or cloned DI particles, it has been consistently observed that papovavirus DI DNA contains two or more initiation sites for DNA synthesis while the standard viral DNA contains only one.

The DI particles isolated by the helper procedure have been deleted much further than those isolated by the complementation procedure (Lee *et al.*, 1975; Griffin and Fried, 1975). The ones that have been analyzed consist of a segment of SV40 or polyoma DNA involving just the nucleotide sequences around the origin of replication; in one case, both reiterated and nonreiterated host cell DNA sequences are also present (Lee *et al.*, 1975). Studies of mass populations of DI particles have also indicated that reiterated as well as nonreiterated host cell sequences can be found in DI particle preparations (Frenkel *et al.*, 1974). The host cell sequences involved are different in different populations of DI particles, but there is some evidence that only a restricted range of host cell DNA is found in the particle (Frenkel *et al.*, 1974).

Martin *et al.* (1974) have isolated an interesting defective particle from a SV40-like virus recovered from human beings (the DAR virus). Their variant, which was effectively isolated by a helper procedure, contains a triplication of a segment of the SV40 genome. Within the monomer portion of the DNA, there is evidence of an inversion of the normal order of sequences in SV40 (Khoury *et al.*, 1974; Martin *et al.*, 1974). To generate this particle, it would appear to have been necessary that a recombinational event occurred fusing DNA sequences that are ordinarily on different strands of the DNA molecule. After this fusion, a deletion must have occurred to generate the monomer-size unit, which was then triplicated (Khoury *et al.*, 1974); Fareed *et al.*, 1974; Davoli and Fareed, 1975).

3.1.4. Mechanism of Synthesis of DI Particles

The previous results show that four types of DNA recombinational events can be found among the various papovavirus DI particles: deletion of DNA, tandem duplication of DNA, acquisition of host cell DNA, and polymerization of small monomer sequences. Special mechanisms must be involved for each of these events.

3.1.4a. Deletion

A deletion, like many of the DNA recombinations involved in DI formation, could involve "illegitimate recombination" (Franklin, 1971). Illegitimate recombination is the fusion of two pieces of DNA in the absence of extensive homology. Thus, we would imagine the deletion occurring by two portions of an SV40 circle coming into apposition and being recombined with each other, generating two circles one of which would lack the origin of replication and so would be a dead end. This model, like all models involving illegitimate recombination, is very easy to draw on paper but difficult to imagine in molecular terms. It is therefore attractive to look for another mechanism which might be involved.

A number of investigators have suggested that deletion might occur during the process of replication of the circular genome. Because replication starts from a single origin and is bidirectional, a partially replicated molecule has a number of 5' and 3' ends as well as a number of continually opening and closing swivel points. Robberson and Fried (1974) and Folk and Wang (1974) have both proposed models of deletion involving the joining of these loose ends. A specific proposal related to the generation of λdv bacteriophage was also made by Chow et al. (1974).

3.1.4b. Duplication

Two pressures lead to selection from a population of deleted DNA molecules of those with a tandem duplication of the region of DNA around the origin of replication. One pressure is the need for a certain minimal size of DNA in order to form a virion; the second is the replication advantage of having two origins for initiation of replication. The mechanism of duplication is totally unclear. The possibility exists for a nascent DNA strand to loop out from the template and then recopy a region of template already copied once.

3.1.4c. Acquisition of Host DNA

The most likely mechanism for the acquisition of host DNA by papovavirus DI particles is covalent integration of host and viral DNA followed by excision of a joint host–viral DNA molecule (Tai et al., 1972). The puzzle posed by the rapid selection of such molecules during sequential high-multiplicity passages is whether the host DNA is an accidental acquisition at high frequency or whether the joint host–viral DNA molecules have an advantage over pure viral DNA molecules.

3.1.4d. Polymerization of Small Monomers

The DI particles found after a number of serial passages are most often polymers (3–9 mers) of either a piece of viral DNA or a joint viral–host DNA piece; in either case, the origin of viral DNA replication is inevitably included. Polymerization of a monomer occurs with great rapidity (Davoli and Fareed, 1975).

3.2. Other DNA Viruses

3.2.1. Herpesvirus

Serial high-multiplicity passage of herpesvirus (or of the related pseudorabies virus) leads to a decreased titer and the formation of DI particles (Bronson *et al.*, 1973; Ben-Porat *et al.*, 1974; Frenkel *et al.*, 1975a; Wagner *et al.*, 1975). Cyclic variation in the ratio of standard to DI particles has been seen (Murray *et al.*, 1975). Both herpes simplex viruses type 1 and type 2 growing in a variety of cells generate DI particles (Bronson *et al.*, 1973) as does *Herpesvirus saimiri* (Fleckenstein *et al.*, 1975). At least some of the DI particles have DNA of a different density than that of standard virions, and much of this DNA involves multiple tandem duplications of a small monomer unit (Rubenstein and Kaplan, 1975; Fleckenstein *et al.*, 1975). A class of DI particles with DNA of standard buoyant density and with limited reiteration has been reported (Kaplan *et al.*, 1975). The DI DNA has the same length as standard DNA (Rubenstein & Kaplan, 1975; Frenkel *et al.*, 1975a). Presumably the generation of the DNA of herpes DI particles involves similar mechanisms as the generation of papovavirus DI particles, although no host DNA has been found in the herpes DI DNA.

Infection of cells with standard virus plus the herpes DI particles containing the highly reiterated DNA generates an excess of a virus-specific polypeptide of molecular weight about 175,000 (Murray *et al.*, 1975; B. Roizman, personal communication). Increasing glycosylation of the virion proteins has also been reported to occur in such cells (Ben-Porat *et al.*, 1975).

3.2.2. Adenovirus

The only candidate adenovirus DI particles are the type 12, low-density particles described by Mak (1971). High-multiplicity serial passages of adenovirus types 2 and 5 have not led to production of detectable DI particles despite a serious effort to find them (P. Sharp, personal communication).

3.3. Reovirus

DI particles among reovirus preparations were first detected when cores were isolated after chymotrypsin treatment of virions from multiple serial passages and found to be of a lower density than cores made from standard virus (Nonoyama *et al.*, 1970). Without chymotrypsin treatment, DI particles and standard virions cannot be separated from each other. Interference with the growth of standard reovirus by its DI particles is so slight that detection of DI particles by this method is often misleading. However, host cells other than L cells may show greater interference.

The frequency of formation of DI particles differs for different reovirus strains. DI particles have been isolated from five different clones of type 3 (Dearing strain) reovirus, and, in each case, they become detectable by the seventh serial high-multiplicity passage. When the double-stranded RNA genomes from the DI cores are analyzed by polyacrylamide gel electrophoresis, all five isolates lack the largest segment (Nonoyama and Graham, 1970). A temperature-sensitive mutant from complementation group C and one from group F show a higher rate of synthesis of DI particles (Schuerch *et al.*, 1974). The genome segment that is missing from these isolates of DI particles is still the largest piece; continued passaging can result in the deletion of a second piece of the double-stranded genome.

The use of reovirus DI particles to probe the functional defects of temperature-sensitive mutants has been exploited by Spandidos and Graham (1975*a,b*). Complementation studies between DI particles and temperature-sensitive mutants show that complementation group C is the only group which fails to complement DI particles or to be complemented by DI particles. Since the largest segment of double-stranded RNA is deleted from DI particles, group C must have a lesion in this same segment. Another group, E, which complements only asymmetrically is interpreted to have a lesion in a *cis*-acting function. These studies demonstrate the potentially powerful tool that DI particles can provide in genetic analyses.

3.4. Picornavirus

The initial poliovirus DI particle appeared in a virus stock that had been passaged for a long time without cloning. Later isolates of poliovirus DI particles were produced by serial high-multiplicity passage of cloned virus. A series of four papers have described the properties of these DI particles and the mechanisms by which they interfere with standard virus and are enriched at the expense of standard virus (Cole *et al.*, 1971; Cole and Baltimore, 1973*a,b,c*).

3.4.1. Nature of the Defect

Poliovirus DI particles contain about 15% less RNA than standard poliovirus particles but have a normal amount of the usual complement of capsid protein. Like standard poliovirus RNA, DI RNA contains a tract of about 70 adenylic residues at the 3′ end (Spector, 1975). Because DI particles have a lower density than standard poliovirus, DI poliovirus particles can be purified to homogeneity using cesium chloride density gradients. Infection of HeLa cells by a purified preparation of DI particles leads to synthesis of a normal amount of virus-specific RNA and to a normal rate of viral protein synthesis. However, there is no formation of progeny virions in cells infected by DI particles alone. New DI particles are produced only when cells are coinfected by standard virions and DI particles.

The inability of DI particles to reproduce themselves is a consequence of the deletion in viral RNA. This deletion results in the synthesis of a capsid protein precursor that is missing about one-third of its amino acid sequence. The precursor is not processed correctly, and none of the four viral capsid proteins is detectable in infected cells.

Analysis of the translation of DI RNA *in vitro* has shown that the maximum size of product formed is 15% smaller than the maximum product synthesized under the direction of standard poliovirus RNA (Villa-Kamaroff *et al.,* 1975). Translation of both standard and DI RNA begins with the same nonapeptide, indicating that the initiation site for protein synthesis is not deleted in DI RNA. Because the capsid protein precursor is encoded by the 5′ half of the viral RNA (Taber *et al.,* 1971; Summers and Maizel, 1971), the deletion must be an internal deletion within the 5′ half of the viral RNA.

3.4.2. Mechanism of Interference

In previous work, it was not possible to examine directly the synthesis of RNA in coinfected cells because no method had been perfected to separate standard from DI RNAs sufficiently to allow their relative quantitation. Recently Hewlett, Rozenblatt, Ambrose, and Baltimore (in preparation) using agarose gel electrophoresis showed directly that when cells are infected by an equal mixture of standard and DI RNA the replicating pool of RNA contains an equal mixture of the two kinds of molecules. The total amount of RNA made by the infected cell is, however, not increased by coinfection, and therefore a constant amount of RNA synthesis must be apportioned between two replicating systems in the coinfected cell. The consequence of this situation is that when cells are infected by standard particles along with DI particles there is less standard RNA synthesis than if the cells were infected only by the standard virus. Because we do not know what limits the total amount of RNA synthesis in infected cells, it is difficult to say exactly why interference occurs.

The consequence of having two types of potential messenger RNA in infected cell is that the polyribosomes of the infected cell are programmed in two different ways. One set of polyribosomes is able to make capsid protein, while the other set, the ones programmed by DI RNA, are unable to synthesize capsid protein. A second consequence of the two types of RNA in the coinfected cell is that the capsid proteins have two RNAs to choose from when virions are formed. The two kinds of RNA are apparently encapsidated with equal efficiency, although DI RNA seems to be encapsidated a little faster than standard RNA (Cole and Baltimore, 1973*b*).

From this analysis, it is evident that interference at the level of RNA synthesis has two consequences: a decreased amount of capsid protein is formed, and part of the capsid protein is utilized to synthesize DI particles rather than standard virions (see Baltimore *et al.*, 1974). The consequence of these two levels of competition is that the yield of standard virions is a function of the square of the percentage of standard virions in the inoculum. Experiments have shown that this relationship holds (Cole and Baltimore 1973*b*). The square relationship makes poliovirus DI particles increasingly potent interfering agents as the proportion of DI particles in the population increases.

3.4.3. Mechanism of Enrichment

The ability of DI particles to become a large percentage of the total virus population requires that they replicate somewhat more efficiently than standard virions; as previously stated, this process is called "enrichment" (Cole and Baltimore, 1973*b,c*). We do not known how enrichment occurs in the poliovirus system, but it is known to be an event that occurs very early during the infectious cycle. Because the poliovirus DI RNA has an internal deletion, it is difficult to see how it could have a different initiation site for viral RNA synthesis than the standard RNA. However, neither does preferential encapsidation explain enrichment in this system. We are therefore left without a clear explanation of enrichment (see Baltimore *et al.*, 1974).

3.5. Togavirus

Most of the phenomena associated with DI particles have been observed with the alphaviruses, Sindbis virus, and Semliki Forest virus. High-multiplicity passage of cloned stocks of either virus leads reproducibly to the generation of stocks of virus having a greatly increased ratio of physical particles to infectious particles. Such stocks interfere with the growth of standard virus and therefore appear to contain DI particles (Inglot and Chudzio, 1972; Schlesinger *et al.*, 1972; Eaton and Faulkner, 1973; Shenk and Stollar, 1973*a*). Cyclic variation of the relative amount of interfering particles with extended passages has been noted (Johnston *et al.*, 1975; Guild and Stollar, 1975). Interference by Sindbis DI virus shows homotypic specificity; Sindbis DI particles interfere better with standard Sindbis virus than they do with Eastern equine encephalitis virus (Shenk and Stollar, 1973*b*), and Semliki Forest virus DI particles do not affect growth of Sindbis virus (Bruton and Kennedy, 1976). A host cell effect on the rate of generation of DI particles has been seen (Levin *et al.*, 1973). Finally, it has been suggested that the relative ease of production of DI particles can explain the relative avirulence of different field isolates of Semliki Forest virus (Woodward and Smith, 1975).

Attempts to fractionate stocks serially passaged at high multiplicity have met with variable success. Originally it was reported that a population of particles more dense than the standard virions has properties of DI particles (Shenk and Stollar, 1973*a*). Later it was reported that the appearance of such particles was variable and that stocks able to interfere often gave rise to only a single density class of virions (Shenk and Stollar, 1973*b*; Guild and Stollar, 1975). Weiss and Schlesinger (1973) were unable to fractionate serial high-multiplicity passage stocks into two classes of particles. Johnston *et al.* (1975) were able to recognize by electron microscopy a class of small particles appearing only in stocks rich in DI particles, but they could not physically separate these from standard virions. Bruton and Kennedy (1976) have reported a fractionation of Semliki Forest virus DI particles by density in CsCl. Because of the difficulty of reproducibly purifying them, most work on putative togavirus DI particles has been done on mixtures of defective and standard particles, a circumstance which has hampered progress.

Togavirus DI particles do not appear to be able to induce RNA synthesis by themselves (Shenk and Stollar, 1973*a;* Weiss *et al.*, 1974).

Coinfection of cells with standard and defective particles, however, modifies the standard pattern of RNA synthesis. Pure togavirus preparation synthesizes mainly two classes of RNA: 49 S virion RNA and a 26 S RNA of the same polarity as the 49 S RNA but only one-third of its length (Simmons and Strauss, 1972a). The 26 S RNA is the messenger RNA for the synthesis of the three major proteins found in the virion (Cancedda and Schlesinger, 1974; Simmons and Strauss, 1974). The major effect of DI particles is to reduce the synthesis of 26 S RNA and to cause the synthesis of a new, quite homogeneous species of RNA sedimenting at 20 S (Shenk and Stollar, 1972; Eaton and Faulkner, 1973; Weiss and Schlesinger, 1973; Levin et al., 1973). For Sindbis virus, the 20 S RNA is about one-half the size of the 26 S RNA, contains some of the 26 S sequences, and, like 26 S RNA, has poly(A) at its 3′ end (Weiss et al., 1974). For Semliki Forest virus, two 20 S species have been reported (Bruton and Kennedy, 1976) and they do not contain sequences of the 26 S RNA (Kennedy, personal communication). Along with the 20 S RNA species appears a 12 S double-stranded RNA species that is probably a duplex of plus and minus 20 S RNAs (Shenk and Stollar, 1972; Eaton and Faulkner, 1973; Levin et al., 1973). During early times of infection, coinfected cells synthesize about the same amount of RNA as cells infected only by standard virus, but RNA synthesis turns off more rapidly in the coinfected cells, leading to a lower yield of RNA (Shenk and Stollar, 1973a; Weiss et al., 1974).

Another major effect of togavirus DI particles is reduction of the amount of nucleocapsid synthesized (Eaton and Faulkner, 1973; Weiss and Schlesinger, 1973). Thus 49 S RNA is made in the coinfected cells but is not efficiently encapsidated, and therefore much of it does not appear in progeny. A likely cause for reduction in nucleocapsid synthesis is the reduction of 26 S RNA synthesis, leading to a deficit of messenger RNA for the synthesis of the virion proteins. Although it was reported that 20 S RNA can be recovered from polyribosomes (Eaton and Faulkner, 1973), a more complete analysis of its potential as a messenger RNA has found that it is inactive by a number of criteria (Weiss et al., 1974). Therefore, it probably does not contribute to the synthesis of virion proteins.

The RNA of Sindbis virus DI particles has been studied both in the dense DI particles of Shenk and Stollar (1973a) and in mixtures of DI and standard particles where DI particles should be in excess. Some RNA is 49 S with a variable amount of lower molecular weight RNA

(Shenk and Stollar, 1973a; Weiss and Schlesinger, 1973; Johnston *et al.*, 1975). For Semliki Forest virus, Bruton and Kennedy (1976) have reported that DI particles have lower molecular weight RNA. They believe that a number of molecules of "20 S" RNA are present in their DI particles, and they show that the 42 S RNA in their preparations is fully infectious.

The existence of 20 S and a corresponding 12 S double-stranded RNA suggests that the 20 S RNA arises by a self-duplication mechanism. For Semliki Forest virus, a replicative intermediate for the 20 S RNA has been seen, and it is thought that the 20 S RNA replicates as an independent entity (Kennedy, personal communication). Therefore, DI RNA seems to be like poliovirus DI RNA—the 20 S RNA appears to be a huge deletion mutant of the 42 S RNA that retains much of the 5′ end of 42 S RNA and a bit of the 3′ end (Kennedy, personal communication). Such a deletion mutant would replicate independently in coinfected cells but would be unable to initiate its own infection because of an inability to code for the needed proteins.

3.6. Rhabdovirus

Because DI particles of VSV were the first to be purified and physically identified as the agents responsible for homologous interference, this virus system, perhaps more than any other, has served to codify our knowledge of DI particles.

3.6.1. Discovery of DI Particles

A "von Magnus-like" effect was first described for VSV by Cooper and Bellett in 1959. Although they could transfer the agent responsible for interference to uninfected cells, they were unable to decide if the interference were due to an incomplete virus particle or to interferon (Bellett and Cooper, 1959). Electron microscopy later showed that virion preparations contained bullet-shaped particles of varying lengths (Reczko, 1960), and biophysical separations indicated another virus particle at a lower density than standard virus which contained all of the constituents of standard virus (Prevec and Whitmore, 1963). Hackett (1964) suggested the possibility that the shorter bullet-shaped particle might be the interfering component described by Bellett and Cooper (1959).

3.6.2. Characterization

Separation of the DI particles from standard VSV was accomplished by rate zonal centrifugation in sucrose gradients (Huang *et al.*, 1966; Crick *et al.*, 1966; Hackett *et al.*, 1967). Preparations containing only standard virus must be made from very recently cloned isolates; DI particles become detectable after only three successive undiluted passages from cloning (Stampfer *et al.*, 1971). The Indiana serotype of VSV makes DI particles of specific sizes; a particular strain of *ts* mutant consistently makes a particular size of DI particle even after cloning (Reichmann *et al.*, 1971). The New Jersey serotype of VSV makes heterogeneous lengths of DI particles (Hackett *et al.*, 1967).

Several DI particles of the Indiana serotype have been isolated and characterized. They range from one-fourth to two-thirds the length of the standard bullet-shaped particle. The length of the bullet is consistent with the size of nucleic acid found in DI particles (Huang and Wagner, 1966b; Brown *et al.*, 1967; Reichmann *et al.*, 1971). The RNA from DI particles contains sequences found in standard RNA and will anneal with different size classes of complementary messenger RNA (Leamnson and Reichmann, 1974). Therefore, deletions appear to arise anywhere in the genome, although with some greater frequency in the region coding for the largest of the messenger RNAs (Schnitzlein and Reichmann, 1976). Whether or not there are single base changes or minor sequence differences between the DI RNAs and standard RNA remains unknown. Recent evidence indicates that some DI RNAs may be cross-linked or contain a great deal of intramolecular complementarity (Lazzarini *et al.*, 1975; Perrault, 1976).

The polypeptides and supramolecular organization of VSV DI particles have been compared to those of standard virus (Huang *et al.*, 1966; Wagner *et al.*, 1969; Perrault and Holland, 1972b). They appear to be identical except for the relative amounts of each polypeptide species, which may be a reflection of the changed surface to volume ratio of shorter DI bullets to standard bullets. The similarity in polypeptides predicts that DI particles also contain the virion-associated RNA-dependent RNA polymerase (Baltimore *et al.*, 1970). DI particles have been found to contain a full complement of polymerase activity, but to demonstrate its existence it must first be solubilized and added to functional templates made from standard virus (Emerson and Wagner, 1972). With the endogenous DI template, there is some polymerase activity which results mainly in the synthesis of poly(A) (Reichmann *et al.*, 1974). Hybridization of DI RNA to isolated messenger RNA for the L protein shows that the sequences found in DI RNA represent only a part of the cistron coding for the messenger RNA (Leamnson and Reichmann, 1974). Therefore, DI RNA contains a binding site for the polymerase and some sequences which signal the addition of poly(A).

A special type of defective VSV particle which is two-thirds the length of standard virions has some properties of DI particles. It contains two-thirds of the normal amount of viral genetic information and does not contain sequences in common with the region coding for the L protein (Leamnson and Reichmann, 1974). It differs from DI particles because it interferes equally well against both serotypes (New Jersey and Indiana) of VSV (Petric and Prevec, 1970; Prevec and Kang, 1972; Prevec, 1973). This heterotypic interference is, however, similar to homotypic interference because it occurs in the presence of actinomycin D and is abolished by ultraviolet light irradiation of the DI particle. Until the homologous sequences between Indiana and New Jersey VSV RNAs are better characterized, it is difficult to classify this long defective particle as a true DI particle.

Other rhabdoviruses besides VSV have been found to produce DI particles (Atanasiu et al., 1963; Murphy et al., 1972; Wild, 1972; Crick and Brown, 1973, 1974, 1975). Because their properties are less well known, we shall present only a discussion of interference by VSV DI particles.

3.6.3. Interference

Virus-specific RNA synthesis of VSV involves two distinct steps: transcription and replication (Baltimore et al., 1970; Printz-Ané et al., 1972; Perlman and Huang, 1973; Soria et al., 1974). The mechanism of interference for VSV DI particles centers around RNA replication rather than RNA transcription. DI particles do not inhibit transcription by standard virus (Huang and Manders, 1972; Perrault and Holland, 1972b), nor do they affect the synthesis of individual species of virus-specific proteins (Kang and Prevec, 1969; Wagner et al., 1969; Little and Huang, unpublished observations). Using temperature-sensitive mutants defective in either transcription or replication as helper standard virus, it was shown that DI-specific RNA synthesis occurs only when RNA replication of the temperature-sensitive helper is not inhibited (Palma et al., 1974). When cycloheximide is used to prevent replication in VSV-infected cells, DI-specific RNA synthesis is similarly affected (Wertz and Levine, 1973; Perlman and Huang, 1973; Palma *et al.*, 1974). When maximum inhibition of standard virus growth occurs in cells coinfected by standard and DI particles, only primary transcription and some synthesis of DI-specific RNA appear to occur (Huang and Manders, 1972; Palma *et al.*, 1974). The lack of any replication of the standard genome RNA into complementary plus and minus 40 S RNA strands suggests that DI RNA successfully competes for replicase molecules.

A current model for VSV RNA replication postulates the modification of a core transcriptase molecule so that it would synthesize full genome size or DI RNA pieces (Perlman and Huang, 1973). This modification appears to be determined by both the virus and the host cell because there is selective synthesis of DI RNA in different host cells (Huang and Baltimore, 1970; Perrault and Holland, 1972*a*) as well as discrimination in the synthesis of different DI RNAs by the replicase provided by a particular standard virus (Reichmann, personal communication). The polypeptides involved in this complex interaction are not well understood; it appears that for DI RNA to successfully compete against standard RNA, the binding site on DI RNA for replicase must be of a higher affinity than that on standard RNA.

3.6.4. Mechanism of Enrichment

A natural consequence of inhibition at the replicative step is the reduction of 40 S RNA templates for the amplification of transcription, the so-called secondary transcription from newly replicated standard virion RNA. Therefore, overall structural protein synthesis is expected to be reduced. If only DI RNA were being replicated, then there would be no competition for the limited pool of structural proteins. Therefore, enrichment appears to be determined by the enhanced ability of DI RNA to replicate in the presence of standard RNA.

Competition for structural proteins during encapsidation and other stages of morphogenesis does not appear to play a role in enrichment of DI particles for the VSV system. There is some evidence to suggest the contrary. During interference, nucleocapsids containing DI RNA and its complementary RNA are preferentially retained in the cytoplasm as if there were a specific block to the further maturation of DI nucleocapsids (Palma and Huang, 1974; Huang and Palma, 1975). Why this occurs is at present unknown.

3.7. Paramyxovirus

A "von Magnus-like" effect with Sendai virus was first seen by Tadokoro (1958) and by Sokol *et al.* (1964). Examination of virion RNA preparations indicated that RNAs smaller than the 50 S standard RNA are more readily found in virions made during undiluted passages (Kingsbury *et al.,* 1970). These smaller RNAs belong to DI particles of Sendai virus. The standard virions can be separated from the DI particles by rate zonal centrifugation (Kingsbury *et al.,* 1970).

Kingsbury and Portner, in a series of papers published between 1970 and 1972, demonstrated the properties of Sendai DI particles. These have been fully reviewed by Kingsbury (1972). It should be pointed out that tissue specificity in the production of DI particles was first reported with the Sendai virus system. Chick lung cells favor the synthesis of standard virus, whereas chick fibroblasts favor the synthesis of DI particles (Kingsbury and Portner, 1970). Also, preexposure of cells to DI particles increases their ability to interfere; this enhancement can occur in the presence of inhibitors of protein synthesis (Portner and Kingsbury, 1971). In an attempt to relate the synthesis of DI-specific RNA to either transcription or replication of the standard RNA, the authors found that both activities appear to be responsible for the synthesis of DI RNA (Portner and Kingsbury, 1972). These results may indicate that Sendai DI particles differ from VSV DI-T particles in their mechanism of interference; it will be interesting to see if Sendai DI RNA contains information capable of coding for a functional polypeptide.

A peculiarity of paramyxoviruses, although not unique to these viruses, is the high proportion of complementary genomes in DI and standard virions (Robinson, 1970; Portner and Kingsbury, 1970). Annealing of these genomes is concentration dependent, suggesting that the complementary sequences reside in separate RNA strands. Similar phenomena have been seen for certain DI particles of vesicular stomatitis virus (Roy *et al.,* 1973). The significance of these findings is at present unknown.

3.8. Orthomyxovirus

The von Magnus phenomenon was first described for influenza viruses (see Sections 1, 4.1, and 4.2). Serial high-multiplicity passage produces particles which contain less nucleic acid (Ada and Perry,

1956; Lief and Henle, 1956). Examination of these RNAs by electrophoretic mobility shows that the largest segment of this multisegmented genome is missing from von Magnus particles and smaller pieces are present when compared to standard preparations (Duesberg, 1968; Pons and Hirst, 1969). Recent electrophoretic separations of standard virion RNAs on urea-agarose polyacrylamide gels demonstrate that influenza RNAs are composed of seven to nine pieces (Palese and Schulman, 1976; Pons, 1975) with the likely possibility that each is unique. Preparations containing a large proportion of von Magnus particles are likely to yield the higher number of RNA pieces (Palese and Schulman, 1976). When the proteins are examined, virions from serial high-multiplicity passages are identical to virions from low-multiplicity passages (Lenard and Compans, 1975). A report on differences in lipid content between virions made by the two types of passages has not been confirmed (Blough and Merlie, 1970).

Undoubtedly, von Magnus particles fit the definition for DI particles. In retrospect, many of the studies on interference with influenza viruses indicate a role for DI particles (Gard and von Magnus, 1947; Henle and Henle, 1949; Schlesinger, 1950). Although there is an indication that certain strains of influenza viruses may produce DI particles that are separable from standard virus by centrifugation (Yoshishita *et al.,* 1959), there has been no clear separation of DI particles from standard influenza virus. Therefore, interpretations of these results are limited. Even diluted passages of influenza virus in eggs or chick embryo fibroblasts contain significant numbers of DI particles. The discovery that growth of influenza virus in Madin–Darby bovine kidney cells leads to preparations containing mostly standard virions has helped to define some of the characteristics unique to standard influenza virus (Choppin, 1969; Choppin and Pons, 1970).

The mechanism of interference remains a mystery for influenza virus DI particles. It is not known whether the loss of a large piece of RNA in the virion preparation is a consistent phenomenon in the generation of DI particles—nor is it known whether the missing segment is totally deleted or appears as a smaller piece of RNA. When intracellular genome RNAs or double-stranded RNAs are examined, the largest segment appears to be synthesized during conditions of interference (Lerner and Hodge, 1969; Sugiura, 1972). The finding that each RNA piece has its own polymerase (Bishop *et al.,* 1972) suggests independence in the synthesis of each piece of RNA. In such a system, it is difficult to imagine how successful competition occurs during replication between a number of pieces of RNA and one larger piece of

RNA. It has been suggested that smaller RNAs are capable of faster replication (Pons, 1975). Nucleocapsids accumulate in the nucleus during interference (Rott and Scholtissek, 1963), a finding which has led investigators to search for a defect in maturation (Nayak, 1972; Meier-Ewert and Dimmock, 1970). Preferential selection during maturation for certain pieces of RNA has been ruled out (Compans *et al.*, 1970).

3.9. Arenavirus

The prototype of the arenavirus group is lymphocytic choriomeningitis (LCM) virus. A great deal of information on the biology of LCM virus has led to the suggestion of DI particles without their actually being isolated or characterized. The early work by Traub (1938) demonstrated an unknown type of interference in persistent LCM virus infections of mice. It was noted that the interference was homologous, involved localized areas, and did not depend on humoral immunity. Later it was demonstrated that interferon did not play a role in this interference (Wagner and Snyder, 1962). *In vitro* studies with macrophages from LCM virus-infected mice showed that these cells are resistant to superinfection by homologous virus, but not by West Nile virus; interferon is not present in these cells, but they are capable of synthesizing it and supporting its action on heterologous virus (Mims and Subrahmanyan, 1966). Persistently infected carrier cultures also show a similar interference phenomenon and appear to produce a transmissible interfering agent (Lehmann-Grube *et al.*, 1969; Staneck *et al.*, 1972).

Characterization of LCM DI particles and a quantitative assay for them based on interference with infectious center formation by standard virus were accomplished by Welsh and Pfau (1972). BHK or L-cell carrier cultures of certain strains of LCM virus will continually produce DI particles with little or no standard virus (Welsh *et al.*, 1972). DI particles of LCM virus sediment with standard virus, have similar antigenic determinants, and cannot be readily separated from standard virus by physical means (Welsh *et al.*, 1972). Examination of their RNAs shows no difference in the relative content of 23 S and 31 S RNA, and smaller RNAs do not appear in preparations containing predominantly DI particles (Welsh *et al.*, 1975). Little can be said about the molecular basis of interference without further biochemical studies on purified populations of DI particles.

The tremendous variability in the interactions of arenaviruses like

LCM virus with host cells both *in vivo* and in carrier cultures (see Hotchin, 1962, 1971) suggests an interesting biological role for DI particles in this system. Even without their purification, an understanding of the interactions between standard LCM virus and DI particles in populations of cells and their effects on host functions will be valuable. For example, studies on some persistently infected cultures show that freezing and thawing can drastically reduce the production of DI particles and induce the synthesis of standard virus (Staneck *et al.,* 1972). Also, although persistently infected, DI-producing neuroblastoma cultures do not show any major difference in growth rates from uninfected cultures, luxury functions such as choline acetyltransferase and acetylcholinesterase activities appear to be reduced (Oldstone *et al.,* 1975). Such approaches offer hope for understanding the complexities of virus–host interactions and the possible role of DI particles in viral pathogenesis.

4. DI PARTICLES AND VIRAL DISEASE

4.1. Are DI Particles Made in the Animal?

In retrospect, all of the early biological studies on influenza virus in mice and embryonated eggs clearly demonstrate *in vivo* synthesis of DI particles (von Magnus, 1954). As long ago as 1950, Bernkopf recognized the need to characterize DI or incomplete particles when he wrote that "A study of chemical differences between such mature [standard] and immature [incomplete] virus may prove extremely interesting if such differences could be correlated with the ability of a particle to multiply and to cause toxic reactions." However, biochemical characterization of DI particles made in animals has been achieved only recently. Holland and Villareal (1975) harvested virus from mouse brains and showed that DI particles of vesicular stomatitis virus made therein are identical to DI particles made in continuous cell cultures.

Attempted correlations of the presence or absence of DI particles with the course of a viral disease in individual animals have been difficult. Studies with mice show that barely detectable amounts of Semliki Forest DI particles are produced in each mouse. Detection of DI particles by their ability to interfere in cell culture assays is often only marginal (Woodward and Smith, 1975). It appears that only pooling of several brains from mice will permit detection of DI particles of vesicular stomatitis virus (Holland and Villareal, 1975). Therefore, data relating the synthesis of DI particles to the course of a disease are likely to contain a great deal of variability until a virus system presents itself that has the property of making large amounts of DI particles in the animal during the normal disease course.

4.2. Protective Effects of DI Particles against Viral Disease

Because of the problems of detecting DI particles in animals during the course of a disease, studies on DI particles in animals have focused on the inoculation of large amounts of DI particles and observation of the subsequent effects of these particles on acute viral disease.

Very early studies done with unpurified preparations of incomplete particles of influenza virus suggested that DI particles can prevent acute disease. Intranasal inoculation of mice with virus from undiluted passages not only does not cause death of the mice but also protects them from the lethal effects of virus from diluted passages (Bernkopf, 1950). Such studies led to the suggestion that incomplete virus preparations could be used as vaccines.

More recently, McLaren and Holland (1974) have shown that vaccine strains of poliovirus contain a large proportion of DI particles. When these authors cloned the standard virus from a vaccine strain, it regenerated DI particles with a much higher frequency than does virulent poliovirus. Human as well as monkey cells generate the DI particles, but they are generated most rapidly in human diploid fibroblasts. It is interesting to speculate that DI particles play a role in the ability of vaccines to confer immunity without causing virulent disease.

Von Magnus (1951) also examined the *in vivo* effects of influenza DI particles in mice. After inoculation of virus from undiluted passages, the surviving mice develop long-term resistance to the acute paralytic disease caused by standard preparations of influenza virus. To show that the protective effect can be diluted out, virus from undiluted passages was inoculated into mice at very low concentrations. This not only failed to protect mice from lethal superinfections but in itself caused acute paralysis and death. These observations could be repeated with sedimentable virus particles (Gard *et al.*, 1952), so that, despite the inability to purify influenza DI particles, the protective effects were attributed by most investigators to the presence of virionlike defective or incomplete particles.

In addition to the early *in vivo* work on diluted and undiluted passages of influenza virus, there is an interesting observation relating

DI particles to the lack of neurotropism of some influenza virus strains. Schlesinger (1950) observed that nonneurotropic influenza virus, when inoculated intracerebrally into mice, produces a very small amount of defective virus which has the same antigenic properties as standard influenza virus. The avirulent strains are thus able to protect mice from subsequent inoculations with highly lethal, but antigenically similar, neuroadapted influenza virus. Unfortunately, this defective virus has not been further characterized.

Although these observations support the hypothesis that the DI particles account for the avirulence of certain viral strains and afford protection against virulent preparations of related viruses, subsequent knowledge about interferon production, antibody synthesis, and cell-mediated immunity has made it extremely difficult to interpret these initial observations made on relatively crude preparations of influenza virus. Instead of the postulation of direct interference with the growth of standard virions by defective particles, complex mechanisms such as interferon induction and immune responsiveness have had to be accommodated in subsequent interpretations of viral interference in animals.

More recent work with better-characterized influenza preparations strongly supports the hypothesis that DI particles not only can delay the onset of acute disease but also can prevent it altogether. Gamboa *et al.* (1975) have shown that the protective effect of influenza virus DI particles is more easily observed in 7-week-old mice than in 3-week-old mice. Also, studies with purified preparations of standard and DI particles of vesicular stomatitis virus show that DI particles afford real protection and can be used prophylactically against the virus-induced, acute encephalitis in mice caused by intracerebral inoculations (Doyle and Holland, 1973).

Although in these more recent studies interferon does not correlate with the resistance to viral infection, nor is heterologous interference detected, the exact mechanism of the protective effect produced by inoculating the animals with DI particles remains obscure. Until selectively immunosuppressed animals are tested with inoculations of standard and DI virus, the protective phenomenon cannot be quantitated in regard to the relative contribution of the direct effect of DI particles on the multiplication of standard virus.

A word of caution about the use of DI particles as vaccines should be inserted here. Although DI particles appear to ameliorate or prevent acute disease and death, our present understanding of DI particles predicts the occurrence of partial viral suppression, so that mixed standard virus and DI particle infections can persist for long periods of time in animals (Huang and Baltimore, 1970). DI particles may initiate and maintain chronic infections, especially in the central nervous system; exacerbations may result from occasional bursts of synthesis of standard virus. Such models and some supporting evidence are discussed in the following section.

4.3. Persistent Viral Disease and DI Particles

Among the many viral diseases which persist in animals, lymphocytic choriomeningitis has been the most thoroughly studied (see Hotchin, 1971). This disease of rodents, in its various manifestations, remains one of the most likely candidates where DI particles will be shown to have an important role. Without isolation of DI particles from the persistently infected animals, the importance of DI particles can only be surmised from the following set of observations: (1) although antibody to LCM virus is present, there is continued production of viral antigens or infectious virus; (2) there is little or no detectable interferon production in these persistently infected animals; (3) the animals are resistant to superinfection with a lethal dose of homologous or closely related virus; and (4) intracellular virus-specific antigen accumulates in some but not all cells of the animals.

Evidence from other virus systems also supports a role for DI particles in establishing and maintaining persistent infections. When animals are inoculated with mixtures of DI and standard particles, the relative concentrations of standard and DI particles in the inoculum are important for the outcome of a viral infection. For instance, inoculation of mice with a mixture of DI and standard particles of vesicular stomatitis virus at a ratio lower than the ratio which protects is found to result in prolonged paralysis and wasting prior to the onset of death (Holland and Doyle, 1973). With measles virus, some strains like the Edmonston strain become more encephalitogenic upon undiluted passage (Janda et al., 1971). These undiluted passages correlate with the development in HeLa cell cultures of a slower cytopathic effect and a high rate of development of carrier cultures (Oddo et al., 1967). On the other hand, diluted passages of the Edmonston strain result in virus which gives a rapid giant-cell-related cytopathic effect. Unfortunately, other variants of measles virus do not exhibit a direct correlation between passage history and either neurotropism or delayed cytopathic effect in cell culture (Janda et al., 1971). The difficulties of isolating DI particles from persistently infected animals suggest that future efforts toward determining the role of DI particles in persistent diseases should concentrate on persistently infected cell cultures.

4.4.   Persistent and Carrier Cultures

Almost every virus group has been found to establish persistently infected cell cultures (see Walker, 1964; Preble and Youngner, 1975). The properties of these infected cultures vary considerably. However, Walker (1968) recognized a particular type of persistent infection which has the following properties: (1) only a fraction of the population of infected cells are productively infected and die; (2) the remaining cells are resistant to superinfection by homologous virus; (3) antibodies to the virus will cure the persistently infected culture; and (4) persistence of the virus in the cell culture can be maintained in the absence of added antibodies or in the absence of interferon production. These properties strongly point to the presence of DI particles for the maintenance of viral persistence in cell culture.

Unfortunately, very few such persistently infected cultures have been characterized in regard to DI particles. Those carrier cultures where DI particles are thought to play a role are cultures infected with lymphocytic choriomeningitis virus (Lehmann-Grube et al., 1969), with Sindbis virus (Inglot et al., 1973), with Parana virus (Staneck and Pfau, 1974), and with rabies virus (Kawai et al., 1975). Other cell cultures persistently infected with measles virus (Rustigian, 1962; Oddo et al., 1967; Raine et al., 1972; Kiley et al., 1974), with encephalomyocarditis virus (Sanders, 1957), with vesicular stomatitis virus (Wagner et al., 1963), and with polyoma virus (Barski and Cornefert, 1962), although not well characterized, may also be maintained by the presence of DI particles. One of the best-characterized carrier cultures was initiated by inoculating cells with a mixture of DI and standard VSV (Holland and Villareal, 1974). Several persistently infected cultures with different wild-type strains and temperature-sensitive mutants of vesicular stomatitis virus have been established this way. The common denominator in all of these cultures is the presence of a particular DI particle, two-thirds the length of standard VSV. The presence of this DI particle is apparently necessary for the initiation and maintenance of each persistently infected cell line. DI particles are detected in each of these carrier cultures, especially if the amplification assay is used to detect DI particles (see Section 2.4.4).

Another approach for studying persistently infected cultures is to establish an artificial carrier culture initiated by adding highly virulent standard virus and DI particles. To maintain continuity, fresh uninfected cells are added to this culture at regular intervals. Palma and Huang (1974) used such cultures infected with VSV to study the synthesis of DI particles by radioactive labeling. Host survival and host resistance to superinfection by homologous standard virus were also monitored. A cyclic overlapping pattern was observed for the production of the two types of viral particles. Depending on the total amount of virus particles and the relative concentrations of DI and standard particles in the preparation, the cells can be rapidly killed or can survive and be resistant to superinfection. In the latter group, some cells are found to retain viral nucleocapsids intracellularly. When preparations of temperature-sensitive mutants of vesicular somatitis virus are similarly passaged, cyclic patterns obtained at the permissive temperature are identical to those of wild-type except that the total virus titers are lower. At the nonpermissive temperature, the infection dies out after a few passages, unless a revertant or a leaky mutant begins to dominate the population of virus particles (Sinarachatanant and Huang, unpublished observations). These findings suggest that temperature sensitivity of the virus may serve to reduce the total amount of progeny synthesized and thus help to establish a persistent infection. Absolute temperature sensitivity, therefore, is not compatible with the persistent state. These studies demonstrate the large variety of cellular responses to viral infection in the absence of the immune system and with only the interplay of standard virus, DI particles, and the host cells.

4.5. Resistance and Susceptibility in Cell Cultures

Another approach toward understanding the role of DI particles in disease is the establishment of primary cell cultures and cell lines directly from animals known to be resistant or susceptible to a certain virus and then the demonstration that resistance is related to the production of DI particles. Darnell and Koprowski (1974) were able to transform fibroblasts from mice which are known to be genetically resistant or sensitive to West Nile virus and to establish these fibroblasts as cell lines. Infection of these cells with West Nile virus showed that the resistant cell line does not produce large amounts of interferon, nor are the cells inordinately sensitive to the action of exogenously added interferon. Standard virus production gradually declines with an increase in the interfering capacity of the preparation upon undiluted passaging in these cells. Biochemical characterization of DI particles made by the resistant cells is, however, still lacking.

Another example of the use of such cell cultures is the demonstration of tissue tropism of Sendai virus; primary cultures of chick lung cells do not support the production of Sendai DI particles as well as chick fibroblasts do (Kingsbury, 1972).

Many viral diseases are evident in the young animal and not in the old animal. Such an age differential of mice to Sindbis virus can be reproduced in cell cultures. Fibroblast cultures of weanling animals produce less standard Sindbis virus than cultures made from newborns (Johnson *et al.*, 1972). Similarly, the resistance of mice chronically infected with lymphocytic choriomeningitis virus can be seen in fibroblast cultures made from these mice (Mims and Subrahmanyan, 1966). Even after four passages, these cells in culture still show resistance. Use of such cultures eliminates many host defense mechanisms and points to the production of DI particles as a possible cause of resistance. Unfortunately, there have been no attempts to identify DI particles in these infected primary cultures.

4.6. Other Hypotheses

Other hypotheses have been put forth to account for viral persistence. These are temperature sensitivity of the virus (Preble and Youngner, 1972, 1975; Fields, 1972) and integration of the viral genetic information into host cell DNA (Mims, 1966; Simpson and Inuma, 1975; Zhdanov *et al.*, 1974). Only further characterization of the virus particles and their life cycle in persistently infected cells will sort out the contributions of these other parameters to the establishment and maintenance of persistence and their possible relationship to the production of DI particles. It should be remembered that these hypotheses are not necessarily mutually exclusive.

ACKNOWLEDGMENTS

We thank Ms. Yvonne Hammond for excellent logistical support. Alice S. Huang is supported by a research career development award from the U.S. Public Health Service, and David Baltimore is an American Cancer Society Professor.

5. REFERENCES

Ada, G. L., and Perry, B. T., 1956, Influenza virus nucleic acid relationship between biological characteristics of the virus particle and properties of the nucleic acid, *J. Gen. Microbiol.* **14:**623.

Atanasiu, P., Lepine, P., Sisman, J., Dauguet, C., and Wetten, M., 1963, Etude morphologique du virus rabique des rues en culture de tissu, *C. R. Acad. Sci. (Paris)* **256:**3219.

Baltimore, D., Huang, A. S., and Stampfer, M., 1970, Ribonucleic acid synthesis of vesicular stomatitis virus. II. An RNA polymerase in the virion, *Proc. Natl. Acad. Sci. USA* **66:**572.

Baltimore, D., Cole, C. N., Villa-Komaroff, L., and Specter, D., 1974, Poliovirus defective interfering particles, in: *Mechanisms of Virus Disease: ICN-UCLA Symposia on Molecular and Cellular Biology*, Vol. 1 (W. S. Robinson and C. F. Fox, eds.), pp. 117–130, Academic Press, New York.

Barski, G., and Cornefert, F., 1962, Response of different mouse cell strains to polyoma infection *in vitro*: Latency and self-inhibition effect in infected cultures, *J. Natl. Cancer Inst.* **28:**823.

Bellett, A. J. D., and Cooper, P. D., 1959, Some properties of the transmissible interfering component of VSV preparations, *J. Gen. Microbiol.* **21:**498.

Ben-Porat, T., Demarchi, J. M., and Kaplan, A. S., 1974, Characterization of defective interfering viral particles present in a population of pseudorabies virions, *Virology* **60:**29.

Ben-Porat, T., Lonis, B., and Kaplan, A. S., 1975, Further characterization of a population of defective interfering pseudorabies virus, *Virology* **65:**179.

Bernkopf, H., 1950, Study of infectivity and hemagglutination of influenza virus in deembryonated eggs, *J. Immunol.* **65:**571.

Bishop, D. H., Roy, P., Bean, W. J., and Simpson, R. W., 1972, Transcription of the influenza ribonucleic acid genome by a virion polymerase. III. Completeness of the transcription process, *J. Virol.* **10:**689.

Blackstein, M. E., Stanners, C. P., and Farmilio, A. J., 1969, Heterogeneity of polyoma virus DNA: Isolation and characterization of non-infectious small supercoiled molecules, *J. Mol. Biol.* **42:**301.

Blough, H. A., and Merlie, J. P., 1970, The lipids of incomplete influenza virus, *Virology* **40:**685.

Bondurant, M. C., Hackett, A. J., and Schaffer, F. L., 1973, Infectivity and RNA patterns as functions of high and low dilution passage of murine sarcoma-leukemia virus: Evidence for autointerference within an oncornavirus population, *J. Virol.* **11:**642.

Brockman, W. W., and Nathans, D., 1974, The isolation of simian virus 40 variants with specifically altered genomes, *Proc. Natl. Acad. Sci. USA* **71:**942.

Brockman, W. W., Lee, T. N. M., and Nathans, D., 1973, The evolution of new species of viral DNA during serial passage of simian virus 40 at high multiplicity, *Virology* **54:**384.

Brockman, W. W., Gutai, M. W., and Nathans, D., 1975, Evolutionary variants of simian virus 40: Characterization of cloned complementing variants, *Virology* **66:**36.

Bronson, D. L., Dreesman, G. R., Biswal, N., and Benyesh-Melnick, M., 1973, Defective virions of herpes simplex virus, *Intervirology* **1:**141.

Brown, F., Martin, S. J., Cartwright, B., and Crick, J., 1967, The ribonucleic acids of the infective and interfering components of vesicular stomatitis virus, *J. Gen. Virol.* **1:**479.

Bruton, C. J., and Kennedy, S. I. T., 1976, Defective interfering particles of Semliki Forest virus: Structural differences between standard virus and defective interfering particles, *J. Gen. Virol.* **31:**383.

Cancedda, R., and Schlesinger, M. J., 1974, Formation of Sindbis virus capsid protein in mammalian cell-free extracts programmed with viral messenger RNA, *Proc. Natl. Acad. Sci. USA* **71:**1843.

Cantell, K., 1961, Mumps virus, *Adv. Virus Res.* **8:**123.

Choppin, P. W., 1969, Replication of influenza virus in a continuous cell line: High yield of infectious virus from cells inoculated at high multiplicity, *Virology* **39:**130.

Choppin, P. W., and Pons, M. W., 1970, The RNAs of infective and incomplete influenza virions grown in MDBK and HeLa cells, *Virology* **42:**603.

Chow, L. T., Davidson, N., and Berg, D., 1974, Electron microscopy study of the structures of λ*dv* DNA's, *J. Mol. Biol.* **86:**69.

Cole, C. N., and Baltimore, D., 1973a, Defective interfering particles of poliovirus. II. Nature of the defect, *J. Mol. Biol.* **76:**325.

Cole, C. N., and Baltimore, D., 1973b, Defective interfering particles of poliovirus. III. Interference and enrichment, *J. Mol. Biol.* **76:**345.

Cole, C. N., and Baltimore, D., 1973c, Defective interfering particles of poliovirus. IV. Mechanisms of enrichment, *J. Virol.* **12:**1414.

Cole, C. N., Smoler, D., Wimmer, E., and Baltimore D., 1971, Defective interfering particles of poliovirus. I. Isolation and physical properties, *J. Virol.* **7:**478.

Compans, R. W., Dimmock, N. J., and Meier-Ewert, H., 1970, *The Biology of Large RNA Viruses* (R. D. Barry and B. W. J. Mahy, eds.), pp. 87–108, Academic Press, New York.

Cooper, P. D., and Bellett, A. J. D., 1959, A transmissible interfering component of vesicular stomatitis virus preparations, *J. Gen. Microbiol.* **21:**485.

Crick, J., and Brown, F., 1973, Interference as a measure of cross relationship in the vesicular stomatitis group of rhabdoviruses, *J. Gen. Virol.* **18:**79.

Crick, J., and Brown, F., 1974, An interfering component of rabies virus which contains RNA, *J. Gen. Virol.* **22:**147.

Crick, J., and Brown, F., 1975, *Negative Strand Viruses* (B. W. J. Mahy and R. D. Barry, eds.), pp. 698–705, Academic Press, London.

Crick, J., Cartwright, B., and Brown, F., 1966, Interfering components of vesicular stomatitis virus, *Nature (London)* **211:**1204.

Darnell, M. B., and Koprowski, H., 1974, Genetically determined resistance to infection with group B arboviruses. II. Increased production of interfering particles in cell cultures from resistant mice, *J. Infect. Dis.* **129:**248.

Darnell, M. B., Koprowski, H., and Lagerspetz, K., 1974, Genetically determined resistance to infection with group B arboviruses. I. Distribution of the resistance gene among various mouse populations and characteristics of gene expression *in vivo*, *J. Infect. Dis.* **129:**240.

Davoli, D., and Fareed, G. C., 1975, Formation of reiterated simian virus 40 DNA, *Cold Spring Harbor Symp. Quant. Biol.* **39:**137.

Doyle, M., and Holland, J. J., 1973, Prophylaxis and immunization of mice by use of virus-free defective T particles protect against intracerebral infection by vesicular stomatitis virus, *Proc. Natl. Acad. Sci. USA* **70:**2105.

Duesberg, P. H., 1968, The RNA's of influenza virus, *Proc. Natl. Acad. Sci. USA* **59**:930.

East, J. L., and Kingsbury, D. W., 1971, Mumps virus replication in chick embryo lung cells: Properties of RNA species in virions and in infected cells, *J. Virol.* **8**:161.

Eaton, B. T., and Faulkner, P., 1973, Altered pattern of viral RNA synthesis in cells infected with standard and defective Sindbis virus, *Virology* **51**:85.

Emerson, S. U., and Wagner, R. R., 1972, Dissociation and reconstitution of the transcriptase and template activities of vesicular stomatitis B and T virions, *J. Virol.* **10**:297.

Enea, V., and Zinder, N. D., 1975, A deletion mutant of bacteriophage f1 containing no intact cistrons, *Virology* **68**:105.

Fareed, G. C., Byrne, J. C., and Martin, M. A., 1974, Triplication of a unique genetic segment in a simian virus 40-like virus of human origin and evolution of new viral genomes, *J. Mol. Biol.* **87**:275.

Fields, B. N., 1972, Genetic manipulation of reovirus—A model for modification of disease, *N. Engl. J. Med.* **287**:1026.

Fleckenstein, B., Bornkamm, G. W., and Ludwig, H., 1975, Repetitive sequences in complete and defective genomes of *Herpesvirus saimiri*, *J. Virol.* **15**:398.

Folk, W. R., and Wang, H. C. E., 1974, Closed circular DNAs with tandem repeats of a sequence from polyoma virus, *Virology* **61**:140.

Franklin, N. C., 1971, Illegitimate recombination, in: *The Bacteriophage Lambda* (A. Hershey, ed.), pp. 175–194, Cold Spring Harbor Laboratory, Cold Spring Harbor.

Frenkel, N., Lavi, S., and Winocour, E., 1974, The host DNA sequences in different populations of serially passaged SV40, *Virology* **60**:9.

Frenkel, N., Jacob, R. J., Honess, R. W., Hayward, G. S., Locker, H., and Roizman, B., 1975a, Anatomy of herpes simplex virus DNA. III. Characterization of defective DNA molecules and biological properties of virus populations containing them, *J. Virol.* **16**:153.

Frenkel, N., Rozenblatt, S., and Winocour, E., 1975b, The repeated sequences in serially-passaged SV40 DNA, in: *Tumor Virus-Host Cell Interaction* (A. Kolber, ed.), pp. 39–58, Plenum Publishing Co., New York.

Fried, M., 1974, Isolation and partial characterization of different defective DNA molecules derived from polyoma virus, *J. Virol.* **13**:939.

Gamboa, E. T., Harter, D. H., Duffy, P. E., and Hsu, K. C., 1975, Murine influenza virus encephalomyelitis. III. Effect of defective interfering particles, *Acta Neuropathol.* **34**:157.

Gard, S., and von Magnus, P., 1947, Studies on interference in experimental influenza. II. Purification and centrifugation experiments, *Arch. Kem. Mineral. Geol.* **24(8)**:1.

Gard, S., von Magnus, P., Svedmyr, A., and Birch Andersson, A., 1952, Studies on the sedimentation of influenza virus, *Arch. Ges. Virusforsch.* **4**:591.

Griffin, B. E., and Fried, M., 1975, Amplification of a specific region of the polyoma virus genome, *Nature (London)* **256**:175.

Guild, G. M., and Stollar, V. S., 1975, Defective interfering particles of Sindbis virus. III. Intracellular viral RNA species in chick embryo cell cultures, *Virology* **67**:24.

Hackett, A. J., 1964, A possible morphologic basis for the autointerference phenomenon in vesicular stomatitis virus, *Virology* **24**:51.

Hackett, A. J., Schaffer, F. L., and Madin, S. H., 1967, The separation of infectious and autointerfering particles in vesicular stomatitis virus preparations, *Virology* **31**:114.

Henle, W., 1950, Interference phenomena between animal viruses: A review, *J. Immunol.* **64**:203.

Henle, W., and Henle, G., 1949, Interference between inactive and active viruses of influenza. II. Factors influencing the phenomenon, *Am. J. Med. Sci.* **207**:717.

Holland, J. J., and Doyle, M., 1973, Attempts to detect homologous autointerference *in vivo* with influenza virus and vesicular stomatitis virus, *Infect. Immun.* **7**:526.

Holland, J. J., and Villareal, L. P., 1974, Persistent non-cytocidal vesicular stomatitis virus infections mediated by defective T particles that suppress virion transcriptase, *Proc. Natl. Acad. Sci. USA* **71**:2956.

Holland, J. J., and Villareal, L. P., 1975, Purification of defective interfering T particles of vesicular stomatitis and rabies viruses generated *in vivo* in brains of newborn mice, *Virology* **67**:438.

Hotchin, J., 1962, The biology of lymphocytic choriomeningitis infection: Virus-induced immune disease, *Cold Spring Harbor Symp. Quant. Biol.* **27**:479.

Hotchin, J., 1971, Persistent and slow virus infections, in: *Monographs in Virology* (J. L. Melnick, ed.), pp. 1–211, Karger, Basel.

Huang, A. S., 1973, Defective interfering viruses, *Annu. Rev. Microbiol.* **27**:101.

Huang, A. S., and Baltimore, D., 1970, Defective viral particles and viral disease processes, *Nature (London)* **226**:325.

Huang, A. S., and Manders, E. K., 1972, Ribonucleic acid synthesis of vesicular stomatitis virus. IV. Transcription by standard virus in the presence of defective interfering particles, *J. Virol.* **9**:909.

Huang, A. S., and Palma, E. L., 1975, Defective interfering particles as antiviral agents, *Perspect. Virol.* **9**:77–88.

Huang, A. S., and Wagner, R. R., 1966a, Defective T particles of vesicular stomatitis virus. II. Biologic role in homologous interference, *Virology* **30**:173.

Huang, A. S., and Wagner, R. R., 1966b, Comparative sedimentation coefficients of RNA extracted from plaque-forming and defective particles of vesicular stomatitis virus, *J. Mol. Biol.* **22**:381.

Huang, A. S., Greenawalt, J. W., and Wagner, R. R., 1966, Defective T particles of vesicular stomatitis virus. I. Preparation, morphology, and some biologic properties, *Virology* **30**:161.

Inglot, A. D., and Chudzio, T., 1972, Incomplete Sindbis virus, in: *Proceedings of the Second International Congress for Virology* (J. L. Melnick, ed.), p. 158, Karger, New York.

Inglot, A. D., Albin, M., and Chudzio, T., 1973, Persistent infection of mouse cells with Sindbis virus: Role of virulence of strains, auto-interfering particles and interferon, *J. Gen. Virol.* **20**:105.

Janda, Z., Zorrby, E., and Marvsyk, H., 1971, Neurotropism of measles virus variants in hamsters, *J. Infect. Dis.* **124**:553.

Johnson, R. T., McFarland, H. F., and Levy, S. E., 1972, Age-dependent resistance to viral encephalitis: Studies of infections due to Sindbis virus in mice, *J. Infect. Dis.* **125**:257.

Johnston, R. E., Jovell, D. R., Brown, D. T., and Faulkner, P., 1975, Interfering passages of Sindbis virus: Concomitant appearance of interference, morphological variants and truncated viral RNA, *J. Virol.* **16**:951.

Kang, C. Y., and Prevec, L., 1969, Proteins of vesicular stomatitis virus. I. Polyacrylamide gel analysis of viral antigens, *J. Virol.* **3**:404.

Kaplan, A. S., Ben-Porat, T., and Rubenstein, A. S., 1976, On the mechanism of herpesvirus DNA replication and the genesis of defective particles, in: *Herpes Viruses* (C. Borek and D. W. King, eds.), Traton Intercontinental Book Corp., in press.

Kawai, A., Matsumoto, S., and Tanabe, K., 1975, Characterization of rabies virus recovered from persistently infected BHK cells, *Virology* **67**:520.

Khoury, G., Fareed, G. C., Berry, K., Martin, M. A., Lee, T. N. H., and Nathans, D., 1974, Characterization of a rearrangement in viral DNA: Mapping of the circular SV40-like DNA containing a triplication of a specific one-third of the viral genome, *J. Mol. Biol.* **87**:289.

Kiley, M. P., Gray, R. H., and Payne, F. E., 1974, Replication of measles virus: Distinct species of short nucleocapsids in cytoplasmic extracts of infected cells, *J. Virol.* **13**:721.

Kingsbury, D. W., 1972, Paramyxovirus replication, *Curr. Top. Microbiol. Immunol.* **59**:1.

Kingsbury, D. W., and Portner, A., 1970, On the genesis of incomplete Sendai virions, *Virology* **42**:872.

Kingsbury, D. W., Portner, A., and Darlington, R. W., 1970, Properties of incomplete sendai virions and subgenomic viral RNA's, *Virology* **42**:857.

Lavi, S., and Winocour, E., 1972, Acquisition of sequences homologous to host deoxyribonucleic acid by closed circular simian virus 40 deoxyribonucleic acid, *J. Virol.* **9**:309.

Lavi, S., Rozenblatt, S., Singer, M. F., and Winocour, E., 1973, Acquisition of sequences homologous to host DNA by closed circular simian virus 40 DNA. II. Further studies on the serial passage of virus clones, *J. Virol.* **12**:492.

Lazzarini, R. A., Weber, G. H., Johnson, L. D., and Stamminger, G. M., 1975, Covalently linked message and anti-message (genomic) RNA from a defective vesicular stomatitis virus particle, *J. Mol. Biol.* **97**:289.

Leamnson, R. N., and Reichmann, M. E., 1974, The RNA of defective vesicular stomatitis virus particles in relation to viral cistons, *J. Mol. Biol.* **85**:551.

Lee, T. N. M., Brockman, W. W., and Nathans, D., 1975, Evolutionary variants of simian virus 40: Cloned substituted variants containing multiple initiation sites for DNA replication, *Virology* **66**:53.

Lehmann-Grube, F., Plenczka, W., and Tees, R., 1969, A persistent and inapparent infection of L cells with the virus of lymphocytic choriomeningitis, *J. Gen. Virol.* **5**:63.

Lenard, J., and Compans, R. W., 1975, Polypeptide composition of incomplete influenza virus grown in MDBK cells, *Virology* **65**:418.

Lerner, R. A., and Hodge, L. D., 1969, Non-permissive infection of mammalian cells: Synthesis of influenza virus genome in HeLa cells, *Proc. Natl. Acad. Sci. USA* **64**:544.

Levin, J. G., Ramseur, J. M., and Grimley, P. M., 1973, Host effect on arbovirus replication: Appearance of defective interfering particles in murine cells, *J. Virol.* **12**:1401.

Lief, F. S., and Henle, W., 1956, Studies on the soluble antigen of influenza virus. III. The decreased incorporation of S antigen into elementary bodies of increasing incompleteness, *Virology* **2**:782.

Mak, S., 1971, Defective virions in human adenovirus type 12, *J. Virol.* **7**:426.

Martin, M. A., Khoury, G., and Fareed, G. C., 1974, Specific reiteration of viral DNA sequences in mammalian cells, *Cold Spring Harbor Symp. Quant. Biol.* **39**:129.

McAllister, P. E., and Pilcher, K. S., 1974, Autointerference in infectious hematopoietic necrosis virus of salmonid fish, *Proc. Soc. Exp. Biol. Med.* **145**:840.

McLaren, L. C., and Holland, J. J., 1974, Defective interfering particles from poliovirus vaccine and vaccine reference strains, *Virology* **60**:579.

Meier-Ewert, H., and Dimmock, N. J., 1970, The role of the neuraminidase of the infecting virus in the production of non-infectious (von Magnus) influenza virus, *Virology* **42**:794.

Mertz, J. E., and Berg, P., 1974, Defective simian virus 40 genomes: Isolation and growth of individual clones, *Virology* **62**:112.

Mertz, J. E., Carbon, J., Herzberg, M., Davis, R. W., and Berg, P., 1975, Isolation and characterization of individual clones of simian virus 40 mutants containing deletions, duplications and insertions in their DNA, *Cold Spring Harbor Symp. Quant. Biol.* **39**:69.

Mims, C. A., 1956, Rift Valley fever virus in mice. IV. Incomplete virus; its production and properties. *Br. J. Exp. Pathol.* **37**:129.

Mims, C. A., 1966, Immunofluorescent study of the carrier state and mechanism of vertical transmission in lymphocytic choriomeningitis virus infection in mice, *J. Pathol. Bacteriol.* **91**:395.

Mims, C. A., and Subrahmanyan, T. P., 1966, Immunofluorescence study of the mechanism of resistance to superinfection in mice carrying the lymphocytic choriomeningitis virus, *J. Pathol. Bacteriol.* **91**:403.

Murphy, F. A., Taylor, W. P., Mims, C. A., and Whitefield, S. G., 1972, Bovine ephemeral fever virus in cell culture and mice, *Arch. Ges. Virusforsch.* **38**:234.

Murray, B. K., Biswal, N., Bookout, J. B., Lanford, R. E., Courtney, R. J., and Melnick, J. L., 1975, Cyclic appearance of defective interfering particles of herpes simplex virus and the concomitant accumulation of early polypeptide VP175, *Intervirology* **5**:173.

Nayak, D. P., 1972, Defective virus RNA synthesis and production of incomplete influenza virus in chick embryo cells, *J. Gen. Virol.* **14**:63.

Nicholson, B. L., and Dunn, J., 1972, Autointerference in the replication of the infectious pancreatic necrosis (IPN) virus of trout, *Bacteriol. Proc.* **72**:196.

Nonoyama, M., and Graham, A. F., 1970, Appearance of defective virions in clones of reovirus, *J. Virol.* **6**:693.

Nonoyama, M., Watanabe, Y., and Graham, A. F., 1970, Defective virions of reovirus, *J. Virol.* **6**:226.

Oddo, F. G., Chiarini, A., and Sinatra, A., 1967, On the hemagglutinating and hemolytic activity of measles virus variants, *Arch. Ges. Virusforsch.* **22**:35.

Oldstone, M. M. A., Welsh, R. M., and Joseph, B. S., 1975, Pathogenic mechanisms of tissue injury in persistent viral infections, *Ann. N.Y. Acad. Sci.* **256**:65.

Palese, P., and Schulman, J. L., 1976, Differences in RNA patterns of influenza A viruses, *J. Virol.* **17**:876.

Palma, E. L., and Huang, A. S., 1974, Cyclic production of vesicular stomatitis virus caused by defective interfering particles, *J. Infect. Dis.* **129**:402.

Palma, E. L., Perlman, S. M., and Huang, A. S., 1974, Ribonucleic acid synthesis of vesicular stomatitis virus. VI. Correlation of defective particle RNA synthesis with standard RNA replication, *J. Mol. Biol.* **85**:127.

Parfanovich, M., Hammarskjöld, B., and Norrby, E., 1971, Synthesis of virus-specific RNA in cells infected with two different variants of measles virus, *Arch. Ges. Virusforsch.* **35**:38.

Perlman, S. M., and Huang, A. S., 1973, Ribonucleic acid synthesis of vesicular stomatitis virus. V. Interactions between transcription and replication, *J. Virol.* **12**:1395.

Perrault, J., 1976, Cross-linked double stranded RNA from a defective vesicular stomatitis virus particle. *Virology* **70**:360.

Perrault, J., and Holland, J. J., 1972*a*, Variability of vesicular stomatitis virus autointerference with different host cells and virus serotypes, *Virology* **50**:148.

Perrault, J., and Holland, J. J., 1972*b*, Absence of transcriptase activity and transcription-inhibiting ability in defective interfering particles of vesicular stomatitis virus, *Virology* **50**:159.

Petric, M., and Prevec, L., 1970, Vesicular stomatitis virus—A new interfering particle, intracellular structures, and virus-specific RNA, *Virology* **41**:615.

Pons, M. W., 1975, Influenza virus RNA(s), in: *The Influenza Viruses and Influenza* (E. D. Kilbourne, pp. 145–170, ed.), Academic Press, New York.

Pons, M., and Hirst, G. K., 1969, The single- and double-stranded RNA's and the proteins of incomplete influenza virus, *Virology* **38**:68.

Portner, A., and Kingsbury, D. W., 1970, Complementary RNA's in paramyxovirions and paramyxovirus-infected cells, *Nature (London)* **228**:1196.

Portner, A., and Kingsbury, D. W., 1971, Homologous interference by incomplete Sendai virus particles: Changes in virus-specific ribonucleic acid synthesis, *J. Virol.* **8**:388.

Portner, A., and Kingsbury, D. W., 1972, Identification of transcriptive and replicative intermediates in Sendai virus-infected cells, *Virology* **47**:711.

Preble, O. T., and Youngner, J. S., 1972, Temperature-sensitive mutants isolated from L cells persistently infected with Newcastle disease, *J. Virol.* **9**:200.

Preble, O. T., and Youngner, J. S., 1975, Temperature-sensitive viruses and the etiology of chronic and inapparent infections, *J. Infect. Dis.* **131**:467.

Prevec, L., 1974, Physiological properties of vesicular stomatitis virus and some related rhabdoviruses, in: *Viruses Evolution and Cancer* (E. Kurstak and K. Mamamorosch, eds.), pp. 677–697, Academic Press, New York.

Prevec, L., and Kang, C. Y., 1970, Homotypic and heterotypic interference by defective particles of vesicular stomatitis virus, *Nature (London)* **228**:25.

Prevec, L., and Whitmore, G. F., 1963, Purification of vesicular stomatitis virus and the analysis of P^{32}-labeled viral components, *Virology* **20**:464.

Printz-Ané, C., Combard, A., and Martinet, C., 1972, Study of the transcription and the replication of vesicular stomatitis virus by using temperature-sensitive mutants, *J. Virol.* **10**:889.

Raine, C. S., Feldman, L. A., Sheppard, R. D., and Bornstein, M. B., 1972, Subacute sclerosing panencephalitis virus in cultures of organized central nervous tissue, *Lab. Invest.* **28**:627.

Reczko, E., 1960, Electronenmikroskopische Untersuchungen am Virus der Stomatitis vesicularis, *Arch. Ges. Virusforsch.* **10**:588.

Reichmann, M. E., Pringle, C. R., and Follett, E. A. C., 1971, Defective particles in BHK cells infected with temperature-sensitive mutants of vesicular stomatitis virus, *J. Virol.* **8**:154.

Reichmann, M. E., Villareal, L. P., Kohne, D., Lesnaw, J. A., and Holland, J. J., 1974,

RNA polymerase activity and poly(A) synthesizing activity in defective T particles of vesicular stomatitis virus, *Virology* **58**:240.

Risser, R., and Mulder, C., 1974, Relative locations of rearrangements in the DNA of defective simian virus 40 (SV40), *Virology* **58**:424.

Robberson, D. L., and Fried, M., 1974, Sequence arrangements in clonal isolates of polyoma defective DNA, *Proc. Natl. Acad. Sci. USA* **71**:3497.

Robinson, W. S., 1970, Self-annealing of subgroup 2 myxovirus RNAs, *Nature (London)* **225**:944.

Rott, R., and Scholtissek, C., 1963, Investigations about the formation of incomplete forms of fowl plague virus, *J. Gen. Microbiol.* **33**:303.

Roy, P., Repik, P., Hefti, E., and Bishop, D. H. L., 1973, Complementary RNA species isolated from vesicular stomatitis (HR strain) defective virions, *J. Virol.* **11**:915.

Rozenblatt, S., Slavi, M., Singer, F., and Winocour, E., 1973, Acquisition of sequences homologous to host DNA by closed circular simian virus 40 DNA. III. Host sequences, *J. Virol.* **12**:501.

Rubenstein, A. S., and Kaplan, A. S., 1975, Electron microscopic studies of the DNA of defective and standard pseudorabies virions, *Virology* **66**:385.

Rustigian, R., 1962, A carrier state in HeLa cells with measles virus (Edmonston strain) apparently associated with noninfectious virus, *Virology* **66**:101.

Sanders, F. K., 1957, Recent advances in the use of tissue culture methods in virus research, *Proc. R. Soc. Med.* **50**:911.

Schlesinger, R. W., 1950, Incomplete growth cycle of influenza virus in mouse brain, *Proc. Soc. Exp. Biol. Med.* **74**:541.

Schlesinger, R. W., 1959, Interference between animal viruses, in: *The Viruses*, Vol. 3 (F. M. Burnet and W. M. Stanley, eds.), pp. 157–194, Academic Press, New York.

Schlesinger, S., Schlesinger, M., and Burge, B. W., 1972, Defective virus particles from Sindbis virus, *Virology* **48**:615.

Schluederberg, A., 1971, Measles virus RNA, *Biochem. Biophys. Res. Commun.* **42**:1012.

Schnitzlein, W. M., and Reichmann, M. E., 1976, The size and the cistronic origin of defective vesicular stomatitis virus particle RNAs in relation to homotypic and heterotypic interference, *J. Mol. Biol.* **101**:307.

Schuerch, A. R., Matsuhisa, T., and Joklik, W. K., 1974, Temperature-sensitive mutants of reovirus. VI. Mutant *ts*447 and *ts*556 particles that lack either one or two genome segments, *Intervirology* **3**:36.

Shenk, T. E., and Stollar, V., 1972, Viral RNA species in BHK-21 cells infected with Sindbis virus serially passaged at high multiplicity of infection, *Biochem. Biophys. Res. Commun.* **49**:60.

Shenk, T. E., and Stollar, V., 1973*a*, Defective interfering particles of Sindbis virus. I. Isolation and some chemical and biological properties, *Virology* **53**:162.

Shenk, T. E., and Stollar, V., 1973*b*, Defective interfering particles of Sindbis virus. II. Homologous interference, *Virology* **55**:530.

Simmons, D. T., and Strauss, J. M., 1972*a*, Replication of Sindbis virus. I. Relative size and genetic content of 26 S and 49 S RNA, *J. Mol. Biol.* **71**:599.

Simmons, D. T., and Strauss, J. M., 1972*b*, Replication of Sindbis virus. II. Multiple forms of double-stranded RNA isolated from infected cells, *J. Mol. Biol.* **71**:615.

Simmons, D. T., and Strauss, J. M., 1974, Translation of Sindbis virus 26 S RNA and 49 S RNA in lysates of rabbit reticulocytes, *J. Mol. Biol.* **86**:397.

Simpson, R. W., and Inuma, M., 1975, Recovery of infectious proviral DNA from mammalian cells infected with respiratory syncytial virus, *Proc. Natl. Acad. Sci. USA* **72**:3230.

Sokol, F., Neurath, A. R., and Vilcek, J., 1964, Formation of incomplete Sendai virus in embryonated eggs, *Acta Virol.* **8**:59.

Sokol, R., Kuwert, E., Wiktor, J. J., Hummeler, K., and Koprowski, H., 1968, Purification of rabies virus grown in tissue culture, *J. Virol.* **2**:836.

Soria, M., Little, S. P., and Huang, A. S., 1974, Characterization of vesicular stomatitis virus nucleocapsids. I. Complementary 40 S RNA molecules in nucleocapsids, *Virology* **61**:270.

Spandidos, D. A., and Graham, A. F., 1975a, Complementation of defective reovirus by *ts* mutants, *J. Virol.* **15**:954.

Spandidos, D. A., and Graham, A. F., 1975b, Complementation between temperature-sensitive and deletion mutants of reovirus, *J. Virol.*

Specter, D. H., 1975, Polyadenylic acid on poliovirus RNA, thesis, Massachusetts Institute of Technology, Cambridge, Mass.

Stampfer, M., Baltimore, D., and Huang, A. S., 1971, Absence of interference during high multiplicity infection by clonally purified vesicular stomatitis virus, *J. Virol.* **7**:409.

Staneck, L. D., and Pfau, C. J., 1974, Interfering particles from a culture of persistently infected Parana virus, *J. Gen. Virol.* **22**:437.

Staneck, L. D., Trowbridge, R. S., Welsh, R. M., Wright, E. A., and Pfau, C. J., 1972, Arenaviruses: Cellular response to long-term *in vitro* infection with Parana and lymphocytic choriomeningitis viruses, *Infect. Immun.* **6**:444.

Sugiura, A., 1972, Influenza viral RNA and ribonucleoprotein synthesized in abortive infection, *Virology* **47**:517.

Summers, D., and Maizel, J. V., Jr., 1971, Determination of the gene sequence of poliovirus with pactamycin, *Proc. Natl. Acad. Sci. USA* **68**:2852.

Taber, R., Rekosh, D. M., and Baltimore, D., 1971, Effect of pactamycin on synthesis of poliovirus proteins: A method for genetic mapping, *J. Virol.* **8**:395.

Tadokoro, J., 1958, Modified virus particles in undiluted passages of HVJ. I. The production of modified particles, *Biken's J.* **1**:111.

Tai, M. T., Smith, C. A., Sharp, P. A., and Vinogoad, J., 1972, Sequence heterogeneity in closed simian virus 40 deoxyribonucleic acid, *J. Virol.* **9**:317.

Thorne, H. V., 1968, Detection of size heterogeneity in the supercoiled fraction of polyoma virus DNA, *J. Mol. Biol.* **35**:215.

Thorne, H. V., Evans, J., and Warden, D., 1968, Detection of biologically defecting molecules in component of polyoma virus DNA, *Nature (London)* **219**:728.

Traub, E., 1938, Factors influencing the persistence of choriomeningitis virus in the blood of mice after clinical recovery, *J. Exp. Med.* **68**:229.

Uchida, S., Watanabe, S., and Kato, M., 1966, Incomplete growth of simian virus 40 in African green monkey kidney culture induced by serial undiluted passages, *Virology* **28**:135.

Villa-Komaroff, L., Guttman, N., Baltimore, D., and Lodish, H. F., 1975, Complete translation of poliovirus RNA in a eukaryotic cell-free system, *Proc. Natl. Acad. Sci. USA* **72**:4157.

von Magnus, P., 1947, Studies on interference in experimental influenza. I. Biological observations, *Mineral. Geol.* **24**(7):1.

von Magnus, P., 1951, Propagation of the RP-8 strain of influenza A virus in chick embryos. III. Properties of the incomplete virus produced in serial passages of undiluted virus, *Acta Pathol. Microbiol. Scand.* **29**:157.

von Magnus, P., 1954, Incomplete forms of influenza virus, *Adv. Virus Res.* **2**:59.

Wagner, M., Skare, J., and Summers, W. C., 1975, Analysis of DNA of defective herpes simplex virus type 1 by restriction endonuclease cleavage and nucleic acid hybridization, *Cold Spring Harbor Symp. Quant. Biol.* **39**:683.

Wagner, R. R., and Snyder, R. M., 1962, Viral interference induced in mice by acute or persistent infection with virus of lymphocytic choriomeningitis (LCM), *Nature (London)* **196**:393.

Wagner, R. R., Levy, A. H., Snyder, R. M., Ratcliff, G. A., and Hyatt, D. F., 1963, Biologic properties of two plaque variants of vesicular stomatitis virus, *J. Immunol.* **91**:112.

Wagner, R. R., Schnaitman, T. A., and Snyder, R. M., 1969, Structural proteins of vesicular stomatitis viruses, *J. Virol.* **3**:395.

Walker, D. L., 1964, The viral carrier state in animal cell cultures, *Prog. Med. Virol.* **6**:111.

Walker, D. L., 1968, Persistent viral infection in cell cultures, in: *Medical and Applied Virology* (Proceedings of the Second International Symposium) (M. Sanders and E. H. Lennett, eds.), pp. 99–110, Warren H. Green, Inc., St. Louis.

Weiss, B., and Schlesinger, S., 1973, Defective interfering passages of Sindbis virus: Chemical composition, biological activity and mode of interference, *J. Virol.* **12**:862.

Weiss, B., Goran, D., Cancedda, R., and Schlesinger, S., 1974, Defective interfering passages of Sindbis virus: Nature of the intracellular defective viral RNA, *J. Virol.* **14**:1189.

Welsh, R. M., and Pfau, C. J., 1972, Determinants of lymphocytic choriomeningitis interference, *J. Gen. Virol.* **14**:177.

Welsh, R. M., O'Connell, C. M., and Pfau, C. J., 1972, Properties of defective lymphocytic choriomeningitis virus, *J. Gen. Virol.* **17**:355.

Welsh, R. M., Burner, P. A., Holland, J. J., Oldstone, M. B. A., Thompson, H. A., and Villareal, L. P., 1975, A comparison of biochemical and biological properties of standard and defective lymphocytic choriomeningitis virus, *Bull. WHO* **52**:403.

Wertz, G. W., and Levine, M., 1973, RNA synthesis by vesicular stomatitis virus and a small plaque mutant: Effects of cycloheximide, *J. Virol.* **12**:253.

Wild, T. F., 1972, Replication of VSV: The effect of purified interfering component, *J. Gen. Virol.* **17**:295.

Woodward, C. G., and Smith, H., 1975, Production of defective interfering virus in the brains of mice by an avirulent, in contrast with a virulent, strain of Semliki Forest virus, *Br. J. Exp. Pathol.* **56**:363.

Yoshiike, K., 1968a, Studies on DNA from low-density particles of SV40. 1. Heterogeneous defective virions produced by successive undiluted passages, *Virology* **34**:391.

Yoshiike, K., 1968b, Studies on DNA from low-density particles of SV40. II. Noninfectious virions associated with a large plaque variant, *Virology* **34**:402.

Yoshiike, K., and Furuno, A., 1969, Heterogeneous DNA of simian virus 40, *Fed. Proc.* **28**:1899.

Yoshino, K., Taniquchi, S., and Arai, K., 1966, Autointerference of rabies virus in chick embryo fibroblasts, *Proc. Soc. Exp. Biol. Med.* **123**:387.

Yoshishita, T., Kawai, K., Fukai, K., and Ito, R., 1959, Analysis of infective particles in incomplete virus preparations of the von Magnus type, *Biken's J.* **2**:25.

Zhdanov, V. M., Bogomolova, N. N., Gavrilov, V. I., Andzhaparidze, O. G., Deryabin, P. G., and Astakhova, A. N., 1974, Infectious DNA of tick-bone encephalitis virus, *Arch. Ges. Virusforsch.* **45**:215.

CHAPTER 3

Virion Polymerases

David H. L. Bishop

Department of Microbiology, The Medical Center
University of Alabama in Birmingham
Birmingham, Alabama 35294

1. INTRODUCTION

The biological importance of virion polymerases to the infection process of viruses can be gauged from the fact that many groups of viruses possess virion nucleic acid polymerases of one form or another. A list of those animal RNA virus groups (and their members) shown to possess RNA-directed RNA polymerases (RNA transcriptases) is given in Table 1. Included are representatives of the arenaviruses, bunyaviruses (and bunyaviruslike viruses), orthomyxoviruses, paramyxoviruses, reoviruses (diplornaviruses), and rhabdoviruses. Oncornaviruses and similar virus types (e.g., visna) possess an RNA- and DNA-directed DNA polymerase, otherwise known as a reverse transcriptase (Table 2). Of the various DNA virus groups, the poxviruses and possibly the icosahedral cytoplasmic deoxyriboviruses possess a virion DNA-instructed RNA polymerase (Table 3). Virus isolated from patients with serum hepatitis appears to possess a DNA-directed DNA polymerase.

The characteristics of the virion polymerases of the various virus groups will be discussed below, not only in relation to what we know from *in vitro* and *in vivo* studies about the polymerase functions but also with regard to the role of polymerase products in the infection process of the different viruses.

TABLE 1

Viruses That Possess Virion RNA-Directed RNA Polymerases[a]

Viruses	References
Lipid-Enveloped	

Rhabdoviruses

    A. *In vitro* assays

Vesicular stomatitis virus	Baltimore *et al.* (1970)
(Indiana serotype) (VSV Indiana)	Aaslestad *et al.* (1971)
Kern Canyon virus (KCV)	Aaslestad *et al.* (1971)
VSV (New Jersey serotype)	Chang *et al.* (1974)
Cocal	Chang *et al.* (1974)
Chandipura	Chang *et al.* (1974)
Piry	Chang *et al.* (1974)
Pike fry rhabdovirus (PFR)	Roy *et al.* (1975)
Spring viremia of carp virus (SVCV)	P. Roy (personal communication)

    B. *In vivo* assays

VSV Indiana	Marcus *et al.* (1971)
	Huang and Manders (1972)
	Flamand and Bishop (1973)
	Wertz and Levine (1973)
VSV New Jersey	Repik *et al.* (1974*b*)
Cocal	Repik *et al.* (1974*b*)
Chandipura	Repik *et al.* (1974*b*)
Piry	Repik *et al.* (1974*b*)
PFR	Roy *et al.* (1975)
Rabies	Bishop and Flamand (1975)
SVCV, Mokola	P. Repik and D. H. L. Bishop (unpublished observations)

Orthomyxoviruses

    A. *In vitro* assays

Influenza A strains	Chow and Simpson (1971)
	Penhoet *et al.* (1971)
	Skehel (1971*a*)
Influenza B strains	Chow and Simpson (1971)
	Schlotissek *et al.* (1971)
	Oxford (1973*b*)

    B. *In vivo* assays

Influenza A strains	Bean and Simpson (1973)
	Repik *et al.* (1974*a*)

Paramyxoviruses

    A. *In vitro* assays

Newcastle disease virus (NDV)	Huang *et al.* (1971)
Sendai virus	Robinson (1971*a*)
	H. O. Stone *et al.* (1971)
Mumps virus	Bernard and Northrop (1974)
SV5	Choppin and Compans (Vol. 4, this series)

    B. *In vivo* assays

Sendai virus	Robinson (1971*c*)

TABLE 1
(*continued*)

Viruses	References
Lipid-Enveloped	
Bunyaviruses	
	A. *In vitro* assays
Lumbo virus	Bouloy *et al.* (1975)
Uukuniemi virus	Ranki and Pettersson (1975)
	B. *In vivo* assays
Snowshoe hare	P. Repik and D. H. L. Bishop (unpublished observations)
La Crosse	D. H. L. Bishop (unpublished observations)
Main Drain	D. H. L. Bishop (unpublished observations)
Bunyamwera	D. H. L. Bishop (unpublished observations)
Arenaviruses	
	A. *In vitro* assays
Pichinde	Carter *et al.* (1974)
	B. *In vivo* assays
	None reported
Coronaviruses	See Tyrrell *et al.* (1975)
Non-Lipid-Enveloped	
Reoviruses: Diplornaviruses	
	A. *In vitro* assays
Reovirus strains	Borsa and Graham (1968)
	Shatkin and Sipe (1968)
Wound tumor virus	Black and Knight (1970)
Cytoplasmic polyhedrosis virus (CPV)	Lewandowski *et al.* (1969)
	Furuichi (1974)
Bluetongue virus	Martin and Zweerink (1972)
	Verwoerd and Huismans (1972)
	B. *In vivo* assays
Reovirus	Watanabe *et al.* (1967a,b)

[a] Virion RNA-directed RNA polymerase activities are detected either (1) by *in vitro* assays or (2) in cells in which protein synthesis is inhibited by cycloheximide or puromycin (see text). This latter test does not rule out the possibility that cellular factors (including enzyme components) may contribute to the observed activities.

What are the advantages, from an evolutionary point of view, for the possession of a virion polymerase? To answer that question, we will have to consider the need for and role of the polymerase transcription products in the infection process. For each of the virus groups listed above, other than the oncornaviruses and the putative hepatitis virus, the function of the virion polymerase is to make messenger RNA.

TABLE 2

RNA Viruses That Possess Virion Reverse Transcriptases

Virus name	References
Avian leukosis virus (MC-29)	Mizutani and Temin (1973)
(Avian induced leukosis virus)	Weber *et al.* (1971)
Avian myeloblastosis virus (AMV)	Garapin *et al.* (1970)
	Fujinaga *et al.* (1970)
	Riman and Beaudreau (1970)
	Smoler *et al.* (1971)
	Spiegelman *et al.* (1970*a*)
Avian reticuloendotheliosis virus	Peterson *et al.* (1972)
Bovine leukemia virus	Kettman *et al.* (1976)
Baboon endogenous viruses (M28, M7)	Benveniste *et al.* (1974)
	Todaro *et al.* (1974*b*)
Chicken syncytial virus (CSV)	Mizutani and Temin (1973)
Duck infectious anemia virus (DIAV)	Mizutani and Temin (1973)
Feline leukemia virus (FeLV)	Hatanaka *et al.* (1970)
(Gardner, Rickard, Theilin strains and	Roy-Burman (1971)
endogenous viruses)	Spiegelman *et al.* (1970*a,b*)
	Todaro *et al.* (1973)
Feline sarcoma-leukemia viruses (FeSV-FeLV)	Fujinaga and Green (1971)
	Hatanaka *et al.* (1970)
	Roy-Burman (1971)
	Scolnick *et al.* (1970*a*)
Gibbon ape lymphoma and related viruses	Kawakami *et al.* (1972)
(GALV)	Scolnick *et al.* (1972)
Guinea pig virus (BUdR, IUdR induced)	Gross *et al.* (1973)
	Hsiung (1972)
	Nayak and Murray (1973)
Guinea pig leukemia-associated virus (L2C)	J. Schlom (personal communication)
Hamster leukemia virus (HaLV)	Hatanaka *et al.* (1970)
Maedi virus of sheep	Lin and Thormar (1972)
Mason-Pfizer monkey virus (M-PMV)	Schlom and Spiegelman (1971)
	Spiegelman *et al.* (1970*a*)
Murine endogenous viruses	Benveniste *et al.* (1974)
	Todaro *et al.* (1975)
	Todaro (1972)
	Sherr *et al.* (1974)
Murine leukemia virus (MuLV)	Baltimore (1970)
(AKR, Maloney, Rauscher, etc.)	Hatanaka *et al.* (1970)
	Scolnick *et al.* (1970*a,b*)
	Spiegelman *et al.* (1970*a*)
Murine mammary tumor virus (MuMTV)	Scolnick *et al.* (1970*a*)
	Spiegelman *et al.* (1970*a*)
Murine sarcoma-leukemia virus (MuSV-MuLV)	Green *et al.* (1970)
	Scolnick *et al.* (1970*a*)
(Harvey, Kirsten, Maloney)	
Murine type A particles	Penit *et al.* (1974)
	Thach *et al.* (1975)
	Wilson and Kuff (1972)
	Yang and Wivel (1973)

TABLE 2

(*continued*)

Virus name	References
Murine xenotropic viruses [S16CL10(I), AT-124]	Benveniste *et al.* (1974)
	Todaro *et al.* (1975)
Progressive pneumonia virus of sheep	L. B. Stone *et al.* (1971*a*)
Rat endogenous viruses (RT21C, CCL38, V-NRK)	Lieber *et al.* (1973)
	Sherr *et al.* (1975)
Rat leukemia virus (RaLV)	Parks *et al.* (1972)
Rat mammary tumor C-type virus (R35)	Chopra and Oie (1972)
RD114	Fischinger *et al.* (1973)
	Livingston and Todaro (1973)
	McAllister *et al.* (1972)
Pig endogenous viruses (MPK, PK)	Sherr *et al.* (1975)
	Todaro *et al.* (1974*a*)
Rous sarcoma virus (RSV)	Bishop *et al.* (1971*c*)
(Schmidt-Ruppin, Prague, B77, etc.)	Coffin and Temin (1971)
	Duesberg and Canaani (1970)
	Duesberg *et al.* (1971*a*)
	Fanshier *et al.* (1971)
	Garapin *et al.* (1970)
	Kang and Temin (1973)
	McDonnell *et al.* (1970)
	Mizutani *et al.* (1970)
	Quintrell *et al.* (1971)
	Robinson and Robinson (1971)
	Spiegelman *et al.* (1970*a*)
	Temin and Mizutani (1970)
Rous-associated viruses (RAV) and Rous sarcoma–RAV-1 mixtures (RSV, RAV-1)	Mizutani and Temin (1971*b*, 1973)
	Robinson and Robinson (1971)
	Spiegelman *et al.* (1970*a*)
Simian foamy virus	Parks *et al.* (1971)
Simian sarcoma virus (SSV)	Harewood *et al.* (1973)
(Woolly monkey virus)	
Simian sarcoma-associated virus (SSAV)	Theilen *et al.* (1971)
Trager duck spleen necrosis virus (SNV)	Mizutani and Temin (1973)
Viper C-type virus	Hatanaka *et al.* (1970)
Visna virus	Lin and Thormar (1971)
	Schlom *et al.* (1971)
	L. B. Stone *et al.* (1971*b*)

For the DNA virus vaccinia, one advantage to possessing a virion RNA polymerase is to allow the infection process to occur (at least principally) in the cell sap and to avoid the necessity for competing with cellular templates for cellular polymerase enzymes. A second potential advantage is the ability of the virus to evolve and possess unique enzyme-binding and transcription initiation sites for messenger RNA synthesis. Such sites can be sites which only the virion

TABLE 3

Viruses That Possess Virion DNA-Directed RNA Polymerases

Virus	References
Poxviruses	
Rabbit poxvirus	Kates and McAuslan (1967)
Vaccinia	Munyon *et al.* (1967)
Yaba virus	Schwartz and Dales (1971)
Euxoa auxiliaris Entomopox virus	Pogo *et al.* (1971)
Amsacta moorei Entomopox virus	McCarthy *et al.* (1975)
Icosohedral cytoplasmic deoxyriboviruses	
Iridescent types 2 and 6 viruses	Kelly and Tinsley (1973)
Frog virus 3	Gravell and Cromeans (1971)

polymerase can recognize. Evolution of preferred binding and initiation sites provides the virus with the capability and delicacies of viral-specified transcription control. Probably, however, the greatest selective advantage to possessing a virion polymerase is the fact that from the onset of the infection process the enzyme is sequestered to the viral genome, and this sequestration allows efficient, repetitive transcription of viral genes into messenger RNA.

For RNA viruses, one can postulate that the possession of an RNA transcriptase confers similar advantages: sequestration of the enzyme to the viral genome, ability to obtain repetitive transcription, transcriptional control, and independence from the need to utilize cellular enzymes. Also, by contrast to those RNA viruses which lack virion polymerases (e.g., the picornaviruses), the virion genome is not committed to cell ribosomes and the hazards of nuclease digestion.

In the case of the oncornaviruses and similar virus types, the possession of a virion RNA-directed DNA polymerase is unique by comparison to all the other virus groups. The presence of this enzyme, the reverse transcriptase, allows the virus to develop its infection process through a DNA intermediate without relying on cellular polymerases or the necessity of translation in the cell's ribosome translation machinery.

This chapter will offer models for possible mechanisms of transcription as well as discuss posttranscriptional modification and evidence for the regulation of transcription. First, however, an overview of the *in vitro* and *in vivo* procedures used for detecting virion polymerases will be presented. This will include descriptions of the effects of inhibitors (such as antibiotics, chemicals, and drugs) or promotors (such as exogenous templates) on the processes. The evidence which suggests that polymerases in virus preparations are in fact located in virus particles and are required for the infection process will be discussed along with procedures which have been used to purify template and polymerases of two virus types. Finally both *in vivo* and *in vitro* product analyses will be discussed in terms of the individual virus groups known to possess virion enzymes.

1.1. *In Vitro* Reaction Conditions Used for Assaying Virion RNA or DNA Polymerases

A discussion of the reaction conditions necessary for obtaining *in vitro* synthesis of RNA by a virion polymerase is provided in the following sections. Many of the basic RNA polymerase reaction conditions are similar for both the poxviruses and the enveloped or non-enveloped RNA virus types; therefore, the following discussion will principally center on those RNA viruses which possess lipid envelopes. This will be followed by a discussion of the particular requirements and properties of reovirus and poxvirus polymerases and finally the reaction conditions and assay procedures for viruses which possess a reverse transcriptase.

1.1.1. RNA Polymerases of RNA Viruses Which Possess a Lipid Envelope

In the initial demonstration by Baltimore *et al.* (1970) that virions of the rhabdovirus vesicular stomatitis virus (VSV) possess an RNA polymerase enzyme activity, it was shown that to achieve *in vitro* RNA synthesis the lipid envelope of the virus had to be solubilized to allow access of reaction ingredients to the enzyme–template complex. Some enveloped RNA virus preparations do not exhibit an absolute requirement for detergent solubilization, presumably because their envelopes are disrupted during the virus purification procedure. Nonionic detergents such as Nonidet P40 and Triton N101, or weakly anionic detergents such as Triton X100, are suitable for solubilizing viral envelopes and are the agents most commonly employed. Sodium deoxycholate, while efficient at solubilizing the viral membrane, is inhibitory to the reaction assay for VSV, Newcastle disease virus (NDV), and the orthomyxovirus influenza A, strain WSN (D. H. L. Bishop, unpublished observations; Hefti *et al.*, 1975; Huang *et al.*, 1971).

TABLE 4

Reaction Ingredients for Assaying RNA Polymerases or DNA Polymerases[a]

RNA polymerase assay		DNA polymerase assay	
ATP	1 mM	dATP	1 mM
CTP	1 mM	dCTP	1 mM
CTP	1 mM	dGTP	1 mM
[³H]UTP[b]	1 mM	[³H]dTTP[b]	1 mM
NaCl (or KCl, etc.)	0.1 M	NaCl (or KCl, etc.)	0.1 M
Tris-hydrochloride (pH 8.0)[c]	0.1 M	Tris-hydrochloride (pH 8.0)[c]	0.1 M
MgCl$_2$[d]	8 mM	MgCl$_2$[d]	8 mM
and, or MnCl$_2$[d]	1 mM	and, or MnCl$_2$[d]	1 mM
Nonidet P40[e]	0.1 % (w/v)	Nonidet P40[e]	0.1 % (w/v)
(or Triton N101, X100, etc.)		(or Triton N101, X100, etc.)	
S-Adenosyl-L-methionine	1 mM		
Dithiothreitol (DTT)	2 mM	Dithiothreitol (DTT)	2 mM

[a] These reaction ingredients are not necessarily applicable to all polymerase assays.
[b] Labeled nucleoside triphosphates are often used at lower concentrations to conserve label; however, as indicated in the text and in Fig. 2, too low a concentration can be detrimental to the rate of synthesis and possibly also the product size. Other labeled triphosphates can also be used, of course.
[c] Other pH values may be optimal for some enzyme systems; see text.
[d] Divalent cations used singly or in combinations must be tested for each system analyzed.
[e] Detergent is used for lipid-enveloped viruses.

The optimal concentrations of the various reaction ingredients have been determined for a variety of virion RNA polymerases, and a typical list is given in Table 4.

1.1.1a. Basic Techniques for Determining the Optimal Reaction Ingredients

The usual criterion for determining the optimal reaction concentration of an ingredient is to determine the maximum rate of *de novo* RNA synthesis as a function of the ingredient concentration by measuring the incorporation of radioactivity into RNA using a labeled nucleoside triphosphate precursor. While the maximum rate of an *in vitro* enzyme reaction may not reflect the maximum fidelity of the transcription process, this procedure for obtaining the best *in vitro* conditions is the one most commonly employed.

The observed *in vitro* virion polymerase activity is frequently expressed as the picomoles of nucleotide incorporated per hour per milligram of viral protein put into the reaction mixture. An alternative procedure is to relate the rate of product RNA synthesis to the amount of template RNA present. If the virus preparation has been grown in the presence of [³H]nucleosides, then the amount of template RNA in the reaction mixture can be measured by the amount of ³H label present and the *de novo* RNA synthesis quantitated by the incorporation of [³²P]nucleotides into acid-insoluble material. The advantages of the latter assay system are numerous but have to be balanced by the cost of performing dual isotope experiments.

Of particular concern to an experimenter setting up an *in vitro* assay of RNA synthesis is the possible presence of nucleases or other agents in a virus preparation which might inhibit progression of the reaction by degrading the template RNA (or the accumulation of product nucleic acids) or directly inhibit the polymerase function in some way and consequently mislead the investigation. The virus purification procedure is one means for removing contaminating nonvirion cellular nucleases. Most virus preparations can be sufficiently purified by combining various procedures to remove nonviral agents. For example, when purifying virus from extracellular fluids, a polyethylene glycol–NaCl precipitation, followed by equilibrium and velocity gradient centrifugations in gradients of sucrose (or glycerol–potassium tartrate combinations), is a technique which has been shown to be successful for most rhabdoviruses (Roy and Bishop, 1972; Obijeski *et al.*, 1974). There has to be for each virus type a careful preliminary exploration of the purification procedures to be employed, whereby the removal of contaminants as well as the effects on virus viability and degree of virion degradation produced by the manipulations are monitored. Purification of intracellular virus from cellular material poses a greater problem for achieving sufficient decontamination, although after mechanical disruption of infected cells the same basic centrifugation procedures can be employed.

The purification procedures commonly used, as outlined above, would not remove "contaminants" present within the virus particle as integral parts of the virus structure. There is an ever-increasing list of enzymes which have been recovered in supposedly highly purified virus (see later) and which are detected only upon detergent solubilization of the virus preparation. Although these may be present in vesicles which copurify with a virus preparation, it seems reasonable to suppose that virus particles contain something other than viral proteins, lipids, nucleic acids, carbohydrates, and water!

In summary, the presence and effect of intraviral as well as extraviral inhibitory agents must be considered in relation to designing *in vitro* experiments to measure viral polymerase activity. In the situation where no *in vitro* polymerase activity is observed, suitable *in vivo* tests can be designed to detect the presence of a virion polymerase, as described later.

1.1.1b. Optimal Detergent Concentration

The effects of different Triton N101 or Nonidet P40 concentrations on the rate of RNA synthesis by the endogenously templated virion polymerases of such rhabdoviruses as VSV or Kern Canyon virus (KCV) are shown in Fig. 1 (Aaslestad *et al.*, 1971). The optimal detergent concentrations for these two viruses as well as other enveloped viruses range from 0.01% to 1%, depending on the virus type and protein concentration in the reaction mixture. At high concentrations of detergent for VSV or KCV, RNA synthesis is inhibited. It is possible that detergents may bind to a polymerase component and affect its enzyme activity directly with regard to both the fidelity of the *in vitro* process and the optimal *in vitro* functioning of the enzyme. This has been suggested for the virion transcriptase of VSV, and could be one reason why the *in vitro* temperature optimum for VSV RNA transcriptase is lower than its *in vivo* optimum (Aaslestad *et al.*, 1971; Szilágyi and Pringle, 1972). One way of overcoming the detergent inhibition is to prepare subviral cores free from detergent.

1.1.1c. Monovalent and Divalent Cation Requirements

For most *in vitro* assays of a virion polymerase, it has been found that monovalent cations are required at concentrations of up to 0.1–0.2 M. Sodium, lithium, potassium, or ammonium ions have been shown to be equally effective for both the VSV or KCV RNA transcriptases (Aaslestad *et al.*, 1971). High salt concentrations (greater than 0.3 M) are inhibitory for many virion polymerases, including VSV; however, this inhibition is reversible by dilution of the salt concentration.

The divalent cations most commonly employed in reaction mixtures are magnesium ($\sim$10 mM) or manganese ($\sim$1 mM) ions, or combinations of both. Manganese ions inhibit the transcription of VSV, even in the presence of magnesium ions (Aaslestad *et al.*, 1971).

For influenza A, strain WSN, a combination of both cations gives an enhanced activity relative to that obtained in the presence of either one (Bishop *et al.,* 1971*a*). The same appears to hold for the bunyavirus Lumbo (Bouloy *et al.,* 1975); however, Uukuniemi virus (Ranki and Pettersson, 1975) is reported to require only manganese ions (3–15 mM). The transcriptase activity of certain NDV strains (H. O. Stone, personal communication) apparently prefers manganese ions, while other NDV strains are reportedly inhibited by manganese substitution (Huang *et al.,* 1971). In the case of KCV transcriptase, cations such as nickel, iron, and calcium were found to be incapable of substituting for magnesium; however, zinc chloride was shown to give some stimulation of the KCV transcriptase activity (Aaslestad *et al.,* 1971).

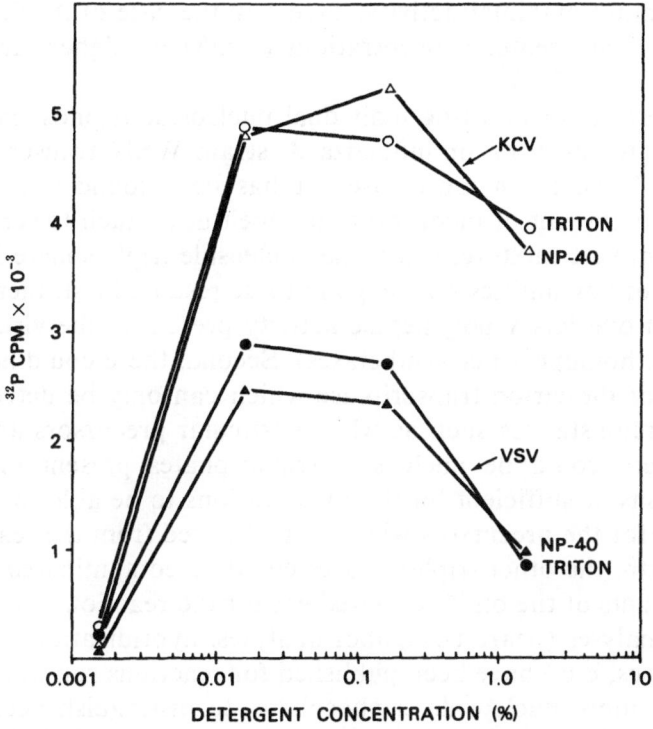

Fig. 1. Effect of nonionic detergents on the RNA polymerase activities of vesicular stomatitis virus (VSV) or Kern Canyon virus (KCV). The rate of incorporation of [^{32}P]uridine monophosphate by VSV virion polymerase or KCV virion polymerase into acid-insoluble product RNA as a function of detergent concentration in the reaction mixture using either Triton N101 or Nonidet P40 was determined. From Aaslestad *et al.* (1971).

1.1.1d. Nucleoside Triphosphate Requirements

In considering the nucleoside triphosphate requirements for demonstrating *in vitro* RNA synthesis, two initial questions have to be considered. Are all four nucleoside triphosphates required for enzyme activity, and, if not, why not? Second, what are the optimal or threshold concentrations for each triphosphate for RNA chain initiation and RNA elongation?

The optimal concentrations for ATP, CTP, GTP, and UTP have been determined for several viral polymerases, and it appears that at least for VSV and influenza A, strain WSN, the concentrations of triphosphates required to give optimal activity vary depending on which triphosphate is being considered (Fig. 2). For both VSV and WSN, reaction concentrations of 0.5–1 mM for each triphosphate are sufficient to obtain maximal activity, except in the case of ATP, for which it appears that greater concentrations (2 mM or higher) are required (Bishop *et al.*, 1971*a*).

The effects of omitting individual nucleoside triphosphates on the reaction rates of VSV or influenza A strain WSN transcriptases are shown in Table 5. In most cases, it has been found that there is a measurable amount of incorporation, albeit at a much lower level than for the complete mixture, when one nucleoside triphosphate is omitted. A variety of possibilities may explain these results. First, there could be more than one RNA polymerase activity present in the virus preparations (e.g., homopolymer synthetases). Second, there could be alternate activities of the virion transcriptase which can only be detected under certain circumstances such as when particular precursors are omitted. Third, there could be nucleoside triphosphates present in the virus preparations at sufficient local concentrations to be able to adequately substitute for the precursors which were omitted from the reaction mixture. Fourth, the other triphosphates could be contaminated with sufficient amounts of the omitted ingredient for the reaction to proceed. No product analyses (nearest-neighbor analyses, hybridizations, or product size analyses, etc.) have been published for reactions run in the absence of one or more nucleoside triphosphate, to distinguish between these possibilities. However, in the case of influenza A, strain WSN, ribonucleoside triphosphates have been isolated from dialyzed, highly purified virus preparations (D. Teninges, personal communication), which means that the presence of virus-associated nucleoside triphosphates could be one reason for the lack of an absolute requirement for all four triphosphates.

Although the omission of one or another triphosphate substantially reduces the incorporation of nucleotides into RNA by the virion polymerase of the paramyxovirus NDV (Huang *et al.,* 1971), the omission of one or more triphosphates from reaction mixtures templated by the paramyxovirus Sendai has been shown to cause either a stimulation of nucleotide incorporation or a three- to twentyfold reduction in enzyme activity (Robinson, 1971*a*; H. O. Stone *et al.,* 1971). Similar results to those obtained with Sendai have been reported for

TABLE 5

Effects of Omission of One Triphosphate from Influenza Virus (WSN) or Vesicular Stomatitis Virus (VSV) Virion Polymerase Reactions[a]

Label	Reaction condition	Incorporation (pmol/h/mg protein)	
		WSN	VSV
$[\alpha\text{-}^{32}P]ATP$	Complete	278	36,200
	−CTP	78	280
	−GTP	29	1,680
	−UTP	64	560
$[\alpha\text{-}^{32}P]GTP$	Complete	220	32,000
	−ATP	8	80
	−CTP	0	32
	−UTP	0	720
$[\alpha\text{-}^{32}P]CTP$	Complete	265	34,308
	−ATP	45	1,286
	−GTP	50	4,680
	−UTP	0	3,326
$[\alpha\text{-}^{32}P]UTP$	Complete	133	18,072
	−ATP	4	14
	−CTP	0	6
	−GTP	0	94

[a] Sixteenfold reaction mixtures were set up lacking manganese, Triton N101, virus, and unlabeled triphosphates but containing $[\alpha\text{-}^{32}P]$adenosine triphosphate (ATP, 0.1 mCi/μmol) at a concentration of 0.8 mM. The mixture was divided into four equal portions, and either guanosine triphosphate (GTP), uridine triphosphate (UTP) and cytidine triphosphate (CTP), or GTP and UTP, or UTP and CTP, or GTP and CTP were added. Each mixture was then divided, and for the WSN set manganese, Trion N101, and [³H]uridine-labeled WSN virus were added, whereas for the VSV set Triton N101 and [³H]uridine-labeled VSV virus were added. The reactions were incubated, and the rates of incorporation were determined and normalized to a constant VSV or WSN ³H recovery. In similar series, the labeled triphosphates employed were $[\alpha\text{-}^{32}P]GTP$ (0.5 mCi/μmol) at 0.4 mM concentration or $[\alpha\text{-}^{32}P]UTP$ (2 mCi/μmol) at 0.2 mM concentration, and similar reaction mixtures were set up in which appropriate triphosphates were lacking.

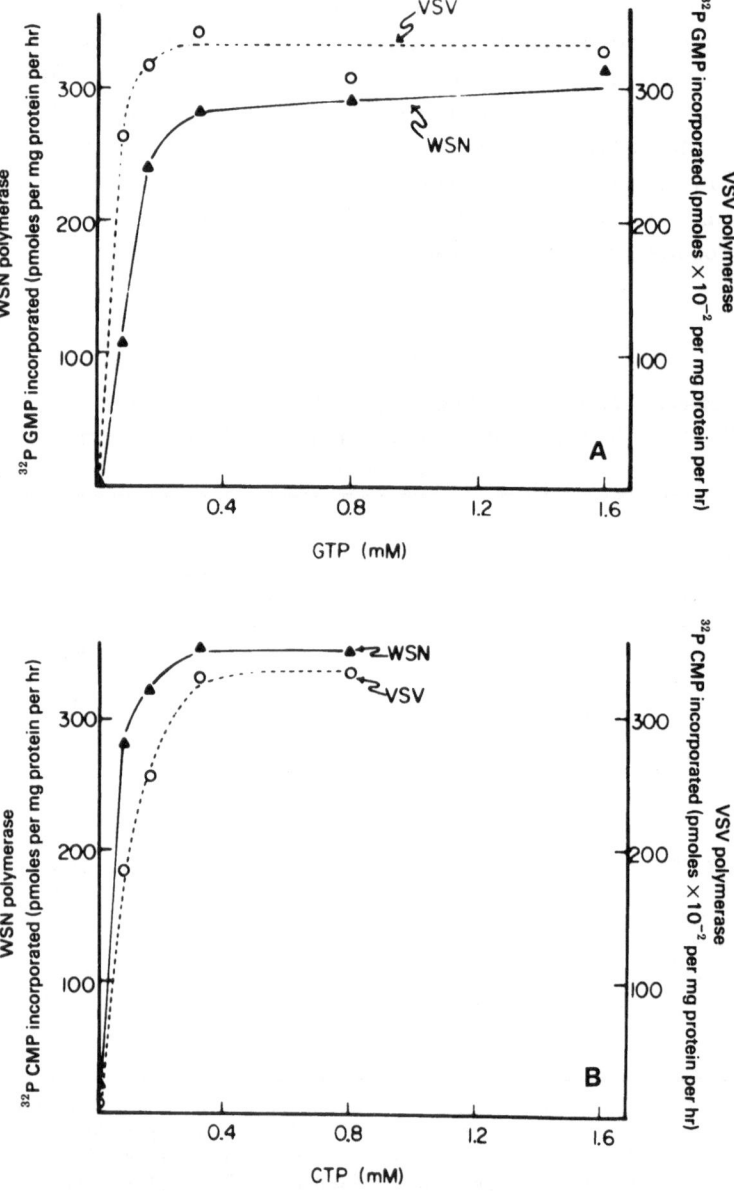

Fig. 2. Threshold requirements for triphosphates by influenza A, strain WSN, or VSV polymerases. A fiftyfold standard reaction was prepared containing 0.5 μmol of one $[\alpha$-$^{32}P]$ribonucleoside triphosphate and 5 μmol of the other three unlabeled ribonucleoside triphosphates, but lacking Triton N101, manganese chloride, or virus. The mixture was divided into ten equal portions. Unlabeled triphosphate (of the same type as the labeled one) was added to the mixture so that the specific activity of the label in pairs of reactions was diluted to 1, 2, 4, 10, or 20 times that of the original mixture. To one set of five mixtures, manganese chloride, Triton N101, and WSN virus were added in the normal WSN reaction proportions, whereas to the other set Triton N101 and VSV

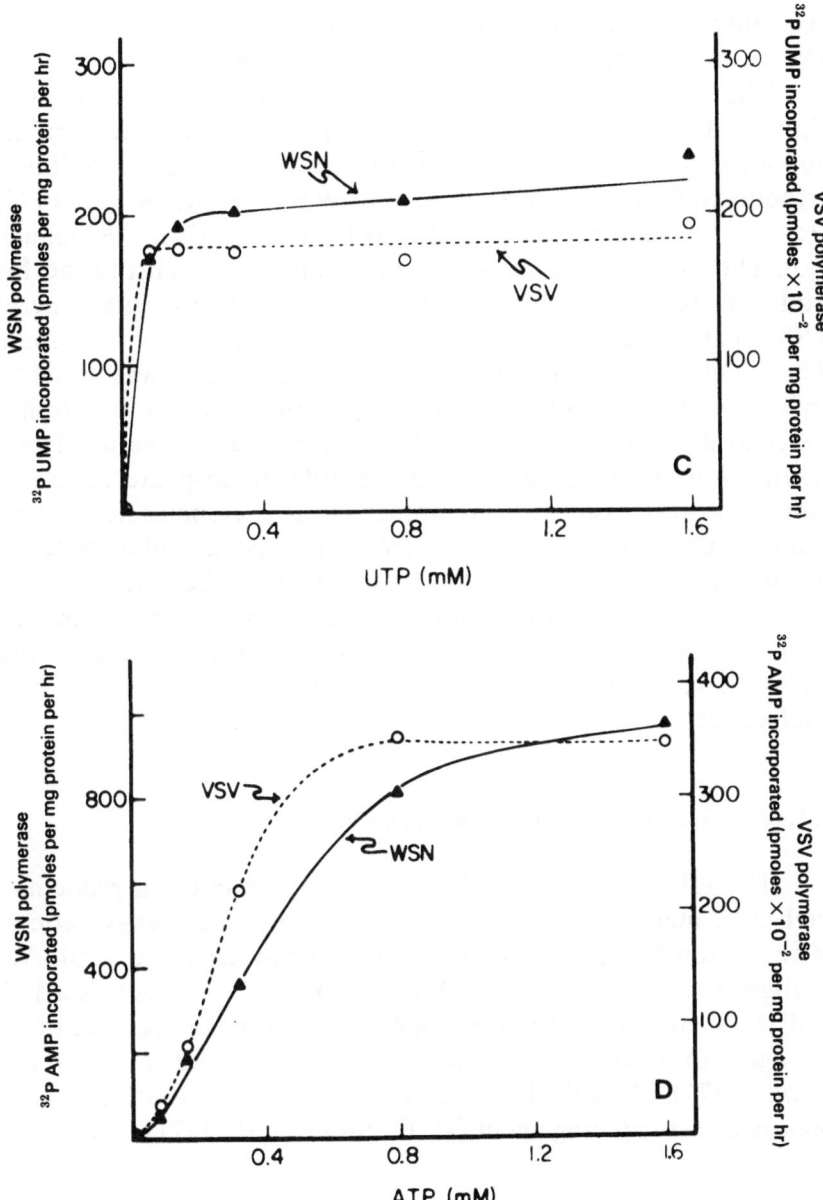

were added in the normal VSV reaction proportions. Portions were removed at 0, 15, 30, 45, 60, 90, and 120 min of incubation at 30°C, and the acid-insoluble radioactivity was determined. The linear rate of incorporation of label (corrected for the differences in triphosphate specific activity and expressed as the picomoles of label per milligram of protein per hour) was plotted against the concentration of the [³²P]triphosphate utilized. The same VSV preparation was used throughout. The WSN preparation in the ATP threshold experiment was not the same as that used in the other WSN experiments. A used [α-³²P]GTP, B used [α-³²P]CTP, C used [α-³²P]UTP, and D used [α-³²P]ATP. From Bishop *et al.* (1971a).

two other paramyxoviruses, mumps (Bernard and Northrop, 1974) and SV5 (Choppin and Compans, Vol. 4, this series).

The ability of one, two, or three (but not four) ribonucleoside diphosphates to substitute for their triphosphate counterparts in reaction mixtures has been shown for preparations of VSV as well as for the orthomyxovirus influenza A, strain WSN (Hefti *et al.*, 1975), and reovirus (Schochetman and Millward, 1972). Ribonucleoside monophosphates were shown to be incapable of substituting for their triphosphate counterparts. In analyzing the transcriptase activity of VSV in the presence of one or more diphosphates, it was found that RNA synthesis occurred after an initial lag. Also, it was found that the amount of RNA synthesis was strongly influenced by the triphosphate concentration (Hefti *et al.*, 1975). It has been suggested from these results that the presence of a nucleoside diphosphate kinase activity (phosphotransferase) in virus preparations is responsible for providing enough substrate triphosphates from the added diphosphates for the transcriptase to function (Hefti *et al.*, 1975). Differential phase extractions of virus preparations can be used to remove the kinase from the transcriptase-active nucleocapsids, which result suggests that the diphosphate kinase activity is not an integral part of the virion nucleocapsid or polymerase function (Hefti *et al.*, 1975).

1.1.1e. Sulfhydryl Group Requirement

The effect of omitting from a reaction mixture a reducing agent, such as 2-mercaptoethanol or dithiothreitol, on the enzymatic activity of the virion RNA polymerases of VSV or influenza A, strain WSN, is negligible (Aaslestad *et al.*, 1971; Bishop *et al.*, 1971a). Kern Canyon and NDV viral transcriptases appear to require the presence of a reducing agent in the *in vitro* reaction mixture (Aaslestad *et al.*, 1971; Huang *et al.*, 1971). Sendai viral transcriptase activity is stimulated by the presence of a reducing agent (H. O. Stone *et al.*, 1971).

1.1.1f. pH Optimum, Buffers, and Other Ingredients

The pH optimum for transcription reaction mixtures using trishydrochloride buffer (0.01 M) has been reported for VSV, KCV, and WSN viral transcriptases to be pH 8.0 ± 0.2 at 31°C. However, NDV appears to have a pH optimum at pH 7.3 (Huang *et al.*, 1971; H. O. Stone, personal communication). Phosphate buffer inhibits the VSV or WSN transcriptase activity; other buffer systems have not been investigated.

A stimulatory effect of *S*-adenosyl-L-methionine (SAM) on the *in vitro* RNA polymerase activity of cytoplasmic polyhedrosis virus has been reported (Furuichi, 1974). A threefold stimulation by 1 mM SAM of the virion RNA polymerase of the rhabdovirus spring viremia of carp virus (SVCV) has been obtained (P. Roy, personal communication). The transcriptase activity of other enveloped viruses (VSV, pike fry rhabdovirus or PFR, influenza A, strain WSN, D. H. L. Bishop, unpublished observations; or NDV and Sendai, H. O. Stone, personal communication) is not stimulated by SAM.

The addition of an ATP-generating system has not been found to be necessary to obtain or enhance rhabdovirus or influenza transcriptase activities (D. H. L. Bishop, unpublished observations).

1.1.1g. Temperature Optimum

The temperature optimum for the maximum initial activity of the virion polymerases of most mammalian rhabdoviruses (Table 1), as well as for influenza A, strain WSN, NDV, and Sendai, has been reported to be 28°–32°C. (Aaslestad *et al.*, 1971; Bishop *et al.*, 1971a; Huang *et al.*, 1971; Robinson, 1971a; H. O. Stone *et al.*, 1971). It has been found that these temperatures also give a more sustained synthetic activity (up to 8 h or longer) than mixtures incubated at 37°C. The 30°C temperature optimum is in contrast to the 36–38°C *in vivo* optimum observed for primary transcription (i.e., transcription of the infecting virion genome by the virion polymerase) for both VSV and influenza A, strain WSN (Flamand and Bishop, 1973; D. H. L. Bishop, unpublished observations). For two fish rhabdoviruses, SVCV and PFR, the temperature optimum for transcriptase action ranges from 18°C to 25°C (Roy *et al.*, 1975; P. Roy, unpublished observations). For IHNV and VHS, an optimum of approximately 12°C has been obtained (P. McAllister and R. Wagner, personal communication).

1.1.1h.   Inhibitors of RNA Synthesis by Viral RNA Polymerases

In those systems tested (VSV, KCV, Sendai, NDV, influenza, and mumps), neither actinomycin D (up to 10 μg/ml) nor deoxyribonuclease (up to 100 μg/ml) inhibits the *in vitro* RNA transcription process (Aaslestad *et al.*, 1971; Baltimore *et al.*, 1970; Bernard and

Northrop, 1974; Chow and Simpson, 1971; Penhoet *et al.*, 1971; Robinson 1971*a*; L. B. Stone *et al.*, 1971*a*). The lack of actinomycin D inhibition for influenza viral transcriptases is in contrast to the inhibition of primary transcription obtained *in vivo* by actinomycin D (Bean and Simpson, 1973), which suggests that the latter is not a direct inhibition of the transcriptase by the drug (see the review by Bishop and Flamand, 1975).

Addition of α-amanitin (8 μg/ml) to reaction mixtures for influenza A, strain NWS, was also shown not to inhibit the transcriptase function (Penhoet *et al.*, 1971). Similar results have been obtained for VSV (D. H. L. Bishop, unpublished observations). Rifampin (100 μg/ml) also does not inhibit RNA transcription for VSV (except apparently for a rifampin-sensitive mutant of VSV, Moreau, 1974) or for influenza and NDV (Baltimore *et al.*, 1970; Chow and Simpson, 1971; Penhoet *et al.*, 1971; Robinson, 1971*a*).

For NDV, pronase treatment was shown to partially inhibit the ability to demonstrate RNA transcription (H. O. Stone *et al.*, 1971). By way of contrast, bromelain pretreatment resulted in little inhibition of influenza transcriptase (Skehel, 1971*b*). In all systems, ribonuclease inclusion in reaction mixtures has been shown to totally inhibit the accumulation of RNA product.

A report by Hunt and Wagner (1975) demonstrated that addition of aurintricarboxylic acid (at 10 mM or higher concentrations) or polyethylene sulfonic acid (at 1 mM or higher concentration) to an *in vitro* reaction mixture before incubation totally inhibited VSV-directed RNA synthesis. A lag in the onset of inhibition of RNA synthesis was observed if either inhibitor was added after the commencement of RNA synthesis. Since both inhibitors are known to prevent initiation of RNA synthesis by Qβ replicase (Blumenthal and Landers, 1973; Kondo and Weissmann, 1972), Hunt and Wagner suggested that the observed inhibition for the VSV polymerase might similarly involve inhibition of transcription initiation rather than chain elongation.

An inhibitory effect of Sendai structural envelope proteins (glycoproteins or matrix protein) on the isolated Sendai virion transcriptive complexes has been reported by Marx *et al.* (1974). Using purified virion transcriptive complexes shown to contain the viral RNA and two of the major virion polypeptides (but not the matrix protein or glycoproteins), they demonstrated that these transcriptive complexes possessed a ninefold greater specific activity by comparison to the initial virus preparation and yet only half the total protein of the original virus. Both purified glycoproteins and matrix protein inhibited the Sendai transcriptive complex, and this inhibition was totally abolished by prior brief heat treatment of the proteins. NDV envelope proteins also inhibited the Sendai transcriptive complexes, although bovine serum albumin, hemoglobin, or the glycoproteins fetuin or ovalbumin did not. Marx and collaborators suggested that the envelope protein of Sendai might interact with the viral RNA–polymerase complex and inhibit its activity during the virus budding–maturation process, although it is difficult to see how a glycoprotein resident on the outer surface of the virus particle can accomplish this, unless some parts of the glycoproteins penetrate the envelope. The specific activities of purified transcriptive complexes of influenza A, strain WSN, or VSV Indiana are the same as their respective unfractionated disrupted virions (Bishop and Roy, 1972; Hefti *et al.*, 1975).

1.1.1i. Promotors of RNA Synthesis by Viral RNA Polymerases

Other than inclusion of standard reaction ingredients, the ability to promote *in vitro* RNA synthesis by other agents has been demonstrated for influenza fowl plague viral transcriptase (McGeoch and Kitron, 1975) and for Sendai viral transcriptase (Stone and Kingsbury, 1973).

For Sendai virus, it has been shown that polyanions (yeast RNA, polyadenylic acid, or polycytidylic acid) at concentrations ranging from 30 to 300 μg/ml will increase RNA synthesis by as much as fivefold (Stone and Kingsbury, 1973; Marx *et al.*, 1974). In all cases, it has been shown that the products made are complementary in composition to the viral genome and not copies of the adjuvant nucleic acid or homopolymers. Predigestion of yeast RNA with alkali was shown to abolish its capacity to stimulate the enzyme. It was found that polyaspartic acid, polyglutamic acid, and DNA (3–300 μg/ml) also stimulate the virion enzymatic activity whereas polyinosinic acid, polyguanylic acid, polyuridylic acid (at similar concentrations), and polyamines such as spermine, cadaverine, and putrescine neither stimulate nor depress the endogenously templated enzyme activity. Dextran sulfate, ammonium sulfate, and polyvinyl sulfate were shown to inhibit Sendai viral transcriptase, depending on the concentrations employed (Stone and Kingsbury, 1973). The mechanism of polyanion stimulation of Sendai transcriptase is not known.

For VSV transcriptase, addition of VSV viral RNA, 28 S HeLa ribosomal RNA, or polio or $Q\beta$ viral RNAs was found to inhibit the endogenously templated viral transcriptase (Bishop and Roy, 1971*a*).

Assays for homopolymeric activity catalyzed by VSV transcriptase and templated by 2 μg of polyinosinic acid, polyguanylic acid, polyadenylic acid, polyuridylic acid, or polycytidilic acid did not detect any polymerase function. Likewise, no synthesis of DNA was detected in endogenously templated VSV reactions or by reactions templated by synthetic heteropolymers or calf thymus DNA (Baltimore, 1970; Bishop and Roy, 1971a). Why polyanions promote one viral polymerase but have no effect on another is not known.

The stimulation of influenza virion polymerase by guanosine, guanosine derivatives, and guanosine-containing nucleotides has been reported (McGeoch and Kitron, 1975). The stimulatory effect by guanosine was found to depend on the influenza strain employed; thus influenza A, strain BEL, was stimulated about sevenfold, influenza A, strain WS, threefold, and influenza B, strain LEE, two-and-a-half-fold. No stimulation by guanosine was detected using the paramyxovirus Sendai.

Although ribose, guanine, and the 2´- or 3´-guanosine monophosphate did not increase the endogenously templated activity of influenza fowl plague virus, guanosine-5´-monophosphate was found to have a stimulatory effect. Various guanosine derivatives (including 7-methylguanosine, 1-methylguanosine, 2´-deoxyguanosine, 8-mercaptoguanosine, 6-mercaptopurine riboside, N^2,N^2-dimethylguanosine, and 8-azaguanosine-5´-monophosphate) did not stimulate fowl plague virion transcriptase. For fowl plague virus, N^2-methylguanosine and 6-mercaptoguanosine and, for both fowl plague and influenza A, strain BEL, viruses, certain dinucleoside monophosphates containing guanosine (GpA, GpG, GpC, but not GpU) were effective in stimulating the viral transcriptase. In a strange result, it was found that with fowl plague virus pretreated with detergent and then dialyzed (to remove virus-associated nucleotides) the incorporation of guanosine did not require the presence of either GTP or UTP (McGeoch and Kitron, 1975).

It was shown using labeled guanosine in complete reaction mixtures that the label was present in virus-complementary RNA in a structure from which it could be totally released as guanosine by venom phosphodiesterase treatment and as guanosine-3´-monophosphate (but not guanosine-containing oligonucleotides) by ribonuclease T_1 digestion. These results argue strongly for the precursor guanosine being incorporated at the 5´ terminus of RNA, presumably the product RNA. Ribonuclease A digestion of reaction product RNA labeled by [^{3}H]guanosine yielded labeled GpCp nucleotides. Neither the mechanism nor the importance of the guanosine-mediated stimulation or incorporation into RNA is understood.

1.1.2. RNA Polymerases of RNA Viruses Which Do Not Possess a Lipid Envelope

Assays for demonstrating the presence of an RNA polymerase associated with virions of the diplornaviruses reovirus and cytoplasmic polyhedrosis virus also require partial dissociation of the viral particles before an enzymatic activity can be obtained. Since neither of these viruses is enveloped in a lipid membrane, the partial dissociation cannot be achieved with a nonionic detergent.

In the initial demonstrations of the presence of an RNA polymerase in reovirus preparations, it was shown that brief heat treatment (e.g., 20 s at 70°C) or predigestion with 50 μg chymotrypsin per milliliter, produced a subviral particle capable of showing RNA polymerase activity (Borsa and Graham, 1968; Shatkin and Sipe, 1968). Since heating, as well as protease treatment, removes proteins from the viral surface (Mayor and Jordan, 1968; Shatkin and Sipe, 1968), presumably activation requires removal of certain structural polypeptides. This is corroborated by the fact that inhibition of activation can be achieved by anticapsid antibody pretreatment (Joklik, Vol. 2, this series).

Of the major virion polypeptides ($\lambda1$, $\lambda2$, $\mu1$, $\mu2$, $\sigma1$, $\sigma2$, and $\sigma3$), it appears that activation requires removal of the $\sigma3$ and some, but not necessarily all, of the $\mu2$ polypeptides (Joklik, 1972). Activation of the viral transcriptase *in vivo* by lysozymal enzymes occurs even in the presence of cycloheximide and appears also to involve removal of certain structural polypeptides including $\sigma3$ (Chang and Zweerink, 1971; Levin *et al.*, 1970*b*; Silverstein and Dales, 1968; Watanabe *et al.*, 1967*a,b*). The exact molecular mechanisms involved in the activation of the dormant enzyme are not known, although presumably whatever protein is removed allows access of precursor nucleotides into the viral core for the transcriptase to function. It has been shown that by adding back reovirus $\sigma3$ protein to activated particles, enzymatic activity can be repressed *in vitro* (Astell *et al.*, 1972). If, as will be discussed later, there are specific exit sites for reovirus mRNA release from activated cores (Gillies *et al.*, 1971), then it is conceivable that $\sigma3$ protein blocks these ports in some as yet undetermined fashion.

The pH optimum for reovirus transcriptase is reported to be pH 8.5 $\pm$ 0.5 in 0.01 M tris-hydrochloride buffer, and the enzyme requires magnesium ions. In some reports it is claimed that manganese ions cannot substitute for magnesium, while other reports mention that either ion can be used (Borsa and Graham, 1968; Joklik, Vol. 2, this series; Shatkin and Sipe, 1968). Like VSV transcriptase, the enzyme activity is inhibited by phosphate buffer (0.01 M) and does not require the presence of 2-mercaptoethanol. Unlike VSV transcriptase, the optimum temperature for *in vitro* transcription is around 50°C (Kapuler, 1970). Transcription of RNA for up to 48 h at 37°C has been obtained. Addition of an ATP-generating system apparently sustains the transcription process (Shatkin and Sipe, 1968).

1.1.3. RNA Polymerases of DNA Viruses

Of the five major groups of animal DNA viruses (poxviruses, papovaviruses, parvoviruses, adenoviruses, and herpesviruses), only members of the poxvirus group have been shown to possess virion DNA-dependent RNA polymerases. In the first demonstration that RNA polymerases were associated with animal viruses, Kates and McAuslan (1967), working with rabbit poxvirus, and Munyon *et al.* (1967), working with vaccinia, showed that an RNA polymerase activity was present in the cores of those two viruses. Later work with Yaba virus confirmed that that poxvirus also possessed a virion DNA-directed RNA polymerase (Schwartz and Dales, 1971). Studies on insect poxviruses have led to the demonstration that they too possess an RNA polymerase (Pogo *et al.*, 1971; McCarthy *et al.*, 1975).

It has been shown, as in the case of other viruses which possess RNA polymerases, that the realization of enzymatic activity requires the presence of all four ribonucleoside triphosphates as well as divalent cations. Magnesium ions at 5–10 mM concentration are preferred over manganese ions in order to obtain optimal activities. In both of the initial reports of poxvirus virion polymerases, it was demonstrated that a reducing agent (e.g., 2-mercaptoethanol) and sonication or trypsin, followed by repurification of the treated material, were necessary to produce enzyme-active cores. Pretreatment of Yaba virus with 0.5% Nonidet P40 and 0.25% 2-mercaptoethanol has also been shown to be efficient in yielding enzyme-active cores (Schwartz and Dales, 1971). The necessity for treating viruses with some disruptive procedure argues that some structural barrier has to be removed before the enzyme can function. A lag of a few minutes before maximal RNA synthesis is achieved can also be interpreted to indicate that the virions in a reaction mixture undergo structural changes (Munyon *et al.*, 1967).

The pH optimum for vaccinia viral polymerase has been reported to be pH 8.4 in 0.01 M tris-hydrochloride buffer (Munyon *et al.*, 1967). Reaction concentrations of around 0.5–1 mM for CTP, GTP, and UTP but 6 mM for ATP have been shown to be necessary to give the highest rates of RNA synthesis (Munyon *et al.*, 1967). Incubation at 35–38°C leads to RNA synthesis for several hours; however, it is not known what effects different temperatures have on the fidelity of transcription or the time course of RNA synthesis. Insect poxviruses possess low temperature optima for their viral transcriptases (Pogo *et al.*, 1971; McCarthy *et al.*, 1975).

1.1.4. DNA Polymerases of RNA Viruses: Virion Reverse Transcriptases

Since the initial demonstration by Baltimore (1970) and Temin and Mizutani (1970) that DNA polymerases were present in virions of Rauscher murine leukemia virus (R-MuLV), the Prague or Schmidt-Ruppin strain of Rous sarcoma virus (RSV), and avian myeloblastosis virus (AMV), DNA polymerases have been identified in virions of a large number of lipid-enveloped oncornaviruses (retraviruses) as well as in viruses of unproven oncogenicity but with similar genome components (Table 2). The term "reverse transcriptase" has been coined for the activity of the enzyme since it directs the synthesis of product DNA from an RNA template, which represents a reversal of the "normal" flow of information from DNA to RNA catalyzed by cellular DNA-templated RNA polymerases.

Studies by Spiegelman *et al.* (1970*a,b,c*) subsequently identified reverse transcriptase activities in eight oncornaviruses: Rauscher murine leukemia virus (R-MuLV), murine sarcoma virus (MuSV), feline leukemia virus (FeLV), the BAl strain A of AMV, a monkey mammary tumor virus, murine and rat mammary tumor viruses, and an RSV–Rous-associated virus type 1 (RSV-RAV 1) mixture. Green *et al.* (1970) demonstrated a reverse transcriptase activity in the Harvey and Moloney strains of murine sarcoma viruses, while Scolnick *et al.* (1970*a,b*) showed that MuMTV, R-MuLV, Moloney-MuLV, Kirsten murine leukemia virus, and the Gardner strain of feline sarcoma virus (FeSV) all possessed reverse transcriptase activities. No reverse transcriptase activity was found in preparations of non-oncornavirus types, e.g., VSV, Sendai, respiratory syncytial virus, lymphocytic choriomeningitis virus, influenza virus, and NDV (Baltimore, 1970; Bishop and Roy, 1971*a*; Scolnick *et al.*, 1970*b*).

With so many analyses of reverse transcriptase activities present in virions of different oncornaviruses, it is hardly surprising that the reported optimal reaction conditions for the various viruses differ even for the same virus type analyzed in different laboratories (see the excellent review by Green and Gerard, 1974).

1.1.4a. Reaction Requirements for Oncornavirus Reverse Transcriptases

It has been shown that the requirement for a nonionic detergent to activate the enzyme is not always absolute (Baltimore, 1970; Temin and Mizutani, 1970). Presumably this is due to partial destruction of the integrity of the viral envelope during the preparative procedures.

The effect of Triton N101 concentration on the integrity of the virus particle of the avian oncornavirus Rous sarcoma virus (RSV) is shown in Fig. 3. It is evident that at high Triton concentrations the RSV genome is released from the virus particle, whereas at low concentrations it remains in a subviral particle which sediments only slightly slower than the untreated virus.

The effect of addition of ribonuclease on RNA-directed DNA synthesis by RSV in reaction mixtures containing high or low Triton N101 concentrations is shown in Fig. 4. The results obtained indicate that at high detergent concentrations the viral genome is more suscepti- ble to degradation by the added ribonuclease than at lower concentra- tions. In point of fact, it has been shown by following the fate of [^{3}H]nucleoside-labeled RSV template RNA during a reaction time course in the absence of added ribonuclease that the RSV template is rapidly degraded by nucleases present in the virus preparation when high Triton concentrations are used, but not when low concentrations are employed (Bishop et al., 1971c; Quintrell et al., 1971). The origin of such nucleases is not known, but these observations emphasize the problem of inhibitory agents present in virus preparations, and should caution an investigator against using too high a detergent concentration to solubilize the viral envelope. Detergents such as Nonidet P40, Triton X100, Triton N101, Tween 40, and Tween 80 at around 0.01% final concentration for viral protein in a reaction mixture of up to 50–100 μg/ml (and proportionately higher for higher viral protein concentra- tions) have been found to be optimal (Garapin et al., 1970; Green et al., 1970). Ether or Tween–ether combinations can also be used to disrupt the viral envelope. Sodium deoxycholate at 1% (w/v) is reported to be inhibitory in reaction mixtures templated by RSV or AMV (Garapin *et al.*, 1970). Stromberg (1972) has reported that of some 60 nonionic detergents tested with AMV, those belonging to the polyoxyethylene alcohol class (e.g., Sterox SL) were the most effective at yielding enzyme-active AMV cores. In that study it was shown that Nonidet P40, which belongs to the polyoxyethylene alkyl phenyl class, was also efficient at giving enzyme-active AMV cores.

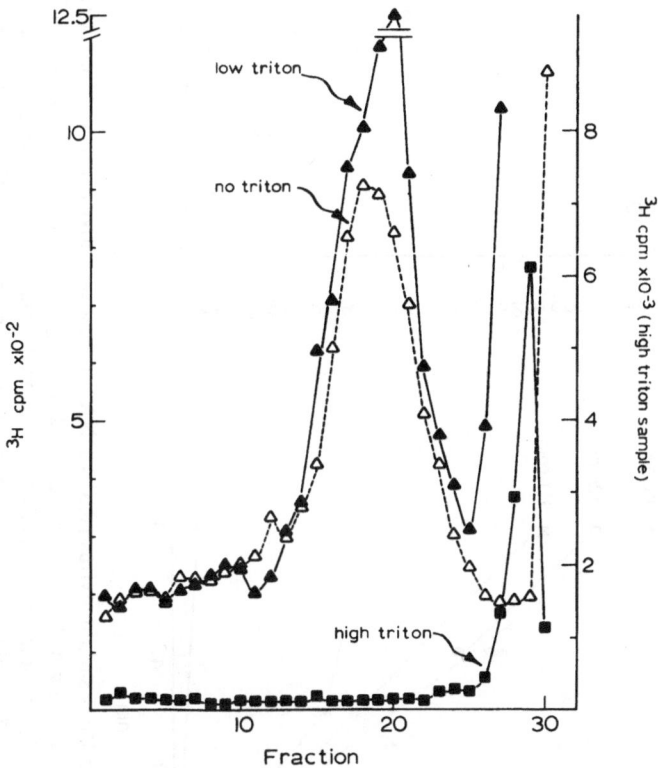

Fig. 3. Sucrose gradient velocity centrifugation of ^{3}H-labeled Rous sarcoma virus (RSV) treated with low or high Triton N101 concentrations. Samples of [^{3}H]uridine-labeled RSV in 0.10 M NaCl and 0.01 M tris-hydrochloride buffer (pH 7.4) were incubated at 37°C for 3 min with Triton N101 at a concentration of 10 μg (low Triton) or 400 μg (high Triton) per 125 μl and centrifuged in 10–30% gradients of sucrose containing 0.15 M NaCl, 0.01 M tris-hydrochloride buffer (pH 7.4), and 0.0002 M dithiothreitol for 60 min at 3°C and 25,000 rev/min in a Spinco SW41 rotor. The distribution of acid-insoluble ^{3}H radioactivity was determined and compared to that of an untreated control.

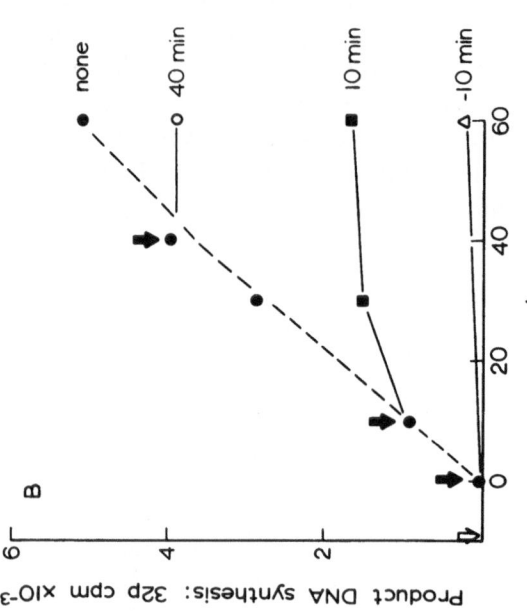

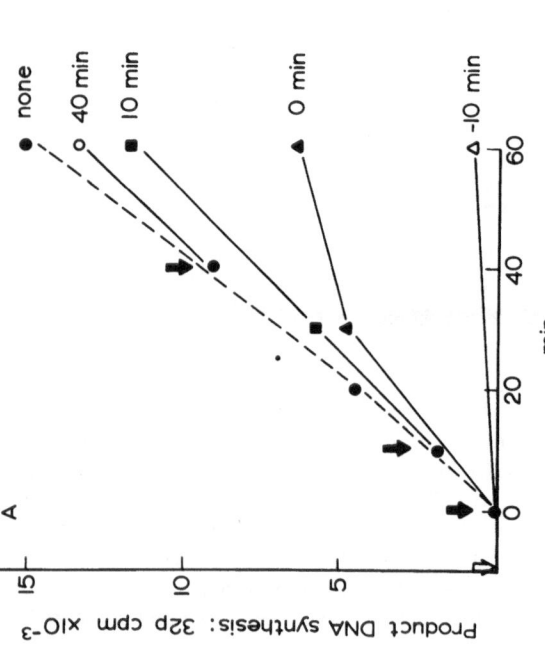

Fig. 4.   Effect of ribonuclease on Rous sarcoma virus (RSV) DNA polymerase activity at two different reaction concentrations of Triton N101. Reaction mixtures containing 10 μg of RSV protein and either 10 μg of Triton N101 (A) or 400 μg Triton N101 (B) per 125 μl reaction volume were incubated at 37°C. At indicated intervals (solid arrows), samples were withdrawn, and 5 μg of ribonuclease A was added and the incubations were continued. The acid-insoluble radioactivity was determined on 50 μl volumes as shown. Pretreatment of 10 μg of virus in 10 μl volume with 5 μg of ribonuclease A for 10 min at 4°C (open arrows) before addition to the reaction mixture is also shown (−10 min, open arrow). Data of Bishop *et al.* (1971c).

The optimal temperature for most virion reverse transcriptases is between 35°C and 45°C depending on the virus type, length of incubation, and laboratory, etc. Most investigators prefer 37°C for mammalian reverse transcriptases, 38–40°C for avian enzymes. The optimal pH is reported to be between pH 7.5 and 8.5, using tris-hydrochloride or tris-acetate buffers at 0.01 M concentration, while acid pH values are inhibitory to the enzyme, as is phosphate buffer (Bishop et al., 1971c; Garapin et al., 1970; Green et al., 1970). Some preference for potassium ions in lieu of (or in addition to) sodium ions has been claimed (Spiegelman et al., 1970a), and NaCl reaction concentrations of 0.06–0.14 M are in the optimal range for most enzyme activities (Scolnick et al., 1970a,b).

Both in the initial demonstrations by Baltimore (1970) and Temin and Mizutani (1970) and in subsequent investigations by others (Green et al., 1970; Garapin et al., 1970), enhanced reverse transcriptase activity was demonstrated in the presence of dithiothreitol or 2-mercaptoethanol (10–20 mM). Glutathione has also been employed as a reducing agent (Riman and Beaudreau, 1970). Magnesium chloride or acetate (1–10 mM, depending on the laboratory) or manganese chloride (usually 1 mM) or combinations of both (8 mM MgCl$_2$, 0.8 mM MnCl$_2$) have been reported to be optimal for the divalent cation requirement depending on the virus type, etc. (Bishop et al., 1971c; Green et al., 1970; Garapin et al., 1970; Scolnick et al., 1970b). Omission of magnesium or inclusion of EDTA or calcium ions drastically inhibits the enzyme activity (Green et al., 1970). It appears that in general the endogenous reaction of the reverse transcriptases of B-type oncornaviruses prefer magnesium ions, whereas those of the C-type oncornavirus prefer manganese ions.

In most endogenously templated reverse transcriptase assays, all four deoxyribonucleoside triphophates are required; ribonucleoside triphosphates cannot substitute for their deoxyribonucleoside counterparts, as shown by Baltimore (1970). The optimal concentrations for the triphosphates vary depending on the virus and laboratory; concentrations of 1 mM for unlabeled deoxyribonucleoside triphosphates (dATP, dCTP, and dGTP) and 0.01 mM for labeled TTP are commonly employed. However, reports suggest that high concentrations of all four triphosphates should be employed (possibly up to 5 mM) in order to obtain large product DNA species (Collett and Faras, 1975; Efstratiadis et al., 1975; Imaizumi et al., 1973; Rothenberg and Baltimore, 1976). Some DNA synthetic activity with detergent-disrupted virus preparations in the absence of dATP or dGTP has been noted (Spiegelman et al., 1970a), but not in the absence of dATP.

Purified reverse transcriptase templated by DNA or RNA requires all four deoxyribonucleoside triphosphates (Kacian *et al.*, 1971).

Reaction mixtures templated by oncornaviruses have been shown to be capable of synthesizing DNA for 12 h or longer under the right conditions (Garapin *et al.*, 1970). A triphosphate-regenerating system (phosphoenolpyruvate and pyruvate kinase) is reported to be helpful in maintaining the rate of DNA synthesis (Temin and Mizutani, 1970; Garapin *et al.*, 1970). Whether nucleoside triphosphates are involved in the polymerization process in ways other than for direct precursor polymerization into DNA is not yet known.

1.1.4b. Template Nucleic Acids Which Can Be Used by Reverse Transcriptases

One characteristic of viral reverse transcriptases which is not shared by the virion polymerases of other RNA viruses is their ability to accept exogenous nucleic acids as templates. Exogenous nucleic acids which have been tested include natural RNA, DNA, or synthetic polymers containing ribonucleotides or deoxyribonucleotides. Both single-stranded and double-stranded nucleic acid species including long single-stranded polymers hydrogen-bonded to short oligonucleotide primers have been extensively investigated using either purified enzymes, disrupted virions, or disrupted virions from which the viral RNA has been removed by preincubation with suitable nucleases, such as pancreatic ribonuclease. From early studies using added native or deoxyribonuclease 1 nicked DNA (activated DNA) as well as in subsequent studies involving RNA or synthetic homopolymers hydrogen-bonded to primer oligonucleotides, it was determined that the presence of a 3'-hydroxyl group was required for the exogenous stimulated DNA polymerase activity (Baltimore and Smoler, 1971; Duesberg *et al.*, 1971*b*; Hurwitz and Leis, 1972; Kiessling and Nieman, 1972; Leis and Hurwitz, 1972*a*; Mizutani *et al.*, 1970; Riman and Beaudreau, 1970; Roy-Burman, 1971; Spiegelman *et al.*, 1970*c*). This conclusion suggests that the polymerase uses a preexisting nucleotide sequence to prime and chain-elongate, a suggestion which has been reinforced by the identification of primers associated with the 60–70 S RNA genome of oncornaviruses (see later).

The exogenously templated reaction using synthetic homopolymer duplexes (DNA, RNA, or DNA:RNA hybrids) yields many times more product DNA than the endogenously templated reaction—as initially shown for the AMV enzyme by Spiegelman *et al.* (1970c). Solely single-stranded homopolymeric DNA or RNA species (i.e., without primer oligonucleotides) were found to be poor templates depending on the constituent nucleotide, although those which possessed, *per se*, some secondary structure did serve as templates for some DNA synthesis. Of the various duplexes tested with the AMV reverse transcriptase in the presence of 12 mM magnesium, those which would template polydeoxyguanylic acid product synthesis [e.g., poly(C):oligo(dI) and poly(dC):oligo(G)] were shown to be excellent templates for the AMV enzyme. Other experimenters have shown that for various mammalian oncornaviruses changing the divalent cation from magnesium to manganese and its concentration to 0.4 mM allows templates such as poly(I):poly(C), poly(dC):poly(G), or poly(A):poly(U) to be transcribed into DNA (Scolnick *et al.*, 1970a).

The nature of product synthesis, primer requirements, and details of the process will be discussed later; suffice it to say that since exogenous templates of the right constitution and under the right experimental conditions give much more product DNA than the endogenously templated reaction, the ability to detect a reverse transcriptase is greatly enhanced by their use and this has greatly facilitated the purification and characterization of reverse transcriptase enzyme proteins for several oncornaviruses.

1.1.4c. Inhibitors of Viral Reverse Transcriptases

In the initial studies by Baltimore (1970) and Temin and Mizutani (1970) with R-MuLV, RSV, and AMV virus preparations, it was shown that ribonuclease A pretreatment (50–1000 μg/ml) potently inhibited DNA synthesis. Lysozyme or cytochrome c pretreatments (at 50 μg/ml) were not inhibitory to enzyme activity and even provided a small increase in DNA synthesis. Ribonuclease inhibition of the endogenously templated activity has been demonstrated for all virion reverse transcriptases so far investigated.

Studies in many laboratories to obtain and characterize drugs or antibiotics which would inhibit viral reverse transcriptases have turned into a bonanza involving analyses of several different drugs and antibiotics. In an initial study in Maurice Green's laboratory (Gurgo *et al.*, 1971), it was shown that for a variety of RNA tumor viruses including MuSV, FeLV, and AMV, rifampicin, which is known to be an effective inhibitor of certain bacterial RNA polymerases, did not inhibit the endogenously templated DNA synthesis by the reverse transcriptases of those viruses. Similar results were obtained in J. Michael Bishop's laboratory using AMV and RSV (Garapin *et al.*, 1970). Green and collaborators subsequently examined the effects of various rifamycin derivatives and then showed that dimethylrifampicin and other derivatives (AF/ABDMP, AF/ABP, AF/013, and C27; see Fig. 5) are in fact effective inhibitors of the RNA-directed synthesis of DNA when used at concentrations greater than 100 μg/ml (Gurgo *et al.*, 1971; Green *et al.*, 1972). These compounds did not, however, inhibit the exogenous DNA-templated synthesis of product DNA, which observation led Green and associates to suggest that in some way these materials acted at the level of RNA-directed, rather than DNA-directed, DNA synthesis.

In systematic studies of various derivatives of rifamycin, involving amine, benzoxazino, and quinoxalino substitutents (at the R position in Fig. 5), Green and collaborators (Green *et al.*, 1972; Gurgo *et al.*, 1971) and others (Yang *et al.*, 1972) demonstrated that rifamycin piperidyl derivatives containing cyclohexyl substitutents were particularly active

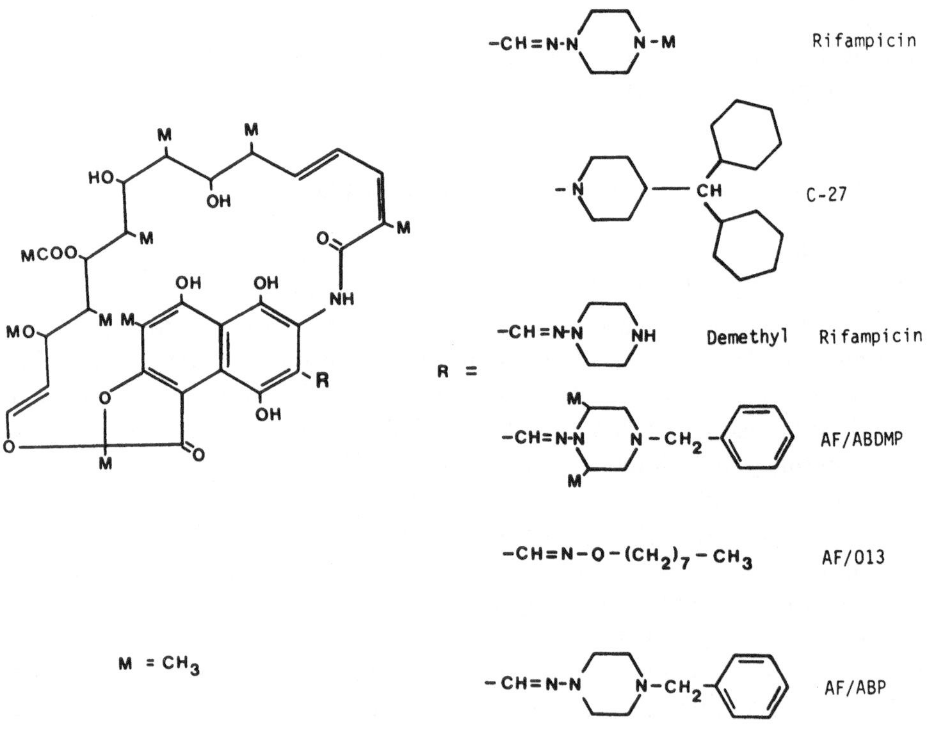

Fig. 5.   Chemical structures of rifampicin and certain derivatives.

in inhibiting certain reverse transcriptase enzymes (50% inhibition at 0.002 mM concentration).

Evidence has been obtained, again principally in Maurice Green's laboratory, which strongly suggests that the site of action of various rifampicin derivatives is the polymerase itself and that in some way the antibiotic thereby inhibits the RNA-directed synthesis of DNA (Gurgo et al., 1971; Green and Gerard, 1974). Results have been obtained which suggest that neither the RNA, DNA, nor nucleoside triphosphate binding sites on the polymerase are affected by AF/ABDMP treatment, and that possibly some allosteric modification of the enzyme results in an inhibition of an early step in DNA synthesis—maybe at the level of the initial induction of DNA polymerization by a primer molecule (see later).

Another set of antibiotics which have been analyzed for their effect on the *in vitro* reverse transcriptase reaction are the streptovaricins A, C, and D, as well as the streptovaricin complex of antibiotics, tolypomycin, streptolidigan, and geldanomycin (Brockman et al., 1971; Gurgo et al., 1971). It has been shown that some inhibition of reverse transcriptase activity can be obtained by treatment with members of the streptovaricin complex or by streptovaricins A, C, and D, but only at relatively high drug concentrations (400 μg/ml). The site where these antibiotics inhibit is not known (Green and Gerard, 1974).

Aminopiperazines (AP4, AP5, and AP8) have been found to be poor inhibitors of MuSV DNA polymerases (Gurgo et al., 1971). However, the same authors reported that actinomycin D, a potent inhibitor of cellular DNA-directed RNA polymerases, was a partial inhibitor (50-70%) of the endogenously templated MuSV(MuLV) reverse transcriptase when used at concentrations of 20-100 μg/ml. They suggested that probably the synthesis of RNA-directed DNA was not inhibited whereas the subsequent synthesis of DNA-directed DNA synthesis was inhibited. This was confirmed when they showed that addition of exogenous DNA to a reaction mixture increased product DNA synthesis sevenfold over the endogenous levels but that this increase was almost completely abolished by addition of 100 μg actinomycin D per milliliter. The sevenfold increase was not, however, inhibited by addition of similar amounts of ribonuclease.

The conclusions drawn by Green and collaborators regarding the mode of inhibition by actinomycin D were corroborated by Manly et al. (1971) using MuLV and confirmed the predictions by J. Michael Bishop and associates concerning the effect of actinomycin D on the RSV reverse transcriptase reaction (Garapin et al., 1970). Since actinomycin D does not prevent the formation of RNA:DNA duplexes by the reverse transcriptase reaction (Garapin *et al.*, 1970; Manly *et al.*, 1971; McDonnell *et al.*, 1970), and the antibiotic is known to intercalate its planar polycyclic ring in duplex DNA between base-paired dinucleotide sequences involving G:C base pairs (Sobell and Jain, 1972), but only poorly to DNA:RNA duplexes, single-stranded RNA or DNA, or double-stranded RNA polymers, it has been concluded that the action of the antibiotic is at the stage of duplex DNA synthesis. How duplex DNA is obtained in the reverse transcriptase reaction is not known, so that the exact site at which actinomycin D inhibits duplex DNA synthesis is still conjectural.

Some inhibition by daunomycin of MuLV, MuSV(MuLV), and RSV DNA polymerase activities has been reported by several laboratories (Mueller *et al.*, 1971; Chandra *et al.*, 1972) using relatively high concentrations (70 μg/ml) of the antibiotic. Daunomycin is also a substance which can intercalate into DNA (Krueger and Mayer, 1970). Effects of the antibiotic on both the endogenous and DNA-templated reverse transcriptase reaction have been reported.

Distamycin and analogues have been shown to be poor inhibitors of the endogenously templated reactions of RSV, MuLV, and MuSV(MuLV), but better inhibitors of those three viral polymerases when templated with certain synthetic templates—particularly those containing deoxyadenosine and thymidine (Chandra *et al.*, 1972; Kotler and Becker, 1971, 1972). Again these antibiotics are believed to inhibit the reaction by intercalating into DNA. Another intercalator, ethidium bromide, has been demonstrated to be a potent inhibitor at 0.01 mM concentration for both the endogenous and exogenously templated reverse transcriptase reactions, again depending (1) the constitution of the template (e.g., particularly with those templates containing deoxyadenosine and thymidine) and (2) the enzyme source (Fridlender and Weissbach, 1971; Hirschman, 1971; Mueller *et al.*, 1971).

A set of drugs which have been shown to be active against the MuSV(MuLV) or MuLV reverse transcriptase reactions (Chandra *et al.*, 1972) are the polycyclic aromatic compounds related to fluoranthrene (e.g., tilorone). Green and associates showed that fluoranthrene derivatives (e.g., anthraquinone, dibenzofuran, fluorene, fluorenone, and xanthenone) at concentrations as low as 0.002 mM inhibited the purified reverse transcriptase of AMV when templated by either synthetic polynucleotides, activated calf thymus DNA, or 70 S oncornavirus RNA (Green and Gerard, 1974). They postulated that these compounds may also derive their antipolymerase activity by intercalating between specific nucleotide sequences on a template nucleic acid.

Other compounds which have been shown to inhibit the polymerase activity are acridine orange, bleomycin, chromomycin, Congo red, histone, olivomycin, protamine, calcium elenolate, thiosemicarbazide, and N-methylisatinthiosemicarbazone (Green and Gerard, 1974; Hirschman, 1972; Levinson *et al.*, 1973). Neither mitomycin C nor chloroquinone inhibits reverse transcriptase activities (Mueller *et al.*, 1971).

Other than antibiotics and drugs, inhibitory effects have been noted with (1) synthetic polynucleotides such as 0.1 mM poly(U) (Tuominen and Kenney, 1971) and poly[(2′-O-methyl)inosinate] (Green and Gerard, 1974), (2) nucleoside triphosphate analogues such as arabinofuranosylcytosine triphosphate (using MuLV and synthetic templates, Tuominen and Kenney, 1971), and (3) salts of cadmium, copper, mercury, nickel, silver, and zinc (Levinson *et al.*, 1973). In this context, it should be borne in mind that the AMV and possibly other viral reverse transcriptases are probably a zinc metalloenzyme containing 1 g-atom zinc per mole of enzyme (Poiesz *et al.*, 1974).

From all these studies on the effects and modes of action of antibiotics or drugs on the reverse transcriptase reaction, several important practical consequences have developed. First, it has been possible to define experimentally the reverse transcriptase reaction into two basic phases, an RNA-directed synthesis of DNA and a DNA-directed synthesis of DNA. Second, by using actinomycin D to block the latter reaction, it has been possible to produce single-stranded DNA and RNA:DNA duplexes (Garapin *et al.*, 1973; Ruprecht *et al.*, 1973). Since RNA can be easily removed from RNA:DNA duplexes by alkali digestion or ribonuclease treatment after heat denaturation, these reaction products have been converted into sources of single-stranded viral complementary DNA (vcDNA) which have been used as probes for determining viral homologies and detecting viral nucleic acid sequences in infected cells, tissues, or tumors (see review by Green and Gerard, 1974).

Since certain antibiotics and drugs are obviously potent inhibitors of the *in vitro* reverse transcriptase reaction, the natural question which has been asked is what are their effects *in vivo*. A detailed discussion of the relevant experiments performed to answer that question is outside the scope of this chapter. However, certain claims have been made which will be summarized briefly below in the context of what is currently known about the oncornavirus infection process.

Although the processes involved in cell transformation induced by an oncornavirus infection are poorly understood, both genetic and other evidence indicate that an early activity of the virion reverse transcriptase is required in an infected cell for transformation to occur. The mutant RSV(α) lacks a functional reverse transcriptase or even any virion proteins identifiable as reverse transcriptase proteins. It does not synthesize DNA in infected cells, nor does it possess the ability to transform chick cells (Hanafusa and Hanafusa, 1971; Hanafusa et al., 1972; Weissbach et al., 1972). Certain temperature-sensitive, conditional lethal mutants of RSV (Linial and Mason, 1973; Mason et al., 1974; Verma et al., 1974a) or R-MuLV (Tronick et al., 1975) have been shown to possess thermolabile reverse transcriptases. These RSV mutants are also defective in their ability to transform cells at nonpermissive temperatures. The inability either of these or of the RSV ts mutants to give transformation has been shown to be an early function so that after a few hours at permissive temperatures, shifting infected cells to nonpermissive temperatures does not inhibit the maintenance or continuation of transformation. Similar results have been obtained for MuSV(tsR-MuLV) pseudotypes (Tronick et al., 1975). Such evidence indicates that the activity of the virion reverse transcriptase is an early, obligatory, function. As will be discussed later in this chapter, it is currently believed that upon infection the viral reverse transcriptase supplies a DNA copy of the viral genome and that this copy is then made into a closed circular DNA which is integrated into, and then replicated along with, the host's genome. The enzymes responsible for replicating viral DNA or transcribing viral RNA from that DNA have not been characterized, but the viral reverse transcriptase does not seem to be one of those enzymes. There is no known function of the reverse transcriptase enzyme late in an infection; however, the door is not shut on the possibility that reverse transcriptases may be involved either as backup viral DNA replication enzymes or in continuously or occasionally providing DNA copies from RNA sequences.

In considering, therefore, whether drugs and antibiotics known to be effective in vitro on the activity of a viral reverse transcriptase might also inhibit the in vivo activity of a viral reverse transcriptase, what has to be considered is the stage at which such potential inhibitors exhibit their maximum effect. Clearly early in an infection is one point at which an antiviral reverse transcriptase inhibitor might be expected to be effective. However, if an antiviral reverse transcriptase inhibitor inhibits an infected cell late in an infection, then there is a good possibility that its effect is not against the viral enzyme but against some other cellular component. Ideally, late in an infection drugs should be used that are capable of selectively inhibiting viral RNA transcription from integrated DNA (or closed circular DNA) sequences, as well as host transformation specific transcription processes.

One further point which should also be borne in mind in these considerations before weighting the importance of the effects of inhibitors on transformed cells is the overall toxicity of the drug to a "normal" cell. When the required concentration of a drug approaches or equals that at which "normal" cells are killed, the value of the inhibitor is obviously slight.

Many of the antibiotics or drugs listed and discussed in preceding paragraphs of this section have been tested *in vivo*, and a variety of claims have been made concerning their effects. As far as the rifampicin derivatives are concerned, in early studies it was reported that rifampicin itself (not an inhibitor of *in vitro* reverse transcriptase activity) inhibited RSV-induced chick cell transformation, focus formation, but not virus replication (Diggelman and Weissman, 1969). The optimal effectiveness of the antibiotic was shown to be between 36 and 60 h postinfection. Vaheri and Hanafusa (1971) obtained similar results except that they showed that following rifampicin treatment cell transformation occurred, but cell division did not. Other investigations have indicated that, depending on the concentration of antibiotic employed, both RSV-induced focus formation and virus replication are inhibited but that at those antibiotic concentrations normal cell growth is also affected (Richert and Balduzzi, 1971; Robinson and Robinson, 1971).

It has been claimed that the rifampicin derivative AF/ABDMP inhibits virus replication, MuSV-directed cell transformation, chemically induced rat tumors, and MuLV-induced cell transformation (Calvin *et al.*, 1971; Green *et al.*, 1972; Hackett and Sylvester, 1972 *a,b*; Joss *et al.*, 1973). Since the levels used were shown to be inhibitory *in vitro* to nonviral polymerases (e.g., normal cell DNA polymerase), it is not known whether these effects represent inhibitions specific to the reverse transcriptase or general inhibitions of one or more normal cellular polymerases (Green *et al.*, 1972). It has also been reported that certain rifampicin derivatives do not inhibit RSV-induced focus formation but do inhibit virus replication in transformed cells (Barlati and Vigier, 1972*a,b*).

The streptovaricin antibiotics have been shown to inhibit MuSV-induced focus formation and spleen enlargement in R-MuLV-inoculated mice (Borden *et al.*, 1971; Carter *et al.*, 1971). Although the drug levels used were lower than those found to be toxic for normal cell growth, the possibility of increased sensitivity of transformed cells being the cause cannot be discounted. Distamycins and fluorenone derivatives have also been claimed to inhibit MuSV-induced focus formation (Chandra *et al.*, 1972; Green and Gerard, 1974; Kruegar and Mayer, 1970).

What future value the anti-reverse transcriptase drugs will have in the treatment of cancer remains to be seen; suffice it to say here that several compounds have been extensively studied *in vitro* (and to some extent *in vivo*) and shown to be effective in one way or another against the reverse transcriptase enzyme and/or against oncornavirus-induced processes. Identification and characterization of specific transcription processes involved in maintenance of the transformed state or viral transcription from DNA (if they exist) are major hurdles to be overcome before these or other drugs can be developed as specific agents which will interfere with the neoplastic state.

1.2. *In Vivo* Assays for Virion RNA or DNA Polymerases

It has been documented in the preceding section that members of eight major groups of viruses (arenaviruses, bunyaviruses, orthomyxoviruses, paramyxoviruses, poxviruses, oncornaviruses, reoviruses, and rhabdoviruses) possess virion polymerases. Do all members of these eight virus groups possess demonstrable quantities of such enzymes? The presumption in that question is that some viral polymerases may require certain cell cofactors before they can function. Such cofactors are not unlikely in view of the recent evidence for mRNA modification (see Shatkin *et al.*, Chap. 1, this volume). Requiring host cofactors could result in a tropism of a virus for one form of differentiated cell but not another. Although the answer to the question of whether all members of the different virus groups possess demonstrable enzyme activities requires investigations of hundreds of viruses—many of which have yet to be cloned, grown, and analyzed—in one particular case, that of rabies virus, the presence of a virion polymerase in preparations of the virus has proved hard to demonstrate. In *in vitro* reaction assays in which various reaction ingredients have been optimized (including the addition of polyanions or SAM), the highest *in vitro* RNA polymerase activity observed in our laboratory with freshly prepared preparations of rabies HEP virus is 5 pmol nucleotide incorporated per hour per milligram of protein. No DNA polymerase activity has been detected. Reaction incubations which yield this small amount of RNA synthesis have a temperature optimum of 33°C and give a linear incorporation of nucleotides into acid-insoluble material for 40 min, but no longer. Mixtures of rabies preparations and preparations of VSV Indiana viruses synthesize RNA in amounts equivalent to those expected for the VSV contribution. It appears, therefore, that no ribonuclease or other general inhibitor is responsible for the low level of demonstrable transcriptase activity in rabies virus preparations. In order to determine whether rabies, a type-member of the rhabdovirus group of viruses, possesses a virion RNA polymerase, another approach has been utilized, namely that of seeking *in vivo* transcription in the presence of inhibitors of *de novo* protein synthesis, puromycin, and/or cycloheximide.

The *in vivo* assay which has been developed (Flamand and Bishop, 1973) allows the detection of small amounts of virus-complementary RNA synthesized in infected cells from the infecting virion genomes (primary transcription). The procedure basically involves using [³H]nucleoside-labeled virus preparations of high specific activity to infect cells at multiplicities as low as 5 PFU per cell, incubating the cells in the presence of inhibitors of protein synthesis (puromycin and/ or cycloheximide at 50–100 μg/ml), extracting the cell nucleic acids, and determining the presence and amounts of virus-complementary RNA therein by self-annealing the extracted nucleic acids and determining the increase in ribonuclease resistance of the ³H label (Fig. 6A). By annealing samples of the infected cell nucleic acids to excesses of added viral [³H]RNA and determining the maximum increase in [³H]ribonuclease resistance, the total quantity of intracellular virus-complementary RNA can be determined.

The results of such *in vivo* analyses for VSV-infected BHK cells are shown in Fig. 6B (Flamand and Bishop, 1974). Such *in vivo* primary transcription assays have demonstrated virus-complementary RNA synthesis in rabies-infected cells in both the presence and the absence of protein synthesis (Fig. 7), and this observation raises the possibility that cell factors or conditions are necessary before rabies-directed RNA transcription can occur (Bishop and Flamand, 1975). Whether rabies utilizes some intracellular protein or polymerase is an open question which needs further investigation.

Certain temperature-sensitive (*ts*) mutants of VSV Indiana belonging to the group I complementation group of *ts* mutants exhibit little or no synthesis of virus-complementary mRNA in infected cells at non-permissive temperatures (usually 39–40°C). Reduced synthetic capability is also apparent for mutants belonging to complementation group IV or some group II mutants, but not for those belonging to group V or group III (see reviews by Pringle, 1975; Bishop and Flamand, 1975;

Fig. 6. Intracellular transcription of VSV viral complementary RNA. In (A), a preparation of [³H]nucleoside-labeled, purified VSV (2.6 × 10⁵ cpm/100 μl Eagle's medium) was added to prewashed cold monolayers of 3 × 10⁶ BHK-21 cells and allowed to adsorb for 30 min at 4°C. One monolayer (the 0 time sample) was washed to remove excess virus, and the cells were recovered and extracted for nucleic acids (containing 1.7 × 10⁴ ³H cpm total). Other infected monolayers were washed with cold medium, then prewashed medium was added followed by incubation at the indicated temperatures for either 20 or 60 min. The cells were then washed and extracted for nucleic acids (1.3 × 10⁴ average ³H cpm/sample). The ³H percentage of ribonuclease resistance before (left side) or after (right side) a self-annealing was determined. In (B), the total amount of virus-complementary RNA synthesized by [³H]nucleoside-labeled VSV at 38°C in BHK-21 cells was determined for infected cells incubated with or without the addition of cycloheximide (to inhibit viral protein synthesis) at different times after infection. The total adsorbed ³H-labeled virus and the number of active virions per cell (representing six per cell) are indicated in the left-hand ordinate. The viral complementary RNA (left-hand panel) was determined by hybridizing infected cell nucleic acids to various amounts of viral [³H]RNA and determining the plateau of [³H]ribonuclease resistance (Flamand and Bishop, 1974). The number of transcriptive intermediates per cell was calculated from the observed linear rates of RNA synthesis established upon cycloheximide addition in comparison to the primary transcription rate developed by the six active virions, and gave values of 6, 40, 400, and 450 for the 0 h, 1 h, 2 h, 3 h, or "no addition" cycloheximide series, respectively. Virus yields from the infected cells are shown in the right-hand panel of B. Data from Flamand and Bishop (1973, 1974).

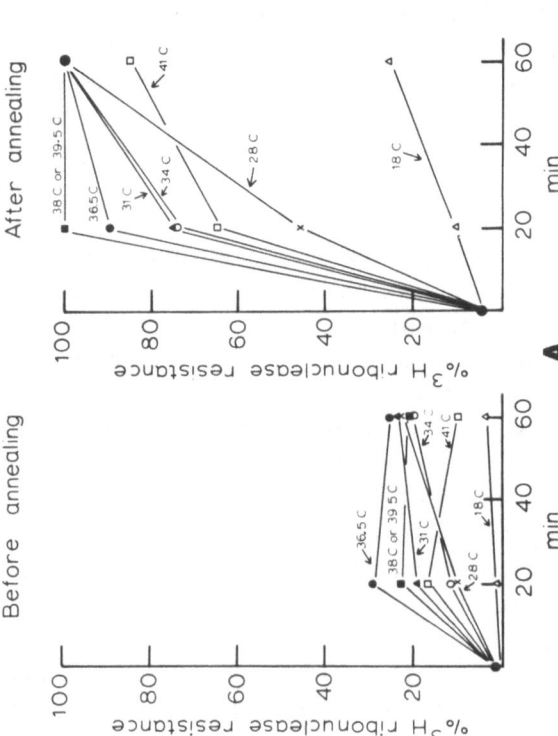

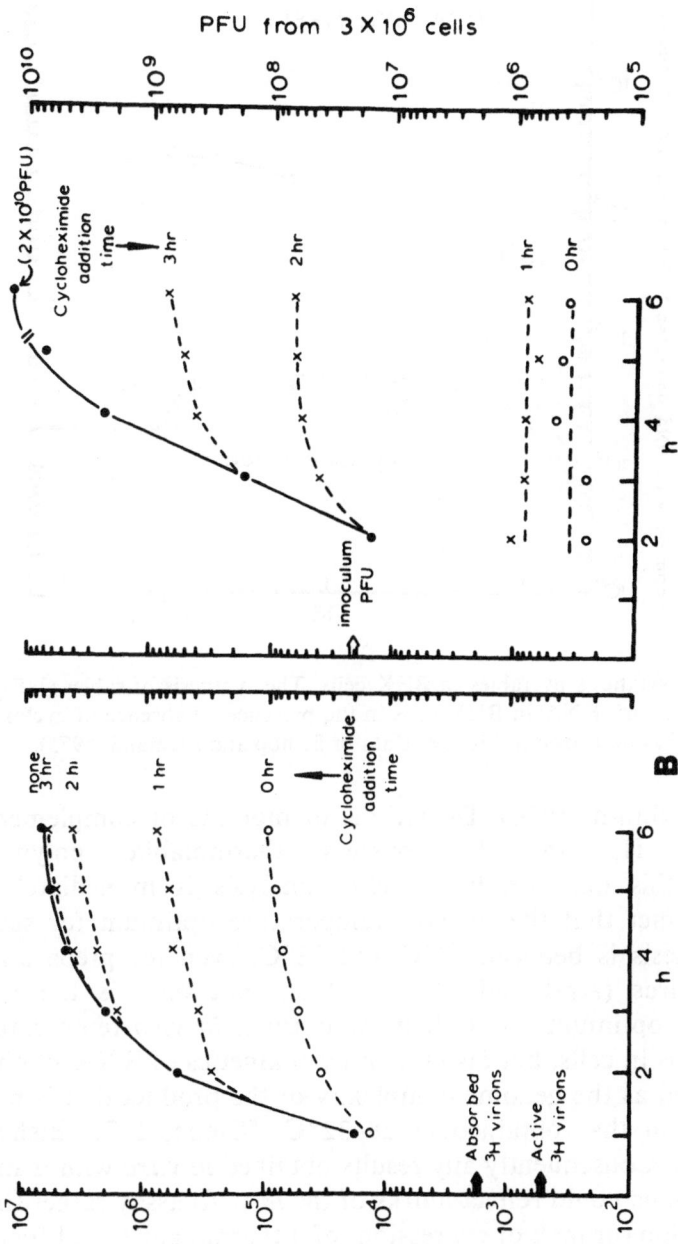

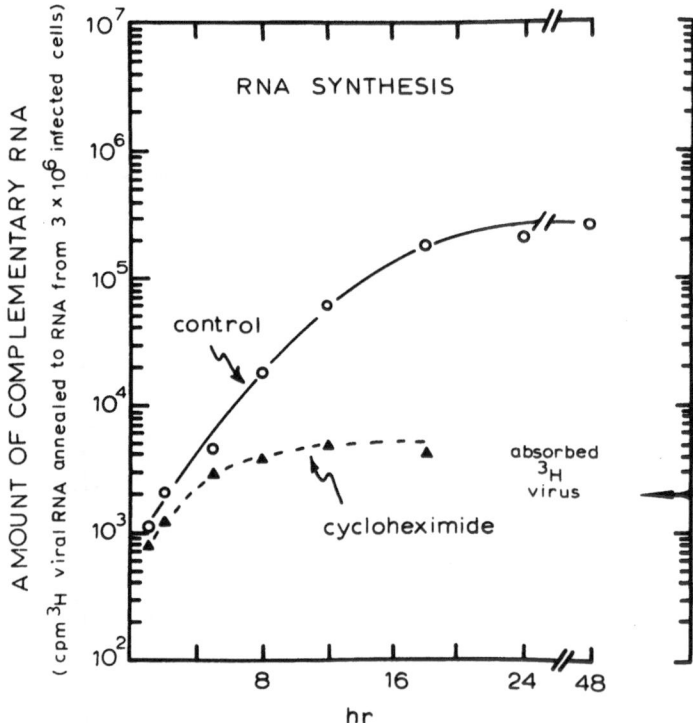

Fig. 7. RNA synthesis by rabies in BHK cells. The synthesis of rabies (HEP strain) virus-complementary RNA in BHK cells in the presence or absence of cycloheximide was determined as described in Fig. 6B. Data of Bishop and Flamand (1975).

Bishop and Smith, 1976). Do virions of mutants of complementation groups I, II, and IV possess thermolabile polymerases? Answering this question by *in vitro* analysis is immediately ruled out by the fact that the *in vitro* temperature optimum for sustained RNA synthesis is between 28°C and 32°C, even for preparations of wild-type virus (Aaslestad *et al.*, 1971). Not only is the *in vitro* temperature optimum lower than the optimal *in vivo* temperature for growing virus in cells, but also the *in vitro* kinetics of RNA synthesis at 37°C, as well as the genome complexity of the product that is made, is different from that obtained at 28–32°C (Bishop, 1971; Bishop and Roy, 1971*b*). Consequently any results obtained *in vitro* with *ts* mutants of VSV Indiana could reflect quirks of the *in vitro* assay rather than the true expression (or lack of expression) of a thermolabile viral function.

Through the use of *in vivo* transcription analyses described above. it has been possible to show that some but not all *ts* mutants of group I, and all VSV Indiana *ts* mutants of group IV so far investigated, perform primary transcription at nonpermissive temperatures, as does the only group II mutant (*ts*II52) so far investigated (Bishop and Flamand, 1975; Bishop and Smith, 1976; Flamand and Bishop, 1973, 1974). It has been concluded, therefore, that some *ts* mutants of VSV Indiana belonging to group I have temperature-sensitive (i.e., thermolabile) polymerase activities and some have polymerases which when synthesized at 31°C are functional (i.e., thermostable) at high temperatures (Flamand and Bishop, 1974).

There are several advantages to the *in vivo* procedure for detecting and analyzing virion polymerase activities. Answers to a variety of questions can be obtained which *in vitro* assays cannot answer or for which *in vivo* product labeling procedures either are too insensitive or are confused by the background of cell-directed nucleic acid synthesis. For instance, is primary transcription influenced by the host cell type? What is the effect of interferon pretreatment (or interferon inducer pretreatment) on the ability of a virus to transcribe nucleic acid? Is transcription complete? What are the rates of viral complementary RNA synthesis? How is the process affected by drugs or temperature changes? Where does primary transcription take place? etc. etc. Answers to some of these questions have been obtained for VSV and other viruses and will be described later.

In vivo assays for detecting virion (as opposed to intracellular virus-induced) RNA polymerases have been applied to a variety of virus types, including members of the reoviruses, poxviruses, rhabdoviruses, bunyaviruses, and orthomyxoviruses. The results obtained will be discussed for each virus type in later sections.

2. PROOF THAT POLYMERASES ARE VIRION COMPONENTS

Evidence that a polymerase activity observed to be associated with a virus preparation is indeed a virion function has been obtained for most viruses by demonstrating (1) that the enzyme activity is present in material possessing the same density as infectious viral particles and (2) that the polymerase is present in the viral core in association with the nucleocapsid. Additional evidence which has been obtained for some, but not all, viruses is (3) that the enzyme activity can be purified from a virus preparation and that it is antigenically or enzymatically distinct by comparison to similar viruses grown in the same cell type, and/or (4) that an enzyme isolated from a purified virus preparation is required for the infectivity of the nucleocapsid. In addition, for some viruses, certain temperature-sensitive mutants have been shown to possess a thermolabile polymerase function (either *in vivo* or *in vitro*). These criteria for concluding that polymerase activities associated with viral preparations are indeed virion functions (and in some cases clearly virus-determined functions) are discussed below.

2.1. Association of Polymerase Activities with Virus Particles

In the initial pioneering studies with poxviruses (rabbitpox and vaccinia), it was shown that the polymerase activity was present in material which could be recovered from an equilibrium gradient in the position where virus particles were present (Kates and McAuslan, 1967; Munyon *et al.*, 1967). This was confirmed later for Yaba poxvirus (Schwartz and Dales, 1971).

Studies with reovirus (Borsa and Graham, 1968; Shatkin and Sipe, 1968), bluetongue virus (Martin and Zweerink, 1972; Verwoerd and Huismans, 1972), cytoplasmic polyhedrosis virus (Lewandowski *et al.*, 1969), and wound tumor virus (Black and Knight, 1970) have also demonstrated that for these diplornaviruses an RNA polymerase activity coincided in an equilibrium gradient with infectious virus particles or was present in preparations identified by electron microscopy to contain virus particles. Analyses of other virus types (rhabdoviruses, arenaviruses, orthomyxoviruses, paramyxoviruses, bunyaviruses, and oncornaviruses) have given similar results.

Another piece of evidence which indicates that polymerase functions reside within virus particles is provided by the demonstration that polymerase activity can often *only* be observed following teatment of a virus preparation with materials which disrupt the virus particle. This has been discussed in preceding sections.

Additional evidence that a polymerase is a virus-associated activity has been provided by the demonstration that subviral nucleocapsids possess some, most, or all of the polymerase functions.

Although, as will be described later, the identification of the polypeptides responsible for poxvirus or reovirus polymerase activities has eluded diligent research, for certain oncornaviruses and rhabdoviruses the polypeptides responsible for the enzyme activity have been purified and shown to be virion components. In the case of influenza virus, or the paramyxoviruses NDV and Sendai, identification of the polymerase protein(s) has been narrowed to a few of the structural polypeptides (Bishop *et al.*, 1972; Hightower *et al.*, 1975; Stone *et al.*, 1972; Robinson, 1971*b*; Zaides *et al.*, 1975). Other virus systems either have not been studied or are in the same category as the myxoviruses.

Various procedures can be used to relate enzyme activity to a subviral core. One simple procedure which we have found to be successful for a variety of lipid-enveloped viruses involves treating a virus preparation with a nonionic detergent in the presence of 1 M NaCl and subjecting the mixture to polyethylene glycol–dextran T500 phase separation. For VSV Indiana and other rhabdoviruses, as well as for the orthomyxovirus influenza A, strain WSN, all of the initial endogenously templated virion transcriptase activity is recovered in the dextran phase while all of the viral glycoprotein and nonglycosylated membrane proteins are recovered in the polyethylene glycol phase (Bishop and Roy, 1972; Hefti et al., 1975). Since for both virus types all of the viral RNA, nucleoproteins as well as some of the minor proteins, is also recovered in the dextran phase, these results suggest that the viral polymerase functions are associated with the nucleocapsid cores of those viruses.

Another procedure which has been successfully applied to influenza virus preparations is to centrifuge the detergent-dissociated material in a velocity gradient and demonstrate (1) that enzyme activity is recovered in association with the viral nucleocapsid and (2) that these nucleocapsids lack the major structural glycoproteins and membrane protein (Bishop et al., 1972). Studies by Emerson and Wagner (1972) involving separation of VSV Indiana nucleocapsids from solubilized viral components by differential centrifugation have also provided definitive evidence that the polymerase resides in the nucleocapsid core.

2.2. Purification of Viral Polymerases

The clearest proof that a polymerase is a component of a virus is the purification of the enzyme polypeptides and a demonstration that not only are those polypeptides always found in viral particles but also are capable of transcribing and rendering infectious the enzyme-stripped nucleocapsid. Such unambiguous proof has yet to be obtained for poxviruses, reoviruses, paramyxoviruses, orthomyxoviruses, bunyaviruses, or arenaviruses. It has been obtained by Emerson and collaborators for the rhabdovirus VSV Indiana (Bishop et al., 1974; Emerson and Wagner, 1972, 1973; Emerson and Yu, 1975).

It has been shown that both purified L protein and NS protein are required to transcribe the genome of VSV Indiana (Emerson and Yu, 1975). The additional fact that removal of the transcriptase proteins results in loss of the ability to productively infect a cell (Szilágyi and Uryvayev, 1973) and restoration of these proteins restores that ability as well as transcriptase activity (Fig. 8 and Table 6) (Bishop et al.,

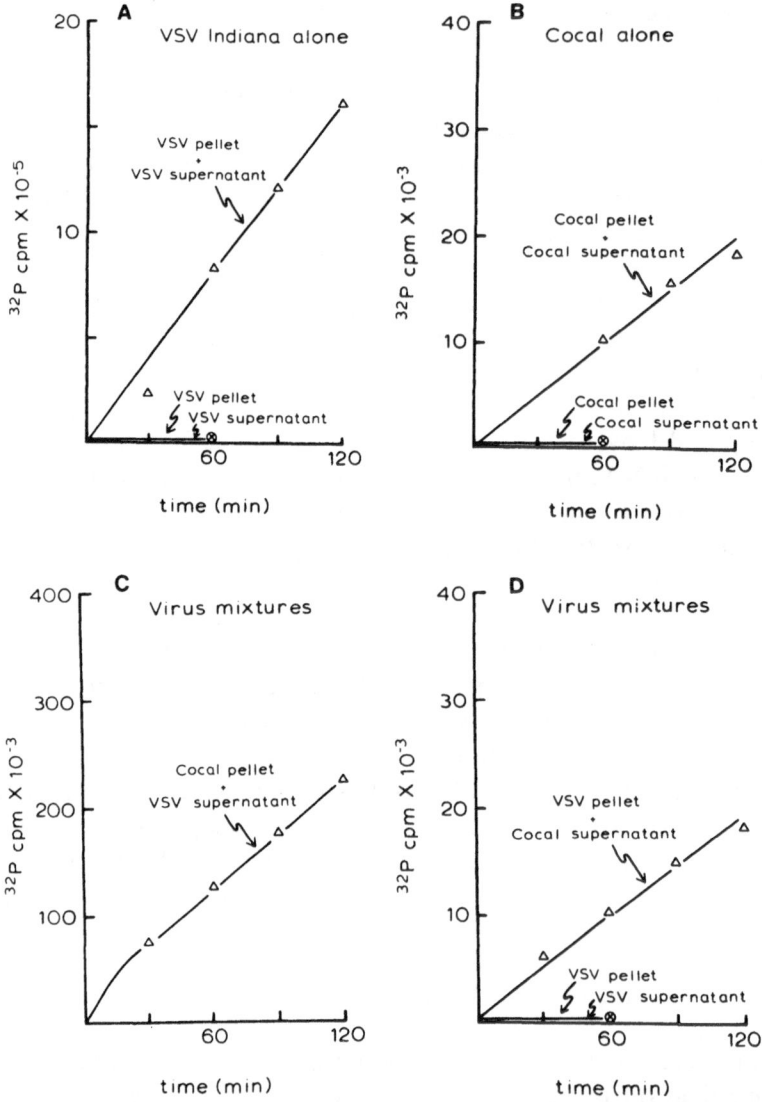

Fig. 8. Reconstitution of transcriptase activity from homologous and heterologous combinations of templates and enzymes of various rhabdoviruses. In (A–D), the reconstitution of transcriptase activity by dissociated components of VSV (Indiana serotype) and Cocal viruses is shown. Purified preparations of [³H]uridine-labeled VSV Indiana and Cocal viruses were dissociated as described by Bishop *et al.* (1974). Samples of each fraction were assayed for transcriptase activity either separately or in the indicated combinations. The initial VSV Indiana virus preparation possessed a specific enzyme activity of 118 pmol [³²P]UMP incorporated per hour per 10^4 ³H cpm viral RNA. The observed reconstituted VSV template plus VSV supernatant specific activity was equivalent to 130 pmol [³²P]UMP incorporated per hour per 10^4 ³H cpm. The initial Cocal virus preparation possessed a specific enzyme activity of 94 pmol [³²P]UMP incorporated per hour per 10^4 ³H cpm. The observed reconstituted Cocal template plus Cocal supernatant specific activity was equivalent to 7.5 pmol [³²P]UMP

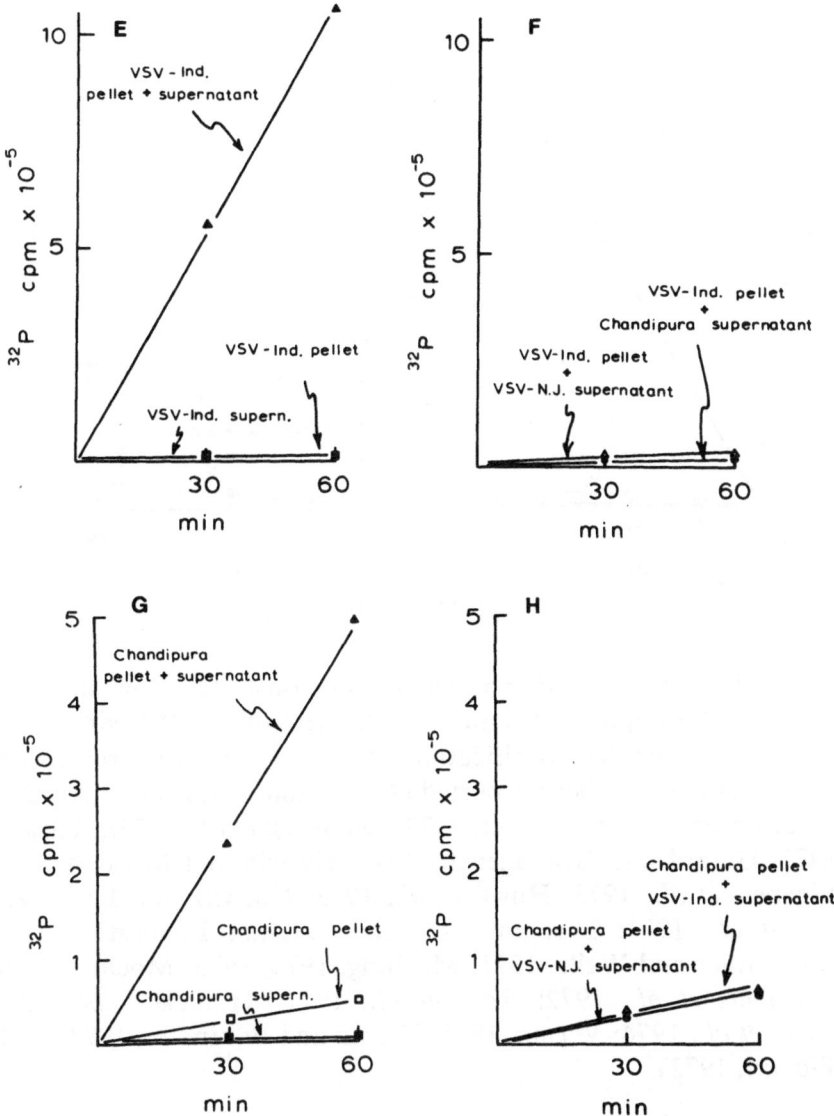

incorporated per hour per 10^4 ^{3}H cpm while that of the Cocal template plus VSV supernatant was 80 pmol per hour per 10^4 ^{3}H cpm. In (E–J), the reconstitution of transcriptase activity by dissociated components of VSV (Indiana serotype), VSV (New Jersey serotype), and Chandipura viruses was attempted. Purified preparations of [^{3}H]uridine-labeled VSV Indiana, VSV New Jersey, and Cocal viruses were dissociated as described by Bishop *et al.* (1974). Samples of each fraction were assayed for transcriptase activity either separately or in the indicated combinations. The initial specific activities of the virus preparations of VSV Indiana, VSV New Jersey, and Chandipura were 95, 54, and 47 pmol [^{32}P]UMP incorporated per hour per 10^4 ^{3}H cpm of virus RNA, respectively. The specific activities of the homologous reconstituted activities of VSV Indiana, VSV New Jersey, and Chandipura viruses were 92, 60, and 50 pmol [^{32}P]UMP incorporated per hour per 10^4 ^{3}H cpm of RNA, respectively. Data of Bishop *et al.* (1974).

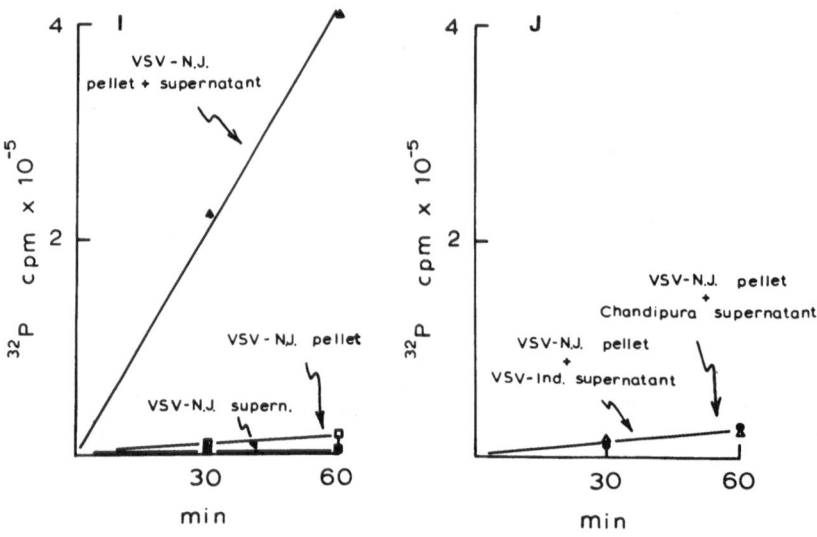

Fig. 8 *(cont'd)*.

1974) is the final proof needed to demonstrate that the viral transcriptase is a required structural component of the VSV virion.

Partial or complete purification of various oncornavirus reverse transcriptases has also been reported (Abrell and Gallo, 1973; Abrell *et al.*, 1975; Baltimore and Smoler, 1972; Duesberg *et al.*, 1971a; Faras *et al.*, 1972; Gerard and Grandgenett, 1975; Gerwin and Milstein, 1972; Grandgenett *et al.*, 1973; Howk *et al.*, 1973; Hurwitz and Leis, 1972; Kacian *et al.*, 1971; Leis and Hurwitz, 1972a,b; Livingston *et al.*, 1972a,b; Miller and Wells, 1971; Moelling, 1974, 1975; Moelling *et al.*, 1971; Robert *et al.*, 1972; Ross *et al.*, 1971; Tronick *et al.*, 1972; Twardzik *et al.*, 1974; Verma, 1975; Verma and Baltimore, 1973; Wells and Fluegel, 1972).

2.3. Antigenic Comparisons of Virion Polymerases from Similar Virus Types

Few serological comparisons of polymerases or polymerase components are available for arenaviruses, bunyaviruses, orthomyxoviruses, paramyxoviruses, poxviruses, rhabdoviruses, or reoviruses.

Only one report exists concerning antigenic comparisons of rhabdovirus transcriptase components, and that is for VSV Indiana and its distant cousin VSV New Jersey. Imblum and Wagner (1975) demonstrated that antibody made against whole VSV Indiana or purified VSV Indiana NS phosphoprotein was unable to inhibit the

TABLE 6

Reconstitution of Infectivity by Dissociated Components of VSV (Indiana Serotype), VSV (New Jersey Serotype), Cocal, and Chandipura Viruses[a]

Component mixtures	PFU	Increase
VSV Indiana pellet	less than $2 \times 10^{2*}$	
VSV Indiana supernatant	less than 20*	
VSV Indiana pellet and VSV Indiana supernatant	6×10^5	.... at least 3000-fold
Cocal pellet	less than 10^{2*}	
Cocal supernatant	less than 17*	
Cocal pellet and Cocal supernatant	1.2×10^3	...... at least 12-fold
VSV New Jersey pellet	3×10^4	
VSV New Jersey supernatant	less than 10^{2*}	
VSV New Jersey pellet and VSV New Jersey supernatant	4×10^5	.... 13-fold
Chandipura pellet	10^4	
Chandipura supernatant	less than 10^{2*}	
Chandipura pellet and Chandipura supernatant	2×10^5	.... 20-fold
VSV Indiana pellet and VSV New Jersey supernatant[b]	2×10^4	.... none[b]
VSV Indiana pellet and Chandipura supernatant[b]	3×10^4	.... none[b]
VSV Indiana pellet and Cocal supernatant	1.5×10^4	.... 75-fold
Cocal pellet and VSV Indiana supernatant	4×10^2	.... at least 4-fold
VSV New Jersey pellet and VSV Indiana supernatant[b]	4×10^4	.... none
VSV New Jersey pellet and Chandipura supernatant	6×10^4	 2
Chandipura pellet and VSV Indiana supernatant[b]	10^4	.... none
Chandipura pellet and VSV New Jersey supernatant	2×10^4	 2

[a] Purified preparations of VSV Indiana, VSV New Jersey, Cocal, and Chandipura viruses were dissociated, and the pellet (= RNA-N) and supernatant (= transcriptase) were assayed separately or in combinations as described by Bishop *et al.* (1974). The PFU observed is expressed as the total obtainable for the complete 0.35 ml fraction (or 0.70 ml combination). In some instances, no plaques were observed at the lowest dilutions plated (*). Homologous reconstitution of infectivity was significantly greater than the component infectivities.

[b] In these combinations, different VSV Indiana pellet and supernatant preparations were used possessing 4×10^4 and less than 1×10^2 PFU, respectively. Other than Cocal–VSV Indiana combinations, heterologous reconstitution of infectivity was not significantly different from the template infectivities; however, small quantities of heterologous reconstitution (less than 10^4) would have been missed.

endogenously templated transcriptase activity of VSV New Jersey under conditions in which either antibody preparation did inhibit chain elongation (82–89%) by VSV Indiana transcriptase. This result points to the degree of evolutionary divergence that those two viruses have experienced (see review by Bishop and Smith, 1976).

Scholtissek *et al.* (1971) showed that in fowl plague convalescent serum antibodies existed that inhibited the polymerase activity of six human, avian, or equine influenza A strains but not the influenza B strain Lee or the paramyxovirus NDV. The similar inhibitions of the various A strains by these antibodies suggest that the polymerase enzymes of influenza A strains are homologous to each other but distinct from those of influenza B, strain Lee.

Antibody raised against purified, semipurified, or unpurified reverse transcriptases of various oncornaviruses has been used to determine the relatedness of those polymerases. Nowinski *et al.* (1972) demonstrated, using a monospecific antiserum raised against purified AMV reverse transcriptase (Watson *et al.*, 1972), that the DNA polymerase activities of both AMV and RSV were inhibited to similar extents by antibody treatment, suggesting that these two enzymes possess at least some similar primary sequences. This has been confirmed by comparative polypeptide mapping (Gibson and Verma, 1974). Antibody against AMV did not neutralize chick cell polymerases or the reverse transcriptases of the avian reticuloendotheliosis virus (REV), murine oncornaviruses (MuMTV, R-MuLV), feline leukemia virus (FeLV), visna virus, or the monkey isolate M-PMV. It did neutralize both Rous sarcoma and Rous-associated viruses of various sorts, and it was shown that the antibody did not react with other structural antigens of AMV. Panet *et al.* (1975a) have developed a sensitive radioimmune assay involving rat antiserum made against the AMV enzyme and ^{125}I-labeled pure enzyme. They confirmed the relatedness of AMV and RSV enzymes and detected a small proportion of antigenic sites on the AMV enzyme which were not present on the RSV enzyme.

Mizutani and Temin (1973) using antibodies raised against AMV also confirmed Nowinski's results and demonstrated that the viral reverse transcriptases of various avian reticuloendotheliosis-type viruses (i.e., CSV, DIAV, REV, TDSNV) were not related to the avian leukemia-sarcoma viruses (e.g., AMV, avian induced leukosis virus, RSV, RAV), nor was the AMV enzyme homologous to two cellular DNA polymerases.

Antibody produced against mammalian C-type oncornaviruses has been shown to be incapable of inhibiting avian virus reverse transcriptases but capable of inhibiting certain, but by no means all, mammalian C-type viral reverse transcriptases (Aaronson *et al.*, 1971; Oroszlan *et al.*, 1971; Parks *et al.*, 1972; Sherr *et al.*, 1975). Antisera against the reverse transcriptase of FeLV and R-MuLV showed reciprocal cross-reactivity but did not significantly inhibit the enzymes of five other C-type viruses (GALV, M7, PK, RD114, and SSAV). Antisera against the monkey SSAV inhibited the enzyme of a gibbon ape virus (GALV) and cross-reacted to lesser extents with the enzyme of pig (PK) virus or with Rauscher murine leukemia virus. Antiserum against GALV gave similar results, although less cross-reactivity was obtained with the murine than with PK enzyme. Antiserum raised against the pig (PK) viral enzyme strongly inhibited SSAV and GALV and to a lesser extent R-MuLV. Interestingly, antiserum against RD114 and the baboon endogenous virus M7 reciprocally cross-reacted but showed no inhibition of the polymerases of the other five viruses tested (R-MuLV, FeLV, SSAV, GALV, or PK). From such data, Sherr and associates concluded that they could identify three sets of interspecies antigenic determinants for polymerases from different groups of mammalian C-type viruses.

3. PRODUCT ANALYSES OF VIRION POLYMERASE REACTIONS

In most analyses of transcription products synthesized by endogenously templated virion polymerases, the product alone is analyzed. In many cases this is sufficient, although analyses involving both labeled template and product have certain advantages, not the least of which is the ability to monitor the fate and integrity of the template species as a function of the reaction time course. In several of the results described below, use has been made of [^{3}H]nucleoside-labeled virus in reaction mixtures containing [α^{32}P]nucleoside triphosphates. The feasibility of having both template and product labeled depends on the system being studied and cost of performing dual isotope experiments. Procedures are available for the synthesis of high specific activity [^{32}P]ribo- or [^{32}P]deoxyribonucleoside triphosphates labeled in the α, $\beta\gamma$, or γ position (Bishop, 1973). Most ^{3}H- or ^{32}P-labeled triphosphates are also available commercially. After a reaction incubation, excess precursor-labeled triphosphates can be conveniently removed from the product nucleic acids by SDS-phenol extraction followed by Sephadex G50 column chromatography (Bishop, 1973).

Product analyses can be performed using a variety of procedures in order to determine (1) its size, using sucrose gradient centrifugation or polyacrylamide or agarose-polyacrylamide gel electrophoresis; (2) the product base ratio and nearest-neighbor ratios (for product labeled by [α-^{32}P]nucleoside triphosphates); (3) the sequences at the 5′ and 3′ ends; (4) the strandedness of the product (single or double stranded) as determined by its resistance to single-strand-specific nucleases; (5) the nature of the transcriptive intermediates; (6) the association of product with the template nucleic acids as monitored by gels, or cesium sulfate centrifugation (e.g., to resolve RNA:DNA duplexes from single-stranded RNA or double-stranded DNA); (7) the presence or absence of virus-complementary nucleic acid sequences as determined by the extent that the product can anneal to viral RNA; (8) the integrity of virus-complementary mRNA as determined by its ability to be translated; (9) the repetitive nature of the transcription process; (10) the completeness of transcription and the proportional representation of the various products. From such analyses, answers to two important questions can be obtained. How is product nucleic acid synthesized? How is its synthesis regulated?

Apart from the obvious interest in determining the mechanism and specificity of mRNA or DNA transcription, the question of mRNA transcriptional control is one which is important in order to understand how a virus regulates the production of its own proteins. For several of the RNA virus types listed in Table 1, the mRNA species are shorter than the viral genome from which they are transcribed; therefore, the questions of whether replication initially involves the same sequence of events as transcription and how transcription of mRNA and viral genome replication are interrelated are also important ones to be answered. What is known about these processes for various virus types which possess virion polymerases will be discussed below in relation to the virus type, and in consideration of both *in vitro* and *in vivo* analyses, as well as in relation to what extent the *in vitro* analyses reflect what occurs *in vivo*.

3.1. Arenaviruses

There is only one published report concerning the presence of a virion RNA polymerase in preparations of the arenavirus Pichinde (Carter *et al.*, 1974). No viral reverse transcriptase has been detected in Pichinde preparations. The reaction requirements for demonstrating polymerase activity and the synthesis of RNA have been shown to be essentially similar to those used for the orthomyxovirus influenza A, strain WSN (e.g., a requirement for magnesium and manganese, dependence on the presence of a nonionic detergent, enhanced synthesis of RNA at 32°C by comparison to 37°C, and no requirement for a reducing agent). The Pichinde virion, like other arenaviruses, possesses not only a segmented RNA genome but also ribosomal particles as well as 4 S RNA (Carter *et al.*, 1973*a*,*b*). Therefore, analyses of Pichinde transcription products are complicated by the fact that there are at least five size classes of virion RNA (28 S and 18 S ribosomal species, 31 S and 22 S viral species, and 4 S RNA of unknown origin). Sucrose gradient centrifugation of Pichinde transcription product has revealed that the product RNA is present in two size classes, one corresponding to 22–26 S and the other 4–6 S. The faster-sedimenting material has been shown to be partially resistant to pancreatic ribonuclease digestion (~40%), whereas the smaller material was shown to be ribonuclease sensitive (i.e., single stranded). No other product analyses have been performed, so that neither the origin nor the constitution of the product is known.

No *in vivo* primary transcription or intracellular mRNA analyses have been published for Pichinde virus, lymphocytic choriomeningitis virus, or other members of the arenavirus group. The coding relationship of the viral segments to the viral proteins (Table 7) is not known.

3.2. Bunyaviruses (and Bunyaviruslike Viruses)

There are more than 140 bunyavirus isolates (including the Bunyamwera supergroup of serologically related viruses and bunyaviruslike viruses not serologically related to the supergroup). Most of these isolates have yet to be adapted to tissue culture, cloned, or analyzed. Several members of the Bunyamwera supergroup, including Bunyamwera itself, snowshoe hare, La Crosse, and Main Drain viruses, appear from *in vivo* studies to perform primary transcription in the absence of *de novo* protein synthesis (see below), and two bunyaviruses, Lumbo and Uukuniemi viruses, have been shown to possess virion RNA-dependent RNA polymerases.

The basic characterization of these viruses is in its infancy; however, it is known for all six viruses listed above that the RNA recovered from purified virions is in three size classes (~3×10^6, 1.8×10^6, and 0.5×10^6 daltons), arranged, at least for Uukuniemi, Lumbo, and La Crosse viruses, in circular RNA–protein complexes (although the RNA therein is not circular, D. H. L. Bishop, unpublished observations). It is also known (Table 8) that the viruses are enveloped viruses possessing two glycoproteins, a major nucleocapsid protein as well as a minor large protein (Bouloy *et al.*, 1974; McLerran and Arlinghaus, 1973; Obijeski *et al.*, 1976; Pettersson and Kaariainen, 1973; Pettersson and von Bonsdorff, 1975; Ranki and Pettersson, 1975; Samso *et al.*, 1975).

3.2.1. *In Vitro* Transcription Analyses of Bunyaviruses

The RNA-dependent RNA polymerase of Lumbo virus was shown by Bouloy *et al.* (1975) to be capable of synthesizing RNA *in vitro* in a linear fashion for 12 h at 32°C or 37°C. The synthesis was shown to require the presence of all four ribonucleoside triphosphates and a nonionic detergent, and it produced more RNA in the presence of both

TABLE 7
Arenavirus Pichinde Structural Components, Viral RNA, and mRNA Species[a]

	Viral proteins	
Species	Mol. wt.	Approximate number per virion
V I	72,000	Not determined
V II	72,000	"
V III	34,000	"
V IV	12,000	"

	Viral RNA	
Species	Approximate mol. wt.	Number per virion
31 S	2×10^6	Not determined
22 S	1.1×10^6	"
28 S (ribosomal)		"
18 S (ribosomal)		"
4 S (cellular)		"

Viral mRNA
Not identified

[a] Polypeptides V II and V III are glycoproteins (Ramos *et al.*, 1972).

TABLE 8

Bunyavirus La Crosse Virion Structural Components, Viral RNA, and mRNA Species[a]

Viral proteins		
Species[b]	Mol. wt.	Approximate number per virion
L	1.8×10^5	50
G1	1.2×10^5	750
G2	0.34×10^5	770
N	0.23×10^5	2440
Viral RNA species		
Species	Mol. wt.	Approximate coding capacity
1	3×10^6	3×10^5
2	1.8×10^6	1.8×10^5
3	0.45×10^6	0.45×10^5
mRNA species		
Unknown		

[a] Information is for La Crosse virus (Obijeski *et al.*, 1976; J. F. Obijeski and D. H. L. Bishop, unpublished observations). Other bunyaviruses (and bunyalike viruses) possess different sizes of viral polypeptides and RNA species (unpublished observations).
[b] L, Large; G1, glycoprotein 1; G2, glycoprotein 2; N, nucleoprotein. G1 and G2 are on the outer surface of virions, N and L are associated with the viral RNA.

magnesium and manganese ions (8 and 1.5 mM, respectively) than in the absence of either or both. The transcription of product RNA was not affected by the addition to a reaction mixture of either deoxyribonuclease, actinomycin D, exogenous RNA, or DNA. All the *in vitro* product accumulation was abolished by ribonuclease treatment. About 60–90% of the product was found to be ribonuclease resistant. Purified single-stranded product was shown to be capable of hybridizing *in toto* to Lumbo viral RNA (but not to Sindbis viral RNA), indicating that it was complementary in composition to the viral RNA species (Bouloy *et al.*, 1975).

Ranki and Pettersson (1975) have demonstrated for Uukuniemi virus that it too possesses an RNA polymerase whose enzyme activity also requires all four triphosphates and a nonionic detergent. The required cation for Uukuniemi was determined to be manganese (3–15 mM); magnesium ions did not substitute for or enhance this manganese requirement. The pH optimum for the Uukuniemi viral polymerase (pH 7.2–8.5) was found to be much broader than that reported for other viral polymerases (e.g., VSV). The highest Uukuniemi enzyme activity was observed for reaction mixtures incubated for 2–3 h at 37–40°C (~30 pmol UMP incorporated per hour per milligram of protein). Neither actinomycin D, rifampin (20 μg/ml), deoxyribonuclease treatment, nor changes in the monovalent cations affected this reaction rate to any significant extent. Studies to characterize the product synthesized by Uukuniemi virus showed that about 20–40% of the product was ribonuclease resistant. Self-annealing total reaction product (template and labeled product) rendered 70–80% of it ribonuclease resistant. Upon sedimentation in a sucrose gradient, some of the product RNA was recovered in the size class of the smaller two viral RNA species, while very little cosedimented with the largest species. Most of the product sedimented as 3–7 S RNA species. Which virion polypeptide(s) constitute the transcriptase is not known.

Whether other bunyaviruses possess virion polymerase activities which can be demonstrated *in vitro* remains to be determined.

3.2.2. *In Vivo* Assays of the Bunyavirus Primary Transcription Process

Primary transcription analyses with certain bunyavirus isolates have been performed in order to detect the synthesis of virus-complementary RNA in infected cells in which *de novo* protein synthesis has been inhibited. The protocol used for such analyses is identical to that described previously for VSV Indiana (Flamand and Bishop, 1973, 1974) and involves the detection by hybridization of virus-complementary RNA in infected cells. The results obtained for the bunyavirus snowshoe hare are shown in Fig. 9; similar results have been obtained for three other bunyaviruses (Main Drain, Bunyamwera, and La Crosse viruses). the conclusions drawn from such results are that these viruses synthesize virus-complementary RNA in the absence of protein synthesis—a result compatible with their possessing virion RNA polymerases. Whether host proteins or factors are involved in the *in vivo* primary transcription will have to await further experimentation. No information is available concerning the size, constitution, or proportionality of bunyavirus mRNA species *in vivo*.

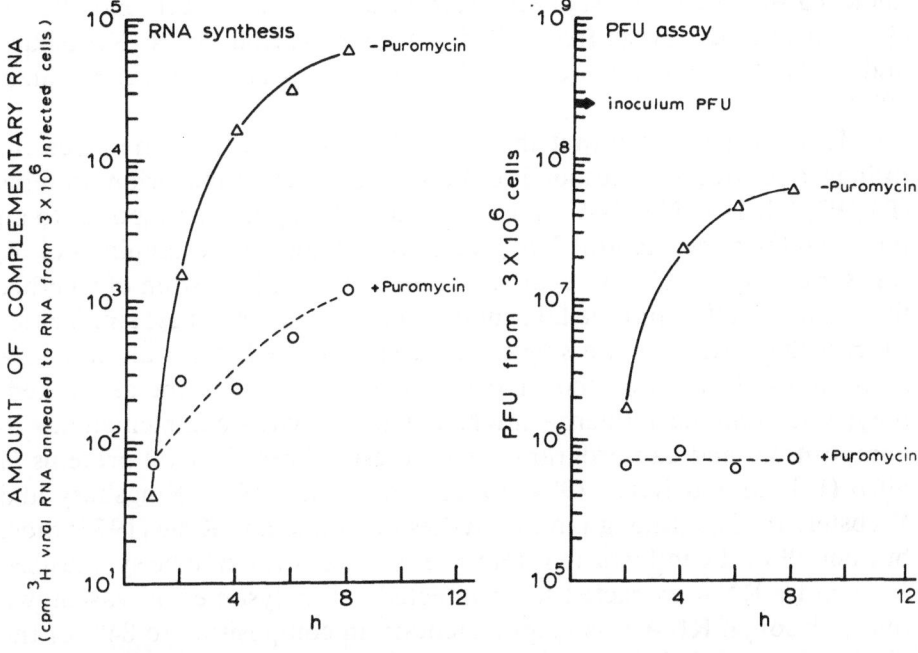

Fig. 9. Synthesis of viral complementary RNA (left-hand panel) in BHK-21 cells infected by [³H]nucleoside-labeled snowshoe hare virus. The synthesis of viral complementary RNA in the presence or absence of 100 μg puromycin/ml was determined as described in Fig. 6 and Bishop and Flamand (1975). The concomitant release of infectious snowshoe hare virus is given in the right-hand panel.

3.3. Orthomyxoviruses: Influenza Viruses

The genetic material of influenza virus has been shown to consist of single-stranded RNA segments which, in dual infections, probably segregate independently of each other and are the explanation for recombinant progeny containing characteristics of both parents (Simpson and Hirst, 1961). Estimates of five to ten for the number of the segments and 1×10^6 to 3×10^5 daltons for their sizes have been made (Bishop *et al.,* 1971a; Duesberg, 1969; Content and Duesberg, 1971; Lewandowski *et al.,* 1971; Skehel, 1971a); eight can be resolved upon extended polyacrylamide gel electrophoresis (W. J. Bean, personal communication). Associated with the RNA is the nucleocapsid protein (NP) as well as one to three minor proteins (P1, P2, and P3). Whether the RNA segments have some structural interrelationship with each other, as suggested by Li and Seto (1971), is not known, although the nucleocapsids can be visualized from virus or infected cell extracts by electron microscopy as free hairpin forms of various sizes (Kingsbury and Webster, 1969; Pons *et al.*, 1969; Compans *et al.*, 1972; Schulze, 1972).

It has been shown that the sizes of seven of the genome pieces of influenza A strains correspond to the sizes of seven of the virion proteins (P1, P2, P3, HA, NP, NA, and MP), while the eighth corresponds to an intracellular nonstructural NS protein (Compans and Choppin, Vol. 4, this series) (see Table 9). Apart from an unresolved question of whether there may be other intracellular nonstructural proteins, these similarities suggest that there is not a single multicistronic mRNA for all influenza A proteins. It is now known from analyses of polysome-associated, poly(A)-containing influenza mRNA that it is virus complementary in composition, and can program the synthesis of certain viral proteins *in vitro* (Etkind and Krug, 1974, 1975; Glass *et al.*, 1975; Kingsbury and Webster, 1973). Although in the studies of Etkind and Krug (1975) most but not all of the influenza proteins were demonstrated to be synthesized *in vitro* by RNA extracted from infected cell polysomes, it was shown that polysomal RNA was complementary in composition to 84% of the viral genome. It is believed that the synthesis of influenza virus-complementary RNA is catalyzed by an RNA-directed RNA polymerase, so that if there is transcriptional control of influenza mRNA synthesis it must be reflected in the ability of the transcriptase(s) to accept the different segments of the genome as templates.

The presence of an RNA polymerase in influenza virus particles has been demonstrated (Chow and Simpson, 1971; Penhoet *et al.*, 1971; Skehel, 1971*a*). Most of the analyses of the *in vitro* or *in vivo* transcription processes have centered on certain laboratory-adapted strains of influenza A viruses (WS, WSN, and fowl plague virus). The basic characteristics and requirements for demonstrating RNA-directed RNA polymerase activities associated with preparations of influenza A strains or influenza B, strain Lee, have been determined in various laboratories; there are no published reports of analyses for enzyme associated with influenza C strains (Bishop *et al.*, 1971*a*; Chow and Simpson, 1971; Penhoet *et al.*, 1971; Scholtissek *et al.*, 1971; Skehel, 1971*b*).

3.3.1. *In Vitro* Transcription Analyses of Influenza A Strains

Both magnesium and manganese ions are required to obtain maximal *in vitro* activity of the virion polymerase (Bishop *et al.*,

1971a). Neither deoxyribonuclease, actinomycin D, rifampicin, nor α-amanitin inhibits the *in vitro* transcription, although phenanthroline derivatives, selenocystamine, and thiosemicarbazones are reported to be inhibitory—an observation commensurate with the presence of zinc ions in virus particles (Oxford, 1973a; Oxford and Perrin, 1974, 1975).

TABLE 9

Orthomyxovirus Influenza A, Strain WSN Virion Structural Components, Viral RNA, and mRNA Species[a]

Viral proteins		
Species	Mol. wt.	Approximate number per virion
P1	0.9×10^5	30
P2	0.8×10^5	30
P3	0.8×10^5	30
HA[b]	0.78×10^5	800
HA1	0.53×10^5	
HA2	0.3×10^5	
NP	0.60×10^5	700
NA	0.55×10^5	170
MP	0.26×10^5	2800

Viral RNA			
Species	Mol. wt.	Approximate coding capacity	Possible protein equivalent
1	1.15×10^6	1×10^5	
2	1.10×10^6	1×10^5	(P1, P2, P3)
3	1.05×10^6	1×10^5	
4	0.82×10^6	0.8×10^5	HA
5	0.70×10^6	0.7×10^5	NP
6	0.58×10^6	0.6×10^5	NA
7	0.35×10^6	0.3×10^5	MP
8	0.32×10^6	0.3×10^5	NS

Viral mRNA

Possibly equivalent in size to the viral RNA species, although this has not been proven for all or any species.

---

[a] Based on Compans and Choppin (Vol. 4, this series) and Bishop *et al.* (1971b and unpublished observations).

[b] HA is the uncleaved precursor of HA1 and HA2 (see Compans and Choppin, Vol. 4, this series).

[c] Protein NS is a nonvirion, intracellular protein found in virus-infected cells. The number of proteins per virion may vary with the host cell used to grow virus.

It has been suggested that different strains of influenza A virus possess different enzyme-specific activities (Chow and Simpson, 1971; Bean and Simpson, 1975). Although at face value this may appear to be true when expressed as the transcription rate per milligram of viral protein, it should be borne in mind that the number of inactive particles in a virus preparation will affect that observation, so that lower specific activities may not reflect inherent or inheritable enzyme differences. That influenza B strains possess a demonstrably different enzyme has been discussed in Section 2.3 (Scholtissek *et al.*, 1971). *In vitro* studies by Oxford (1973*b*) have led to similar conclusions.

Analyses of early reaction products templated by ^{3}H-labeled influenza A, strain WSN, in the presence of $[\alpha$-^{32}P]UTP have shown that by 10 min of incubation at 30°C all of the product RNA is associated with template RNA species of every size class (Bishop *et al.*, 1971*b*). Product synthesis was demonstrated to be linear through 2 h of incubation, and by the 2 h point most of the product was present in RNA size classes which migrated on polyacrylamide gels either (1) in complexes with template viral species which moved slower than free single-stranded viral RNA or (2) in complexes which migrated in the position where single-stranded viral RNA was recovered. Some template-free products were obtained, and these migrated faster than the viral RNA species. Melted reaction product nucleic acids were found, for the most part, to possess electrophoretic mobilities somewhat greater than that of the smallest viral RNA segments (i.e., they possessed sizes of from 3×10^5 to 1×10^4 daltons), indicating that the transcription process was incomplete. At the 2 h point, 60% of the total product RNA was shown to be ribonuclease resistant, although it could be converted into a ribonuclease-sensitive form by heat denaturation. Annealing experiments demonstrated that all of the influenza product RNA was complementary to the viral RNA. These annealing experiments, which were performed on the *total* mixture of ^{3}H template and ^{32}P product labeled reaction nucleic acids, also demonstrated that when all of the ^{32}P was rendered ribonuclease resistant only 14% of the ^{3}H label was converted into ribonuclease-resistant structures (Bishop *et al.*, 1971*b*). Similar experiments performed with influenza A, strain WS, have shown that for reaction product nucleic acids obtained from an 8 h incubation, 85% of the template was rendered ribonuclease resistant when the reaction products were self-annealed, indicating that at least 85% of the influenza genome was transcribed by the viral polymerase (Bishop *et al.*, 1972). Since under those annealing conditions some 70% of the product RNA was rendered ribonuclease resistant and possessed size ranges upon melting comparable to those of the viral segments, these results indicated that the transcription process was repetitive and essentially complete (D. H. L. Bishop, unpublished observations; Bishop *et al.*, 1972). Why so much of the product remains associated with the template is not known, although this observation raises the possibility that the *in vitro* process may lack some host factor which is required for mRNA release or efficient transcription.

Additional information that the virion polymerase transcribes most if not all of the viral segments has been obtained by partially resolving (Pons, 1971) the various nucleocapsids of influenza A, strain WS, and demonstrating that transcriptase enzyme activity is associated with all of the various nucleocapsid size classes (Bishop *et al.*, 1972). The various nucleocapsids were shown to possess the major virion protein NP and minor proteins P (probably P1, P2, and P3) but not the major viral matrix protein MP, the glycoproteins HA (HA1 and HA2), or NA. Whether the proteins P are the polymerase is not known (Table 9). In unpublished observations, we have been able to separate P proteins from RNA-NP complexes and concomitantly lose enzyme activity; however, we have yet to achieve reconstitution of the enzyme activity (D. H. L. Bishop, unpublished observations).

Evidence has been presented using polyethylene glycol–dextran T500 phase-separated nucleocapsids of influenza A viruses that the initiation of transcription involves pppGpCp . . . sequences (Hefti *et al.*, 1975). Since the 3′-nucleosides of the RNA segments of influenza A virus are reported to be . . . pU_{OH} (Lewandowski *et al.*, 1971), this result suggests that *in vitro* transcription does not start at the terminal nucleoside of the 3′ end of the template RNA. Where it does start or terminate is not known. On the assumptions that viral RNA *replication* does start at the 3′-terminal nucleoside and that these *in vitro* observations reflect what happens *in vivo*, it is possible that transcription and the initial phase of replication are not identical (as suggested for the rhabdovirus VSV Indiana; see below). The relationship of guanosine incorporation in fowl plague transcription assays (McGeoch and Kitron, 1975) to the initiation of transcription is, as discussed earlier, not known.

3.3.2. *In Vivo* Primary Transcription Analyses of Influenza A, Strain WSN

Primary transcription analyses of influenza A, strain WSN, infected cells have been undertaken (Bean and Simpson, 1973; Repik *et al.*, 1974*a*). It has been shown that primary transcription occurs in a linear fashion through 14 h in the presence of cycloheximide and/or puromycin (50–100 μg/ml) but not in the presence of actinomycin D (Bean and Simpson, 1973; D. H. L. Bishop and A. Flamand, unpublished observations; Repik *et al.*, 1974*a*).

Primary transcription is not inhibited in interferon or interferon inducer pretreated cells using chick embryo fibroblasts or mouse L cells, although the same materials totally inhibit secondary transcription (i.e., that which occurs in the presence of *de novo* protein synthesis, Repik *et al.*, 1974*a*). This result indicates that, as far as influenza is concerned, the action of the intracellular mediators induced by interferon is not at the level of RNA-directed virus-complementary RNA synthesis.

It is noteworthy that primary transcription in mouse L cells is on a par masswise (but possibly not in terms of genome complexity) with that observed in chick embryo fibroblasts, even though the L-cell infection is almost nonproductive, and secondary transcription is severely curtailed by comparison to that of the chick cell host (Bishop and Flamand, 1975; Repik *et al.*, 1974*a*). The reason for this cell tropism is not known but points to a requirement for some host organization or factors for productive influenza infections.

How much of the viral RNA is transcribed during primary transcription? Using [³H]nucleoside-labeled influenza A, strain WSN, virus, it appears that primary transcription is developed by only a low number of the virus particles which adsorb to a cell. This has been shown by the slight (3%) increase in the [³H]ribonuclease resistance of the viral RNA (before annealing) during the infection process (Bean and Simpson, 1973; Repik *et al.*, 1974*a*). Because of the low number of genomes participating in primary transcription, it has been difficult to obtain enough intracellular, virus-complementary RNA in cycloheximide-treated cells to determine by self-hybridization if all the genome is transcribed. A minimum figure of 35% has been obtained for influenza virus grown in chick embryo fibroblasts (Repik *et al.*, 1974*a*). However, this value was not a plateau value and can only be considered as a minimal estimate.

It has been claimed that primary transcription is blocked by treating cells with UV light, actinomycin D, α-amanitin, cordycepin, and high concentrations of mitomycin C (W. J. Bean, Ph.D. thesis). We have confirmed the actinomycin D effect. although it has been claimed that cordycepin inhibits primary transcription, studies with fowl plague virus have indicated that replication of this virus is not inhibited by cordycepin (Mahy *et al.*, 1973). It has been proposed that the effect of cordycepin may vary from cell type to cell type, as well as when the drug is added, or depending on what virus is used (Rochovansky and Pons, 1975). Cordycepin does not inhibit the cytoplasmic transcription processes of VSV, although it has been shown that it does inhibit host nucleolar RNA synthesis in HeLa and other cell types (Penman *et al.*, 1970). In view of the observation that the transcriptase activity *in vitro* is not inhibited by actinomycin D (Chow and Simpson, 1971), the *in vivo* results with that antibiotic are puzzling, if one *just* considers that the antibiotics are inhibitors of nucleic acid synthesis and processing.

3.3.3. The Putative Nuclear Involvement in Influenza Transcription

Where does primary transcription take place? Because of the relevance of the abovementioned antibiotics to cell nuclear functions and the observation that influenza virus does not productively infect enucleated cells (Follett *et al.*, 1974), the possibilities must be considered that primary transcription either involves DNA synthesis or requires a nuclear function (e.g., a cell protein or RNA) or occurs within the nuclear organization. Evidence has been recently obtained that one or other of these possibilities may well be the case. No viral proteins or virus-complementary RNA can be detected in enucleated BHK cells (Kelly *et al.*, 1974). Altogether, the evidence is mounting which suggests that influenza viruses have some nuclear involvement. Both viruslike RNA and NP and NS proteins have been isolated from nuclear preparations (Krug, 1972; Krug and Etkind, 1973; Etkind and Krug, 1974; Lazarowitz *et al.*, 1971; Taylor *et al.*, 1970).

Preparations of nuclei from infected cells have been shown by comparison with uninfected cells to contain an RNA-dependent RNA polymerase activity, which is insensitive *in vitro* to actinomycin D (Hastie and Mahy, 1973; Mahy *et al.*, 1975). The time course of appearance of this nuclear activity is important; studies have shown that it occurs earlier (although in lesser amounts) than a cytoplasmic enzyme activity which has also been isolated from infected cells and which is maximally active 5 h or later postinfection (Compans and Caliguiri, 1973; Hastie and Mahy, 1973; Scholtissek, 1969). Also, the nuclear enzyme activity appears to decrease from 3 h postinfection. Comparative analyses of known host nuclear and host cytoplasmic enzymes, in relation to the distribution of this RNA-dependent RNA polymerase activity, have given reasonably clear evidence that it is probably not due to cytoplasmic contamination (Hastie and Mahy, 1973). At least part of the product of the nuclear activity has been shown to be virus complementary, although whether viruslike RNA sequences are synthesized is not known.

As if this abundance of evidence for nuclear-related activities during influenza viral infections is not enough, it has been suggested that by comparison with uninfected cells there is, in addition to the RNA-dependent synthesis, a stimulation of nuclear DNA-dependent RNA polymerase activity in influenza infected cells (Mahy et al., 1972, 1975). This poses the question of what host function is being turned on. The nuclear DNA-dependent synthesis reaches a maximum about 90 min postinfection. Addition of α-amanitin (a known inhibitor of host DNA-dependent RNA polymerase II) to an infected cell culture inhibits the first but not the second phase of nuclear RNA synthesis. It has been suggested therefore that there are at least two operationally distinct nuclear phases of RNA synthesis: one host DNA dependent (early, peaking around 90 min) and the other virus dependent (later, peaking around 2–3 h). Secondary transcription starts before 1 h postinfection (Repik et al., 1974a). This means that primary transcription and the earliest round of replication also start before 1 h postinfection. The nuclear and cytoplasmic enzyme activities have to be considered in this light. Is primary transcription influenced by a product of host DNA-dependent RNA polymerase activity? The actinomycin D results would support this view.

It should also be mentioned that although actinomycin D is a complete and efficient inhibitor of the replication cycle when added early in an influenza infection time course (Barry, 1964; Rott et al., 1965; Pons, 1967; Gregoriades, 1970), it does not significantly inhibit virus yield if added after 2 h postinfection. However, it has been claimed that actinomycin D does inhibit virus-complementary RNA synthesis to some extent when added later. Whereas α-amanitin is a less efficient inhibitor (Mahy et al., 1972; Rott and Scholtissek, 1970), it apparently does not inhibit RNA synthesis at all when added later.

In recent experiments, Armstrong and Barry (1974, 1975) have obtained evidence to suggest that virus-induced RNA synthesis in vivo can be detected only in the nucleus, not in the cytoplasm. Their experiments involved pulse-labeling infected cells in the presence or absence of antibiotics followed by autoradiographic examination. Although these analyses may not be sensitive enough to detect the site of primary transcription, they suggest that both replication and secondary transcription may be nuclear events.

In relation to all these considerations, the glaring observation which needs an explanation is that primary transcription proceeds in puromycin- or cycloheximide-treated cells but not in the presence of actinomycin D (Bean and Simpson, 1973; Repik *et al.,* 1974*a*). If a cell protein is needed for developing the viral infection and its availability requires host cell DNA-directed mRNA synthesis (sensitive to actinomycin D and possibly α-amanitin, mitomycin C, ultraviolet light, and cordycepin), then why is influenza primary transcription not likewise inhibited by cycloheximide or puromycin? It could be suggested that a newly synthesized protein is not involved, and that whatever host RNA is synthesized, if related to primary transcription, it is involved as an RNA and not as a protein translational product. For instance, the host RNA could act as primer molecule for RNA synthesis or influence transcription, as demonstrated for Sendai by the polyanion stimulation of transcription. Although some transcription initiation has been demonstrated *in vitro* to involve *de novo* pppGpCp . . . initiations (Hefti *et al.,* 1975) or guanosine initiations (McGeoch and Kitron, 1975), it is possible that another mechanism of RNA synthesis exists involving RNA primers (as in the oncornaviruses) or that the putative host cell RNA species is involved in another phase of the transcription process.

Other explanations which could be advanced are that DNA synthesis from the influenza genome occurs *in vivo* and that this DNA synthesis is actinomycin D sensitive. Alternatively, it could be postulated that cycloheximide or puromycin treatment of a cell in arresting the ribosomal translation machinery thereby makes available a host protein which influences the influenza virus-complementary RNA synthesis by stabilizing or releasing the transcripts from the transcriptive complexes. This would promote repetitive transcription and give rise to the observed intracellular primary transcription in the presence of those drugs. In the absence of cycloheximide or puromycin, this theory would have to predict that there is no protein available and that the normal source of the host protein is by translation of newly synthesized, actinomycin D-sensitive messenger RNA formed in the nucleus by the DNA-dependent RNA polymerase II. Such host messenger RNA synthesis might normally be turned on by the association with the nucleus of an infecting virion component.

Another possible explanation of the actinomycin D effect relates to transport questions concerning the viral genome. Let us suppose that for primary transcription the infecting virion genome has to enter or associate with the cell nucleus. It has been shown that certain antibiotics, including those which inhibit influenza replication, cause some physical change in the nuclear structure—particularly the nucleolus (Schoefl, 1964; Busch and Smetana, 1970; Marinozzi and Fiume, 1971). If these physical changes involve a subsequent prohibition of the normal nuclear association of the infecting virion transcription complexes, then this might also be an explanation of the observations.

Another possibility which has been advanced by Scholtissek *et al.* (1969) is that the various inhibitors which are known to block viral infection (actinomycin D, mitomycin C, and ultraviolet light, to which one could add α-amanitin and for some cases cordycepin) do so by stimulating or liberating a nuclease which destroys the viral RNA (or RNA transcripts as they are being synthesized during primary transcription). Since these drugs do not inhibit VSV or certain paramyxoviruses, presumably the activity of this nuclease would be a local restricted one (e.g., unique to a nuclear location).

Clearly the interrelationships of virus-directed or host-mediated control of influenza RNA transcription and replication need further investigation to unravel some of these problems, especially in relation to understanding the host nuclear involvement and determining the actinomycin D effect on primary transcription.

3.4.   Paramyxovirus Viral Transcriptases

The genome of the paramyxoviruses consists of a single, negative-stranded 57 S RNA of molecular weight 5-6 $\times$ 10^6 associated with a nucleocapsid protein. Other RNA species frequently found in virus preparations are believed to be of host origin; however, complementary viral RNA species (including some of full length) have been obtained from purified virus preparations (Kolakofsky *et al.,* 1974; Robinson, 1970).

As in the case of VSV, the virus-complementary messenger RNA synthesized both *in vitro* and *in vivo* is smaller than the viral genome (Table 10) (Barry and Bukrinskaya, 1968; Blair and Robinson, 1968; Bratt and Robinson, 1967; Collins and Bratt, 1973; Kingsbury, 1966; Pridgen and Kingsbury, 1972; Weiss and Bratt, 1974). Since the viral genome is not segmented, this suggests that either messenger RNA is transcribed from the genome as a single molecule and then cleaved or it is transcribed in sections with particular initiation signals (and possibly particular termination signals) for transcription. Although most investigators tend to support the latter contention, no definitive evidence has been obtained to rule out one or the other possibility.

TABLE 10

Paramyxovirus Sendai Virion Structural Components, Viral RNA, and mRNA Species

	Viral proteins[a]		
Protein species	Mol. wt.	Approximate number per virion	
L	1.0×10^5	10	
P	0.74	250	
H,N	0.70	450	
F0	0.63×10^5	600	
(F)	(0.45×10^5)	950	
NP	0.60×10^5	2000	
M	0.40×10^5	2000	

	Viral RNA (5.4×10^6 mRNA species)[a]		
Species	Mol. wt.	Approximate coding capacity	Possible protein equivalents
1 28 S	1.6×10^6	1.6×10^5	L
2 18 S	8.7×10^5	0.9×10^5	P
3 18 S	8.1×10^5	0.8×10^5	HN
4 18 S	7.2×10^5	0.7×10^5	F
5 18 S	5.8×10^5	0.6×10^5	NP
6 18 S	4.1×10^5	0.4×10^5	M

[a] Information supplied by K. Glazier and D. Kingsbury (personal communication); see also Zaides *et al.* (1975). Basically similar data are available for NDV and SV5 (see Choppin and Compans, Vol. 4, this series; Hightower *et al.*, 1975). Protein molecules per virion, their size, and their number vary depending on the virus strain, its form, conditions of growth, etc. H,N is a hemagglutinin–neuraminidase glycoprotein; F0 (and its derivative F) is a glycoprotein believed to be responsible for fusion; NP is a nucleocapsid protein; M is an internal matrix or membrane protein; the functions of P and L are not known but may be related to the virion transcriptase (D. W. Kingsbury, personal communication).

3.4.1. *In Vitro* Analyses of the Transcription Process

Subviral complexes possessing transcriptase activity have been isolated from Sendai virus (Marx *et al.*, 1974) and NDV (Meager and Burke, 1973), and shown to consist principally of the viral RNA, a major nucleocapsid protein and minor amounts of a large viral protein (Marx *et al.*, 1974). Similar complexes have been obtained with SV5 virus preparations (Choppin and Compans, Vol. 4, this series). No reconstitution of enzyme or infectivity experiments have been performed, presumably because of the low specific enzyme activity of paramyxovirus virion transcriptases (Huang *et al.*, 1971; Robinson, 1971*a*; H. O. Stone *et al.*, 1971; Bernard and Northrop, 1974). The identification of the transcriptase polypeptides has not yet been achieved, although it is presumed that the large protein is a polymerase component (Table 10).

An issue discussed in Section 1.1.1d which has complicated *in vitro* product analysis experiments, and which has been apparent for preparations of paramyxoviruses of all types, relates to the RNA polymerase activity that is obtained in reaction mixtures lacking one or more of the precursor nucleoside triphosphates. Substantial residual RNA polymerase activity can be found in many paramyxovirus preparations when one or another triphosphate is omitted (H. O. Stone *et al.*, 1971; Bernard and Northrop, 1974). The nature and product of this reaction are not known, nor is it clear if this is due to a viral function, a contaminant cellular enzyme, or virion triphosphates.

Other interesting aspects of the *in vitro* transcription process of Sendai virus are (1) the stimulation of the enzyme activity by polyanions (Stone and Kingsbury, 1973) and (2) the inhibition by viral membrane protein or glycoproteins (Marx *et al.*, 1974). These aspects have been discussed in Sections 1.1.1h and 1.1.1i. Whether viral proteins or cellular polyanions have regulatory roles in intracellular transcription is not known.

In vitro transcription product size analyses for NDV have shown that the transcripts possess a size range comparable to that of the various NDV messenger RNA species isolated from infected cells and, like their *in vivo* counterparts, have 3′ sequences of polyadenosine nucleotides (Pridgen and Kingsbury, 1972; Weiss and Bratt, 1974). No poly(U) tracts are present in NDV viral RNA (Marshall and Gillespie, 1972), so that it appears that poly(A) addition is an untemplated function of some virion enzyme, possibly the polymerase itself or a cellular enzyme incorporated into virions.

It has been demonstrated that mRNA, extracted from infected cells, programs the synthesis of viral proteins *in vitro* (Kingsbury, 1973) and that the size ranges of these RNA species are approximately those expected for monocistronic messenger RNA species (Bratt *et al.*, 1975; Collins and Bratt, 1973).

Using the Beaudette C strain of NDV, Colonno and Stone (1975) demonstrated that the *in vitro* 18 S viral complementary transcripts synthesized in the presence of *S*-adenosyl-L-methionine (SAM) possess 5′ sequences with blocked and methylated components. Methylation was shown to occur only in the presence of RNA synthesis (i.e., none was obtained in the absence of magnesium or manganese ions, CTP, Triton N101, NaCl, or virus). The presence of SAM neither stimulated nor depressed the RNA transcription rate. In the presence of [³H]SAM, the kinetics of methylation and RNA synthesis were observed to be similar. Methylation could be inhibited 90–100% by the presence of 0.1–0.3 mM S-adenosyl-L-homocysteine in addition to [³H]SAM, although again the rate of overall RNA synthesis was not inhibited. These results indicate that NDV virions possess methylation enzyme(s) which is not an obligatory component of the virion transcriptase.

It has been concluded that the only site of methylation for NDV transcripts is at the 5′ terminus of the product RNA and that on average 1.6 methyl residues are incorporated per 18 S molecule (Colonno and Stone, 1975). The 5′ sequence is believed to be m⁷G⁵′ppp⁵′Gp(Py), possibly also ribose-methylated (Colonno and Stone, 1975; H. O. Stone, personal communication). Similar 5′ termini have been identified for NDV *in vivo* messenger RNA species (H. O. Stone, personal communication). It is not known if the methylation of NDV transcripts is obligatory for subsequent translation as shown for reovirus transcripts (see Shaktin *et al.*, Chap. 1, this volume).

Nothing has been published concerning the extent or proportional representation of the various *in vitro* transcripts for NDV or Sendai viruses (or other paramyxovirus).

With respect to the question of whether the viral mRNA synthesized *in vitro* is produced as a single unit which is then cleaved into messenger-size pieces, the evidence obtained so far does not exclude the possibility.

3.4.2. *In Vivo* Analyses of the Transcription Process

It is known that the mRNA isolated from polyribosomes extracted from NDV- or Sendai virus-infected cells is complementary to their respective viral genomes, and possesses 3′-poly(A) sequences. The extent of the genome that is represented therein has been shown to be of the order of at least 65–70% (Bratt and Robinson, 1967; Kaverin and Varick, 1974; Portner and Kingsbury, 1970; Pridgen and Kingsbury, 1972; Weiss and Bratt, 1974; Bratt *et al.*, 1975). In studies with Sendai virus-infected cell extracts, Roux and Kolakofsky (1975) claimed that there is a 33 S intracellular RNA species which is complementary to

40% of the viral genome and an 18 S set of RNA species complementary to the remaining 60% of the viral RNA. The 33 S component, sufficient to code for the viral L protein, is a minor cellular component by comparison to the 18 S mixture. This suggests that at least in the cell all of the genome is transcribed into viral complementary RNA and that there is some form of transcriptional control whereby some parts of the genome are transcribed more frequently than others. Such observations argue that these mRNA species do not represent cleavage products of a whole-length precursor (from which one would expect to obtain equivalent molar proportions of the cleavage products). This sort of evidence, however, is the only evidence to date for separate initiation and transcription of the various mRNA species.

Primary transcription for Sendai virus has been demonstrated (Robinson, 1971c) and shown to lead to the synthesis of both 18 S and 35 S RNA species but not virion 57 S species. The ability of paramyxoviruses to replicate in the presence of actinomycin D or other inhibitors of DNA synthesis, or in UV-irradiated cells, differentiates these viruses from the influenza viruses (for review, see Choppin and Compans, Vol. 4, this series).

3.4.3. An RNA-Dependent DNA Polymerase Activity Associated with a Mutant NDV Obtained from a Persistent Infection

Thacore and Youngner (1969, 1970) obtained from L cells originally infected with the Herts strain of NDV a persistently infected cell culture which (1) continuously produced low levels of a mutant NDV, (2) produced some interferon, and (3) was resistant to superinfection by VSV. The mutant NDV (NDV_{pi}) was shown to be more thermolabile than the original virus (NDV_0), and gave small plaques in chick embryo fibroblast cultures. It also was more sensitive to interferon inhibition than the original strain.

In 1973, Furman and Hallum reported that NDV_{pi} grown in 10-day-old embryonated eggs or chick embryo fibroblasts possessed an RNA-directed DNA polymerase activity which was not present in the original NDV_0 strain. The requirements for the reverse transcriptase assay were shown to be a nonionic detergent (Triton X100), magnesium ions, and all four deoxyribonucleoside triphosphates. The *in vitro* synthesis of DNA was totally abolished by inclusion of ribonuclease in the reaction mixture. It was shown that the product of the reaction was DNA on the basis of its sensitivity to deoxyribonuclease and density in cesium sulfate.

The reverse transcriptase activity of NDV_{pi} was demonstrated to be associated with particles in the virus preparation which possessed a chick erythrocyte hemagglutination property (i.e., like other strains of NDV) and a density equivalent to that of NDV (or other enveloped viruses). Hybridization studies revealed that the product made by the NDV_{pi} particles was capable of hybridizing to RNA extracted from either NDV_{pi} or NDV_0 virus particles, but not to VSV viral RNA. No investigation was undertaken to rigorously prove that the DNA would hybridize to purified 57 S viral RNA (as opposed to other viral species).

Although the possibility of stable heterozygotes with some form of passenger oncornavirus derived from the L cells could not be rigorously excluded, a search for the presence of RNA tumor-type antigens, as detectable by immunodiffusion tests, was negative (Furman and Hallum, 1973). These results, taken together, argued strongly for a reverse transcriptase associated with the NDV_{pi} particles and raise the intriguing question that possibly during a persistent infection a reverse transcriptase activity was endowed upon the NDV mutant. Whether an NDV DNA copy was present in infected cells (either L cells or chick cells) was not determined; however, Simpson and Iinuma (1975), using another paramyxovirus (respiratory syncytial virus), have obtained evidence for a DNA copy in persistently infected cultures. The DNA they isolated was capable of transfecting permissive cells, and this transfection was deoxyribonuclease sensitive but not ribonuclease sensitive.

3.5. Reoviridae (Diplornavirus) Transcriptases

Characteristically, the genomes of various Reoviridae (diplornaviruses) consist of 10–12 double-stranded RNA segments named, for convenience, "large" (L1, L2, etc.), "medium" (M1, M2, etc.), and "small" (S1, S2, etc.), and possessing size ranges varying from 0.3×10^6 to 2.8×10^6 daltons (see Table 11). Virus particles of some members (e.g., reovirus) but not others (e.g., bluetongue virus) also contain a substantial mass of short oligonucleotides of unknown function but possessing particular nucleotide sequences of various types (Joklik, Vol. 2, this series).

For reovirus, the ten double-stranded segments are in equimolar proportions and add up to a total RNA mass per virion of 1.5×10^7 daltons (i.e., 7.5×10^5 dalton equivalents of protein, Joklik, Vol. 2, this series). The reovirus particle is known to contain seven major polypeptides (outer shell: $\mu2$, $\sigma1$, and $\sigma3$; inner core: $\lambda1$, $\lambda2$, $\mu1$, and

TABLE 11

Reovirus Virion Structural Components, Viral RNA, and mRNA Species

Viral proteins[a]		
Protein species	Mol. wt.	Approximate number per virion
λ1	1.55×10^5	110
λ2	1.40×10^5	90
μ1	0.80×10^5	20
(μ2)	(0.72×10^5)	550
σ1	0.42×10^5	30
σ2	0.38×10^5	200
σ3	0.34×10^5	900
Viral double-stranded RNA species		
RNA species	Mol. wt.	Number per virion
L1	2.8×10^6	1
L2	2.7×10^6	1
L3	2.6×10^6	1
M1	1.6×10^6	1
M2	1.55×10^6	1
M3	1.4×10^6	1
S1	0.9×10^6	1
S2	0.8×10^6	1
S3	0.65×10^6	1
S4	0.6×10^6	1
Viral mRNA species		
RNA species	Mol. wt.	Possible virion protein equivalents
l1	1.4×10^6	
l2	1.35×10^6	(λ1, λ2)
l3	1.3×10^6	
m1	0.8×10^6	
m2	0.78×10^6	μ1 (μ2)
m3	0.7×10^6	
s1	0.45×10^6	
s2	0.4×10^6	
s3	0.23×10^6	σ1, σ2, σ3

[a] Information based on Joklik (Vol. 2, this series); see also Shatkin *et al.* [Chap. 1, this volume (Table 2)]; for the protein assignment to mRNA species, refer to those two reviews for possible assignments. Protein μ2 is a cleavage product of μ1.

$\sigma2$). Of these polypeptides, $\mu2$ is a cleavage product of $\mu1$ and exists in the form of two small polypeptide chains held together by disulfide bonds. From the known sizes of the ten individual genome segments of reovirus and their coding capacity, it is likely (Table 11) that six of the virion proteins are the primary gene products of six of the ten virion RNA species (i.e., not counting the $\mu2$ protein).

Various types of cores, lacking some or all of the outer shell polypeptides, can be obtained from preparations of certain reovirus strains by heat treatment or chymotrypsin digestion (see Section 1.1.2). Chymotrypsin, depending on the conditions, removes the $\sigma3$, $\mu2$, and $\sigma1$ polypeptides in sequence, leaving a chymotrypsin-resistant core. Given a source of nucleoside triphosphates, magnesium ions, or manganese ions (etc.), chymotrypsin-derived cores (i.e., virions lacking $\sigma3$ and at least some of the $\mu2$ polypeptides) possess the capacity of supporting the *in vitro* transcription of the viral genome (Joklik, 1972, and Vol. 2, this series). Such transcription results in the synthesis of single-stranded RNA which is complementary to only one strand of each duplex and which is also equivalent to the reovirus messenger RNA found in infected cells (Hay and Joklik, 1971; Watanabe *et al.*, 1968; Zweernik and Joklik, 1970).

It is not known which of the reovirus core polypeptides comprise the transcriptase. However, an elegant electron microscopic examination of cores involved in producing transcripts has shown that up to nine strands of product RNA can be simultaneously extruded from an active core, each from a separate port in the core particle (Fig. 10) (Gillies *et al.*, 1971). This observation suggests that not only are there structural sites for product extrusion but also there are several transcriptase enzyme polypeptides in the virion particle.

3.5.1. *In Vitro* Transcription Analyses of Reovirus and Other Diplornaviruses

Transcriptase enzyme activities have been demonstrated in bluetongue virus, cytoplasmic polyhedrosis virus, reovirus strains, and wound tumor viruses (Table 1). Other than that for the reovirus strains, the temperature optimum for three of those four virus types is between 25°C and 30°C. For reovirus, the maximum *in vitro* activity is obtained between 45°C and 55°C (Kapuler, 1970). The preferred divalent cation for the bluetongue viral transcriptase is reported to be manganese,

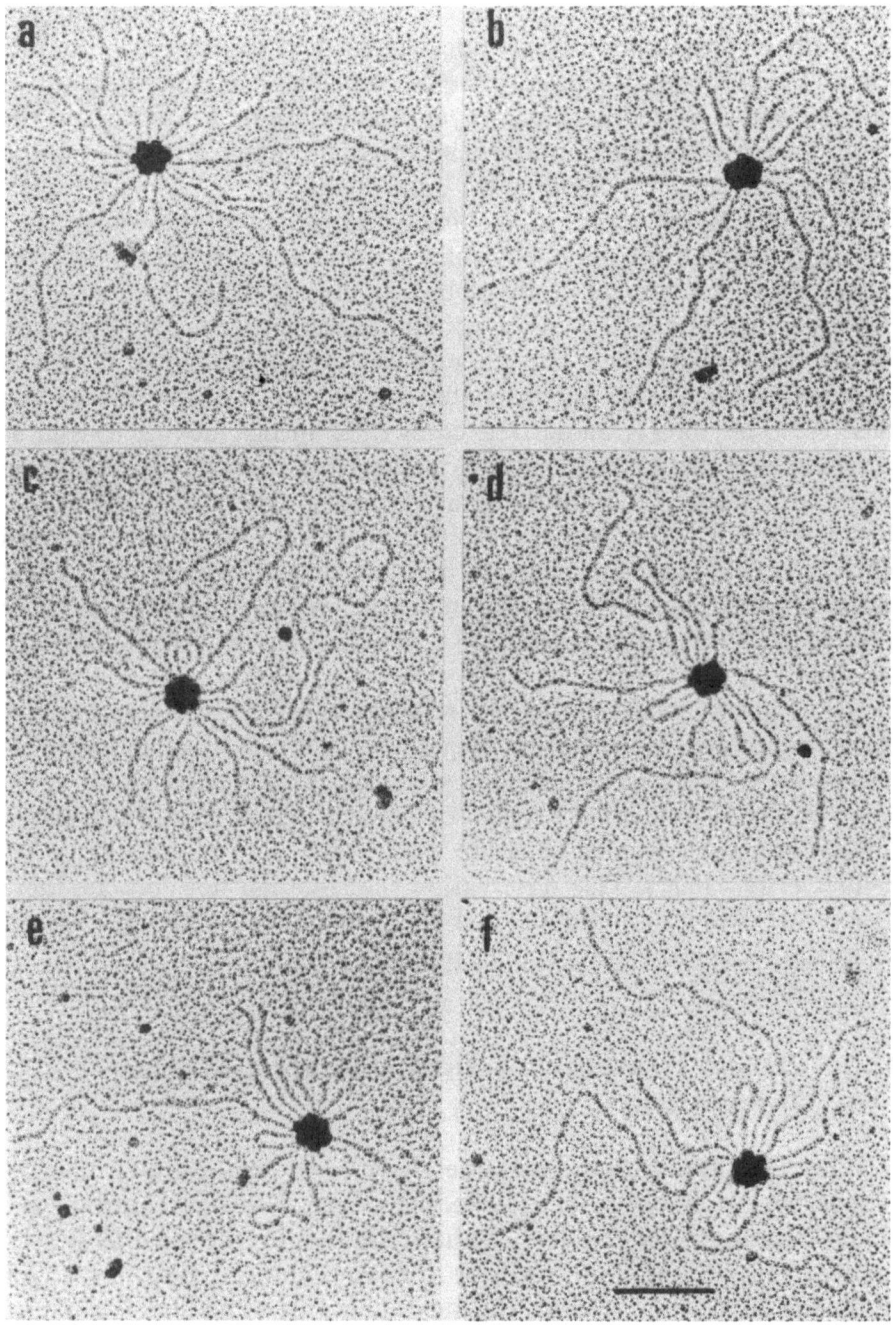

Fig. 10. Release of reovirus transcripts from reovirus cores is shown in an electron micrograph of transcription-active particles of reovirus. Courtesy of R. Bellamy.

while that for the other three viruses is magnesium. Bluetongue virus and reoviruses require activation by removal of certain outer virion polypeptides. Some preparations of cytoplasmic polyhedrosis viruses also have to be activated (e.g., by extraction with difluorodichloro-methane, Shimotohno and Miura, 1973); others do not (Donoghue and Hayashi, 1972; Lewandowski *et al.,* 1969). Wound tumor virus preparations apparently do not require activation (Black and Knight, 1970). As mentioned previously, capsomer removal is a required procedure for activating reovirus to a transcription functional form. The removal can, however, be blocked by anticapsid antibody treatment, which stabilizes the outer shell (Joklik, Vol. 2, this series).

It has been shown that the single-stranded reovirus transcripts synthesized *in vitro* possess (1) the same length as RNA obtained by denaturation of the double-stranded genome (Banerjee and Shatkin, 1970; Skehel and Joklik, 1969) and (2) the same length as reovirus mRNA isolated from infected cells (Hay and Joklik, 1971; Watanabe *et al.,* 1968; Zweernik and Joklik, 1970).

For reovirus the *in vitro* (or *in vivo* mRNA) transcripts can be sized into three major classes represented by decreasing order of size as large (l1, l2, l3), medium (m1, m2, m3), and small (s1, s2, s3, s4).

Under optimal conditions, the *in vitro* rates of synthesis of the individual RNA transcripts appear to be equivalent (Skehel and Joklik, 1969), therefore resulting in essentially similar masses of the various species but different molecular ratios due to the size differences between the largest and smallest species (Table 11). This observation indicates that chain initiation under these conditions is not a rate-limiting process. Changing the *in vitro* reaction conditions (e.g., lowering the magnesium concentration from 10 to 1 mM) not only reduces the rate of RNA synthesis but also changes the mass proportions of the various RNA species synthesized. For instance, the large (l) species are less well represented at low magnesium concentrations (Joklik, Vol. 2, this series). Presumably chain initiation can be rate limiting under particular circumstances, and this could be a way in which transcription regulation is achieved *in vivo* (see below). Reducing the ATP concentration to trace amounts in the reaction mixture and incubating at 31°C result in a curtailment of all RNA synthesis except certain small (s) RNA species, and this has been used to selectively obtain s RNA for nucleotide sequencing purposes (Nichols *et al.,* 1972). The 5′ sequence of the product obtained by such procedures was found to be (p)ppGCCAUUUUUGCUCUUCCAGACGUUG.... It is not known if this sequence represents the normal transcript 5′ sequence;

however, it is noteworthy that it does not contain any initiating AUG codons for protein synthesis.

An interesting characteristic of the *in vitro* (and *in vivo*) transcription process is its conservative nature (Levin *et al.,* 1970*a*); Shatkin and Rada, 1967; Skehel and Joklik, 1969). Neither viral strand is released as single-stranded RNA during the transcription process. It is not known how transcription (which involves some form of minus strand templated plus strand synthesis) occurs on a duplex RNA without displacement of the parental plus strand. Presumably, if plus strand displacement occurs, it is a local, transitory phenomenon which is reversed as an area of the template is transcribed.

The fact that only one strand of each duplexes is copied repetitively *in vitro* has been established by showing (1) that purified product cannot self-anneal and (2) that it can anneal to the melted viral duplexes and thereafter be resolved into ten duplexes by polyacrylamide gel electrophoresis (Skehel and Joklik, 1969).

Kapuler (1971) was able to use ethidium bromide to measure fluorometrically the synthesis of RNA *in vitro* due to an increased fluorescence obtained as the reaction was incubated. Since to fluoresce the drug has to bind to double-stranded RNA regions, this was interpreted to indicate that the product species (and in fact all size classes of product species) contain highly ordered structures even though they are linear single strands (Warrington *et al.,* 1973).

The rate of reovirus RNA synthesis has been estimated to involve the incorporation of between 10 and 60 nucleotides per second (Banerjee and Shatkin, 1970; Skehel and Joklik, 1969), with all classes of RNA product present by 1–2 min of reaction incubation. Reactions can be incubated productively for 2 days or more (Levin *et al.,* 1970*b*).

Furuichi (1974) has recently shown that addition of *S*-adenosyl-L-methionine (SAM) to reaction mixtures templated by cytoplasmic polyhedrosis virus (CPV) stimulates the endogenous transcription rate more than tenfold and that there is a methylation coupled transcription of CPV product RNA. The 5′-terminal nucleotide sequences of the various double-stranded RNA segments of CPV have been determined (Furuichi and Miura, 1975; Miura *et al.,* 1974) and shown to be m^{7}GpppAmpGpUp... for all the plus strands (mRNA) and ppGpGpCp ... for all the minus strands (see Fig. 11). The 3′ sequences are all ... pXpCpC$_{OH}$ or ... pXpCpU$_{OH}$, respectively (X = purine). The 5′-terminal sequences of CPV *in vitro* product RNA have also been shown to be m^{7}G$^{5'}$p$\overset{\alpha}{-}$pp$\overset{\beta\alpha}{^{5'}}$AmpGpUp (Fig. 11). In this structure, the m^{7}G is 7-methylguanosine, Amp is 2′-*O*-methyladenylic acid, and the origins of

Cytoplasmic polyhedrosis virus: m⁷GpppAmpGpUp..............pXpCpC_OH   plus strand.

                                       _HO_UpCpXp..............pCpGpGpp   minus strand.

                              mRNA: m⁷GpppAmpGpUp....................

Reovirus: viral   m⁷GpppGmpCpUp..............pApUpC_OH   plus strand.

                                       _HO_CpGpAp..............pUpApGpp   minus strand.

                              mRNA: m⁷GpppGmpCpUp...................

X = purine

Y = pyrimidine

m⁷GpppAmpGp.... =

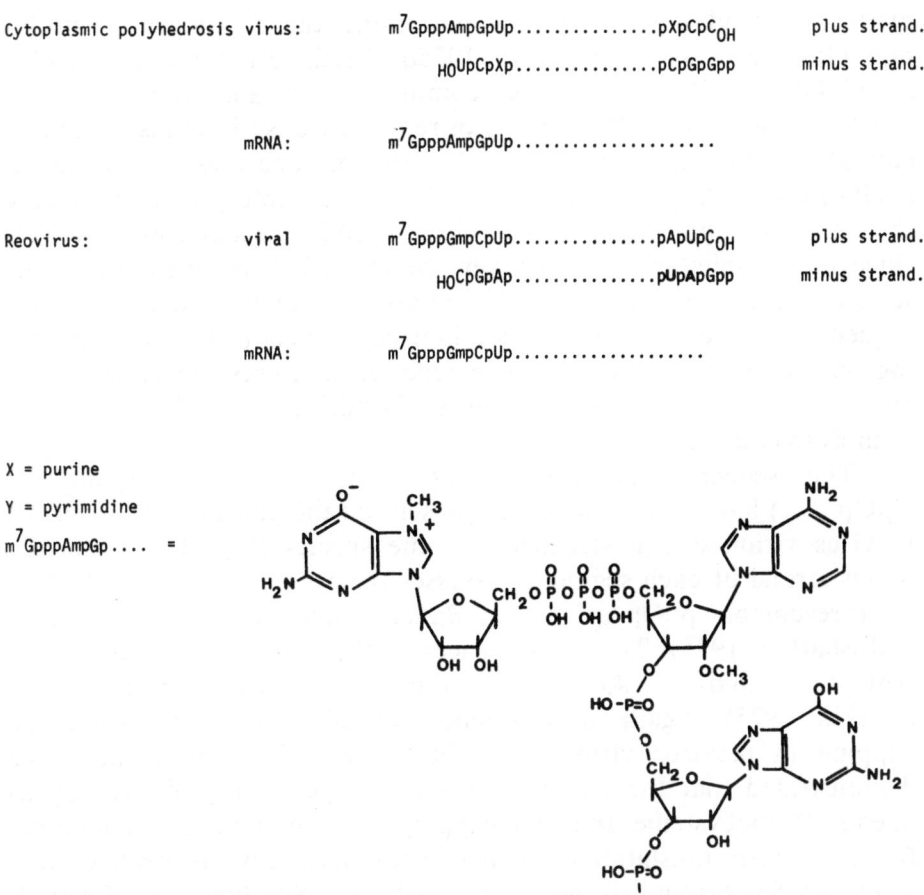

Fig. 11.   The 3′- and 5′-nucleoside sequences of cytoplasmic polyhedrosis virus and reovirus.

the phosphates in the bridge are indicated. The implications of these observations are that during transcription an inverted guanosine nucleotide (the cap) is added to the product strand and this guanosine is methylated at the 7 position, while the next nucleotide is methylated on the ribose 2′ position. The enzymes responsible for the capping and methylation activities of CPV are not known. Nor is it known if the m⁷G is added from a methylated precursor GTP or is methylated *in situ* (see discussion below).

It has been shown that reovirus cores (in the absence of SAM) synthesize product RNA species possessing the 5′ sequence ppGpCp ... (Banerjee *et al.*, 1971: Levin *et al.*, 1970a). Recent studies have shown that when given a source of SAM, reovirus cores also synthesize methylated and capped species possessing the 5′ sequence $m^7G^{5′}$-$ppp^{5′}GmpCp$ . . . (Furuichi *et al.*, 1975*a*). Methylation does not involve an m^7GTP (A. Shatkin, personal communication) and probably occurs *in situ*. Unlike with CPV, the rate of reovirus transcription is not markedly altered by the presence of SAM. Previous analyses by Levin *et al.* (1970*a*) using $[\beta,\gamma$-$^{32}P]GTP$ (and no SAM) were interpreted to indicate the additional presence of $\overset{\gamma\beta}{ppp^{5′}}Gp$ . . . product sequences; however, since no phosphatase analyses were performed, it is possible that the sequence they obtained was $G^{5′}p\overset{\beta}{pp^{5′}}Gp$ . . . , and in fact this latter sequence has been identified as a minor component (up to 30% depending on the reaction conditions) in reaction mixtures incubated in the absence of SAM (A. Shatkin and Y. Furuichi, personal communication; Both *et al.*, 1975*d*).

The sequence $XpppGmpCpUp$ . . . (presumably $m^7GpppGmp$-$CpUp$ . . .) has been shown to be present on the plus strands of all the reovirus virion double-stranded genome species (Fig. 11), while the minus strand of each segment possesses the 5′ sequence $ppGpPupPyp$ . . . (presumably $ppGpApUp$ . . . ; Banerjee and Shatkin, 1970; Chow and Shatkin, 1975). The 3′ sequences of the two strand types are . . . $ApUpC_{OH}$ and . . . $ApGpC_{OH}$, respectively (Muthukrishnan and Shatkin, 1975). Again the enzymes responsible for methylation and capping in reovirus virions are not known. However, it has been demonstrated that the methylation can be specifically blocked by *S*-adenosylhomocysteine and that capping and methylation are required for the *in vitro* translation of *in vitro* synthesized reovirus mRNA (Both *et al.*, 1975*b*; Muthukrishnan *et al.*, 1975; see Shatkin *et al.*, Chap. 1, this volume).

Studies with wound tumor virus have indicated that its *in vitro* RNA transcripts possess the 5′ sequence $m^7G^{5′}ppp^{5′}AmpXp$ . . . (A. Banerjee, personal communication).

Other than the transcriptase enzyme activity, it has been shown that reovirus cores catalyze an exchange reaction between inorganic pyrophosphate and all four ribonucleoside triphosphates, or just GTP, e.g., $[\beta,\gamma$-$^{32}P]GTP + PP_i \rightleftharpoons GTP + ^{32}PP_i$ (Wachsman *et al.*, 1970). Also, reovirus cores possess a nucleoside triphosphatase enzyme activity which is capable of catalyzing the reaction $[\gamma$-$^{32}P]GTP \rightarrow GDP + ^{32}P_i$, with ATP > GTP > CTP > UTP being the order of preferred substrates (Borsa *et al.*, 1970; Kapuler *et al.*, 1970). The virion polypeptides (or their origins, i.e., viral or cellular) responsible for these enzyme activities are not known. How these reactions relate to the initiation of transcription (see Fig. 12) or the transcriptase polypeptide(s) is also not known.

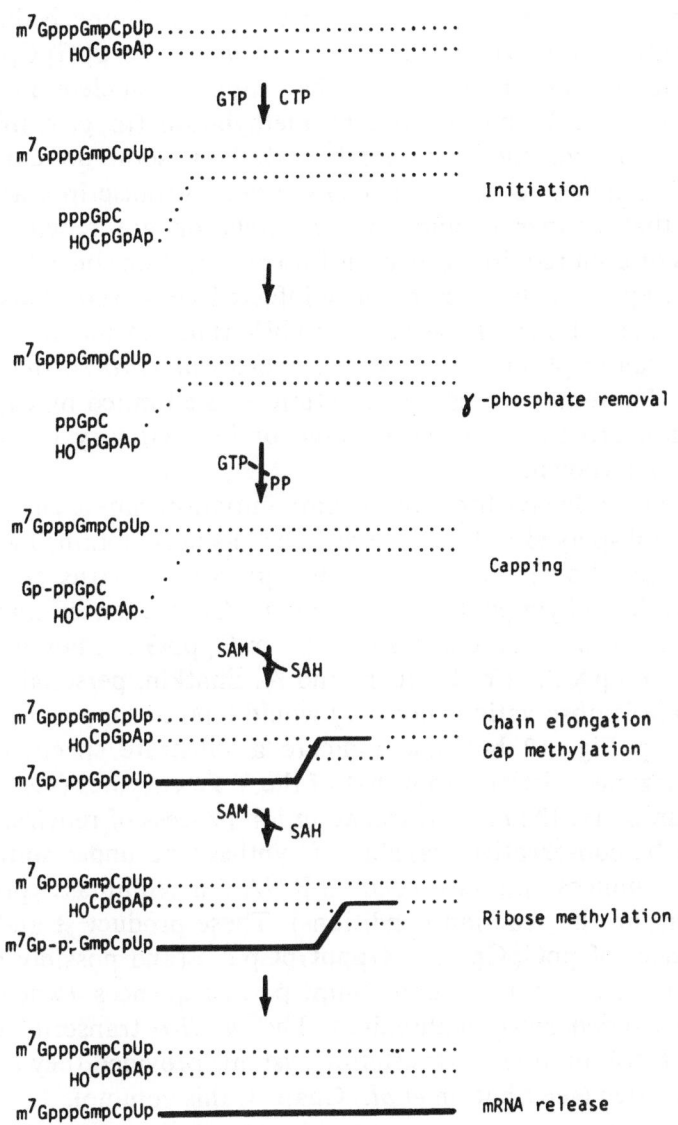

Fig. 12. A model for the sequence of transcription and 5′ modification of reovirus mRNA transcripts. In this model, the transcription process is indicated as being conservative and involving a product transiently possessing a 5′-triphosphate (not definitely identified). Also, capping is represented as occurring prior to transcription elongation (Y. Furuichi, S. Muthukrishnan, J. Tomasz, and A. Shatkin, personal communication). Capping and modification of released transcripts are alternative possibilities (see text). Scheme of capping and modification courtesy of A. Shatkin.

In terms of possible transcription initiation mechanisms, if the *in vitro* transcription process involves an initial synthesis of product pppGpCp . . . sequences (not yet proven), then it is conceivable that the triphosphatase catalyzes their conversion *in situ* to ppGpCp . . . , termini suitable for addition of the capping nucleotide (to give $G^{5'}ppp^{5'}GpCp . . .$) and subsequent methylation (to give $m^7G^{5'}ppp^{5'}$-GpCp . . .). If so, then one might ask how soon do capping and methylation have to occur after the initiation of transcription? It is conceivable that neither capping nor methylation are reovirus-specified functions or even required processes for transcription since the necessary enzymes appear to be present in uninfected cells (see Shatkin *et al.,* Chap. 1, this volume). However, for CPV (but not reovirus) methylation and transcription are coupled processes *in vitro* so that this suggests that for some viruses transcription is rate-limited by capping and methylation processes. At what stage or in what way the process is limited is not known.

Various pathways for transcription initiation can be envisaged, one of which is displayed in Fig. 12. Evidence has been obtained which indicates (1) that ^{32}ppGpCp dinucleotides given to reovirus cores can be capped and methylated to give Gp^{32}ppGpCp sequences and (2) that reovirus cores convert GpppGp . . . to m^7GpppG . . . but not GpppA . . . to m^2GpppA . . . (Y. Furiuchi and A. Shatkin, personal communication). These observations not only would tend to support the pathway indicated in Fig. 12 but also indicate a substrate specificity of the reovirus enzymes distinct from that of the VSV enzymes (see later).

In summary, the *in vitro* transcription process of reovirus has been shown to be conservative, capable of synthesizing under some but not all circumstances plus-stranded mRNA in equimass proportions (depending on the reaction conditions). These product strands possess 5′ sequences of ppGpCp . . . , GpppGpCp . . . (and possibly pppGpCp . . .) sequences, or $m^7G^{5'}ppp^{5'}GmpCp . . .$ sequences (when given a source of *S*-adenosyl-L-methionine). The *in vitro* transcript can function as mRNA in *in vitro* translation systems provided they are capped and methylated (see Shatkin *et al.,* Chap. 1, this volume).

3.5.2. *In Vivo* Transcription Analyses of Reovirus and Other Diplornaviruses

Following absorption, reovirions are converted by lysozomal enzymes to subviral particles (SVP) by 1 h postinfection, and this process involves the removal of the polypeptide $\sigma 3$ and a 8000 dalton fragment of $\mu 2$ (Chang and Zweerink, 1971); Silverstein *et al.*, 1970, 1972). The SVP particles exhibit transcriptase activity *in vivo* or when isolated from infected cells *in vitro*, and are responsible for the synthesis of early reovirus mRNA (Levin *et al.*, 1971, Watanabe *et al.*, 1967a,b). Reovirus mRNA is principally found in infected cells in association with polysomes.

By hybridizing pulse-labeled intracellular mRNA (extracted from cells 2 h postinfection or later) to unlabeled double-stranded RNA, then separating the hybrids by gel electrophoresis and determining the radioactivity in each hybrid (Watanabe *et al.*, 1968; Zweerink and Joklik, 1970), it has been shown that the *in vivo* mRNA synthesis is neither in equimass nor in equimolar proportions (Table 11). This is in contrast to certain *in vitro* results described above. Although Zweerink and Joklik (1970) found that all transcripts were present in similar proportions throughout the infection cycle (Table 11), Watanabe *et al.* (1968) found that only l3, m3, s3, and s4 mRNA species were found early (or when cycloheximide was used to inhibit protein synthesis) but that all ten mRNA species were found late in a productive infection. Similar results have been obtained by Shatkin and LaFiandra (1972) using chymotrypsin-treated virions (i.e., all ten transcripts are synthesized in productive infections but predominantly only s3, s4, and m3 in cycloheximide-treated cells).

It is not known if the particular mRNA species seen in cycloheximide-treated cells are artifacts of the drug treatment or a true expression of early transcriptional control. If the latter, then one could reasonably postulate that some cellular protein structure or local condition suppresses the transcription of some of the reovirus segments and that a newly synthesized viral or cell protein is needed to release that inhibition. Although the *in vivo* subviral particle is not exactly comparable to the *in vitro* particle, it has been shown that recovered subviral *in vivo* particles synthesize all ten mRNA species *in vitro*.

Recent analyses have shown that the *in vivo* reovirus mRNA species, like their *in vitro* counterparts, also possess capped and methylated 5′ sequences (Furuichi *et al.*, 1975b). No 3′-polyadenylic sequences have been identified on reovirus mRNA species (Stoltzfus *et al.*, 1973).

The infection process of reovirus, involving different phases of mRNA transcription (i.e., early primary transcription and late secondary transcription—which are dependent on *de novo* protein synthesis and replication), has been reviewed by Joklik (Vol. 2, this series).

Although it is believed that plus strand mRNA species serve as the templates for minus strand synthesis, it is not known if or how the enzyme(s) responsible for synthesizing minus strand RNA is related to the virion transcriptase. The conservative natures of the transcription and replication cycles are noteworthy features of reovirus infections, and it has been reported that there is a release of the parental infecting virion together with progeny virions late in an infective time course (Joklik, Vol. 2, this series). For a comprehensive review and discussion of these and other aspects of the infection process of reovirus and other diplornaviruses, the reader is referred to Joklik's chapter in Volume 2 of *Comprehensive Virology*.

3.6. Rhabdovirus Transcriptases

It has been demonstrated that rhabdoviruses are negative-stranded viruses which possess a virion-associated RNA-directed RNA polymerase.(Baltimore *et al.,* 1970). Certain characteristics of rhabdovirus *in vivo* or *in vitro* transcription processes have been discussed in Section 1 and 2; others will be described below.

For discussions of the basic structural features of rhabdoviruses, the reader is referred to reviews by Wagner (Vol. 4, this series) and Bishop and Smith (1976). The bullet-shaped enveloped viruses, possessing an outer glycoprotein and envelope-associated membrane protein(s), contain a helical nucleocapsid core which consists of an RNA molecule in association with a nucleoprotein and other proteins. For VSV Indiana there are three major virion proteins, i.e., a single glycoprotein, G, a single membrane protein, M, and a nucleoprotein, N (Table 12). Other rhabdoviruses may possess one or two membrane proteins. There is no evidence to date which suggests that the VSV Indiana G, M, or envelope components affect the virion-associated transcriptase function either *in vitro* (Bishop and Roy, 1972; Emerson and Wagner, 1972) or *in vivo* (see Section 1.1.1h).

The nucleocapsids of several rhabdoviruses, e.g., VSV Indiana, VSV New Jersey, Cocal, Chandipura, and Piry, contain not only RNA and N protein but also (Table 12) small amounts of a large protein, L, and a phosphoprotein, NS (Obijeski *et al.,* 1974; Sokol *et al.,* 1974; Wagner *et al.,* 1972). Rabies appears to possess some phosphorylated N protein as well as an L protein but no separate NS protein (Sokol and Clark, 1973; Sokol and Koprowski, 1975).

Although for VSV Indiana the known structural polypeptides correspond to almost all the genetic information of the virus (Table 12), other proteins and various enzyme activities have been observed to be present in rhabdovirus preparations (see later). In our laboratory, we consistently find in VSV Indiana and other rhabdovirus preparations

TABLE 12

Structural Components of the Rhabdovirus Vesicular Stomatitis Virus (VSV) Indiana Serotype

Viral proteins[a]		
Protein species	Mol. wt.	Approximate number per virion
L	1.5×10^5	52
G	0.67×10^5	1800
N	0.52×10^5	2000
NS	$0.4–0.25 \times 10^{5b}$	200
M	0.25×10^5	4000

Virus RNA[c]		
	3.8×10^6	

Virus mRNA species[d]			
Species	RNA size	Mol. wt.	Protein specified
1	28 S	1.7×10^6	L
2	17.5 S	0.70×10^6	G
3	15 S	0.55×10^6	N
4	12.5 S	0.28×10^6	M
5	12.5 S	0.28×10^6	NS

[a] Data from Bishop and Roy (1972) and Bishop and Smith (1976).
[b] NS protein is a phosphoprotein whose electrophoresis mobility varies with pH (Obijeski *et al.*, 1974). G protein is a glycoprotein; N is the nucleocapsid protein; M is a membrane or matrix protein; NS and L form the transcriptase (Wagner, Vol. 4, this series). Other virus protein species (e.g., A and B observed by Bishop and Roy, 1972) are regularly observed in association with nucleocapsids of virus preparations, but may be of host origin (see text).
[c] Data from Repik and Bishop (1973).
[d] Data from Rose and Knipe (1975) and Both *et al.* (1975*b,c* and unpublished observations).

minor but distinct protein bands possessing an electrophoretic mobility slightly faster than that of the N protein. Such proteins, A and B, described by Bishop and Roy (1972), are associated with the nucleocapsid when viral cores are made (i.e., lacking lipid, G, and M proteins). Other investigators have observed similar bands (J. F. Obijeski, personal communication). We do not know what function these protein species have or if they are derived from other viral structural proteins or are host proteins packaged during maturation.

The genome of rhabdoviruses consists of a linear, single-stranded RNA possessing a size of 3.8×10^6 daltons (Repik and Bishop, 1973). Associated with this RNA is about 100×10^6 daltons of N protein (Bishop and Roy, 1972). The 5′-terminal sequence of VSV Indiana viral RNA has been shown to be pppApCpGp . . . (Hefti and Bishop, 1975a). It is neither capped nor methylated (Hefti and Bishop, 1975a,b). The ratio of N protein to RNA in the nucleocapsid is not only sufficient to protect the RNA from ribonuclease digestion but also even prevents the removal of the 5′ terminal pppApCp sequence by pancreatic ribonuclease treatment (Hefti and Bishop, 1975b). Therefore, it can be concluded that the viral N protein completely covers the RNA and, from the size of the whole nucleocapsid, probably maintains it in an extended form (Wagner, Vol. 4, this series). How the RNA–N protein complex is synthesized or how the N protein interacts with the RNA is not known.

For rhabdoviruses, like other viruses, the productive infection of a cell involves the synthesis of virus-specified products in sufficient quantities, location, and timeliness to execute the necessary procedures to produce infectious progeny. Following uncoating, the next critical step to a successful infection is the synthesis of viral messenger RNA (plus strands) by transcription of the virion genome (minus strand). How and in what form these mRNA species are synthesized will be discussed below in relation to evidence gleaned from both *in vitro* and *in vivo* rhabdovirus studies. Since certain viral proteins are evidently required before others, a second set of questions which has to be asked is whether there is any regulation of viral messenger RNA synthesis *in vivo* (or as seen *in vitro*) to give unequal amounts of the various mRNA species at one time or another so that through translation certain proteins species are available in greater amounts than others. Evidence pertaining to the answer of some of these questions will be discussed below.

As mentioned previously, transcription from the infecting rhabdovirus genome is termed "primary transcription," and transcription of mRNA from progeny replicas of the infecting genome is termed "secondary transcription." In order to obtain progeny genome RNA species, it has been shown by molecular and genetic procedures that both protein synthesis and replication must occur (reviewed by Bishop and Flamand, 1975; Bishop and Smith, 1976; Wagner, Vol. 4, this series). Clearly, the processes of mRNA transcription and replication must somehow involve the same template. However, these processes are not necessarily conflicting, and this could be an important attribute of the infection process evolved by rhabdoviruses, as will be proposed later in this chapter. The interrelationships of primary transcription, replication, and secondary transcription will be discussed using evidence gleaned from both *in vivo* and *in vitro* analyses.

3.6.1. *In Vitro* Studies on the Viral Transcriptases of Rhabdoviruses

The properties of the *in vitro* transcription process of VSV Indiana serotype have been studied extensively subsequent to the initial demonstration by Baltimore *et al.* (1970) that VSV Indiana virions possess an RNA-directed RNA polymerase. Since the virus nucleocapsid is enclosed in a membrane, in order to assay the synthesis of RNA *in vitro* it is necessary to solubilize the structure to allow access of labeled nucleoside triphosphates to the enzyme–template complex as discussed in Section 1.1.1. The optimal reaction ingredients have been determined for various rhabdoviruses (see Section 1.1.1) including VSV Indiana, VSV New Jersey, Chandipura, Piry, Cocal, and Kern Canyon viruses, and, apart from KCV, each of the other five rhabdoviruses appears to have similar requirements to that of VSV Indiana (Aaslestad *et al.*, 1971; Chang *et al.*, 1974).

It is noteworthy that for all rhabdoviruses so far investigated except the fish rhabdoviruses, the optimal temperature for the *in vitro* transcription reaction is between 28°C and 32°C (Aaslestad *et al.*, 1971; D. H. L. Bishop, unpublished observations; Huang *et al.*, 1971; Roy *et al.*, 1975). This is in clear opposition to the results obtained *in vivo*, where the optimal temperature for primary transcription was shown for VSV Indiana to be 38°C (Flamand and Bishop, 1973). Why is the *in vitro* temperature optimum so low? The reason is not definitely known and could reflect a quirk of the *in vitro* reaction conditions. It has also been shown that the *in vitro* transcription process is not sustained at 37°C (Aaslestad *et al.*, 1971) and that it is also selective and incomplete (Bishop and Roy, 1971b). Within an infected cell, the primary transcription process is both sustained and complete at 38°C, or even 40°C (Flamand and Bishop, 1973). Some evidence has been presented which suggests that the observed *in vitro* inhibition may be due to the presence of a nonionic detergent (see Section 1.1.1b). However, it is known for other enzyme systems, such as that of the oncornavirus Rous sarcoma virus, that the *in vitro* RNA-directed DNA synthesis in the presence of Triton N101 is greater at 38°C than at 31°C (see Section 1.1.4a).

The optimal temperature for the *in vitro* assay of the virion transcriptase of pike fry rhabdovirus is around 20°C (Roy *et al.*, 1975), and that of SVCV or Egtved viruses around 18°C (P. Roy and D. H. L. Bishop, unpublished observations).

The endogenously templated virion transcriptase enzyme-specific activities (assayed in terms of pmoles of nucleoside monophosphate incorporated into product RNA per hour per milligram of viral protein, or per milligram of viral RNA) vary considerably from one rhabdo-virus to another. The highest rhabdovirus enzyme-specific activity we have observed was obtained with preparations of VSV Indiana, giving values of about 30,000 pmol GMP per hour per milligram of protein (Bishop *et al.*, 1971a). Other rhabdoviruses exhibit ten- to a hundred-fold lower *in vitro* specific activities: VSV New Jersey, Cocal, Chan-dipura, Piry (Chang *et al.*, 1974), KCV (Aaslested *et al.*, 1971), PFR (Roy *et al.*, 1975), SVCV, Egtved, and rabies viruses (P. Roy and D. H. L. Bishop, unpublished observations). However, their lower activities do not correlate with the observed *in vivo* primary transcription rates (P. Repik and D. H. L. Bishop, unpublished observations). As mentioned previously (Section 1.2), rabies virus preparations exhibit very little, if any, *in vitro* transcriptase activity (usually less than 5 pmol UMP incorporated per hour per milligram of protein), while the *in vivo* primary transcription rate for rabies is on a par with that obtained for influenza virus—which possesses an *in vitro* transcriptase specific activity of around 1000 pmol per milligram of protein per hour (Bishop and Flamand, 1975; D. H. L. Bishop, unpublished observations). It is not known if the rabies intracellular activity reflects an interaction with some specific factors in the cell which facilitate transcription.

3.6.2. Other Enzymes Associated with Rhabdovirus Preparations

Apart from the virion transcriptase, certain other enzymes have been consistently observed in preparations of all rhabdoviruses so far investigated. One of these enzyme activities is a nucleoside diphosphate kinase activity (nucleoside triphosphate phosphotransferase). This activity is activated by a nonionic detergent and is capable of catalyzing reactions such as $GTP + CDP \rightleftharpoons GDP + CTP$ (Roy and Bishop, 1971). It has been shown that, provided an excess of at least one nucleoside triphosphate is supplied, the *in vitro* transcription of RNA by VSV Indiana preparations can function in the presence of the three other nucleoside diphosphates following their conversion into triphosphates (see Section 1.1.1d).

Nucleoside triphosphatase enzyme activities (e.g., ATPase) are also present in all rhabdovirus preparations so far investigated (Roy and Bishop, 1971). These enzymes appear to be inhibited by Triton N101 and probably do not inhibit the *in vitro* transcription process unless limiting concentrations of triphosphates are present. Whether they are involved in the transcription initiation process is not known.

The presence of a protein kinase enzyme activity associated with the nucleocapsids of VSV Indiana and other rhabdoviruses has been demonstrated (Sokol and Clark, 1973; Strand and August, 1971). The preferred endogenous substance for the VSV protein kinase is the resident NS phosphoprotein, although the viral M protein and other exogenous substances can be phosphorylated. When VSV Indiana virion nucleocapsids are prepared from virus particles by nonionic detergent extraction, the protein kinase activity is not completely removed (Imblum and Wagner, 1974; Moyer and Summers, 1974). In fact, in unpublished studies we have performed, all of the endogenously templated protein kinase activity remains with the nucleocapsids even when 95% of the viral G, M, and lipids are removed by detergent and polyethylene glycol–dextran T500 phase separation (D. H. L. Bishop, unpublished observation). Parenthetically, it has been shown that such detergent extraction separates into the PEG phase the G, M, lipids, some L protein, as well as all the triphosphatase and phosphotransferase enzyme activities, while the nucleocapsid core (RNA-N, NS and L, protein kinase plus the A and B proteins) is removed in the dextran phase (Bishop and Roy, 1972; Hefti *et al.,* 1975). Which proteins constitute the protein kinase activity or these other enzyme activities is not known, although it has been shown that neither highly purified RNA-N complexes nor L or NS proteins possess protein kinase activities (Imblum and Wagner, 1974). Since the viral protein kinase specific activity varies depending on the host cell selected for virus growth, this is suggestive evidence that the enzyme may be a host protein (Imblum and Wagner, 1974).

3.6.3. Template and Enzyme Components of VSV Indiana and Other Rhabdovirus Transcriptases

The RNA-N complex serves as the template for the VSV Indiana virion transcriptase; naked VSV viral RNA will not function in lieu of that complex (D. H. L. Bishop and S. U. Emerson, unpublished observations; Bishop *et al.*, 1974; Emerson and Yu, 1975; Emerson and Wagner, 1972, 1973). The viral N protein is not permanently removed from the viral RNA during transcription. It is not recovered in association with the product RNA species (Bishop and Roy, 1972).

As discussed in Section 2.2.1, S. U. Emerson and R. R. Wagner have conclusively shown in an elegant series of experiments that the VSV Indiana transcriptase consists of the viral L protein and NS phosphoprotein (Emerson and Yu, 1975; Emerson and Wagner, 1972, 1973; Imblum and Wagner, 1975). The number of transcriptase enzymes per virion are not known, although from the number of L and NS proteins per virion it appears that there are several (Bishop and Roy, 1972). Purified template (RNA-N) possesses no transcriptase activity and is not infectious *per se*. Purified enzyme preparations or enzyme components also exhibit no transcriptase activity, nor are they infectious *per se*. Suitable combinations of template and enzyme (RNA-N plus NS and L; but not RNA-N plus L or RNA-N plus NS) both exhibit transcriptase activity *in vitro* and are infectious (Bishop *et al.*, 1974; Emerson and Yu, 1975; S. Emerson and D. H. L. Bishop, unpublished observations). The molecular interrelationships of NS with L and the template RNA-N complex are not known.

It has been shown that enzyme preparations of VSV Indiana can transcribe the template components of Cocal virus (a rhabdovirus known by the criteria of serology, Federer *et al.*, 1967, and by RNA genome homology, Repik *et al.*, 1973, to be closely related to VSV Indiana), and render the Cocal template infectious (Table 6 and Fig. 8). Likewise, the reverse heterologous reconstitution of transcriptase and infectivity between Cocal enzyme and VSV Indiana template has been demonstrated (Bishop *et al.*, 1974). In contrast, VSV Indiana enzyme shows little if any ability to transcribe or render infectious the VSV New Jersey template (Table 6 and Fig. 8), even though VSV New Jersey is a distant cousin of VSV Indiana (again based on the criteria of serology, Federer *et al.*, 1967, or RNA homology, Repik *et al.*, 1973). The VSV New Jersey enzyme will transcribe and reconstitute infectious entities only with VSV New Jersey template (Table 6 and Fig. 8).

The results of these experiments suggest that the enzyme of one rhabdovirus is specific for its own template and will not accept a template of a distantly related (or presumably unrelated) virus. Whether this specificity is due to an RNA sequence or relates to the recognition processes of one or all of the transcriptase proteins for the template remains to be determined. However, it should be noted that these results provide the clearest proof that virion transcriptases are required functions for a productive infection by a rhabdovirus.

It is not known whether the transcriptase enzyme components of all rhabdoviruses are composed of L proteins and NS phosphoproteins. Although all rhabdoviruses so far investigated have been shown to possess L proteins, some appear to possess phosphorylated N proteins and others both phosphorylated N and other proteins of unknown function. The functional or evolutionary significance of these differences is not known (see review by Bishop and Smith, 1976).

3.6.4. Transcription Product Analyses for VSV Indiana

It has been shown that the *in vitro* transcription process for VSV Indiana is complete at 31°C under optimal conditions (Bishop, 1971). This has been demonstrated by showing that the product is complementary to the viral genome and on annealing can render the genome almost completely ribonuclease resistant (Bishop, 1971). No viruslike product sequences have been detected in *in vitro* transcription analyses (Aaslestad *et al.*, 1971). It has been shown that the *in vitro* transcription process is repetitive, sequential, and disproportionate in that some sequences are transcribed more frequently than others (Bishop and Roy, 1971*b*; Roy and Bishop, 1972). This has been demonstrated by hybridizing the product RNA either to VSV viral RNA or to VSV defective T-particle RNA (representing a unique one-third of the viral genome). By such studies it was shown that (1) much less than one-third of the product could hybridize to the defective particle RNA for any timepoint examined and (2) those product sequences which did hybridize to the T-particle RNA could only be detected late in a reaction time course and not at all in reactions conducted at 37°C (Bishop and Roy, 1971b). Product RNA species are, for the most part, smaller than the viral genome (Bishop and Roy, 1971*b*; Roy and Bishop, 1972).

3.6.5. Direction of Product Synthesis, Transcription Initiation

The direction of product RNA synthesis has been shown by [γ-^{32}P]ribonucleoside triphosphate incorporation studies and pulse-chase experiments to occur in a 5′ to 3′ mode (Roy and Bishop, 1973).

In early experiments, a variety of 5′ initiation sequences were identified among the *in vitro* transcription reaction products. These sequences were shown to possess either pppAp . . . or, to lesser extents, pppGp . . . 5′ nucleotides (Roy and Bishop, 1973; Chang *et al.*, 1974). No product initiations have been obtained commencing with pyrimidine nucleotides, and this has been documented for VSV Indiana, VSV New Jersey, Chandipura, Cocal, and Piry viruses (Chang *et al.*, 1974). Of the product RNA sequences initiated *in vitro* with purine nucleotides, four were partially sequenced for VSV Indiana, and gave sequences pppApCpGp . . . , pppApApPypXpGp . . . , pppGpCp . . . , and pppGpGpPyp . . . (Roy and Bishop, 1973; Chang *et al.*, 1974). What roles do these initiation sequences have in the process of *in vitro* mRNA transcription or viral RNA replication? Before proposing answers to these questions, let us first consider the evidence for specific posttranscriptional modification of VSV mRNA species.

The transcription process of VSV RNA can be delineated into several phases: (1) initiation, (2) 5′ modification involving capping and methylation, (3) chain elongation, and (4) release and polyadenylation of the 3′ terminus. The temporal relationship of 5′ modification to initiation or chain elongations is not known, nor is it definitively known if there are unique or multiple sites for mRNA initiation (see later).

The results obtained on transcription initiation with cores of VSV Indiana isolated by polyethylene glycol–dextran phase separation to remove the phosphotransferase from the transcriptase-active complexes (RNA–N protein, NS and L proteins) and utilizing [γ-^{32}P]ribonucleoside triphosphates (Roy and Bishop, 1973) indicated that initiation involved purine nucleotides (Fig. 13). The ^{32}P-labeled terminal nucleotides can be recovered from the whole product RNA by alkali hydrolysis or, for [γ-^{32}P]GTP-labeled product RNA, by ribonuclease T$_1$ digestion (Roy and Bishop, 1973; Hefti and Bishop, 1975*b*, 1976). The label in these termini is sensitive to alkaline phosphatase (Fig. 13), and is present on uncapped, unmodified sequences. When the [γ-^{32}P]ATP- or [γ-^{32}P]GTP-labeled product is digested by pancreatic ribonuclease and resolved by DEAE-cellulose column chromatography at pH 5.5, four distinguishable termini can be identified (Roy and Bishop, 1973). Nearest-neighbor analyses using [α-^{32}P]ribonucleoside triphosphates as well as analyses involving ribonuclease T$_1$ digestion have been used to characterize these termini as pppApCpGp . . . , pppApApCpXpGp . . . , pppGpAp . . . , pppGpCp . . . , and pppGpGpPyp . . . (Chang *et al.*, 1974; Hefti and Bishop, 1975*b*, 1976; Roy and Bishop, 1973). The proportions of the different termini usually obtained for VSV reactions incubated for extended periods (up to 8 h) are given in Table 13.

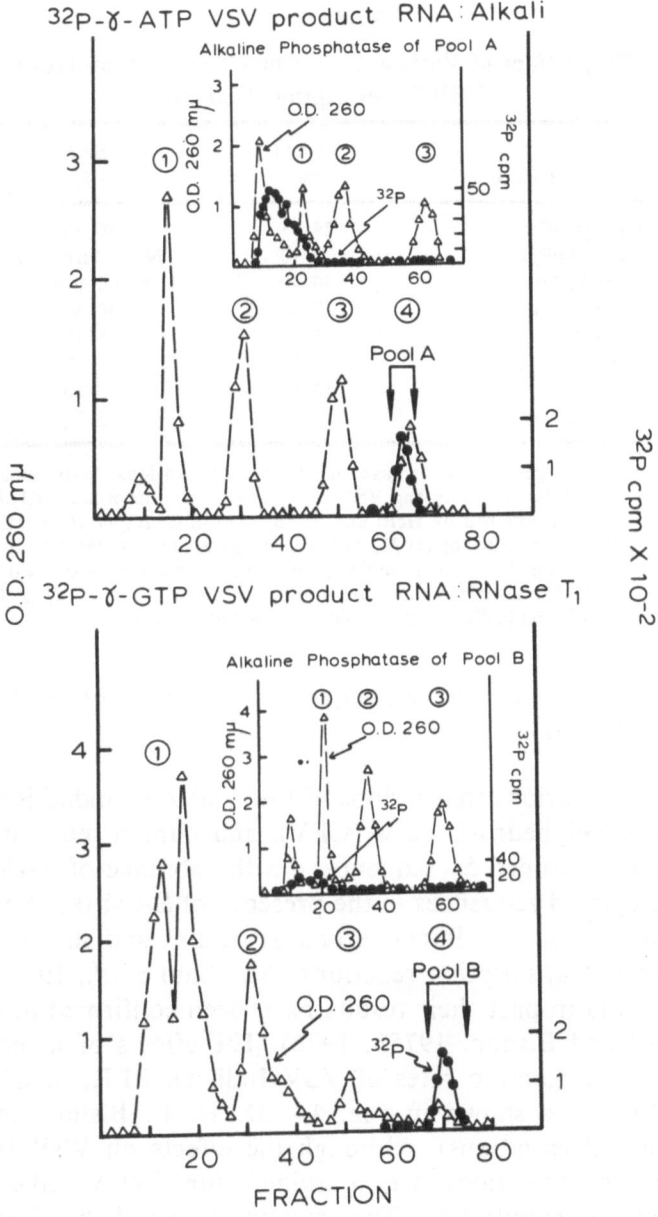

Fig. 13.   Viral complementary transcripts of the rhabdovirus VSV Indiana which possess 5′ sequences starting with pppAp . . . or pppGp. . . . VSV *in vitro* transcription product RNA labeled by either [γ-³²P]ATP or [γ-³²P]GTP was purified and digested with alkali or ribonuclease T₁ and the labeled nucleotides were resolved by DEAE-cellulose column chromatography with suitable marker nucleotides. The respective ³²pppAp and ³²pppGp nucleotides were recovered (pool A or B), digested with alkaline phosphatase, and shown to give in both cases labeled phosphate (inserts). Data of Hefti and Bishop (1975*b*).

TABLE 13

Proportions of Various 5′ Product Termini Observed in VSV Transcription Analyses[a]

Sequence	−SAM (%)	+SAM (%)
pppApCpGp . . .	25–35	20–30
pppApApPyp . . .	5–10	Not determined
GpppApAp . . .	30–40	Not detected
m⁷GpppAmpApCp . . .	Not detected	40–50
pppGpCp . . .	5–15	5–15
pppGpAp . . .	5–15	5–10
pppGpGp . . .	10–15	5–10
Unidentified		5–15

[a] The ranges of percentage occurrence of various 5′-terminal nucleotides found among VSV *in vitro* transcription product RNA are data obtained by Hefti and Bishop (1976), Chang *et al.* (1974), and Roy and Bishop (1973). Other investigators have so far detected only GpppAp . . . and m⁷GpppAmpAp . . . sequences (see text). Triphosphate-terminating nucleotides have been identified by both [γ-³²P]- and [α-³²P]ribonucleoside triphosphate precursors.

3.6.6. Modification of the 5′ Sequences of Transcripts by Capping and Methylation

The observations that virions of the double-stranded RNA viruses, cytoplasmic polyhedrosis virus (CPV), and human reovirus synthesize capped and uncapped 5′ sequences (in the absence of SAM) or additional methylated sequences in the presence of SAM (see Section 3.5.1) have been extended by Banerjee and associates and shown to apply to VSV Indiana transcription reactions (Abraham *et al.,* 1975; Rhodes *et al.,* 1974), and in part their results have been confirmed in our laboratory (Hefti and Bishop, 1975*b*, 1976). The effects of a methyl donor (SAM) on the reaction rates of VSV Indiana, PFR, or SVCV virion transcriptases are shown in Fig. 14 (D. H. L. Bishop and P. Roy, unpublished observations). Although the effects on VSV Indiana and PFR transcription rates are marginal, for SVCV (like CPV) the stimulation is significant. The results obtained by Banerjee and associates, or in our laboratories, indicate that the VSV Indiana transcript methylation involves a posttranscriptional modification of certain RNA species synthesized by the virion transcriptases (Abraham *et al.,* 1975; Hefti and Bishop, 1975*b*, 1976). Whether this modification is obligatory or occurs while product RNA is being completed or after it is released is not known, although the observations for SVCV (Fig.

14) that the reaction rate is stimulated by the presence of SAM argue, as in the case of CPV, for the former.

The termini analyzed in the experiments discussed in the previous section represent only those sequences which have retained their γ-phosphates. When transcription initiation studies are performed in the presence of $[\alpha\text{-}^{32}\text{P}]$ribonucleoside triphosphates ($\pm$SAM), a substantial number of capped termini can be demonstrated (Table 13) possessing the sequence GpppApA . . ., which in the presence of SAM are quantitatively converted into termini possessing the sequence m^{7}GpppAmp-ApCpXpGp . . . (Abraham *et al.*, 1974; Hefti and Bishop, 1975*b*, 1976; Rhodes *et al.*, 1974). In the analyses reported from Banerjee's laboratory, only capped ($-$SAM) or capped and methylated ($+$SAM) sequences were detected, whereas in our analyses both uncapped and capped ($-$SAM) or uncapped and capped and methylated ($+$SAM) sequences have been detected (Table 13). The explanation for the differences observed between these two laboratories is not immediately apparent, but as shown in Fig. 15 it is not because the $[\gamma\text{-}^{32}\text{P}]$ATP-labeled product is smaller than the mRNA species (12–28 S, Table 12). As will be discussed later, Rose (1975), has obtained evidence which indicates that VSV messenger-size RNA species isolated from infected cells possess a variety of capped and methylated 5′ sequences as well as a small proportion of uncapped termini possessing pppAp . . . or pppGp . . . sequences. One possible explanation for the difference between the results obtained by Banerjee and by us is that there are strain or growth differences for various VSV Indiana isolates and such

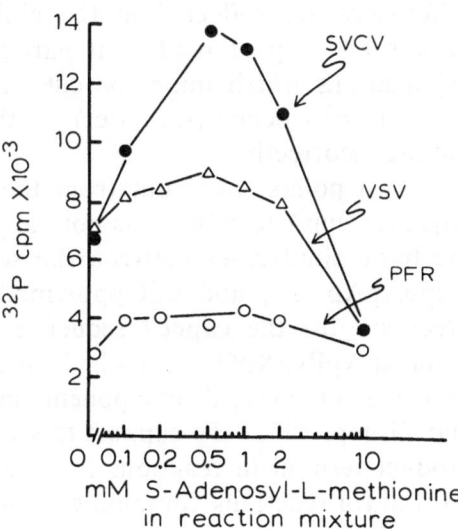

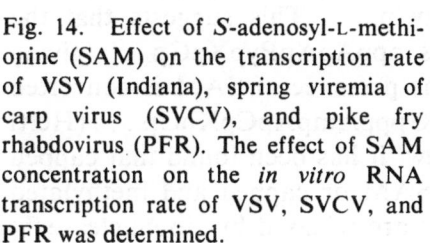

Fig. 14. Effect of *S*-adenosyl-L-methionine (SAM) on the transcription rate of VSV (Indiana), spring viremia of carp virus (SVCV), and pike fry rhabdovirus (PFR). The effect of SAM concentration on the *in vitro* RNA transcription rate of VSV, SVCV, and PFR was determined.

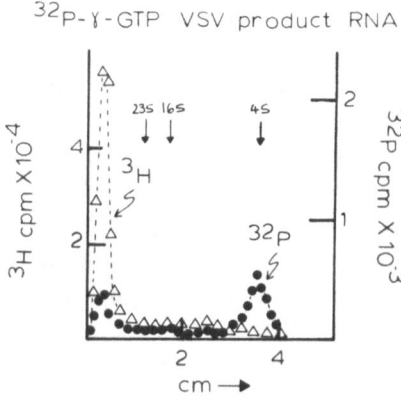

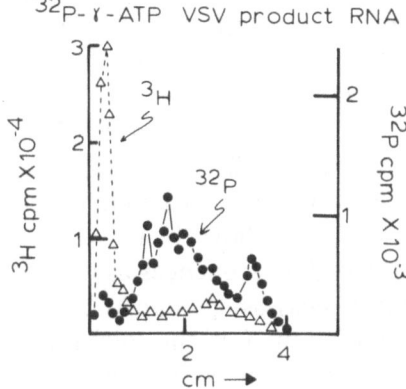

Fig. 15.   Size of VSV *in vitro* transcription product RNA labeled by [γ-³²P]GTP and [γ-³²P]ATP. Product RNAs were purified and resolved by polyacrylamide gel electrophoresis. A parallel electrophoresis of *E. coli* 23 S, 16 S, and 4 S RNA species was used to provide size markers.

differences are reflected in the ability of the virus transcriptase to initiate and cap or modify at particular sites on the template. Other explanations which might be entertained are that *in vitro* replication attempts can occur (see later) or that *in vitro* the transcriptase can initiate incorrectly.

Two points stand out from the results obtained: one is that the sequence pppApCpGp . . . is not capped or methylated and the other is the basic similarities between the sequences pppApApPypXpGp . . . , GpppApAp . . . , and m⁷GpppAmpApCp. . . . This suggests that the precursor for the capped sequence is pppApApPypXpGp . . . (giving GpppApApPypXpGp . . .) which in the presence of SAM is converted into the methylated component m⁷GpppAmpApCpXpGp . . . (Hefti and Bishop, 1976). To support this view, it has been found that capped product termini in reactions lacking SAM or capped and methylated termini (in reactions containing SAM) are released by ribonuclease T_1

digestion of suitably labeled reaction products and these can be recovered from DEAE chromatograms in positions two charges greater than those isolated by pancreatic ribonuclease digestion (D. H. L. Bishop, unpublished observation). Although this evidence is not in itself sufficient, since no pulse-chase experiments have been performed, it does suggest one plausible mechanism for initiation, capping, and methylation of VSV product termini as exemplified in Fig. 16. Other mechanisms will be discussed below.

Banerjee and associates have been able to resolve the different *in vitro* transcription product RNA species into distinct RNA species (28 S; 17.5 S; 14.5 S and 12 S RNA; see Table 12) and show that when synthesized in the presence of SAM the 17.5–12 S species possess the 5' sequence $m^7G^{5'}\overset{\alpha\beta}{pp}-\overset{\alpha}{p}AmpAp$ (Abraham *et al.*, 1975; Moyer *et al.*, 1975*b*).

If there are uncapped and/or capped 5' sequences of VSV product RNA formed *in vitro* and destined to become capped and methylated $m^7GpppAmpAp \ldots$ mRNA species, then what are the functions, if any, of the uncapped species which do not appear to become capped or methylated? The major one of these is the $pppApCpGp \ldots$ sequence; others detected involve $pppGp \ldots$ sequences. The relevance of these other observed sequences to the transcription process producing VSV mRNA is not known. Possibly they represent alternate initiations either for the 28 S mRNA or for the other mRNA species. Since they represent both variable and minor proportions of the various 5' sequences, these questions will be hard to resolve. It has been noted that a second capped and methylated sequence can be detected among VSV Indiana product sequences (Hefti *et al.*, 1976). This sequence is bigger than the $m^7GpppAmpAp \ldots$ but occurs in minor quantities. It is possible that it represents a $m^7GpppAmpAmpCp \ldots$ sequence (like some of the methylated and capped 18–12 S VSV mRNA species detected *in vivo*, Rose, 1975) or another sequence related to the guanosine initiations.

The occurrence *in vitro* of the $pppApCpGp \ldots$ sequence is of interest since it has been demonstrated among transcription products of several rhabdoviruses (Chang *et al.*, 1974; Roy and Bishop, 1973) and is the same sequence found at the 5' end of VSV viral RNA. Some evidence has been obtained in our laboratory which suggests that the 3' sequence of VSV viral RNA is . . . $pPypGpU_{OH}$, i.e., possibly the complement of the $pppApCpGp \ldots$ sequence. Do the *in vitro* $pppApCpGp$ . . . sequences represent initiation at the 3' terminus of the viral RNA, and if so how does this relate to mRNA transcription initiation? The

Scheme 1. Capping and methylation post-transcription.

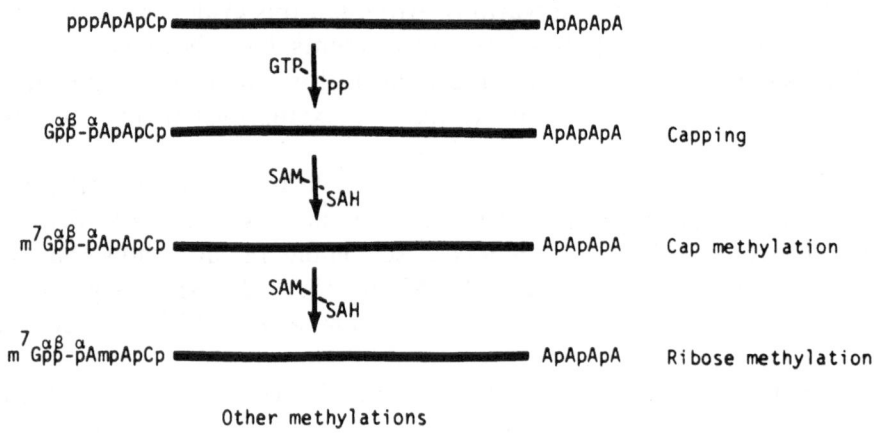

Other methylations

· ·

Scheme 2. Capping prior to chain elongation.

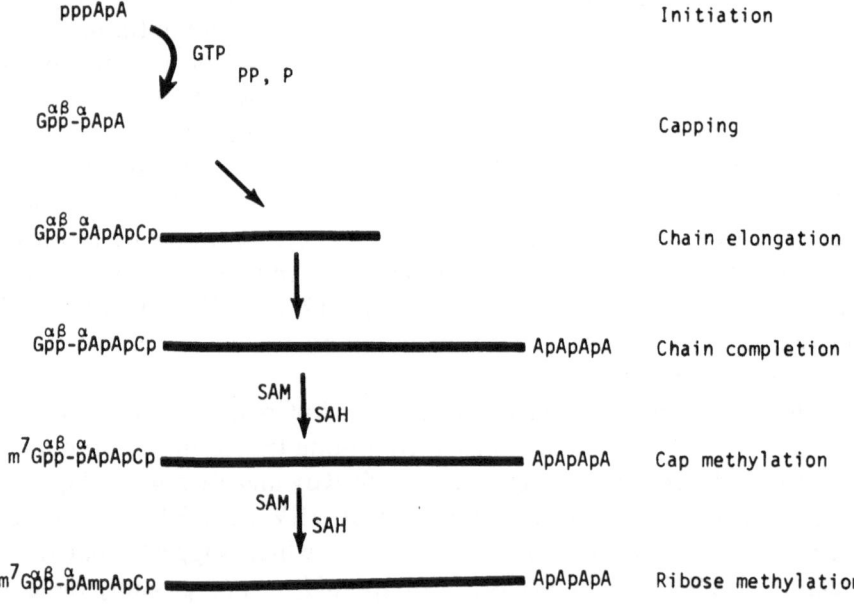

Fig. 16. Schemes for the 5′ modification of VSV viral complementary mRNA. In these schemes, the 5′ capping of VSV mRNA is indicated as occurring either after transcription or prior to chain elongation (alternate schemes are discussed in the text).

answers to these questions are not known at present, but two concepts can be considered. The first is that they are related to mRNA synthesis and the second is that they relate to *in vitro* replication attempts. These possibilities will be discussed in the models presented in Sections 3.6.13 and 3.6.14.

In summary, then, it appears that *in vitro* transcripts are initiated by purine nucleotides and of the various sequences initiated *in vitro* probably at least one type (pppApApCpXpGp . . .) is subsequently capped ($G^{5'}ppp^{5'}ApApCpXpGp$) and methylated in the presence of SAM ($m^7G^{5'}ppp^{5'}AmpApCpXpGp$ . . .) and represents the 5′ sequence found on most if not all VSV mRNA species. Another *in vitro* 5′ sequence does not appear to become capped or methylated (pppApCpGp . . .), and whether other 5′ sequences are capped or methylated is not known.

3.6.7. Transcription Termination, Polyadenylation

Are there particular stop sites for transcription termination? The answer to this question is not known, but possibly it is in the affirmative since mRNA species of particular sizes can be obtained (Moyer *et al.,* 1975*b*). What is clear is that polyadenylation of the mRNA product can occur at the 3′ end of the RNA, producing poly(A) stretches of up to 200 nucleotides (Banerjee and Rhodes, 1973; Moyer *et al.,* 1975*b*; Villareal and Holland, 1973). This is also true for the VSV *in vivo* mRNA species (Ehrenfeld, 1974; Ehrenfeld and Summers, 1972; Galet and Prevec, 1973; Moyer *et al.,* 1975*b*; Soria and Huang, 1973). Whether polyadenylation occurs on most or all product RNA strands is not certain. In our laboratories, we have observed less polyadenylation of VSV transcription products than that reported by others (D. H. L. Bishop, unpublished observations). This may be due to our reaction conditions or to the virus or cell strains employed. There is no decisive evidence to indicate whether polyadenylation is an attribute of the virion transcriptase or another enzyme present in minor amounts. No long stretches of poly(U) have been identified in the viral RNA sequences (Ehrenfeld, 1974; Ehrenfeld and Summers, 1972), so presumably the poly(A) elongation proceeds as a nontemplated function. The poly(A) stretches are heterogeneous in size, consisting of some 50–200 nucleotides (Moyer *et al., 1975b),* and this may account for the observed heterogeneity of the majority of the transcription product RNAs (12–18 S). It is conceivable that poly(A) synthesis is initiated from a short stretch of viral poly(U) (say, five to six nucleotides) and that, through slippage or nontemplated activity, polyadenylation is continued.

3.6.8.   Transcriptional Control as Evidenced by the *in Vitro* Analyses

The possibility that there may be some form of transcriptional control opreative during *in vitro* analyses stems from the observation that the *in vitro* product RNA does not equally represent all parts of the viral genome (Roy and Bishop, 1972). The mRNA obtained *in vivo* also is disproportionate, as shown by a variety of procedures including hybridization of total cellular RNA to VSV complete or defective RNA species or *in vivo* labeling procedures (Huang and Manders, 1972; Flamand and Bishop, 1974). Despite the obvious problem that the *in vitro* analyses may not represent *in vivo* conditions, it is worthwhile to speculate about potential control mechanisms which are suggested by these analyses.

One possible control point could be the transcription initiation sequences. If, as all evidence currently suggests, the individual messages are transcribed in a 5′ to 3′ direction involving separate 5′ initiations, then the preference of the transcriptase to pick one initiation point over another would result in selective transcription. Other than template sequence considerations, how could one initiation point be preferred over another? A first possibility would involve a role for the NS phosphoprotein, whereby at particular locations on the RNA–N protein complex the NS protein promotes initiation at greater efficiency than at other locations. The relative efficiencies may reflect the number of NS molecules per location or the degree of protein kinase-catalyzed NS phosphorylation. However, we do not know if all NS proteins are equally phosphorylated or if some are selectively phosphorylated during transcription.

Second, transcription control mechansims could reflect differences in the rate of initiation as opposed to the rate of RNA elongation. Alternatively, there could be differences in the elongation rate of different transcripts whereby longer transcripts are transcribed at a slower rate due to the intrinsic sequences involved, or by requiring greater or more sustained unwinding of the RNA–N protein complex.

Third, although stop or prevention signals in transcription have no supportive evidence at this time, it can be suggested that such signals may exist. If so, then they could function in controlling transcription by holding up the completion or release of some but not all transcripts. Again, this could be a possible role for the NS protein, or involve differences in the poly(A) fabrication machinery.

As a fourth possibility for transcriptional control, a sequential synthesis mechanism can be envisaged whereby there is an interdependence of transcript syntheses, transcript A being synthesized before transcript B synthesis is triggered, with the efficiency of triggering B being less than 100%, etc.

All these hypotheses assume that transcription of VSV mRNA involves the separate initiations of each mRNA species; if, however, the individual mRNA species are derived from a common precursor, then other forces of transcriptional control might be entertained.

3.6.9. The Size and Messenger RNA Capability of VSV *in Vitro* (and *in Vivo*) Transcripts

It has been shown that the VSV messenger RNA species isolated from infected cells consist of a viral complementary 28 S RNA and a mixture of species which are broadly recovered in the 11–18 S region of a sucrose gradient (Huang *et al.*, 1970; Mudd and Summers, 1970). The heterogeneities observed in the 11–18 S region are probably a function of differences in the possession and sizes of the poly(A) 3' sequences present on the individual mRNA species. If, however, VSV mRNA species possessing poly(A) 3' tails are selected from the infected cell nucleic acids through their ability to bind to oligo(dT)-cellulose columns, recovered, and separated on sucrose gradients, then it is possible to resolve the 11–18 S mixture of RNA species into three reasonably well-defined mRNA size classes (Both *et al.*, 1975c; Moyer *et al.*, 1975a). The same results can be obtained for *in vitro* synthesized VSV transcripts (Moyer *et al.*, 1975b). Alternative procedures which have been used to separate *in vivo* mRNA species involve recovering RNA species following their separation by gel electrophoresis run under denaturing conditions (Knipe *et al.*, 1975).

By using the *in vitro* (or *in vivo*) VSV mRNA species to prime wheat germ or cell-free Krebs II mouse ascites cell extracts, it has been conclusively demonstrated that the 17.5 S VSV mRNA codes for a nonglycosylated G protein, the 14.5 S for the VSV N protein and the 12 S species for both M and NS proteins (see Table 12) (Both *et al.*, 1975b,c; Knipe *et al.*, 1975). Evidence has been presented from nucleotide sequence complexity studies that the *in vivo* 12 S mRNA probably contains two species of RNA of similar size (Rose and Knipe, 1975). 28 S mRNA is usually recovered in small quantities (either from *in vitro* transcription analyses or from extracts of infected cells). There is only one report using *in vivo* derived material indicating that it codes for the L protein (Morrison *et al.*, 1974). The requirement for the

methylation of capped mRNA species before these species can be
translated has been demonstrated (Both *et al.*, 1975*b*) and is discussed
in Chapter 1 of this volume.

3.6.10. *In Vivo* Transcription Analyses of Rhabdoviruses

Within an infected cell, the initial synthesis of VSV messenger
RNA occurs by transcription from the parental genome (primary
transcription). This synthesis can be monitored using either cycloheximide or puromycin to inhibit cellular and viral protein synthesis,
thereby preventing the appearance and activity of new transcriptase or
replicase enzymes (Flamand and Bishop, 1973; Huang and Manders,
1972; Marcus *et al.*, 1971; Wertz and Levine, 1973).

The numbers of adsorbed virus particles which participate in
primary transcription has been estimated from the percentage of virus
RNA molecules which become ribonuclease resistant as a result of the
infection process (Flamand and Bishop, 1973). Values of about one out
of five are commonly observed for VSV, although occasional values of
one out of two have been obtained (Bishop and Flamand, 1975). As far
as can be ascertained, the inactive particles remain inactive throughout
an infection cycle even during high-multiplicity infections since they do
not appear capable of being "rescued" by active particles (Flamand
and Bishop, 1974). It does not seem likely that the inactive virions
represent particles which have not penetrated (since most of the
adsorbed viruses become uncoated, Bishop and Smith, 1976). Probably
their defect is in some structural or enzymatic attribute.

By hybridization experiments, the rate, completeness, and extent
of VSV Indiana primary transcription have been measured, as
described in Section 1.2. It has been demonstrated that primary
transcription can occur at a linear rate for up to 6 h postinfection
(Flamand and Bishop, 1974). The fact that the transcripts represent the
entire viral genome has been demonstrated by self-annealing of nucleic
acids extracted from cells infected by [³H]nucleoside-labeled virus,
which renders the [³H]RNA therein completely ribonuclease resistant
(Fig. 6). Flamand and Bishop (1973, 1974) showed that primary
transcription proceeds at a greater rate at 38°C than at 31°C (or
40°C). From the number of active viruses per cell and the rate of virus-
complementary RNA synthesis, they calculated that minimally one
genome mass equivalent of virus-complementary RNA was synthesized
every 90 s. If the 28 S viral messenger RNA represents half the

genome, and the other four messenger RNA species about 12% each, then this rate corresponds to a synthesis time of 45 s for 28 S RNA and about 6 to 18 s for the other species.

As far as primary transcription is concerned, it has been shown that neither actinomycin D, cycloheximide, puromycin, interferon, nor poly(rI):poly(rC) pretreatments affect primary transcription, although the latter four totally inhibit secondary transcription, presumably by inhibiting or preventing viral protein synthesis (Flamand and Bishop, 1973, 1974; Repik *et al.*, 1974*a*). The rate of primary transcription developed in interferon-treated cells was observed to be comparable to that found in cycloheximide- or interferon-plus-cycloheximide-treated cells. These results do not preclude the possibility that the transcripts are less stable in the interferon-treated cells because of induced nucleases or lack of proper processing or other factors. However, they do indicate that the transcriptase itself is not directly affected by interferon-induced products.

It has been shown that from 45 min to 1 h postinfection, the rate of virus-complementary VSV RNA synthesis dramatically increases, and, although this increase is not affected by treating the cells with actinomycin D, it is inhibited by cycloheximide, puromycin, interferon, or poly(rI):poly(rC) treatments. Neither secondary transcription nor viruslike RNA has been observed in cells infected with group IV or group I temperature-sensitive mutants grown at nonpermissive temperatures for virus production (reviewed by Bishop and Smith, 1976; Pringle, 1975). The conclusion to be reached from those results is that secondary transcription and the synthesis of viral RNA require the synthesis and expression of new viral gene products.

In an experiment which involved adding cycloheximide at various times postinfection and measuring the linear rates of RNA transcription established thereafter (by comparison with that obtained during primary transcription), the number of intracellular transcriptive intermediates has been estimated. Such estimates have given values of around 450 transcriptive intermediates per infected BHK-21 cell by 3–6 h postinfection (Flamand and Bishop, 1974). No increase in the number of intermediates was observed subsequent to 3 h postinfection, although progeny virus continued to be liberated.

The total amount of virus-complementary RNA produced per infected BHK cell has been calculated. Values of 2×10^4 genome mass equivalents per cell (i.e., about 1×10^5 individual mRNA species) have been obtained. Similar experiments involving a thirtyfold higher multiplicity of infection have indicated that by 3–6 h postinfection, equivalent amounts of viral complementary RNA are obtained to that observed for the lower multiplicity of infection, even though the initial rate of RNA synthesis was thirtyfold higher (Flamand and Bishop, 1974). These results suggested that there is some finite limit to the amount of virus-complementary RNA which can be synthesized in an infected cell and also some form of host-modulated regulation of transcription.

Comparative experiments employing the same batch of virus and different cell types (e.g., BHK-21, mouse L cells, and secondary chick embryo fibroblasts) have indicated that both the yield of infectious virus per cell and the accumulated amount of virus-complementary RNA during primary or secondary transcription are dependent on the host cell type. For VSV, the BHK cell line is one of the most permissive host cell types we have so far investigated. Which host cell factors influence this permissiveness is not known.

It is noteworthy that secondary transcription induced by VSV in BHK cells starts at a time when the first progeny virus and newly synthesized viruslike RNA can be detected. These observations strongly suggest that the templates for secondary transcription are the newly synthesized viruslike RNA species. These observations also suggest that once formed, even at such an early stage of the infection cycle (*viz.,* the first rounds of replication), the progeny viruslike RNA species have a possibility of becoming progenitors of infectious virus particles or templates for secondary transcription (Flamand and Bishop, 1974).

At later stages of the infection cycle when the number of transcriptive intermediates reaches a plateau, a majority of the viruslike RNA species give rise to progeny virus particles. Why? What interrelationships exist between primary transcription and replication, and between replication and secondary transcription? At the moment we do not know. Since the template for primary transcription must be the template for replication (there is no other, unless one proposes stitching together the messenger RNA species), what form of conversion from transcription to replication can be envisaged?

3.6.11. Evidence for the Intracellular Control of Transcription

Evidence suggesting that some form of transcription control is operative within the infected cell, at least late in an infection, comes from two sources. First, it has been shown by hybridization studies that there is a disproportionate representation (on a molar basis) of the various VSV messenger RNA species (Flamand and Bishop, 1974). Which species are involved is not known, although it is probable that there are fewer 28 S mRNA species than other species. The reason for this judgment has come from hybridization experiments using defective T-particle RNA (VSV-111) and viral RNA to determine the relative molar representation of the different mRNA sequences. On a molar basis, less than one-third of the total mRNA annealed to the T-particle RNA by comparison to that which annealed to the viral RNA (Flamand and Bishop, 1974). It has been shown that the standard T-particle RNA is probably equivalent to part of the 28 S messenger RNA (Leamnson and Reichmann, 1974). Furthermore, examination of the relative proportions of messenger RNA species obtained by labeling experiments has indicated that there is always less 28 S RNA than the other mRNA species (Mudd and Summers, 1970; Perlman and Huang, 1973; Rose and Knipe, 1975). It is not known it some of the other mRNA species in the 11–18 S size range are represented unequally.

In addition to the suggestions discussed previously (Section 3.4.8) of how transcription might be controlled, the possibility of an interrelationship between transcription and translation should not be overlooked. Are RNA transcripts picked by the translational machinery with an equal probability despite sequence, size, and conformational differences? If there is a selection (or selective processing procedure), then there could also be a feedback mechanism. Until these questions can be answered, we will not fully understand the transcriptional control mechanisms or their ramifications. Whether transcriptional control is operative during both primary and secondary transcription is not known.

3.6.12. Characterization of the VSV mRNA Species Formed *in Vivo*

The VSV mRNA species have been identified as a 28 S RNA which translates to give the viral L protein, a 17–18 S RNA which can be translated into G protein, a 14–15 S RNA which makes N protein, and a 11–12 S RNA which codes for both NS and M proteins (see Section 3.6.9). These mRNA species possess polyadenylated 3′ sequences, which facilitates their isolation through the use of oligo(dT)-cellulose columns. It is not known if all VSV mRNA species possess equal-size polyadenylated 3′ sequences.

In work from the laboratory of J. Rose (1975), it appears that most VSV Indiana mRNA species are capped and methylated, although some uncapped sequences were also found. In Banerjee's laboratory, only one capped sequence was identified (Moyer *et al.*, 1975*a*). Rose found that each size class of VSV Indiana messenger RNA contained a variety of 5′ sequences. The majority (65–70%) were either m⁷G⁵′ppp⁵′AmpAp . . . or m⁷G⁵′ppp⁵′mAmpAp . . . (where mA is a base-methylated adenosine). Of the other 5′ termini, 20% were in m⁷G⁵′ppp⁵′AmpmAmpCp . . . and m⁷G⁵′ppp⁵′mAmpmAmpCp . . . sequences. The remaining 10–15% of the terminal sequences possessed pppAp . . . or pppGp . . . nucleotides and were on mRNA size transcripts not associated with ribosomes. Clearly from these results it is evident that both ribose and base methylations are variable functions, both of which may be required before selection for translation by ribosomes. That adenosine and guanosine triphosphate 5′ termini were also found is an interesting observation and may tie in with some of the previously reported *in vitro* results.

3.6.13. A Summary and Model of VSV Transcription Processes

What is known about the *in vitro* (and *in vivo*) RNA transcription process for VSV Indiana is as follows:

1. VSV virions possess several copies of the RNA-instructed RNA polymerase enzyme per virion which are able to transcribe the whole genome (*in vitro* or *in vivo*) into viral complementary RNA.

2. The viral RNA–N protein complex is the template for the transcriptase (L plus NS). At best, transcription involves only a tempory lifting of N from the RNA.

3. The transcription process is highly repetitive, possibly selective with regard to which RNA species are transcribed most frequently, and possibly also sequential.

4. The direction of RNA transcription is 5′ to 3′ and *in vitro* involves the production of various uncapped sequences and one capped sequence (−SAM) or various uncapped sequences and at least one capped, methylated sequence (+SAM). The principal capped, methylated sequence is m⁷G⁵′ppp⁵′-AmpApCpXpGp. . . . The principal uncapped sequence is pppApCpGp. . . .

5. VSV mRNA transcripts are complementary to the viral genome, usually possess 3′ poly(A) tails (50–200 nucleotides)

and to function. as mRNA must be modified (to give m⁷GpppAmpAp . . . 5′ sequences or other methylated derivatives of that sequence).

A model for how transcription occurs is presented in diagrammatic form in Fig. 17. On the presumption that it involves the separate initiations of the different transcripts (mRNA), it can be suggested that the process possesses the following attributes:

1. There are a limited number of specific transcription initiation sites on the template complex.
2. There are specific transcription stop regions, e.g., short poly(U)sequences which trigger slippage, release, and polyadenylation of the 3′ end of the mRNA.
3. The N proteins are contiguous. They are hydrogen-bonded to each other and to the phosphate backbone of the viral RNA. Such N protein associations protect the 5′ end as well as internal regions of the RNA (and possibly the 3′ end) from nuclease digestion. This sequestering function of N protein is in addition to its role in maintaining the helical structure of the nucleocapsid in the virion, as well as its function in transcription and replication.
4. The function of the NS phosphoprotein is to lift the N protein off the RNA during product RNA synthesis by substituting its phosphate for the phosphate of the template RNA. This creates a bubble in the N protein framework and allows access of the transcriptase to the RNA. It is not necessary for this function that the NS molecules be integral parts of the transcriptase; they could just initiate a bubble by attaching to N protein at particular sites. Either way, this can constitute what can be termed the *bubble hypothesis* of RNA transcription.
5. Ribosomes can attach to transcripts after capping and methylation. This may occur either prior or subsequent to the completion of RNA transcription, depending on the efficiency of the capping and methylation processes.

As an alternative to the bubble hypothesis can be postulated a *groove hypothesis*. In this case, it can be suggested that the attributes of the RNA-N complex are functionally the same as those described in the list, except that the bases of the RNA are exposed in a groove and accessible to the transcriptase. During transcription, there would be no liftoff of the N protein. Possibly local structural variations of the

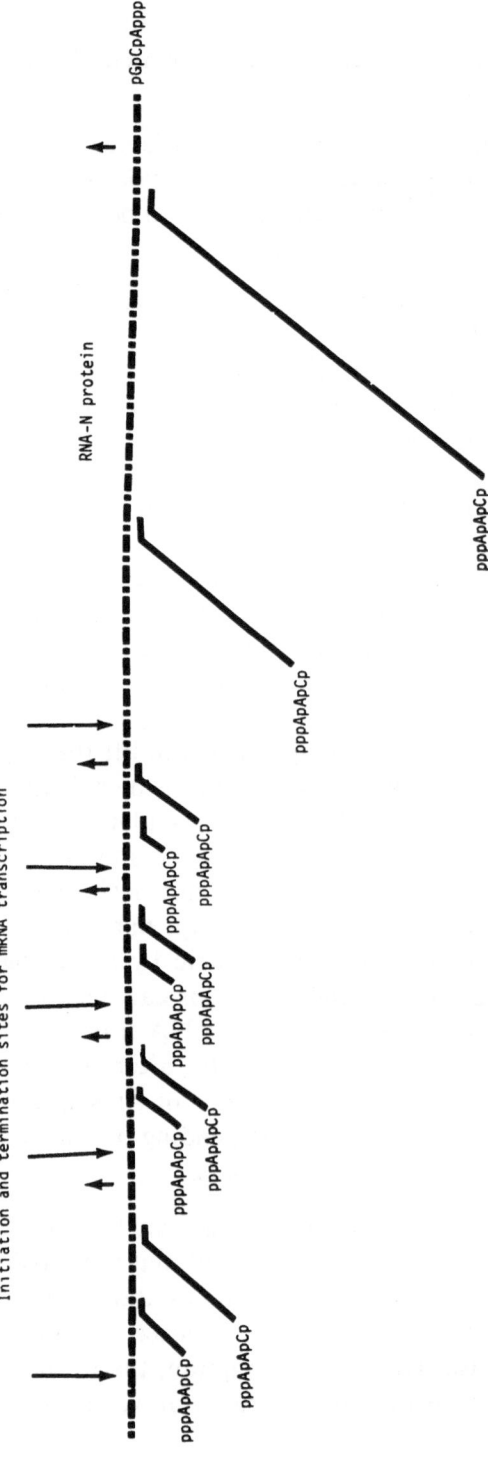

Fig. 17. A model of the process of VSV mRNA transcription. The synthesis (5′ to 3′) of viral complementary mRNA by the VSV transcriptase is depicted as involving specific internal initiation (↓) and termination (↑) signals for individual mRNA transcripts. The template is the RNA–N protein complex, the enzymes are NS and L, and the product is shown as naked RNA. How the viral proteins interact is not known. Although in this model *no* 5′ modification is depicted as occurring *during* transcription, this is not critical to the scheme (see text). Transcription is envisaged as involving a temporal N protein bubble or alternatively a groove mechanism (see text). The gene sequence order is not known, although some evidence that the L mRNA may be toward the 5′ end of genome has been obtained (Roy and Bishop, 1972).

groove or enzyme–NS interactions with it would designate transcription initiation or termination sites.

An alternate model to the one just proposed is one in which transcription is initiated at the 3′ end of the viral RNA and the various mRNA species are excised from that transcript by a restriction endonuclease specific for RNA sequences. The excision products would have to possess 5′ termini with 5′-monophosphates (e.g., pApApCp- . . .), which are then capped and methylated to give the modified 5′ sequences of the eventual mRNA. Transcription control could reflect the efficiency by which the original transcript completed the process. Although this model cannot be rigorously excluded by the available data, if pppApApPypXpGp . . . sequences are the precursors of capped sequences, and thereafter capped and methylated sequences, their occurrence would tend not to support this model. However, without pulse-chase or other experimental results, the model cannot be excluded at present.

3.6.14. The Interrelationships of Transcription and Replication

There is very little evidence concerning the replication processes of rhabdoviruses. In part, this is due to the fact that there is more transcription of mRNA in infected cells than viruslike RNA synthesis and also to the fact that no *in vitro* systems have been devised to study RNA replication. Reviews of what is known about the replication processes are available (Bishop and Smith, 1976; Wagner, Vol. 4, this series). The information obtained by molecular and genetic studies will not be reiterated here. However, since replication probably uses in its initial phase (step 1) the viral genome to form a complete viral complementary RNA (unless the individual mRNA pieces are stitched together), the interrelationships of transcription and replication are worth considering, particularly with respect to the virion polymerase. Two observations made in Alice Huang's laboratory pertain to this question: The first is that it has been shown that synthesis of viral RNA requires ongoing protein synthesis—because viral RNA synthesis stops upon the addition of cycloheximide, while mRNA transcription does not (Perlman and Huang, 1973). The second important observation is that intracellular 42 S plus strand RNA occurs in RNA–N protein complexes (Soria *et al.,* 1974). We have found similar structures in infected cells (unpublished observations), and this has led us, like Huang, to believe that replication involves an RNA–N protein intermediate which contains a 42 S plus strand RNA. The intriguing question concerning the replication process is its interrelationship with transcription. Temperature-sensitive mutants of group I and group IV do not synthesize viruslike RNA at nonpermissive temperatures, but many of their members do perform primary transcription at 39.5°C (see Section 1.2). Mutants of group III and group V, on the other hand, do synthesize viruslike RNA at high temperatures (see reviews by Bishop and Smith, 1976; Pringle, 1975; Wagner, 1975). Since it is clear, therefore, that the synthesis of viruslike RNA depends on certain gene products becoming available through translation of the primary mRNA transcripts, which products are involved and what do they do to the virion transcriptase?

Replication can be envisaged in two steps. The first step of replication is the synthesis of a complete 42 S virus-complementary, plus strand RNA. The second step is the synthesis of 42 S viral minus strand RNA from the 42 S plus strand template. The first step is the intriguing one. How does it relate to the ongoing transcription process? In order to provide working hypotheses, two models can be suggested for the first stage of VSV replication.

The first model which can be proposed is a *simultaneous model for transcription and replication*. This is exemplified in Fig. 18. The essential features of this model are as follows:

1. Transcription and replication can occur on the same template without restriction of each other.
2. Replication starts at the 3′ end of the viral RNA; transcription may also initiate there but possibly does not (Fig. 18; as discussed in Section 3.6.13).
3. The replication process is envisaged as employing the same bubble or groove methods of synthesis that are postulated to be used in primary transcription.
4. In this model, it can be proposed that the 42 S plus strand becomes coated with N protein as it is synthesized. This is the essential feature of this proposal. Whether 42 S replication is held up until N protein or a replicase is available is not critical to the model. Either possibility may in fact regulate the process. If the virion transcriptase is the replicase, then one could postulate that naked 42 S plus strand RNA may be made by the transcriptase only at relatively low frequency until N protein is available. Alternatively, if the replicase is a modified transcriptase (such as one possessing more NS molecules, or super- or dephosphorylated NS species, or even an additional, newly synthesized gene product), then it could

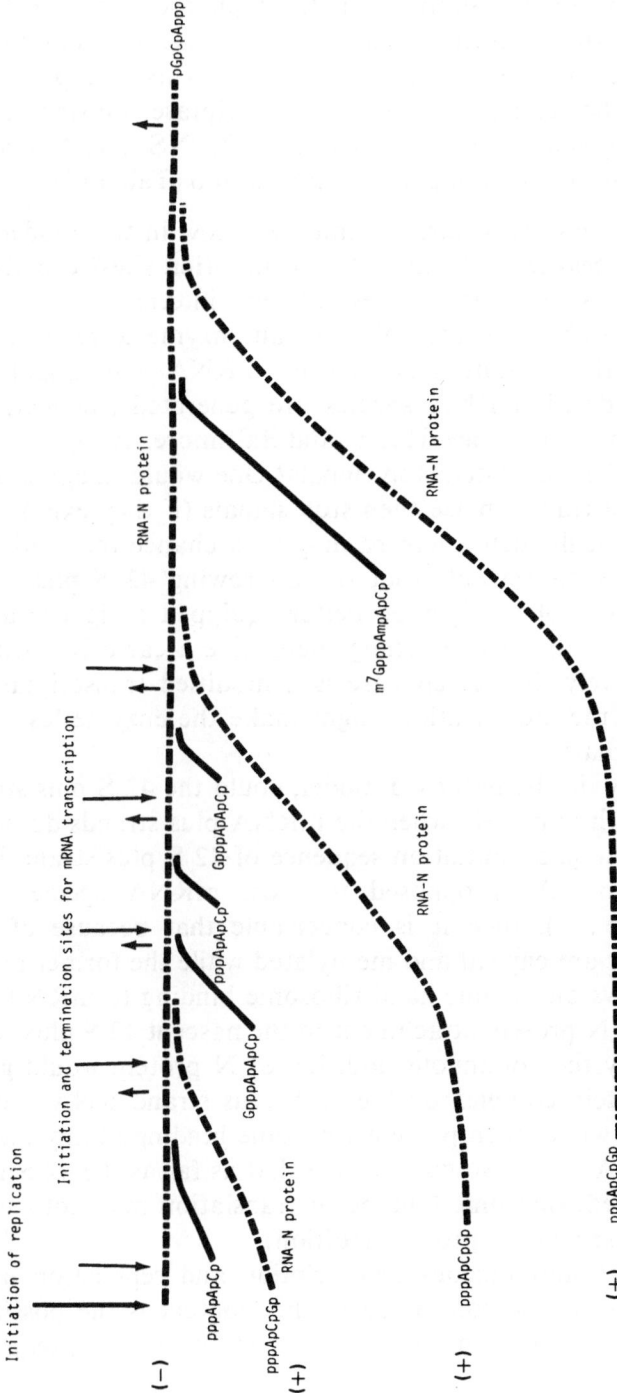

Fig. 18. Simultaneous transcription and the initial replication step. A scheme for simultaneous, individual mRNA transcription and complete viral complementary (+) RNA synthesis is presented. In this scheme, the two processes are shown as being noncompetitive and the complete plus strand product is envisaged as being stabilized by addition of N protein to give RNA(+)-N protein complexes. By contrast, the mRNA species are naked. In this scheme, mRNA 5′ modification is depicted as occurring during transcription (see text).

be suggested that replication would be held up until they became available. Similarly, if the replicase is not related to the transcriptase then one might expect that replication would await the availability of the replicase. This last possibility, wherein the replicase is not the transcriptase, we find hard to believe, in view of the fact that L, G, N, NS, and M account for almost all the viral genetic information (Table 12).

If mRNA species are separately initiated, how in this model does the replication process relate to the mRNA initiation sites? Clearly, for an RNA in process of being synthesized, any internal initiation site would pose little problem to the ability of an enzyme to read through that site. Replication initiation at internal mRNA sites might be a means by which deletion RNA species are generated (replicating to give defective progeny genomes, Huang and Baltimore, 1970).

What about the postulated stop signals? One would imagine that if the replicase is the transcriptase then stop signals (if they exist) would be a hazard. Although such a hazard may be a chance the replication process takes, it is conceivable that the ongrowing 42 S plus strand RNA–N protein complex might be better equipped to ignore a stop signal because of its distinctive arrangement (like a car driver with no brakes). Alternatively, if the replicase is a modified transcriptase (as suggested above), the modification might make the enzyme less sensitive to the stop signals.

How and why, in the proposed model, could the 42 S plus strands become coated with N protein when the mRNA plus strands do not? If there is a difference in 5′ initiation sequence of 42 S plus strand RNA (e.g., pppApCpGp . . .) as opposed to most mRNA species (e.g., pppApApCpXpGp . . .), then it is conceivable that because of their sequence the latter are capped and methylated while the former are not and that this allows on the one hand ribosome binding to mRNA, and on the other hand N protein attachment to the nascent 42 S plus strand RNA. Presumably the continuous addition of N protein would give a contiguous N protein complex on the 42 S plus strand RNA. This, it can be suggested, would then prevent ribosome binding at any internal site along the RNA (it is also conceivable that as far as the 42 S RNA species is concerned, ribosome binding or translation may not even be detrimental to subsequent N protein addition).

This model of simultaneous transcription and replication makes certain predictions. To be considered in this context is the possibility that the *in vitro* initiations involving pppApCpGp . . . sequences represent *in vitro* replication attempts by the virion polymerase. These are currently being investigated experimentally.

The second alternate model which can be proposed is a *nonsimultaneous transcription and replication* process, whereby either the viral genome template is used in transcription (e.g., primary transcription) or it is used in replication. A possible scheme is shown in Fig. 19. In this case, it can be suggested that internal mRNA initiation sites and stop signals are blocked. As discussed above, the virion transcriptase could function as a replicase (with or without prior modification) or the replicase could be a new gene product. Blocking in this scheme could be achieved by NS proteins complexing with the N protein at mRNA initiation or stop sites, causing permanent liftoff or changes in the groove structure. Other site-blocking mechanisms can also be envisaged.

Step 2 replication is defined as the synthesis of viral minus strand RNA from the 42 S plus strand RNA template. Huang and associates (Soria *et al.,* 1974) have suggested that the template for this phase of the replication process is a 42 S plus strand RNA–N protein complex and not a double-stranded or naked RNA replicative intermediate. We would support that proposal as suggested in Fig. 20. Without internal initiation sites present on the plus strand RNA, the virion transcriptase could presumably function as the polymerase. If synthesis of viral RNA leads directly into progeny virion packaging, then the enyzme might even become enclosed within the virus particle. Alternatively, a modified transcriptase might function as a replicase (as in the simultaneous transcription–replication scheme described above).

This step 2 replication process would again employ either the bubble or the groove mechanism of enzyme–template interaction. The essential distinctive feature of this process, like that of the nonsimultaneous step 1 replication scheme, is that N protein covers both the template plus strand and the product minus strands. It would be in keeping with this scheme of events that N protein availability through ongoing protein synthesis controlled this RNA replication.

Other mechanisms of RNA synthesis could be and have been proposed; however, we believe that the simultaneous transcription–replication and step 2 replication procedures outlined above are more plausible mechanisms. It is to be hoped that in the not too distant future it will be possible to obtain evidence which will lead to more definitive models for all the mechanisms involved in VSV RNA synthesis (transcription and replication).

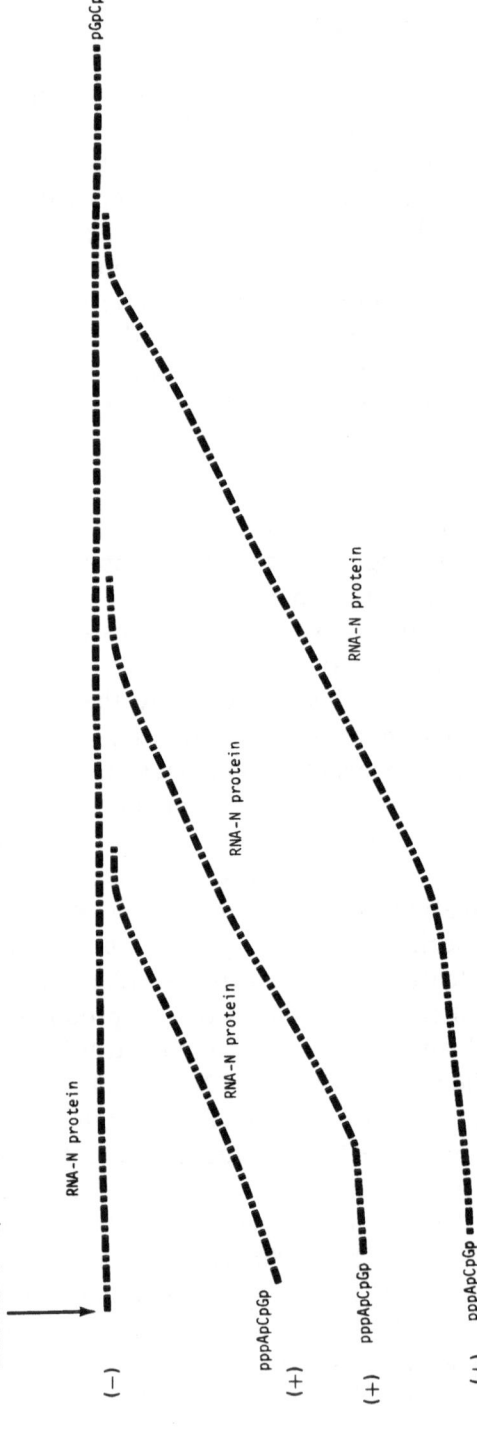

Fig. 19. Nonsimultaneous transcription and step 1 replication. In this scheme, the synthesis of complex viral complementary (+) RNA has usurped the synthesis of individual mRNA species. Although complete RNA plus strands are shown as RNA(+)–N protein complexes, this may not be a necessary feature (see text).

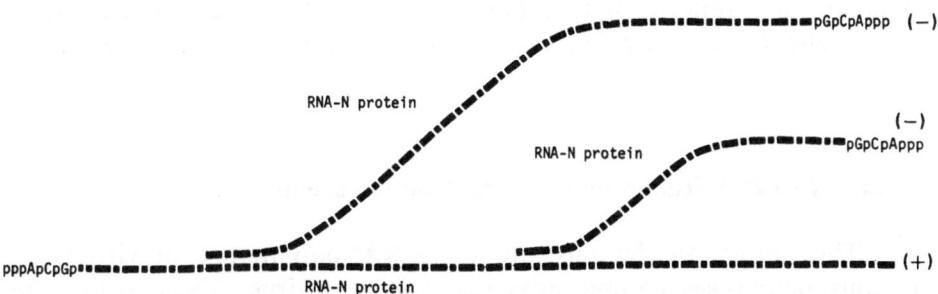

RNA-N protein

pGpCpAppp (−)

(−)
pGpCpAppp

RNA-N protein

pppApCpGp

(+)

RNA-N protein

Fig. 20.   Replication step 2: the synthesis of viruslike RNA. Viruslike (−) RNA synthesis is shown as being templated by RNA(+)-N protein complexes and producing RNA(−)-N protein complexes.

3.7.   Poxvirus Transcriptases

Like the enveloped RNA viruses, poxviruses require a membrane-disrupting treatment before an *in vitro* DNA-directed RNA polymerase can be demonstrated (Section 1.1.3). The presence of other enzyme activities with one or another poxvirus has also been demonstrated. These include (1) a protein kinase activity (Downer *et al.,* 1973; Kleiman and Moss, 1973, 1975*a,b*; Paoletti and Moss, 1972), (2) a poly(A) polymerase activity (Kates and Beeson, 1970*b*; Moss and Rosenblum, 1974; Moss *et al.,* 1973, 1975), (3) various nucleoside triphosphatase activities (Gold and Dales, 1968; Munyon *et al.,* 1968; Paoletti and Moss, 1974; Paoletti *et al.,* 1974), (4) both acid and alkaline deoxyribonucleases (Pogo and Dales, 1969; Pogo *et al.,* 1971), (5) a guanylytransferase activity (Ensinger, *et al.,* 1975; Martin *et al.,* 1975), and (6) two mRNA methyltransferases (Martin *et al.,* 1975; Martin and Moss, 1975; Wei and Moss, 1974, 1975). All of these enzyme activities have been shown to remain associated with the virus particles after extensive purification and exhibit enhanced activities after dissociation or removal of the outer viral proteins.

Although the poxviruses were the first animal viruses shown to possess an RNA polymerase, there have been few studies published on their RNA polymerases. In part, this is due to the complex nature of the genome and virions of these DNA viruses (genome approximately 140×10^6 daltons of double-stranded DNA). No doubt, though, the virion polymerase of poxviruses, as well as a variety of other aspects of the poxvirus infection process, will receive greater attention from molecular virologists in the future. Most of the reported studies have used vaccinia virus as the prototype virus. For a discussion of the

structural and functional aspects of poxviruses, the reader is referred to
the excellent review by Bernard Moss in Volume 3 of *Comprehensive
Virology*.

3.7.1. *In Vitro* Studies on the Viral Polymerase of Poxviruses

The ability to demonstrate an RNA polymerase in virions of
various poxviruses (rabbit poxvirus, vaccinia virus, Yaba virus, and
various insect poxviruses; see Table 2) has been reported to require the
prior dissociation of the virus envelope (discussed in Section 1.1.3). The
requirements of the enzyme for divalent cations and all four ribonucleoside triphosphates have been demonstrated (Section 1.1.3). For vaccinia virus, the *in vitro* synthesis of RNA under optimal conditions
proceeds at a linear rate for several hours. Although the RNA
transcription process is repetitive, it is limited in scope to some 14% of
the genome's length (Kates and Beeson, 1970*a*). The RNA formed is,
however, equivalent to the early mRNA found in infected cells. How
transcriptional control is effected *in vivo* or *in vitro* is not known,
although it is probable that one or more of the proteins synthesized
from the early mRNA species are the mediators of the *in vivo* switch to
total mRNA synthesis.

In early studies reported by Kates and Beeson (1970*a*), it was
shown that the principal $[\gamma\text{-}^{32}P]$ribonucleoside triphosphate incorporated into product RNA was ATP, although some labeled CTP was
also incorporated into product RNA (approximately one-fifth that of
the ATP).

More recent studies have indicated that the mRNA transcripts
made *in vitro* by vaccinia cores are for the most part either capped or
possess 5′-diphosphate groups and become methylated when given a
source of SAM (Wei and Moss, 1974, 1975; B. Moss, personal communication). SAM does not appear, however, to stimulate the overall
in vitro DNA-templated RNA transcription rate. The methylation and
capping enzyme activities have been purified and/or identified (see
later). For the endogenous transcription process using whole cores, it
has been shown that the methylation process requires and parallels the
ongoing synthesis of RNA. No methylation is obtained in the presence of
only one of the four ribonucleoside triphosphates or magnesium ions,
etc. It is also inhibited ($\sim 90\%$) by the presence of actinomycin D in the
reaction mixture. In the preliminary experiments reported by Wei and
Moss (1974, 1975), it was shown that exogenous substrates which were

unable to be methylated by vaccinia cores were yeast RNA, *Escherichia coli* tRNA, rRNA, poly(A), poly(U), poly(C), and poly(G). This indicated that these RNAs possess no terminal or internal sequences suitable for being methylated by the vaccinia enzyme (see Section 3.7.2). When given unmethylated vaccinia *in vitro* RNA transcripts, it was demonstrated that the vaccinia enzyme was capable of using them as substrates. It was also shown that the methylated RNA formed by vaccinia cores possessed a size range of approximately 10–20 S, annealed to poly(U)-sepharose [i.e., possessed poly(A) sequences—as shown previously for vaccinia RNA transcripts by Kates (1970)], and also annealed to immobilized, denatured, vaccinia DNA. Other experiments which showed that the methylated products were in fact vaccinia RNA transcripts included demonstration that the methylated products were rendered acid soluble by KOH, hot TCA, and ribonuclease (but not deoxyribonuclease or protease treatments). It was calculated that two or three methyl groups were present per 1000 nucleotides of product RNA.

Subsequent analyses by Wei and Moss (1975) and B. Moss (personal communication) indicated that in the presence of SAM the principal product $5'$ sequences were $m^7G^{5'}ppp5'Gmp\ldots$ and $m^7G^{5'}\overset{\alpha}{p}\text{-}\overset{\beta\alpha}{pp}Amp.\ldots$ In the absence of SAM, but presence of $[\beta\gamma\text{-}{}^{32}P]GTP$, various alkali-resistant $5'$ sequences were observed, one of which possessed alkaline phosphatase-sensitive phosphate groups, and presumably represented an uncapped sequence involving guanosine nucleotides (later determined to be a ppGp... sequence, B. Moss, personal communication). Similar experiments using labeled ATP have detected uncapped ppAp... sequences. The relative amount of the various $5'$ termini ($\pm$SAM) apparently depends on the reaction conditions and varies with varying ATP and GTP reaction concentrations (B. Moss, personal communication).

Kates and Beeson (1970*a*) showed that RNA synthesis by vaccinia cores exhibits a lag in the initial synthetic rate but that this lag could be abolished by suitable preincubation with particular ribonucleoside triphosphates (e.g., CTP and/or ATP). In determining the K_m values for the four ribonucleoside triphosphates, it has been shown (Kates and Beeson, 1970*a*; Munyon *et al.,* 1967) that the K_m for ATP ($\sim 1 \times 10^{-3}$) is about twenty-fold higher than that of UTP, GTP, or CTP (~ 5–7×10^{-5}). It has been postulated from kinetic analyses of the amount of core-bound product RNA which is resistant or sensitive to ribonuclease digestion that initially during RNA transcription the product RNA is held within some core structure which protects it from ribonuclease digestion; however, later it is extruded and becomes sensitive to ribonuclease treatment. Both multiple RNA transcription initiation sites and extrusion sites have been postulated to be present as well as an ATP-dependent extrusion mechanism (Kates and Beeson, 1970a). Although the structural features of the core have yet to be worked out, the existence of sequestered genome and enzymes may be one reason that core preparations do not respond to added DNA as a template (Moss, Vol. 3, this series).

The protein(s) responsible for the DNA-directed RNA polymerase activity of vaccinia are not known. In part, this is due to the complicated architecture of poxviruses and poxvirus cores (reviewed by Moss, Vol. 3, this series) and also to the fact that the polymerase is inhibited by sodium deoxycholate—a detergent which is, however, capable of releasing other virion enzymes in an active form from vaccinia cores (see below).

3.7.2. Other Enzyme Activities Associated with Poxviruses

The presence of various enzymes associated with virus and core preparations of certain poxviruses has been documented, and some of these enzyme activities are described in detail in the chapter by Moss on poxviruses in *Comprehensive Virology,* Volume 3. These activities include a poly(A) polymerase, triphosphate phosphohydrolases, deoxyribonucleases, and a protein kinase, a guanylyltransferase, and two methyltransferases.

The demonstration that a poly(A) polymerase is associated with vaccinia cores (Kates and Beeson, 1970b) and vaccinia-infected cells (Brakel and Kates, 1974a,b) as well as other poxviruses (McCarthy et al., 1975) has raised the question of whether this enzyme activity is the RNA polymerase itself or is a component of the polymerase enzyme. This question is also relevant to other virus types which are known to possess virion RNA polymerases and which produce transcripts possessing poly(A) 3′ sequences (e.g., VSV, influenza, NDV).

For vaccinia, investigations to determine if the poly(A) polymerase is, or is related to, the virion RNA polymerase have been undertaken, and differences were detected between the two enzyme activities in their pH optima, divalent cation requirements, and heat stabilities (Brown et al., 1973).

The virion-associated poly(A) polymerase has been shown by Kates and Beeson (1970b) to (1) utilize ATP as a sole nucleoside triphosphate precursor (and not be active with GTP, UTP, or CTP); (2)

be insensitive to actinomycin D, partially sensitive to ethidium bromide (~44%), and largely inhibited by proflavin (~90%), under conditions in which one of these totally inhibits the RNA polymerization process; and (3) synthesize poly(A) (approximately 150 nucleotides in length) for a limited time period only (~5 min) when given just ATP. In the presence of all four ribonucleoside triphosphates, poly(A) tracts (approximately 100 nucleotides in length) have been shown to be present on heteropolymeric transcript RNA species of various sizes.

Studies on a poly(A) polymerase activity purified 340-fold from vaccinia virus cores by affinity chromatography and other procedures (Moss *et al.*, 1973, 1975) have shown that it is comprised of two polypeptides (35,000 and 51,000 daltons), prefers manganese ions over magnesium, possesses a pH optimum of 8.6, and is inhibited only to minor extents by the presence of other ribonucleoside triphosphates at equimolar concentrations. Sodium chloride concentrations (greater than 0.1 M) were shown to inhibit the enzyme activity. Effective exogenous substrates were poly(C), short poly- and oligodeoxyribonu-cleotides of dT, dC, and dI but not dA or dG, as well as short poly- and oligoribonucleotides of various sorts other than poly(G). Although long homopolyribonucleotides [other than poly(C)] or RNA were poor primers, after digestion and alkaline phosphatase treatment their priming capacity could be increased. Using uridylate oligomers, the minimum effective primer length was found to be four to six nucleotide residues (Moss *et al.*, 1975). Other analyses have indicated that the purified enzyme utilizes ATP but not CTP, UTP, or GTP (Moss and Rosenblum, 1974). Nearest-neighbor analyses using labeled ATP and polyribonucleotides indicate that the enzyme activity works by extending existing polynucleotides possessing 3′ hydroxyl groups rather than as a transcriptase making complementary polynucleotide copies.

No evidence has been obtained which indicates that the poly(A) polymerase is the virion transcriptase, and, although the possibility that it is one of the polymerase polypeptides is hard to rigorously exclude, based on current information a reasonable conjecture is that the enzyme is one of a complex of enzymes in vaccinia cores associated with the production and modification of RNA transcripts.

The presence and 250-fold purification of a protein kinase enzyme activity from vaccinia virions using DNA affinity and DEAE-cellulose chromatography has also been reported (Downer *et al.*, 1973; Kleiman and Moss, 1973, 1975*a,b*; Paoletti and Moss, 1972). The purified enzyme (62,000 daltons) requires protamine or other basic proteins as an activator, ATP, magnesium ions, and alkaline pH values (pH 9.5–

10.5) in order to catalyze the phosphorylation of two viral proteins (38,500 and 11,700 dalton polypeptides). The enzyme activity is not stimulated by cyclic mononucleotides (Downer *et al.,* 1973; Kleiman and Moss, 1975*b*; Paoletti and Moss, 1972). The relevance of the protein kinase activity to the RNA polymerase is not known.

Deoxyribonucleases have been demonstrated in vaccinia, Yaba, and insect poxviruses (Pogo *et al.,* 1971; Pogo and Dales, 1969; Schwartz and Dales, 1971), and one nuclease has been purified from vaccinia virus (Rosemond-Hornbeak *et al.,* 1974, Rosemond-Hornbeak and Moss, 1974). The enzyme has an acid pH optimum (pH 4.4), and is specific for single-stranded DNA, exhibiting both endonucleolytic and exonucleolytic activities. In vaccinia, more than one deoxyribonuclease activity may be present, based on pH optima analyses (Aubertin and McAuslan, 1972; Pogo and Dales, 1969). Again the relationship (if any) of these enzymes to the virion RNA polymerase is not known.

Nucleoside triphosphatases in various poxviruses have been described (Gold and Dales, 1968; Munyon *et al.,* 1968; Pogo *et al.,* 1971; Schwartz and Dales, 1971). For vaccinia, two immunologically distinct activities have been partially resolved and shown to prefer as substrates either (1) ATP (or dATP), and be stimulated by DNA, or (2) any ribo- or deoxyribonucleoside triphosphate, and be stimulated by DNA or RNA (Paoletti *et al.,* 1974*a,b*; Paoletti and Moss, 1974).

The copurification from vaccinia cores of a guanylyltransferase enzyme activity (responsible for capping vaccinia mRNA) and an mRNA (guanine-7-)methyltransferase activity (responsible for cap methylation) has been reported (Martin *et al.,* 1975). Chromatography of virus extracts on DNA-agarose separated these two enzymes from mRNA (nucleoside-2′-)methyltransferase enzyme(s), but further chromatography on poly(U)-sepharose or poly(A)-sepharose, gel filtration, or sucrose gradient centrifugation failed to separate the capping from the cap-methylation enzyme activities, suggesting that these two enzymes are associated in some way. It was found that the active complex (~127,000 daltons) possessed two polypeptides as determined by denaturing SDS-polyacrylamide gel electrophoresis (95,000 and 31,400 daltons).

Systematic analyses performed by Martin and Moss (1975) with the purified polypeptides demonstrated that the capping enzyme was capable of capping the 5′ end of viral mRNA or synthetic poly(G) or poly(A) sequences provided that they possessed 5′-diphosphates (not monophosphates). The reaction was shown to require magnesium ions and GTP (and not ATP, CTP, or UTP) and be optimally effective at pH 7.8. It was shown that the reaction was reversible and that inorganic pyrophosphate was an inhibitor. It was proposed that in essence the enzyme catalyzed the reaction

$$ppN\text{-} + GTP \rightleftharpoons G^{5'}pppN\text{-} + PP_i$$

Other analyses reported by Martin and Moss, 1975, indicated that the associated mRNA (guanine-7-)methyltransferase (cap-methylation enzyme) promoted, in the presence of SAM, methylation of capped GpppAp . . . and GpppGp . . . sequences in a reaction which was not dependent on GTP or divalent cations. The pH optimum for this reaction was in a broad range around neutrality and the reaction was inhibited by S-adenosylhomocysteine (SAH). It was proposed that the enzyme catalyzed the reaction

$$G^{5'}pppN\text{-} + SAM \rightarrow m^7G^{5'}pppN\text{-} + SAH$$

From the foregoing discussions, it can be appreciated that several of the ancillary enzyme activities responsible for modification of the RNA polymerase products [poly(A) polymerase, capping and two methylation enzymes] have been identified; others possibly involved in some aspect of either RNA synthesis, DNA synthesis, or transcriptional control (triphosphatases, deoxyribonucleases, and the protein kinase) have also been identified. These results suggest that the RNA transcription and processing procedures are a complex interwoven series of events. How the polypeptides involved are arranged in vaccinia cores is an interesting question which will have to await *in vitro* reconstruction attempts. It is not known how many of the ancillary enzyme activities are coded by the vaccinia genome; although there is ample genetic information available, it is conceivable that these enzymes are obtained from the host cell.

3.8. Possible RNA Polymerases Associated with the Icosahedral Cytoplasmic Deoxyriboviruses

The icosahedral cytoplasmic deoxyriboviruses include frog virus 3 (FV$_3$), the iridescent viruses, the lymphocystitis viruses, and African swine fever viruses (Kelly and Robertson, 1973). McAuslan and associates (and others) have demonstrated in frog virus 3 the presence of nucleotide phosphohydrolases (Aubertin *et al.*, 1971; Vilaginès and McAuslan, 1971), a deoxyribonuclease activity (Kang and McAuslan, 1972; Palese and McAuslan, 1972), single- and double-stranded ribonucleases (Palese and Koch, 1972), and a protein kinase activity (Gravell and Cromeans, 1972). RNA polymerase activities have been claimed to be present in preparations of the iridescent type 2 and 6 viruses (Kelly and Tinsley, 1973) and postulated to exist in frog virus 3 (Gravell and Cromeans, 1971). Cells infected with FV_3 synthesize virus-specific mRNA species in the cell cytoplasm in the presence of an inhibitor of protein synthesis. Whether this is in fact due to a virion polymerase or represents transcription by a cellular polymerase is not known for sure. Although the DNA of African swine fever virus is infectious, this does not preclude a virion polymerase being present in virions of that or other icosahedral cytoplasmic DNA viruses.

3.9. Herpesviruses, Adenoviruses, Papovaviruses, and Parvoviruses

No RNA polymerase activity has been reported for herpesviruses, adenoviruses, papovaviruses, and parvoviruses, which are known to possess infectious DNA, although other virion-associated enzymes have been claimed to be present from some of them (see reviews by Philipson and Lindberg, Roizman and Furlong, Rose, and Salzman and Khoury in Vol. 3, this series).

3.10. DNA-Directed DNA Polymerase Activity of the Putative Hepatitis B Virus

Two reports from W. S. Robinson's laboratory (Greenman and Robinson, 1974; Kaplan *et al.*, 1973) have indicated that associated with the candidate human hepatitis B virus there is a DNA-directed DNA polymerase activity. It was shown that the enzyme was a component of the 42 nm Dane particle and also the 28 nm core structures prepared from Dane particles by nonionic detergent treatment (NP40). The DNA reaction products were shown to remain associated with the Dane core structures, from which they could be released by SDS and resolved as 15 S circular double-stranded DNA (approximately 0.78 μm in length, approximate molecular weight 1.6×10^6) as determined by cesium chloride centrifugation followed by electron microscopy (Robinson *et al.*, 1974). Similar structures have been obtained from unincubated Dane particles, as well as additional linear 0.5–12 μm DNA molecules. The interrelationships of these DNA species (homologies, helper functions, defectiveness, etc.) or of the polymerase to the virus infection and replication processes are not known and need further study.

3.11. RNA Transcriptases of Coronaviridae

In a recent presentation of the features of various Coronaviridae, there are references to unpublished observations from several sources of RNA-dependent RNA polymerases associated with avian infectious bronchitis virus and murine hepatitis virus (Tyrrell *et al.*, 1975). The information to support these claims has yet to be published.

3.12. The Reverse Transcriptases of Oncornaviruses and Similar Virus Types

The presence and basic requirements for demonstrating a reverse transcriptase (RNA-directed DNA polymerase) activity in preparations of oncornaviruses (and viruses with similar genome components) have been described in Section 1.1.4 and Table 2, together with some discussion of the utility of exogenous natural or synthetic RNA or DNA substrates as well as the effect of a variety of inhibitors on the enzyme activity. Purification procedures and antigenic similarities and dissimilarities for various reverse transcriptases are discussed in Sections 2.2.4 and 2.3.

In this part of the chapter, some discussion of the constitution of reverse transcriptases and possible mechanisms of product synthesis will be presented; for more extensive treatises, readers are referred to the reviews by Bader (Vol. 4, this series) and Green and Gerard (1974).

3.12.1. Polypeptide Composition of Viral Reverse Transcriptases

The reverse transcriptase activity obtained from crude AMV preparations has been purified several hundredfold and can be separated into two active fractions by chromatography on phosphocellulose (Grandgenett *et al.*, 1973; Kacian *et al.*, 1971). The major enzyme isolated from AMV (designated $\alpha\beta$) has a molecular weight of 160,000 and is comprised of two polypeptides possessing molecular weights of 65,000–69,000 (α) and 105,000–110,000 (β) (Gibson and Verma, 1974; Grandgenett *et al.*, 1973; Hurwitz and Leis, 1972; Kacian *et al.*, 1971; Moelling, 1975; Verma and Baltimore, 1973). Equal molar proportions of the α and β subunits are usually found for the $\alpha\beta$ complex except on storage at $-20°C$, when the proportion of the α form rises (Moelling, 1975). The minor AMV DNA polymerase activity (approximately 10–20% of the total original virus activity) is reported to possess a size of about 90,000 daltons and a subunit molecular weight of 65,000; it is comprised of either one or two α polypeptides (Grandgenett *et al.*, 1973). The minor component can be obtained not only from the $\alpha\beta$ complex by rechromatography of the latter on phosphocellulose but also apparently by protease degradation of the β subunit using trypsin or subtilisin (Moelling, 1975). Extraction of AMV in the presence of a protease inhibitor gives an enzyme preparation possessing mostly β subunits. Purified AMV $\alpha\beta$ enzyme preparations when iodinated after storage at $-20°C$ have been shown to possess not only relatively more α subunits but also a small polypeptide ($\sim 40,000$ daltons) presumed to be the other cleaved portion (γ) of the β subunit (Moelling, 1975). Polypeptide analyses of the AMV α and β moieties confirm the relationship of α and β proteins (Gibson and Verma, 1974). Taken together, these results suggest that the α subunit is a cleavage product of β (along with the γ moiety) and that the α (or $\alpha\alpha$) enzyme and the $\alpha\beta$ (and possibly β or $\beta\beta$) enzyme possesses the reverse transcriptase capability. It has been claimed that the uncleaved β (? one β or $\beta\beta$) enzyme is more efficient at transcribing natural viral RNA than the α or $\alpha\beta$ enzyme (Moelling, 1975). Furthermore, it has been suggested that the β enzyme has a better affinity for template or enhances the affinity of α for the template or substrate (Panet *et al.*, 1975*b*). Green and associates have also claimed that the $\alpha\beta$ complex is more efficient at synthesizing DNA than the α subunit and that the β subunit possibly influences the mechanism of action of the ribonuclease H activity associated with AMV DNA polymerase (see below) (Grandgenett *et al.*, 1973; Green and Gerard, 1974). Whether this is so remains to be determined in view of the recent finding that the β subunit is the precursor of the α protein. Whether the reverse transcriptase of infectious virus consists of β ($\beta\beta$) or α ($\alpha\alpha$) or $\alpha\beta$ complexes is not known.

Similar minor and major components and shared polypeptide sequences for RSV DNA polymerase have been identified ($\alpha\beta$ complex as determined by gel filtration or centrifugation of approximately 110,000 daltons, although by SDS gel electrophoresis the α subunit is approximately 65,000 daltons and the β subunit is approximately 105,000 daltons) (Duesberg *et al.*, 1971*a*; Faras *et al.*, 1972). In view of the antigenic and primary sequence similarities of the AMV and RSV enzymes (Gibson and Verma, 1974), it is probable that for RSV both β and α polypeptides possess reverse transcriptase activity.

The sizes of the DNA polymerases of Tragger duck spleen necrosis virus (SNV) and reticuloendotheliosis virus are reported to be 68,000–

75,000 daltons based on Sephadex gel filtration and glycerol gradient centrifugation (Mizutani and Temin, 1974, 1975). Again, whether these proteins represent active components derived from some larger precursor enzyme is not known.

The reverse transcriptase activities of several mammalian C-type viruses, including R-MuLV (Ross *et al.,* 1971, Abrell and Gallo, 1973), SSV (Abrell and Gallo, 1973), FeLV (Tronick *et al.,* 1972), and GaLV˙ (Mondal *et al.,* 1975), are reported to possess sizes of approximately 70,000 daltons by both gel filtration and sedimentation analyses of the active enzyme. Polypeptide analyses of purified reverse transcriptase enzyme preparations using SDS gel electrophoresis have identified either three components for M-MuSV(MuLV) ($\sim$ 82,000, 68,000, and 60,000, Gerard and Grandgenett, 1975), two components for Friend MuLV (67,000 and 51,000 Weimann *et al.,* 1974), two components for HaLV (68,000 and 53,000, Verma *et al.,* 1974*b*), or one component for M-MuLV (80,000, Verma, 1975) and one component for SSV and GaLV (70,000, Abrell and Gallo, 1973; R. Gallo, personal communication). The relevance and interrelationships of the various polypeptides identified in these preparations to the reverse transcriptase enzyme activity of the respective viruses are not known. Protease treatment of F-MuLV reverse transcriptase does not cleave the 70,000 polypeptide (Moelling, 1975), although whether the 70,000 polypeptide is a cleavage product of some larger functional virion polypeptide is also not known.

Mason-Pfizer monkey virus, a virus derived from a rhesus monkey mammary tumor, possesses a reverse transcriptase which has been shown, based on sedimentation and SDS-polyacrylamide gel electrophoresis, to contain a single polypeptide of 110,000 daltons (Abrell and Gallo, 1973).

Milk from R111 mice, known from electron microscopic analyses to contain both B- and C-type oncornaviruses, has yielded reverse transcriptase enzyme activities possessing sizes of approximately 110,000 daltons (B-type virus) and approximately 70,000 daltons (C-type virus) (Howk *et al.,* 1973). However, Dion *et al.* (1974) detected only the 110,000 component in R111 milk virus, possibly because less C-type virus was present in the milk they processed.

The DNA polymerase of the Russell's viper C-type virus, based on glycerol gradient centrifugation, possesses a molecular weight of 109,000. No polypeptide analyses have been reported, so that if subunits or other functional polypeptides exist, their size is not known.

Although visna, maedi, and progressive pneumonia virus (PPV) are known to possess virion reverse transcriptases (Table 2), the sizes of their reverse transcriptase polypeptides are not known for certain, although a size of approximately 125,000 daltons has been suggested for visna, based on analyses of virus lysates (Lin and Thormar, 1973). The major structural polypeptides of these viruses (including Zwoegerziekte virus) have been determined (Haase and Baringer, 1974; Mountcastle *et al.*, 1972), and the presence of 60–70 S RNA for some of these viruses has also been documented (Harter *et al.*, 1971; Haase *et al.*, 1974*a*; Lin and Thormar, 1971, 1972; Stone *et al.*, 1971*b*).

The size and polypeptide composition of the reverse transcriptase associated with the various foamy viruses (Table 2) have not been reported.

3.12.2. Ribonuclease H Activity Associated with Viral Reverse Transcriptases

The presence in oncornaviruses of one or more exonucleolytic ribonuclease activities which degrade RNA from RNA:DNA hybrids but not the DNA strand of the hybrid or double- or single-stranded RNA (or DNA) was initially demonstrated for AMV virus preparations (Moelling *et al.*, 1971) and then for other oncornaviruses (Grandgenett *et al.*, 1972). Purified AMV α and $\alpha\beta$ complexes possess both reverse transcriptase and ribonuclease H activities (Baltimore and Smoler, 1972; Grandgenett *et al.*, 1973; Keller and Crouch, 1972; Watson *et al.*, 1973). Some evidence has been presented that the two enzymes may be present on different polypeptides (Leis *et al.*, 1973), although this now appears to be doubtful (see below). It has been demonstrated that the AMV $\alpha\beta$ enzyme (1) requires free termini to be present on a substrate RNA in an RNA:DNA hybrid since it is unable to remove circular poly(A) complexed to poly(dT) (Leis *et al.*, 1973), (2) digests the RNA in a processive manner in that once attached it does not appear to switch substrates upon substrate dilution, and also (3) proceeds from either the 3′ or 5′ end of the RNA to give nucleotides with 3′-hydroxyl groups and 5′-phosphate groups (Baltimore and Smoler, 1972; Keller and Crouch, 1972; Leis *et al.*, 1973). Using the purified α subunit of AMV reverse transcriptase, it has been shown that its ribonuclease H activity is a random rather than processive exonuclease (Grandgenett and Green, 1974). Whether the β AMV subunit functions in the same manner or confers the capability of processiveness is not known.

The observation that a temperature-sensitive mutant of RSV possesses both thermolabile reverse transcriptase and thermolabile ribonuclease H activities (Verma *et al.,* 1974*a*; Panet *et al.,* 1975*a*) argues strongly for both enzyme activities residing in one polypeptide. Ribonuclease H activity has also been identified in purified preparations of SNV.

Some evidence has been presented which indicates that upon purification of the reverse transcriptase activity of mammalian C-type viruses it too exhibits ribonuclease H activity (Grandgenett *et al.,* 1972). Gerard and Grandgenett (1975) demonstrated that two ribonuclease H activities are present in crude lysates of Moloney MuSV(MuLV). The larger enzyme (RNase H-I), which is present in smaller amounts (approximately 10% of the total activity), was shown to copurify with the viral reverse transcriptase enzyme activity, require manganese (optimum concentration approximately 2 mM using as substrate [^{3}H]poly(A):poly(dT), and possess an apparent molecular weight of about 70,000. The smaller enzyme (RNase H-II), which is present in greater quantities (approximately 90% of the total activity), was found to possess no detectable DNA polymerase activity, prefer magnesium (10–15 mM optimum) over manganese (5–10 mM optimum) for the degradation of [^{3}H]poly(A):poly(dT), and possess an apparent molecular weight of about 30,000. In addition, it was shown that the two enzymes exhibit different substrate preferences whereby RNase H-II can degrade [^{3}H]poly(A):poly(dT) 6 times faster than [^{3}H]poly(C):poly(dG) in the presence of manganese (10 mM), while RNase H-I digests the same two substrates at about equal rates in the presence of manganese (2 mM). Gerard and Grandgenett (1975) also noted that under limited substrate digestion using [^{3}H]poly(A):poly(dT) the two enzymes released either 10 and 40 nucleotide-length oligonucleotides (RNase H-I) or oligonucleotides only 10 nucleotides in length (RNase H-II). In this respect, the RNase H-I resembles the α and $\alpha\beta$ activities of the AMV enzyme (Grandgenett and Green, 1974). It was not determined whether the two MuSV(MuLV) enzymes were exo- or endoribonucleases.

Although it was shown by Gerard and Grandgenett (1975) that the RNase H-I activity copurified with the RNA-directed DNA polymerase activity of MuSV(MuLV), SDS-polyacrylamide analysis of the most highly purified preparations of that DNA polymerase showed them to contain three polypeptides (82,000, 68,000, and 60,000 daltons), so that which polypeptide represented the ribonuclease H activity could not be determined (see below).

Weimann *et al.* (1974) found that the ribonuclease H activity in preparations of Friend leukemia virus also copurified with the DNA polymerase activity. It was present in material which contained two polypeptides of molecular weights 67,000 and 51,000, although it was reported that the Friend leukemia virus enzyme was present in a complex corresponding to a size of 123,000 daltons.

Analyses of Rauscher MuLV reported by Robert Gallo's laboratory have indicated that cores of that virus exhibit reverse transcriptase activity but no ribonuclease H activity (Wu *et al.*, 1974). The properties of the ribonuclease H enzymes isolated by them were different from those of the Moloney MuSV(MuLV) activity discussed above. The major RNase H activity of R-MuLV (or K-MuSV) did not bind at 0.2 M KCl to phosphocellulose [the RNAase H-II of M-MuSV(MuLV) elutes only at 0.5 M KCl (Gerard and Grandgenett, 1975)]; also, the DNA polymerase activity recovered from R-MuLV (and K-MuSV) was shown to possess very little RNase H activity. No or undetectable amounts of ribonuclease H activities have also been reported for hamster leukemia virus (Verma *et al.*, 1974*b*) and K-MuSV(MuLV) and M-MuSV(MuLV) (Wang and Duesberg, 1973). Whether these results are due to insensitive reaction conditions or a lack of a ribonuclease H activity is not known.

Verma (1975) using cloned Moloney MuLV demonstrated that its purified reverse transcriptase (one polypeptide of approximately 80,000 daltons) possessed a ribonuclease H activity that acted as a random exonuclease, giving large-size digestion products possessing 3′-hydroxyl and 5′-phosphate groups. These studies as well as those reported by Gerard and Grandgenett (1975) suggest that MuLV at least possesses a ribonuclease H activity associated with the reverse transcriptase polypeptide. Whether the lack of activity obtained for other mammalian C-type viruses reflects the presence of an inhibitor, insensitive assay procedures, or incorrect reaction conditions remains to be determined. Some evidence that a temperature-sensitive MuLV mutant possesses a thermolabile reverse transcriptase and ribonuclease H activity has been obtained (I. Verma, personal communication).

In summary, it is clear that the avian oncornavirus ribonuclease H activity is associated with subunits of the viral reverse transcriptase activity; also, some evidence for the association of ribonuclease H with the mammalian oncornavirus reverse transcriptase has been obtained. It appears that there is often more than one ribonuclease H activity present in preparations of the mammalian C-type viruses, raising the possibility that either a cellular enzyme or another viral coded function is present.

3.12.3. Substrates Which Can Be Used by Purified Viral Reverse Transcriptases

As pointed out in Section 1.1.4, synthetic polynucleotides have been widely used to determine the presence of a reverse transcriptase activity and also aid in its purification. By definition, the reverse transcriptase possesses the capability of transcribing RNA into DNA, and this capacity needs to be verified before ascribing a DNA polymerase activity to the reverse transcriptase category. As far as oncornaviruses are concerned, the capacity of a viral reverse transcriptase to transcribe the viral genome into DNA is a criterion which should be satisfied, and which will be considered first. The capability of purified enzymes to transcribe other RNA species and synthetic polymers will be briefly discussed.

The abilities of both avian reverse transcriptases and those of mammalian oncornavirus origin to utilize 60–70 S oncornavirus RNA species as templates have been demonstrated. Studies involving the purified enzyme of both AMV (Grandgenett *et al.*, 1973; Kacian *et al.*, 1971; Leis and Hurwitz, 1972*a*) and RSV (Duesberg *et al.*, 1971*a,b*) have shown that either enzyme is able to synthesize DNA using an avian oncornavirus genome as a template. The reaction conditions for the AMV RNA-templated AMV reverse transcriptase have been reported to require all four deoxyribonucleoside triphosphates (0.2 mM dATP, dCTP, and dGTP and 0.04 mM [³H]dTTP) and 6 mM magnesium (Kacian *et al.*, 1971). Enhanced activity was observed in the presence of KCl (0.1 M) and 0.4 mM dithiothreitol (Kacian *et al.*, 1971). Leis and Hurwitz (1972*a*) obtained similar results and demonstrated that reactions involving manganese and magnesium mixtures (0.2 and 10 mM, respectively) were better than reactions involving one or the other cation alone. Heat-denatured (80°C for 3 min) AMV RNA preparations (or RSV RNA) were less effective as templates for the AMV enzyme than undenatured RNA, the loss of templating capability depending on whether magnesium or manganese–magnesium cation combinations were employed in the reaction mixture (Leis and Hurwitz, 1972*a*). It was also shown that reactions templated by AMV RNA (1) incorporated all four deoxyribonucleotides at approximately equal rates, (2) incorporated more nucleotides at pH 8.2 than at other pH values, and (3) were totally inhibited by ribonuclease or deoxyribonuclease.

Other RNA species which have been effectively used as templates for the AMV enzyme include RSV RNA, R-MuLV RNA, bacteriophage f2, MS2, or Qβ RNA, as well as bulk *E. coli* tRNA (Leis and Hurwitz, 1972*a*). The relative transcription efficiencies varied with the source of RNA, and it was noted that the AMV viral RNA (on a molar basis) was a preferred substrate. Other studies have demonstrated that in addition to oncornaviral and bacteriophage RNAs, globin messenger RNA, influenza viral RNA, ribosomal, tobacco mosaic viral RNA, poliovirus RNA, and *E. coli* 5 S RNA species can be used as templates by the avian reverse transcriptases (Duesberg *et al.*, 1971*a,b*; Efstratiadis *et al.*, 1975; Faras *et al.*, 1972; Goodman and Spiegelman, 1971; Kacian *et al.*, 1972; Ross *et al.*, 1972; Spiegelman *et al.*, 1971; Taylor *et al.*, 1973; Verma *et al.*, 1972).

Of concern in studying the ability of a reverse transcriptase to utilize RNA as a template are the size and fidelity of the transcript made. Both full-length and partial transcripts of globin mRNA have been obtained, and it has been claimed that the amount of the former among the product nucleic acids can be increased by employing high concentrations of deoxyribonucleoside triphosphates in the reaction mixture (Collett and Faras, 1975; Efstratiadis *et al.*, 1975; Imaizumi *et al.*, 1973; Rothenberg and Baltimore, 1976).

Analyses with purified reverse transcriptase derived from certain mammalian C-type viruses have also demonstrated an ability to transcribe oncornaviral and RNA species (Abrell *et al.*, 1975; Gerard and Grandgenett, 1975; Verma, 1975). The purified reverse transcriptases used were derived from R-MuLV, SSV, GaLV, M-MuLV, and M-MuSV(MuLV). The synthesis of DNA obtained was on a par (or nearly so) with that observed for the AMV enzyme. These results lay to rest earlier observations from a variety of laboratories indicating that purified mammalian C-type virus reverse transcriptases are unable to copy heteropolymeric RNA (see review by Green and Gerard, 1974). Purified reverse transcriptase from Mason-Pfizer monkey virus (M-PMV), a virus which has both B-type and C-type properties, also has been shown to use 60–70 S oncornaviral RNA as a template (Abrell *et al.*, 1975).

Using M-MuLV enzyme templated with M-MuLV 60–70 S RNA, magnesium (10 mM) has been reported to be a better divalent cation than manganese (1 mM), and a similar result was obtained when AMV RNA was used to template the reaction (Verma, 1975). Addition of an oligo(dT) primer was shown to increase tenfold the AMV RNA transcription rate. Other RNA species which have been shown to function as templates for the M-MuLV enzyme include 10 S rabbit reticulocyte globin mRNA [with an oligo(dT) primer] and 18 S slime mold ribosomal RNA [with a oligo(dC) primer].

In experiments involving M-MuSV(MuLV) enzyme templated by AMV 60–70 S viral RNA, manganese (0.5 mM) was found to be the preferred divalent cation (Gerard and Grandgenett, 1975). Experiments with the M-PMV enzyme using AMV 60–70 S RNA have been reported (Abrell *et al.*, 1975) in which manganese (1 mM) was the preferred cation by comparison to magnesium (10 mM). Dion *et al.* (1974) with a purified mouse mammary tumor virus enzyme (MuMTV) showed that it could transcribe MuMTV 60–70 S RNA in the presence of magnesium (10 mM, although manganese was not examined), with a notable increase in transcription when an oligo(dT) primer was added.

The ability of the various purified reverse transcriptases to synthesize product when given native calf thymus DNA, activated DNA, or heat-denatured DNA has been demonstrated, as has their ability to use synthetic templates of various sorts and sizes (see review by Green and Gerard, 1974). It has been noted that synthetic polymer-templated reactions may show a different divalent cation preference to that found using native RNA.

In summary, purified reverse transcriptases of avian and mammalian origin (including both C-type and B-type viruses) can use heteropolymeric RNA, DNA, and synthetic polynucleotides of various sorts (see Section 1.1.4b) as templates to synthesize DNA. Whether all mammalian C-type viral enzymes can transcribe RNA remains to be determined (e.g., the hamster HaLV enzyme which was reported to be unable to perform an endogenous reaction). The meaning of the divalent cation preference of an enzyme when using one but not another template is obscure. Since in some cases combinations of manganese and magnesium enhance the transcription rate by comparison to that obtained with either magnesium or manganese (Leis and Hurwitz, 1972a), the reported preferences for one or the other divalent cation may not reflect the optimal reaction conditions. However, until the conditions for synthesizing a biologically competent product DNA are found, we will not know whether one or another or both cations are preferred for faithful transcription.

3.12.4. The Nature of the Viral Genome and Associated RNA Species

Analyses of the nucleic acids associated with preparations of oncornaviruses have demonstrated the presence of a 60–70 S RNA species, 28 S and 18 S RNA (which are probably cellular ribosomal RNA species), 7 S RNA of host origin, 4 S RNA possessing the capability of accepting amino acids, and small amounts of DNA, also apparently of host origin (Bishop *et al.*, 1970*a,b*; Carnegie *et al.*, 1969; Duesberg, 1968; Erikson, 1969; Erikson and Erikson, 1970, 1971, 1972; Erikson *et al.*, 1973; Faras *et al.*, 1973*a,b*; Levinson *et al.*, 1970; Montagnier *et al.*, 1969; Randerath *et al.*, 1971; Riman and Beaudreau, 1970; Robinson *et al.*, 1965; Rosenthal and Zamecnik, 1973*a,b*; Sawyer and Dahlberg, 1973; Trávníček, 1968, 1969; Wang *et al.*, 1973).

Although the exact nature of the 60–70 S RNA remains to be elucidated, the following characteristics have been observed: (1) The 60–70 S RNA is a complex of RNA molecules. (2) By electron microscopy using a variety of Kleinschmidt spreading techniques, including the use of T4 gene 32 protein, or glyoxal–formamide or urea–formamide to denature and extend the 60–70 S viral RNA and its subunits, the following sizes for oncornavirus RNA species have been obtained. For the 60–70 S RNA of RSV-B, a size of $6.2 \pm 0.4 \times 10^6$ was observed (Delius *et al.*, 1975). For the 35 S RNA derived from the Prague strain of RSV and a transformation-defective variant (*td* Pr-RSV-C), sizes of $3.3 \pm 0.2 \times 10^6$ and $2.9 \pm 0.2 \times 10^6$, respectively, were observed (Kung *et al.*, 1975*b*). Similar results were found for the 35 S RNA of Prague RSV-B (2.7 ± 0.1, Delius *et al.*, 1975) and RSV-D ($3.1 \pm 0.3 \times 10^6$, Jacobson and Bromley, 1975). For FeLV a similar size of $3.3 \pm 0.2 \times 10^6$ was observed, and although RD114 initially gave a size of approximately 5×10^6 further analyses using more stringent denaturing conditions gave values of 2.8×10^6 (Kung *et al.*, 1975*a*). It has been suggested from these latter studies of the RD114 RNA that the 35 S subunits are hydrogen-bonded at their 5′ ends (see below). Whether this is true for all oncornaviruses remains to be determined.

Analyses of the sequence complexity of cloned Prague RSV RNA involving determining the proportion of the total labeled RNA that individual labeled oligonucleotides represent has given values of $2.5 \pm 0.6 \times 10^6$ to $3.5 \pm 0.3 \times 10^6$ daltons for the 60–70 S RNA of that virus (Duesberg and Vogt, 1973; Duesberg *et al.*, 1975). Analyses of AMV RNA by kinetic hybridization studies have also indicated that its 60–70 S RNA possesses a sequence complexity of approximately 3.3×10^6 daltons (Baluda *et al.*, 1975).

The subunits of avian leukosis viruses and those of sarcoma viruses are distinguishable. Cloned naturally occurring leukosis viruses or nontransforming derivatives of transforming viruses contain a single RNA subunit type (*b*), whereas viruses capable of transforming cells in tissue culture contain a larger-size subunit (*a*) (Duesberg and Vogt, 1973; Stone *et al.*, 1975). Oligonucleotide fingerprint analyses have suggested that the *b* subunit is a deletion derivative of the *a* type (Lai *et al.*, 1973). Whether virus strains can upon repeated passage contain both *a* and *b* subunit types is a question which has not been completely resolved (Stone *et al.*, 1975), although evidence that the change from cloned transforming virus to transformation-defective strains results in a conversion of all *a* to all *b* RNA subunits argues not only for polyploidy but also for the lack of stable *ab* heterozygotes (Duesberg and Vogt, 1973; Stone *et al.*, 1975; Vogt, 1973).

Taken together, it appears that the viral 60–70 S RNA of avian oncornaviruses is composed of identical or, at least so far, indistinguishable subunits of approximately 3×10^6 daltons size. These results therefore suggest that the genome is polyploid (presumably *aa,* or *bb,* although *ab* heterozygotes may also occur). For mammalian C-type viruses, a similar situation probably exists, although no evidence for distinguishable *a* and *b* subunits has been obtained.

In addition to the major 35 S subunits, low molecular weight tRNA has been demonstrated to be present in the 60–70 S complex. For reasons which will be apparent later, this tRNA is known as a primer RNA. Although some 15–30% of the RNA in avian and other oncornaviruses is small RNA, of the virus-associated 4 S tRNA species it appears that principally only one type (a tryptophan tRNA) is present in the 60–70 S RNA complex in any significant amount. By heat denaturation, the primer Trp tRNA can be separated from the high molecular weight complex and resolved by two-dimensional gel electrophoresis (Dahlberg *et al.*, 1975; Elder and Smith, 1973; Erikson and Erikson, 1971; Faras *et al.*, 1974; Gallagher and Gallo, 1973; Ikemura and Dahlberg, 1973; Nichols and Waddell, 1973; Rosenthal and Zamecnik, 1973a,b; Sawyer and Dahlberg, 1973). Recovery analyses of the free and 60–70 S complexed tRNA indicate that there are one or two copies of Trp tRNA species associated with the 60–70 S complex and some six to eight copies of the same tRNA present in the virion as free tRNA species (Bishop *et al.*, 1973; Dahlberg *et al.*, 1974, 1975; Faras *et al.*, 1974). Although minor amounts of other tRNA species have also been recovered from the 60–70 S complex, these results suggest that the Trp tRNA is the principal species.

A 3′-terminal adenosine nucleoside and 3′ sequence of poly(A) (approximately 150–200 nucleotides) are present ʻon the 35 S RNA subunits of AMV RNA as well as M-MuSV(MuLV) and other oncornavirus 60–70 S RNA species (Rho and Green, 1974; Stephenson *et al.*, 1973).

3.12.5. The Mechanism of DNA Synthesis by Reverse Transcriptases

3.12.5a. The Synthesis of Viral Complementary DNA

All DNA polymerases that have been studied so far synthesize DNA by chain extension; the reverse transcriptase appears to be no exception to this rule. It has been shown that for an endogenously templated reverse transcriptase reaction, the initial DNA is covalently attached in the 3′ terminus of the small primer RNA molecules present in the 60–70 S RNA complex (Bishop *et al.*, 1973; Canaani and Duesberg, 1972; Faras *et al.*, 1973*b*). Purified reverse transcriptases templated by purified 60–70 S RNA frequently synthesize more DNA when supplied with exogenous oligonucleotide primers. It has been shown that these primers need free 3′-hydroxyl groups and that they are also chain-extended at their 3′ ends (Leis and Hurwitz, 1972*a*).

Is primer extension the only way oncornaviruses synthesize DNA? Some evidence has been presented that for reticuloendotheliosis virus (REV) there is little *in vitro* endogenously templated DNA synthesis until some RNA synthesis has occurred (Mizutani and Temin, 1975). The ribonucleotides incorporated by REV were shown to be principally UMP, with lesser amounts of AMP and CMP and very little GMP. It was shown that the RNA synthesis required the presence of a nonionic detergent to rupture the viral envelope, a divalent cation (8 mM magnesium), and one or more ribonucleoside triphosphates. The endogenously templated DNA synthetic rate was stimulated fivefold by the inclusion of four ribonucleoside triphosphates, and threefold by the inclusion of ATP or UTP. Nearest-neighbor analyses indicated that some of the newly synthesized DNA was covalently attached to RNA mostly in . . . pUpdAp . . . linkages with some . . .pAp-dAp . . . bonds also present. Alkali degradation of [³H]ribonucleoside triphosphate-labeled product gave some [³H]uridine and less [³H]adenosine nucleosides, as expected for *de novo* RNA-DNA bridges. Although the size of the RNA portion was reported to be 4 S, it was not determined if the newly synthesized RNA was *de novo* initiated RNA or primer chain extended RNA (e.g., involving a nucleotidyltransferase). The studies reported by Mizutani and Temin (1975) suggest therefore another manner in which DNA synthesis might be obtained other than by just DNA chain extension of a preexisting template-associated RNA primer.

Whether other oncornaviruses possess RNA as well as DNA polymerases remains to be seen. The observations obtained for reticuloendotheliosis virus and related viruses need to be confirmed; however, they raise the possibility that *de novo* RNA initiations or RNA chain extension may be involved in the mechanism of DNA synthesis. What enzyme performs this function needs to be determined.

Mizutani and Temin (1971*b*) have demonstrated that in purified preparations of various oncornaviruses there are not only a variety of associated enzyme activities but also nucleotides of various sorts (dTTP, ATP, dATP, CTP, dCTP, UTP, GDP, ADP, UDP, GMP, AMP, dGMP). The virion-associated enzymes identified by them and other investigators in addition to the endogenous reverse transcriptase and ribonuclease H activities (core) include a DNA ligase (core), hexokinase (core), protein kinase and phosphoprotein phosphatase activities, lactic dehydrogenase (noncore), aminoacyl tRNA synthetase, single- and double-strand specific ribonucleases, ATPase (noncore), nucleotide kinases, methylases, DNA exonuclease and DNA endonuclease activities, nucleotide kinases, methylases, DNA exonuclease and DNA endonuclease activities, nucleotidyltransferases, and nucleoside triphosphate phosphotransferases (Erikson and Erikson, 1972; Gantt *et al.*, 1971, 1973; Grandgenett *et al.*, 1972; Hung, 1973; Hurwitz and Leis, 1972; Mizutani *et al.*, 1970*a*, 1971; Mizutani and Temin, 1971*a*; Moelling *et al.*, 1971; Roy and Bishop, 1971; Trávníček and Riman, 1973). Whether any of the enzyme activities are functional in the DNA polymerization process or are fortuitously present in virions is not known. Although many of these enzymes may be regarded as host enzymes picked up at the time of budding (or in vesicles copurified with the virus particles), it should be remembered that for vaccinia, VSV, and NDV several of the enzymes associated with virus preparations are in fact operative in transcript nucleic acid modification, and at least for VSV and NDV are clearly host derived. For those viruses, mRNA modification is evidently required for eventual mRNA translation.

Whether the presence of host enzymes in oncornavirus preparations is pertinent to the infection process remains to be determined. No evidence for virion RNA polymerase activities necessary for subsequent DNA synthesis has been reported for oncornaviruses other than for REV. It will be of interest therefore to know if this activity is unique to REV and also what its relevance is to the transcription of REV DNA. Other investigators have documented the endogenous synthesis of DNA by REV preparations (Kang, 1975; Waite and Allen, 1975), although they have not investigated the question of *de novo* RNA involvement.

Let us consider the reported involvement of a preexisting RNA primer in DNA synthesis. Where is the primer located? Evidence has been obtained which indicates that the primer is located near the 5′ end of the 35 S RNA subunit(s) (Taylor and Illmensee, 1975). Some evidence suggesting that it is located approximately 110 nucleotides from the 5′ end has also been obtained, as well as evidence that DNA chain extension initially goes from the 3′ end of the primer to the 5′ end of the template (J. Taylor, personal communication). If so, then how is the whole viral complementary DNA strand made?

It has been suggested by electron microscopy of RD114 RNA that the two 35 S subunits are hydrogen-bonded together near or at their 5′ ends (Kung *et al.*, 1975*a*). If this is true for avian and other oncornaviruses and it is true that the primers for each of these viruses are located near the 5′ ends of the 35 S species, then this raises a problem in terms of a 5′ to 3′ direction of product DNA synthesis. What happens to DNA synthesis when primer extension reaches the 5′ end of the 35 S template? Had the 35 S RNA species possessed hydrogen-bonded 3′ to 5′ termini (or protein-linked ends), then one might propose that in some way the primer–product DNA chain is extended over the gap and thence around the circle. To date, however, there is no evidence for hydrogen-bonded circular template species.

In order to obtain a complete viral complementary DNA transcript, one can envision several alternative methods. For example, the following procedures might be proposed:

1. DNA is synthesized in a 5′ to 3′ mode, starting at the 3′ end of the template and proceeding in a continuous fashion to the 5′ end of the template. Either a hairpinlike folded poly(A) sequence at the 3′ end of the 35 S RNA or preexisting oligonucleotides or other RNA species could serve as primers.

2. DNA is synthesized in a 5′ to 3′ mode in separate portions, initially using the Trp tRNA primer and then other primer oligonucleotides. The DNA copies are eventually ligased together to form a linear, continuous, virus-complementary DNA strand. This model requires the presence and availability of other primer oligonucleotides.

3. DNA is synthesized first by chain-extending of the primer Trp RNA until the DNA reaches the 5′ end of the 35 S template. The primer–single-stranded DNA product is displaced by a succeeding primer and a new DNA strand is made. The initial primer–DNA molecule then hydrogen-bonds to the 3′ end of the 35 S RNA [on the poly(A) tail if the DNA contains some poly(T) sequence]. The DNA is then chain-extended to the 5′ terminus of the template. This model requires either that some of the initial RNA sequence near or at the 5′ end of the tem-

plate is reiterated at the 3′ end or that near the 5′ end of the RNA there are sufficient tandem adenylic nucleotides to generate a virus-complementary poly(T) sequence which could be transposed to some part of the polyadenylate 3′ sequence of the viral RNA.

4. DNA synthesis involves one or more *de novo* RNA initiation sequences which are chain-extended by the product DNA and ligased together after removal of the RNA.

5. Synthesis of DNA, starting at the primer, extends over the 5′ end of the RNA to the 3′ end and thence around the RNA until the primer is again reached. After removal of the primer, the DNA is ligased together to give a complete circle. In this model (Fig. 21), the 5′ and 3′ ends [not, though, the 3′ poly(A) tail] are held in juxtaposition by virtue of being hydrogen-bonded to another stretch of RNA (depicted as a sequence on the other 35 S strand in Fig. 21, although an internal region of the same RNA could also function in that capacity). Although particular RNA sequences at the 5′ and 3′ ends are presented in the model, they are for illustration only, and these sequences—other than the 3′ poly(A) and 5′ m⁷GpppGmpCp . . . (Furuichi *et al.*, 1975c)—are not known. It should be noted that in this model no genetic information is lost in making the complete circle of DNA.

One final point should be made before considering the next phase of DNA synthesis: no evidence has been obtained to prove that the genome-associated Trp tRNA is the primer *in vivo*. If evidence can be obtained, then this would strengthen the credibility of the *in vitro* observations.

3.12.5b. The Synthesis of Viruslike DNA Strands

The subsequent synthesis of viruslike DNA is not understood at all. Again, various models can be proposed which involve nicking or degrading the RNA portion of the initial RNA:DNA hybrid, generating 3′-hydroxyl groups, and then chain-extending the RNA fragments by DNA and eventual DNA ligation (Fig. 21). Whether ribonuclease H is involved is not known, and in fact a recent article by Collett and Faras (1976) suggests that ribonuclease H plays no part. An alternative procedure which can be proposed involves using a second primer–DNA product viral complementary strand to initiate the viruslike DNA syn-

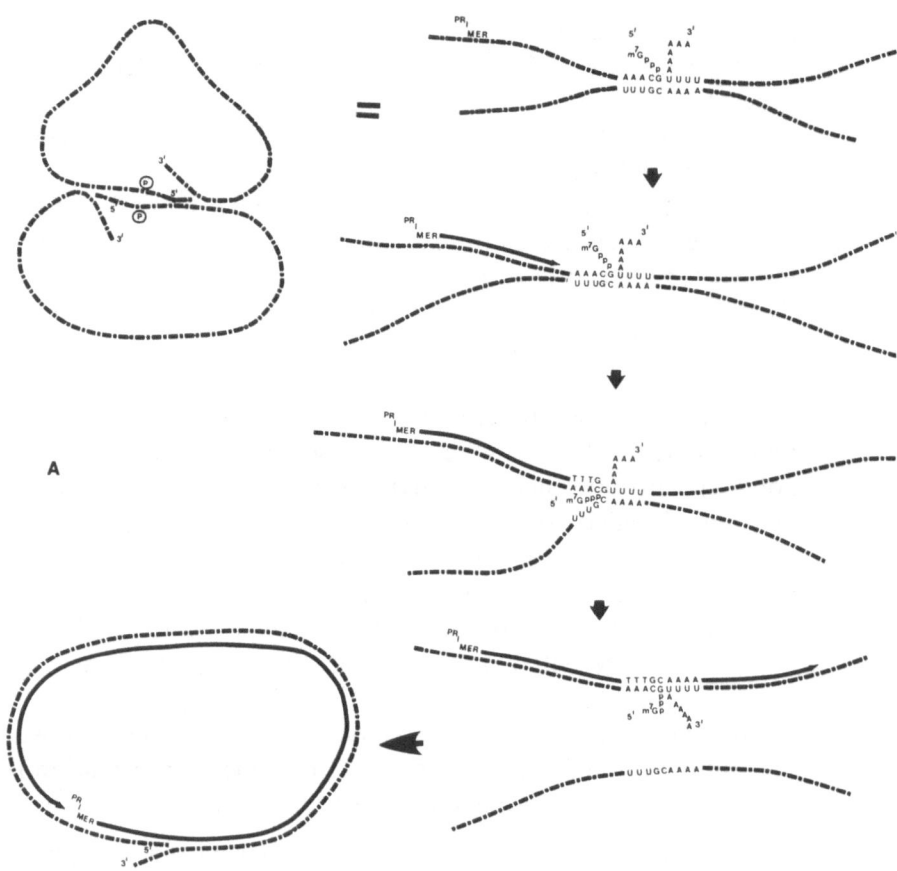

Fig. 21. A hypothetical model for the synthesis of closed circular DNA for a 60–70 S RNA genome. In the first phase of DNA synthesis (A), the two 35 S RNA subunits are envisaged as being hydrogen-bonded together, near their 5′ ends (Kung *et al.*, 1975a). The juxtaposition of the 5′ and 3′ ends [not including the 3′ poly(A) tail] is shown as being hydrogen-bonded to another RNA sequence (in this figure involving the other 35 S subunit—not obligatory to the model). By primer 3′ extension, *de novo* DNA synthesis is shown as coming to, then bridging the 5′, 3′ gap and proceeding around the RNA template to the primer region. Both 35 S RNA subunits could be transcribed simultaneously in this model. Following removal of the primer (B), the DNA (minus strand) could be circularized by a ligase and a complementary DNA (plus strand) synthesized either by displacement of the RNA or by nicking and RNA 3′ extension (as shown). Final ligation would produce a closed circular DNA possessing all the genetic information of the original RNA. The nucleotide sequences shown [other than the poly(A) 3′ tail or, for RSV, the 5′ mGpppGmpCp] are hypothetical (see text). This model of DNA synthesis is hypothetical and originates from suggestions and discussions with Drs. R. W. Compans and E. Hunter. Note that at the prefinal step, recombination between the two DNA derivatives of the 35 S subunits could give rise to high-frequency recombination (E. Hunter, personal communication).

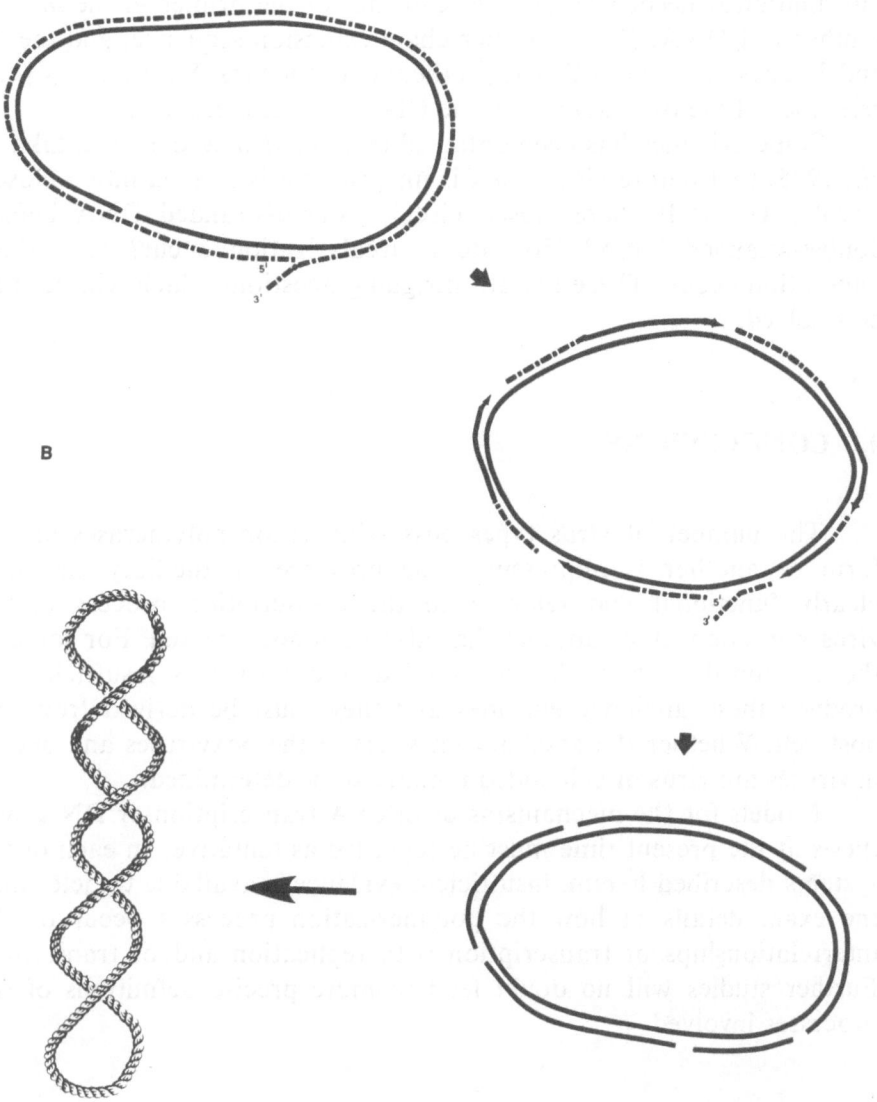

B

thesis on another region of the virus-complementary DNA. For instance, if the 5′ end of the viral RNA is a self-complementary hairpin, then this method could start viruslike DNA synthesis at the 3′ end of the virus-complementary DNA sequence following annealing of the primer–DNA to the virus-complementary DNA sequence.

Clearly, before any reasonable models can be proposed for the synthesis of virus-complementary DNA or viruslike DNA, more information is needed on (1) the relevance of the primer to the *in vivo* synthesis of DNA, (2) the primer chain extension sequence, and the 5′ and 3′ ends of the viral RNA, (3) the fate of the viral RNA, and (4) the relevance of the two subunits to the DNA synthetic process.

Some evidence has been obtained from *in vivo* studies (Guntaka *et al.*, 1975) that before viral DNA is integrated it is formed into a closed circular DNA. Is there closed circular single-stranded DNA before double-stranded DNA? How does circularization occur? How does integration occur? These are all intriguing questions which will need to be resolved.

4. CONCLUSIONS

The number of virus types possessing virion polymerases of one form or another is impressive. The presence of ancillary enzymes, clearly functional and relevant to the transcription process of the viruses in one way or another, has also been documented. For some of the enveloped viruses, the virus-coded information is insufficient to produce these ancillary enzymes and they must be derived from the host cell. Whether the ancillary enzymes of the poxviruses and oncornaviruses are virus or cell coded remains to be determined.

Models for the mechanisms of mRNA transcription or DNA synthesis at the present time must be regarded as tentative. In each of the systems described herein, insufficient evidence is available to determine the exact details of how the polymerization processes occur or the interrelationships of transcription with replication and/or translation. Further studies will no doubt lead to more precise definitions of the processes involved.

ACKNOWLEDGMENTS

I thank Drs. R. Gallo, B. Moss, A. Shatkin, H. Stone, and J. Taylor for permission to use unpublished material and Drs. R. W. Compans and E. Hunter for reading sections of this chapter. Studies conducted in our laboratories and reported herein were aided by U.S. Public Health Service Grant Al 13402.

5. REFERENCES

Aaronson, S. A., Parks, W. P., Scolnick, E. M., and Todaro, G. J., 1971, Antibody to the RNA-dependent DNA polymerase of mammalian C-type RNA tumor viruses, *Proc. Natl. Acad. Sci. USA* **68**:920.

Aaslestad, H. G., Clark, H. F., Bishop, D. H. L., and Koprowski, H., 1971, Comparison of the ribonucleic acid polymerase of two rhabdoviruses, Kern Canyon virus and vesicular stomatitis virus, *J. Virol.* **7**:726.

Abraham, G., Rhodes, D. P., and Banerjee, A. K., 1975, The 5′-terminal structure of the methylated mRNA synthesized *in vitro* by vesicular stomatitis virus, *Cell* **5**:51.

Abrell, J. W., and Gallo, R. C., 1973, Purification, characterization and comparison of the DNA polymerases from two primate RNA tumor viruses, *J. Virol.* **12**:431.

Abrell, J. W., Reitz, M. S., and Gallo, R. C., 1975, Transcription of 70 S RNA by DNA polymerases from mammalian RNA viruses, *J. Virol.* **16**:1566.

Armstrong, S. J., and Barry, R. D., 1974, The topography of RNA synthesis in cells infected with fowl plague virus, *J. Gen. Virol.* **24**:535.

Armstrong, S. J., and Barry, R. D., 1975, The detection of virus-induced RNA synthesis in the nuclei of cells infected with influenza viruses, in: *Negative Strand Viruses*, Vol. 1 (B. W. J. Mahy and R. D. Barry, eds.), pp. 491–499, Academic Press, New York.

Astell, C., Silverstein, S. C., Levin, D. H., and Acs, G., 1972, Regulation of the reovirus RNA transcriptase by a viral capsomere protein, *Virology* **48**:648.

Aubertin, A. M., and McAuslan, B. R., 1972, Virus-associated nucleases: Evidence for endonuclease and exonuclease activity in rabbitpox and vaccinia viruses, *J. Virol.* **9**:554.

Aubertin, A., Palese, P., Tan, K. B., and McAuslan, B. R., 1971, Proteins of a polyhedral cytoplasmic deoxyvirus. III. Structure of frog virus 3 and localisation of virus-associated adenosine triphosphate phosphohydrolase, *J. Virol.* **8**:643.

Baltimore, D., 1970, Viral RNA-dependent DNA polymerase, *Nature (London)* **226**:1209.

Baltimore, D., and Smoler, D. F., 1971, Primer requirement and template specificity of the DNA polymerase of DNA tumor viruses, *Proc. Natl. Acad. Sci. USA* **68**:1507.

Baltimore, D., and Smoler, D. F., 1972, Association of an endoribonuclease with the avian myeloblastosis virus DNA polymerase, *J. Biol. Chem.* **247**:7282.

Baltimore, D., Huang, A. S., and Stampfer, M., 1970, Ribonucleic acid synthesis of vesicular stomatitis virus. II. An RNA polymerase in the virion, *Proc. Natl. Acad. Sci. USA* **66**:572.

Baluda, M. A., Shoyab, M., Markham, P. D., Evans, R. M., and Drohan, W. N., 1975, Base sequence complexity of 35 S avian myeloblastosis virus RNA determined by molecular hybridization kinetics, *Cold Spring Harbor Symp. Quant. Biol.* **39**:869.

Banerjee, A. K., and Rhodes, D. P., 1973, *In vitro* synthesis of RNA that contains polyadenylate by virion-associated RNA polymerase of vesicular stomatitis virus, *Proc. Natl. Acad. Sci. USA* **70**:3566.

Banerjee, A. K., and Shatkin, A. J., 1970, Transcription *in vitro* by reovirus-associated ribonucleic acid-dependent polymerase, *J. Virol.* **6**:1.

Banerjee, A. K., Ward, R. L., and Shatkin, A. J., 1971, Initiation of reovirus mRNA synthesis *in vitro*, *Nature (London) New Biol.* **230**:169.

Barlati, S., and Vigier, P., 1972a, Effect of two rifamycin derivatives on the Rous sarcoma virus transformation system, *J. Gen. Virol.* **17**:221.

Barlati, S., and Vigier, P., 1972b, Selective inhibition of Rous sarcoma virus production in transformed chick fibroblasts by two rifamycin derivatives, *FEBS Lett.* **24**:343.

Barry, R. D., 1964, The effects of actinomycin D and ultraviolet light irradiation on the production of fowl plague virus, *Virology* **24**:563.

Barry, R. D., and Bukrinskaya, A. G., 1968, The nucleic acid of Sendai virus and ribonucleic acid synthesis in cells infected by Sendai virus, *J. Gen. Virol.* **2**:71.

Bean, W. J., and Simpson, R. W., 1973, Primary transcription of the influenza virus genome in permissive cells, *Virology* **56**:646.

Bean, W. J., Jr., and Simpson, R. W., 1975, Virion-associated transcriptase activity of influenza recombinant and mutant strains, *J. Virol.* **16**:516.

Benveniste, R. E., Lieber, M. M., Livingston, D. M., Sherr, C. J., Todaro, G. J., and Kalter, S. S., 1974, Infectious type C virus isolated from a baboon placenta, *Nature (London)* **248**:17.

Bernard, J. P., and Northrop, R. L., 1974, RNA polymerase in mumps virion, *J. Virol.* **14**:183.

Bhattacharyya, J., Xuma, M., Reitz, M., Sarin, P. S., and Gallo, R. C., 1973, Utilization of mammalian 70 S RNA by a purified reverse transcriptase from human myelocytic leukemic cells, *Biochem. Biophys. Res. Commun.* **54**:324.

Bishop, D. H. L., 1971, Complete transcription by the transcriptase of vesicular stomatitis virus, *J. Virol.* **7**:486.

Bishop, D. H. L., 1973, RNA-dependent polymerases, in: *Methods in Molecular Biology,* Vol. 4 (A. I. Laskin and J. A. Last, eds.), pp. 1–62, Dekker, New York.

Bishop, D. H. L., and Flamand, A., 1975, Transcription processes of animal RNA viruses, in: *Control Processes in Virus Multiplication* (D. C. Burke and W. C. Russell, eds.), pp. 95–152, Cambridge University Press, Cambridge.

Bishop, D. H. L., and Roy, P., 1971a, Kinetics of RNA synthesis by vesicular stomatitis virus particles, *J. Mol. Biol.* **57**:513.

Bishop, D. H. L., and Roy, P., 1971b, Properties of the product synthesized by vesicular stomatitis virus particles, *J. Mol. Biol.* **58**:799.

Bishop, D. H. L., and Roy, P., 1972, Dissociation of vesicular stomatitis virus and relation of the virion proteins to the viral transcriptase, *J. Virol.* **10**:234.

Bishop, D. H. L., and Smith, M., 1976, Rhabdoviruses, in: *The Molecular Biology of Animal Viruses* (D. Nayak, ed.), Dekker, New York.

Bishop, D. H. L., Obijeski, J. F., and Simpson, R. W., 1971a, Transcription of the influenza ribonucleic acid genome by a virion polymerase. I. Optimal conditions for *in vitro* activity of the ribonucleic acid-dependent ribonucleic acid polymerase, *J. Virol.* **8**:66.

Bishop, D. H. L., Obijeski, J. F., and Simpson, R. W., 1971b, Transcription of the influenza ribonucleic acid genome by a virion polymerase. II. Nature of the *in vitro* polymerase product, *J. Virol.* **8**:74.

Bishop, D. H. L., Ruprecht, R., Simpson, R. W., and Spiegelman, S., 1971c, Deoxyribonucleic acid polymerase of Rous sarcoma virus: Reaction conditions and analysis of the reaction product nucleic acids, *J. Virol.* **8**:730.

Bishop, D. H. L., Roy, P., Bean, W. J., Jr., and Simpson, R. W., 1972, Transcription of influenza ribonucleic acid genome by a virion polymerase. III. Completeness of the transcription process, *J. Virol.* **10**:689.

Bishop, D. H. L., Emerson, S. U., and Flamand, A., 1974, Reconstitution of infectivity and transcriptase activity of homologous and heterologous viruses: Vesicular stomatitis (Indiana serotype), Chandipura, vesicular stomatitis (New Jersey serotype), and cocal viruses, *J. Virol.* **14**:139.

Bishop, J. M., Levinson, W. E., Quintrell, N., Sullivan, D., Fanshier, L., and Jackson, J., 1970a, The low molecular weight RNAs of Rous sarcoma virus. I. The 4 S RNA, *Virology* **42**:182.

Bishop, J. M., Levinson, W. E., Sullivan, D., Fanshier, L., Quintrell, N., and Jackson, J., 1970b, The low molecular weight RNAs of Rous sarcoma virus. II. The 7 S RNA, *Virology* **42**:927.

Bishop, J. M., Tsan-Deng, C., Faras, A. J., Goodman, H. M., Levinson, W. E., Taylor, J. M., and Varmus, H. E., 1973, Transcription of the Rous sarcoma virus genome by RNA-directed DNA polymerase, in: *Virus Research* (C. F. Fox, ed.), Academic Press, New York.

Black, D. R., and Knight, C. A., 1970, Ribonucleic acid transcriptase activity in purified wound tumor virus, *J. Virol.* **6**:194.

Blair, C. D., and Robinson, W. S., 1968, Replication of Sendai virus. I. Comparison of the viral RNA and virus-specific RNA synthesis with Newcastle disease virus, *Virology* **35**:537.

Blumenthal, T., and Landers, T. A., 1973, The inhibition of nucleic acid binding proteins by aurintricarboxylic acid, *Biochem. Biophys. Res. Commun.* **55**:680.

Borden, E. C., Brockman, W. W., and Carter, W. A., 1971, Selective inhibition by streptovaricin of splenomegaly induces Rausher leukaemia by virus, *Nature (London) New Biol.* **232**:214.

Borsa, J., and Graham, A. F., 1968, Reovirus: RNA polymerase activity in purified virions, *Biochem. Biophys. Res. Commun.* **33**:895.

Borsa, J., Grover, J., and Chapman, J. D., 1970, Presence of nucleoside triphosphate phosphohydrolase activity in purified virions of reovirus, *J. Virol.* **6**:295.

Both, G. W., Banerjee, A. K., and Shatkin, A. J., 1975a, Methylation-dependent translation of viral messenger RNAs *in vitro*, *Proc. Natl. Acad. Sci. USA* **72**:1189.

Both, G. W., Moyer, S. A., and Banerjee, A. K., 1975b, Translation and identification of the mRNA species synthesized *in vitro* by the virion-associated RNA polymerase of vesicular stomatitis virus, *Proc. Natl. Acad. Sci. USA* **72**:274.

Both, G. W., Moyer, S. A., and Banerjee, A. K., 1975c, Translation and identification of the viral mRNA species isolated from subcellular fractions of vesicular stomatitis virus-infected cells, *J. Virol.* **15**:1012.

Both, G. W., Furuichi, Y., Muthukrishnan, S., and Shatkin, A. J., 1975d, Ribosome binding to reovirus mRNA in protein synthesis requires 5′ terminal 7-methylguanosine, *Cell* **6**:185.

Bouloy, M., Krams-Ozden, S., Horodniceanu, F., and Hannoun, C., 1974, Three-segmented RNA genome of Lumbo virus (bunyavirus), *Intervirology* **2**:173.

Bouloy, M., Colbere, F., Krams-Ozden, S., Vialat, P., Garapin, A. C., and Hannoun, C., 1975, Activité RNA polymérasique associée à un bunyavirus (Lumbo), *C. R. Acad. Sci. Ser. D* **280**:213.

Brakel, C., and Kates, J. R., 1974a, Poly(A) polymerase from vaccinia virus-infected cells. I. Partial purification and characterization, *J. Virol.* **14**:715.

Brakel, C., and Kates, J. R., 1974b, Poly(A) polymerase from vaccinia virus-infected cells. II. Product and primer characterization, *J. Virol.* **14**:724.

Bratt, M. A., and Robinson, W. S., 1967, Ribonucleic acid synthesis in cells infected with Newcastle disease virus, *J. Mol. Biol.* **23**:1.

Bratt, M. A., Collins, B. S., Hightower, L. E., Kaplan, J., Tsipis, J. E., and Weiss, S. R., 1975, Transcription and translation of Newcastle disease virus RNA, in: *Negative Strand Viruses,* Vol. 1 (B. W. J. Mahy and R. D. Barry, eds.), pp. 387–408, Academic Press, New York.

Brockman, W. W., Carter, W. A., Li, L.-H., Reusser, F., and Nichol, F. R., 1971, Streptovaricins inhibit RNA dependent DNA polymerase present in an oncogenic RNA virus, *Nature (London)* **230**:249.

Brown, McK., Dorson, J. W., and Bollum, F. J., 1973, Terminal riboadenylate transferase: A poly(A) polymerase in purified vaccinia virus, *J. Virol.* **12**:208.

Busch, H., and Smetana, K., 1970, Effect of drugs and other agents on the nucleolus, in: *The Nucleolus,* Academic Press, New York.

Calvin, M., Joss, V. R., Hackett, A. J., and Owens, R. B., 1971, Effect of rifampicin and two of its derivatives on cells infected with Moloney sarcoma virus, *Proc. Natl. Acad. Sci. USA* **68**:1441.

Canaani, E., and Duesberg, P., 1972, Role of subunits of 60–70 S avian tumor virus ribonucleic acid in its template activity for the viral deoxyribonucleic acid polymerase, *J. Virol.* **10**:23.

Carnegie, J. W., Deeney, A. O. C., Olsen, K. C., and Beaudreau, 1969, An RNA fraction from myeloblastosis virus having properties similar to transfer RNA, *Biochim. Biophys. Acta* **190**:274.

Carter, M. F., Biswall, N., and Rawls, 1973*a,* Characterization of nucleic acid Pichinde virus, *J. Virol.* **11**:61.

Carter, M. F., Murphy, F. A., Brunschwig, J. P., Noonan, C., and Rawls, W. E., 1973*b,* Effects of actinomycin D and ultraviolet and ionizing radiation on Pichinde virus, *J. Virol.* **12**:33.

Carter, M. F., Biswal, N., and Rawls, W. E., 1974, Polymerase activity of Pichinde virus, *J. Virol.* **13**:577.

Carter, W. A., Brockman, W. W., and Borden, E. C., 1971, Streptovaricins inhibit focus formation by MSV (MLV) complex, *Nature (London) New Biol.* **232**:212.

Chandra, P., Zunino, F., Goetz, A., Gericke, D., Thorbeck, R., and DiMarco, A., 1972, Specific inhibition of DNA-polymerases from RNA tumor viruses by some new daunomycin derivatives, *FEBS Lett.* **21**:264.

Chang, C.-T., and Zweerink, H. J., 1971, Fate of parental reovirus in infected cells, *Virology* **46**:544.

Chang, S. H., Hefti, E., Obijeski, J. F., and Bishop, D. H. L., 1974, RNA transcription by the virion polymerase of five rhabdoviruses, *J. Virol.* **13**:652.

Chopra, H. C., and Oie, H. K., 1972, Possible etiological role of virus particles detected in rat and monkey mammary tumors, *J. Natl. Cancer Inst.* **48**:1059.

Chow, N., and Shatkin, A. J., 1975, Blocked and unblocked 5′ termini in reovirus genome RNA, *J. Virol.* **15**:1057.

Chow, N., and Simpson, R. W., 1971, RNA dependent RNA polymerase activity associated with virions and subviral components of myxoviruses, *Proc. Natl. Acad. Sci. USA* **68**:752.

Coffin, J. M., and Temin, H. M., 1971, Comparison of Rous sarcoma virus-specific deoxyribonucleic acid polymerases in virions of Rous sarcoma virus and in Rous sarcoma virus-infected chicken cells, *J. Virol.* **7**:625.

Collett, M. S., and Faras, A. J., 1975, *In vitro* transcription of DNA from the 70 S RNA of Rous sarcoma virus: Identification and characterization of various size classes of DNA transcripts, *J. Virol.* **16**:1220.

Collett, M. S., and Faras, A. J., 1976, *In vitro* transcription of 70 S RNA by the RNA-directed DNA polymerase of Rous sarcoma virus: Lack of influence of RNase H, *J. Virol.* **17**:291.

Collins, B. S., and Bratt, M. A., 1973, Separation of the messenger RNA's of Newcastle disease virus by gel electrophoresis, *Proc. Natl. Acad. Sci. USA* **70**:2544.

Colonno, R. J., and Stone, H. O., 1975, Methylation of messenger RNA of Newcastle disease virus *in vitro* by a virion-associated enyzme, *Proc. Natl. Acad. Sci. USA* **72**:2611.

Compans, R. W., and Caliguiri, L. A., 1973, Isolation and properties of an RNA polymerase from influenza virus-infected cells, *J. Virol.* **11**:441.

Compans, R. W., Content, J., and Duesberg, P. H., 1972, Structure of the ribonucleo-protein of influenza virus, *J. Virol.* **10**:795.

Content, J., and Duesberg, P. H., 1971, Base sequence differences among the ribonu-cleic acids of influenza virus, *J. Mol. Biol.* **62**:273.

Dahlberg, J. E., Sawyer, R. C., Taylor, J. M., Faras, A. J., Levinson, W. E., Goodman, H. M., and Bishop, J. M., 1974, Transcription of DNA from the 70 S RNA of Rous sarcoma virus. I. Identification of a specific 4 S RNA which serves as primer, *J. Virol.* **13**:1126.

Dahlberg, J. E., Harada, F., and Sawyer, R. C., 1975, Structure and properties of an RNA primer for initiation of Rous sarcoma virus DNA synthesis *in vitro, Cold Spring Harbor Symp. Quant. Biol.* **39**:925.

Delius, H., Duesberg, P. H., and Mangel, W. F., 1975, Electron microscope measure-ments of Rous sarcoma virus RNA, *Cold Spring Harbor Symp. Quant. Biol.* **39**:835.

Diggelman, H., and Weissman, C., 1969, Rifampicin inhibits focus formation in chick fibroblasts infected with Rous sarcoma virus, *Nature (London)* **224**:1277.

Dion, A. S., Vaidya, A. B., Fout, G. S., and Moore, D. H., 1974, Isolation and characterization of RNA-directed DNA polymerase from a B-type RNA tumor virus, *J. Virol.* **14**:40.

Donoghue, T. P., and Hayashi, Y., 1972, Cytoplasmic polyhedrosis virus (CPV) of *Malacosoma disstria*: RNA polymerase activity in purified free virions, *Can. J. Microbiol.* **18**:207.

Downer, D. N., Rogers, H. W., and Randall, C. C., 1973, Endogenous protein kinase and phosphate acceptor protein in vaccinia virus, *Virology* **52**:13.

Duesberg, P. H., 1968, Physical properties of Rous sarcoma virus RNA, *Proc. Natl. Acad. Sci. USA* **60**:1511.

Duesberg, P. H., 1969, Distinct subunits of the ribonucleoprotein of influenza virus, *J. Mol. Biol.* **42**:485.

Duesberg, P. H., and Canaani, E., 1970, Complementarity between Rous sarcoma virus (RSV) RNA and the *in vitro*-synthesized DNA of the virus-associated DNA polymerase, *Virology* **42**:783.

Duesberg, P. H., and Vogt, P. K., 1973, RNA species obtained from clonal lines of avian sarcoma and from avian leukosis virus, *Virology* **54**:207.

Duesberg, P., Helm, K. V. D., and Canaani, E., 1971*a*, Properties of a soluble DNA polymerase isolated from Rous sarcoma virus, *Proc. Natl. Acad. Sci. USA* **68**:747.

Duesberg, P. H., Helm, K. V. D., and Canaani, E., 1971*b*, Comparative properties of

RNA and DNA templates for the DNA polymerase of Rous sarcoma virus, *Proc. Natl. Acad. Sci. USA* **68**:2505.

Duesberg, P., Vogt, P. K., Beemon, K., and Lai, M., 1975, Avian RNA tumor viruses: Mechanism of recombination and complexity of the genome, *Cold Spring Harbor Symp. Quant. Biol.* **39**:847.

Efstratiadis, A., Maniatis, T., Kafatos, F. C., Jeffrey, A., and Vournakis, J. N., 1975, Full length and discrete partial reverse transcripts of globin and chorion mRNA's, *Cell* **4**:367.

Ehrenfeld, E., 1974, Polyadenylation of vesicular stomatitis virus messenger RNA, *J. Virol.* **13**:1055.

Ehrenfeld, E., and Summers, D. F., 1972, Adenylate-rich sequences in vesicular stomatitis virus messenger ribonucleic acid, *J. Virol.* **10**:683.

Elder, K. T., and Smith, A. E., 1973, Methionine transfer ribonucleic acids of avian myeloblastosis virus, *Proc. Natl. Acad. Sci. USA* **70**:2823.

Emerson, S. U., and Wagner, R. R., 1972, Dissociation and reconstitution of the transcriptase and template activities of vesicular stomatitis B and T virions, *J. Virol.* **10**:297.

Emerson, S. U., and Wagner, R. R., 1973, L protein requirement for *in vitro* RNA synthesis by vesicular stomatitis virus, *J. Virol.* **12**:1325.

Emerson, S. U., and Yu, Y.-H., 1975, Both NS and L proteins are required for *in vitro* RNA synthesis by vesicular stomatitis virus, *J. Virol.* **15**:1348.

Ensinger, M. J., Martin, S. A., Paoletti, E., and Moss, B., 1975, Modification of the 5′-terminus of mRNA by soluble guanylyl and methyl transferases from vaccinia virus, *Proc. Natl. Acad. Sci. USA* **72**:2525.

Erikson, R. L., 1969, Studies on the RNA from avian myeloblastosis virus, *Virology* **37**:124.

Erikson, E., and Erikson, R. L., 1970, Isolation of amino acid acceptor RNA from purified avian myeloblastosis virus, *J. Mol. Biol.* **52**:387.

Erikson, E., and Erikson, R. L., 1971, Association of 4 S ribonucleic acid with oncornavirus ribonucleic acids, *J. Virol.* **8**:254.

Erikson, E., and Erikson, R. L., 1972, Transfer ribonucleic acid synthetase activity associated with avian myeloblastosis virus, *J. Virol.* **9**:231.

Erikson, E., Erikson, R. L., Henry, B., and Pace, N. R., 1973, Comparison of oligonucleotides produced by RNase T_1 digestion of 7 S RNA from avian and murine oncornaviruses and from uninfected cells, *Virology* **53**:40.

Etkind, P. R., and Krug, R. M., 1974, Influenza viral mRNA. *Virology* **62**:38.

Etkind, P. R., and Krug, R. M., 1975, Purification of influenza viral complementary RNA: Its genetic content and activity in wheat germ cell-free extracts, *J. Virol.* **16**:1464.

Fanshier, L., Garapin, A. C., McDonnell, J., Faras, A., Levinson, W., and Bishop, J. M., 1971, DNA polymerase associated with avian tumor viruses: Secondary structure of the deoxyribonucleic acid product, *J. Virol.* **7**:77.

Faras, A. J., Taylor, J. M., McDonnell, J. P., Levinson, W. E., and Bishop, J. M., 1972, Purification and characterization of the DNA polymerase associated with RSV, *Biochem.* **11**:1334.

Faras, A. J., Garapin, A. C., Levinson, W. E., Bishop, J. M., and Goodman, H. M., 1973a, Characterization of the low molecular weight RNAs associated with the 70 S RNA of Rous sarcoma virus, *J. Virol.* **12**:334.

Faras, A. J., Taylor, J. M., Levinson, W. E., Goodman, H. M., and Bishop, J. M., 1973*b*, RNA-directed DNA polymerase of Rous sarcoma virus: Initiation of synthesis with 70 S viral RNA as template, *J. Mol. Biol.* **79**:163.

Faras, A. J., Dahlberg, J. E., Sawyer, R. C., Harada, F., Taylor, J. M., Levinson, W. E., Bishop, J. M., and Goodman, H. M., 1974, Transcription of DNA from the 70 S RNA of Rous sarcoma virus. II. Structure of a 4 S RNA primer, *J. Virol.* **13**:1134.

Federer, K. E., Burrows, R., and Brooksby, J. B., 1967, Vesicular stomatitis virus— The relationship between some strains of the Indiana serotype, *Res. Vet. Sci.* **8**:103.

Fischinger, P. J., Peebles, P. T., Nomura, S., and Haapala, D. K., 1973, Isolation of an RD-114-like oncornavirus from a cat cell line, *J. Virol.* **11**:978.

Flamand, A., and Bishop, D. H. L., 1973, Primary *in vivo* transcription of vesicular stomatitis virus and temperature-sensitive mutants of five vesicular stomatitis virus complementation groups, *J. Virol.* **12**:1238.

Flamand, A., and Bishop, D. H. L., 1974, *In vivo* synthesis of RNA by vesicular stomatitis virus and its mutants, *J. Mol. Biol.* **87**:31.

Follett, E. A. C., Pringle, C. R., Wunner, W. H., and Skehel, J. J., 1974, Virus replication in enucleate cells: Vesicular stomatitis and influenza virus, *J. Virol.* **13**:394.

Fridlender, B., and Weissbach, A., 1971, DNA polymerases of tumor virus: Specific effect of ethidium bromide on the use of different synthetic templates, *Proc. Natl. Acad. Sci. USA* **68**:3116.

Fujinaga, K., and Green, M., 1971, Mechanism of carcinogenesis by RNA tumour viruses: The RNA- and DNA-dependent DNA polymerase activities of Feline sarcoma virus, *J. Gen. Virol.* **12**:85.

Fujinaga, K., Parson, J. T., Beard, J. W., Beard, D., and Green, M., 1970, Mechanism of carcinogenesis by RNA tumor viruses. III. Formation of carcinogenesis by RNA tumor viruses. III. Formation of RNA-DNA complex and duplex DNA molecules by the DNA polymerase(s) of avian myeloblastosis virus, *Proc. Natl. Acad. Sci. USA* **67**:1432.

Furman, P. A., and Hallum, J. V., 1973, RNA-dependent DNA polymerase activity in preparations of a mutant of Newcastle disease virus arising from persistently infected cells, *J. Virol.* **12**:548.

Furuichi, Y., 1974, Methylation-coupled transcription by virus-associated transcriptase of cytoplasmic polyhedrosis virus containing double-stranded RNA, *Nucleic Acids Res.* **1**:802.

Furuichi, Y., and Miura, K.-I., 1975, A blocked structure at the 5′ terminus of mRNA from cytoplasmic polyhedrosis virus, *Nature (London)* **253**:374.

Furuichi, Y., Muthukrishnan, S., and Shatkin, A. J., 1975*a*, 5′-Terminal m⁷G(5′)ppp(5′)Gᵐp *in vivo:* Identification in reovirus genome RNA, *Proc. Natl. Acad. Sci. USA* **72**:742.

Furuichi, Y., Morgan, M., Muthukrishnan, S., and Shatkin, A. J., 1975*b*, Reovirus messenger RNA contains a methylated, blocked 5′-terminal structure: m⁷G(5′)ppp(5′)GᵐpCp-, *Proc. Natl. Acad. Sci. USA* **72**:362.

Furuichi, Y., Shatkin, A. J., Stavnezer, E., and Bishop, J. M., 1975*c*, Blocked, methylated 5′-terminal sequence in avian sarcoma virus RNA, *Nature (London)* **257**:618.

Galet, H., and Prevec, L., 1973, Polyadenylate synthesis by extracts from L cells infected with vesicular stomatitis virus, *Nature (London) New Biol.* **243**:200.

Gallagher, R. E., and Gallo, R. C., 1973, Chromatographic analyses of isoaccepting tRNAs from avian myeloblastosis virus, *J. Virol.* **12**:449.

Gantt, R., Stromberg, K. J., and Montes De Oca, F., 1971, Specific RNA methylase associated with avian myeloblastosis virus, *Nature (London)* **234**:35.

Gantt, R., Smith, G. H., and Julian, B., 1973, Base specific methylase activity in RNA tumor viruses: Avian leukosis virion-associated RNA methylase(s), *Virology* **52**:584.

Garapin, A. C., McDonnell, J. P., Levinson, W., Quintrell, N., Fanshier, L., and Bishop, J. M., 1970, Deoxyribonucleic acid polymerase associated with Rous sarcoma virus and avian myeloblastosis virus: Properties of the enzyme and its product, *J. Virol.* **6**:589.

Garapin, A. C., Varmus, H. E., Faras, A. J., Levinson, W. E., and Bishop, J. M., 1973, RNA-directed DNA synthesis by virions of Rous sarcoma virus: Further characterization of the templates and the extent of their transcription, *Virology* **52**:264.

Gerard, G. F., and Grandgenett, D. P., 1975, Purification and characterization of the DNA polymerase and RNase H activities in Moloney murine sarcoma-leukemia virus, *J. Virol.* **15**:785.

Gerwin, B. I., and Milstein, J. B., 1972, An oligonucleotide affinity column for RNA-dependent DNA polymerase from RNA tumor viruses, *Proc. Natl. Acad. Sci. USA* **69**:2599.

Gibson, W., and Verma, I. M., 1974, Studies on the reverse transcriptase of RNA tumor viruses: Structural relatedness of two subunits of avian RNA tumor viruses, *Proc. Natl. Acad. Sci. USA* **71**:4991.

Gillies, S., Bullivant, S., and Bellamy, A. R., 1971, Viral RNA polymerases: Electron microscopy of reovirus reaction cores, *Science* **174**:694.

Glass, S. E., McGeoch, D., and Barry, R. D., 1975, Characterization of the mRNA of influenza virus, *J. Virol.* **16**:1435.

Gold, P., and Dales, S., 1968, Localization of nucleotide phosphohydrolase within vaccinia virus, *Proc. Natl. Acad. Sci. USA* **60**:845.

Goodman, N. C., and Spiegelman, S., 1971, Distinguishing reverse transcriptase of an RNA tumor virus from other known DNA polymerases, *Proc. Natl. Acad. Sci. USA* **68**:2203.

Grandgenett, D. P., and Green, M. 1974, Different mode of action of ribonuclease H in purified α and $\alpha\beta$ RNA-directed polymerase from avian myeloblastosis virus, *J. Biol. Chem.* **249**:5148.

Grandgenett, D. P., Gerard, G. F., and Green, M., 1972, Ribonuclease H: A ubiquitous activity in virions of ribonucleic acid tumor viruses, *J. Gen. Virol.* **10**:1136.

Grandgenett, D. P., Gerard, G. F., and Green, M., 1973, A single subunit from avian myeloblastosis virus with both RNA-directed DNA polymerase and ribonuclease H activity, *Proc. Natl. Acad. Sci. USA* **70**:230.

Gravell, M., and Cromeans, T. L., 1971, Mechanisms involved in nongenetic reactivation of frog polyhedral cytoplasmic deoxyribovirus: Evidence for an RNA polymerase in the virion, *Virology* **46**:39.

Gravell, M., and Cromeans, T., 1972, Virion-associated protein kinase and its involvement in non-genetic reactivation of frog polyhedral cytoplasmic deoxyribovirus, *Virology* **48**:847.

Green, M., and Gerard, G. F., 1974, RNA-directed DNA polymerase—Properties and functions in oncogenic RNA viruses and cells, *Prog. Nucleic Acid Res. Mol. Biol.* **14**:187.

Green, M., Rokutanda, M., Fujinaga, K., Ray, R. K., Rokutanda, H., and Gurgo, C., 1970, Mechanism of carcinogenesis by RNA tumor viruses. I. An RNA-dependent DNA polymerase in murine sarcoma viruses, *Proc. Natl. Acad. Sci. USA* **67**:385.

Green, M., Bragdon, J., and Rankin, A., 1972, 3-Cyclic amine derivatives of rifamycin: Strong inhibitors of the DNA polymerase activity of RNA tumor viruses, *Proc. Natl. Acad. Sci. USA* **69**:1294.

Green, M., Grandgenett, D., Gerard, G., Rho, H. M., Loni, M. C., Robins, M., Salzberg, S., Shanmugam, G., Bhaduri, S., and Vecchio, G., 1975, Properties of oncornavirus RNA-directed DNA polymerase, the RNA template, and the intracellular products formed early during infection and cell transformation, *Cold Spring Harbor Symp. Quant. Biol.* **39**:975.

Greenman, R. L., and Robinson, W. S., 1974, DNA polymerase in the core of human hepatitis B virus candidate, *J. Virol.* **13**:1231.

Gregoriades, A., 1970, Actinomycin D and influenza virus multiplication in the chick embryo fibroblasts, *Virology* **42**:905.

Gross, P. A., Fong, D. K. Y., and Hsiung, G. D., 1973, Characterization of guinea pig C-type virus, *Proc. Soc. Exp. Biol. Med.* **143**:367.

Guntaka, R. V., Bishop, J. M., and Varmus, H. E., 1975, Synthesis and integration of Rous sarcoma virus proviral DNA in duck embryo fibroblast cells, in: *Fundamental Aspects of Neoplasia* (A. A. Gottlieb, O. J. Plescia, and D. H. L. Bishop, eds.), pp. 315–323, Springer-Verlag, New York.

Gurgo, C., Ray, R. K., Thiry, L., and Green, M., 1971, Inhibitors of the RNA and DNA dependent polymerase activities of RNA tumor viruses, *Nature (London) New Biol.* **229**:111.

Haase, A. T., and Baringer, J. R., 1974, The structural polypeptides of RNA slow viruses, *Virology* **57**:238.

Haase, A. T., Garapin, A. C., Faras, A. J., Taylor, J. M., and Bishop, J. M., 1974, A comparison of the high molecular weight RNAs of Visna virus and Rous sarcoma virus, *Virology* **57**:259.

Hackett, A. J., and Sylvester, S. S., 1972*a*, Cell line derived from Balb/3T3 that is transformed by murine leukaemia virus: A focus assay for leukaemia virus, *Nature (London) New Biol.* **239**:166.

Hackett, A. J., and Sylvester, S. S., 1972*b*, Inhibition of MLV-induced transformization in Balb/3T3 derived cells, *Nature (London) New Biol.* **239**:164.

Hanafusa, T., and Hanafusa, H., 1971, Noninfectious RSV deficient in DNA polymerase, *Virology* **43**:313.

Hanafusa, H., Baltimore, D., Smoler, D., Watson, K. F., Yaniv, A., and Spiegelman, S., 1972, Absence of polymerase protein in virions of alpha-type Rous sarcoma virus, *Science* **177**:1188.

Harewood, K. R., Vidrine, J. G., Larson, D. L., Wolff, J. S., Schidlovsky, G., and Mayyasi, S. A., 1973, Biochemical and morphological studies of simian sarcoma virus type 1, *Biochim. Biophys. Acta* **308**:252.

Harter, D. H., Schlom, J., and Spiegelman, S., 1971, Characterization of Visna virus nucleic acid, *Biochim. Biophys. Acta* **240**:435.

Hastie, N. D., and Mahy, B. W. J., 1973, RNA-dependent RNA polymerase in nuclei of cells infected with influenza virus, *J. Virol.* **12**:951.

Hatanaka, M., Huebner, R. J., and Gilden, R. V., 1970, DNA polymerase activity associated with RNA tumor viruses, *Proc. Natl. Acad. Sci. USA* **67**:143.

Hay, A. J., and Joklik, W. K., 1971, Demonstration that the same strand of reovirus genome RNA is transcribed *in vitro* and *in vivo, Virology* **44**:450.

Hefti, E., and Bishop, D. H. L., 1975*a*, The 5′ nucleotide sequence of vesicular stomatitis viral RNA, *J. Virol.* **15**:90.

Hefti, E., and Bishop, D. H. L., 1975*b*, The 5′ sequence of VSV viral RNA and its *in vitro* transcription product RNA, *Biochem. Biophys. Res. Commun.* **66**:785.

Hefti, E., and Bishop, D. H. L., 1976, The 5′ sequences of VSV *in vitro* transcription product RNA (±SAM), *Biochem. Biophys. Res. Commun.* **68**:393.

Hefti, E., Roy, P., and Bishop, D. H. L., 1975, The initiation of transcription by influenza virion transcriptase, in: *Negative Strand Viruses,* Vol. 1 (B. W. J. Mahy and R. D. Barry, eds.), pp. 307–326, Academic Press, New York.

Hightower, L. E., Morrison, T. G., and Bratt, M. A., 1975, Relationships among the polypeptides of Newcastle disease virus, *J. Virol.* **16**:1599.

Hirschman, S. Z., 1971, Inhibitors of DNA polymerases of murine leukemia viruses: Activity of ethidium bromide, *Science* **173**:441.

Hirschman, S. Z., 1972, Inactivation of DNA polymerases of murine leukaemia viruses by calcium elenolate, *Nature (London) New Biol.* **238**:277.

Howk, R. S., Rye, L. A., Killeen, L. A., Scolnick, E. M., and Parks, W. P., 1973, Characterization and separation of viral DNA polymerases in mouse milk, *Proc. Natl. Acad. Sci. USA* **70**:2117.

Hsiung, D. G., 1972, Activation of guinea pig C-type virus in cultured spleen cells by 5-bromo-2′deoxyuridine, *J. Natl. Cancer Inst.* **49**:567.

Huang, A. S., and Baltimore, D., 1970, Defective viral particles and viral disease processes, *Nature (London)* **226**:325.

Huang, A. S., and Manders, E., 1972, Ribonucleic acid synthesis of vesicular stomatitis virus. IV. Transcription by standard virus in the presence of defective interfering particles, *J. Virol.* **9**:909.

Huang, A. S:, Baltimore, D., and Stampfer, M., 1970, Ribonucleic acid synthesis of vesicular stomatitis virus. III. Multiple complementary messenger RNA molecules, *Virology* **42**:946.

Huang, A. S., Baltimore, D., and Bratt, M. A., 1971, Ribonucleic acid polymerase in virions of Newcastle disease virus: Comparison with the vesicular stomatitis virus polymerase, *J. Virol.* **7**:389.

Hung, P. P., 1973, Ribonucleases of Rous sarcoma virus, *Virology* **51**:287.

Hunt, D. M., and Wagner, R. R., 1975, Inhibition by aurintricarboxylic acid and polyethylene sulfonate of RNA transcription of vesicular stomatitis virus, *J. Virol.* **16**:1146.

Hurwitz, J., and Leis, J. P., 1972, RNA-dependent DNA polymerase activity of RNA tumor viruses. I. Directing influence of DNA in the reaction, *J. Virol.* **9**:116.

Ikemura, T., and Dahlberg, J. E., 1973, Small RNAs of *E. coli*. I. Characterization by polyacrylamide gel electrophoresis and fingerprint analysis, *J. Biol. Chem.* **248**:5024.

Imaizumi, T., Diggelman, H., and Scherrer, K., 1973, Demonstration of globin messenger sequences in giant nuclear precursors of messenger RNA of avian erythroblasts, *Proc. Natl. Acad. Sci. USA* **70**:1122.

Imblum, R. L., and Wagner, R. R., 1974, Protein kinase and phosphoproteins of vesicular stomatitis virus, *J. Virol.* **13**:113.

Imblum, R. L., and Wagner, R. R., 1975, Inhibition of viral transcriptase by immunoglobulin directed against the nucleocapsid NS protein of vesicular stomatitis virus, *J. Virol.* **15**:1357.

Jacobson, A. B., and Bromley, P. A., 1975, Determination of the molecular weight of the RNA subunits of Rous sarcoma virus by electron microscopy, *Cold Spring Harbor Symp. Quant. Biol.* **39**:845.

Joklik, W. K., 1972, Studies on the effect of chymotrypsin on reovirions, *Virology* **49**:700.

Joss, V. R., Hughes, A. M., and Calvin, M., 1973, Effect of dimethylbenzyldesmethylrifampicin (DMB) on chemically induced mammary tumours in rats, *Nature* (*London*) *New Biol.* **242**:88.

Kacian, D. C., Watson, K. F., Burny, A., and Spiegelman, S., 1971, Purification of the DNA polymerase of avian myeloblastosis virus, *Biochim. Biophys. Acta* **246**:365.

Kacian, D. L., Spiegelman, S., Bank, A., Terada, M., Metafora, S., Dow, L., and Marks, P. A., 1972, *In vitro* synthesis of DNA components of human genes for globins, *Nature* (*London*) *New Biol.* **235**:167.

Kang, C.-Y., 1975, Characterization of endogenous RNA-directed DNA polymerase activity of reticuloendotheliosis viruses, *J. Virol.* **16**:880.

Kang, C.-Y., and Temin, H. M., 1973, Lack of sequence homology among RNAs of avian leukosis-sarcoma viruses, reticuloendotheliosis viruses, and chicken endogenous RNA-directed DNA polymerase activity, *J. Virol.* **12**:1314.

Kang, H. S., and McAuslan, B. R., 1972, Virus associated nucleases: Location and properties of deoxyribonucleases and ribonucleases in purified frog virus 3, *J. Virol.* **10**:202.

Kaplan, P. M., Greenman, R. L., Gerin, J. L., Purcell, R. H., and Robinson, W. S., 1973, DNA polymerase associated with human hepatitis B antigen, *J. Virol.* **12**:995.

Kapuler, A. M., 1970, An extraordinary temperature dependence of the reovirus transcriptase, *Biochemistry* **9**:4453.

Kapuler, A. M., 1971, Reovirus core transcriptase and ethidium bromide: A continuous fluorimetric assay for polynucleotide synthesis based on the secondary structure of mRNA, *Biochim. Biophys. Acta* **238**:363.

Kapuler, A. M., Mendelsohn, N., Klett, H., and Acs, G., 1970, Four base specific 5′ triphosphatases in the subviral core of reovirus, *Nature* (*London*) **225**:1209.

Kates, J., 1970, Transcription of the vaccinia virus genome and the occurrence of polyriboadenylic acid sequences in messenger RNA, *Cold Spring Harbor Symp. Quant. Biol.* **35**:743.

Kates, J., and Beeson, J., 1970*a*, Ribonucleic acid synthesis in vaccinia virus. I. The mechanism of synthesis and release of RNA in vaccinia cores, *J. Mol. Biol.* **50**:1.

Kates, J., and Beeson, J., 1970*b*, Ribonucleic acid synthesis in vaccinia virus. II. Synthesis of polyriboadenylic acid, *J. Mol. Biol.* **50**:19.

Kates, J. R., and McAuslan, B. R., 1967, Poxvirus DNA-dependent RNA polymerase, *Proc. Natl. Acad. Sci. USA* **58**:134.

Kaverin, N. I., and Varich, N. L., 1974, Newcastle disease virus-specific RNA: Polyacrylamide gel analysis of single-stranded RNA and hybrid duplexes, *J. Virol.* **13**:253.

Kawakami, T. G., Huff, S. E., Buckley, P. M., Dungworth, D. C., Snyder, S. P., and Gilden, R. V., 1972, C-type virus associated with gibbon lymphosarcoma, *Nature* (*London*) *New Biol.* **235**:170.

Keller, W., and Crouch, R., 1972, Degradation of DNA-RNA hybrids by ribonuclease H and DNA polymerases of cellular and viral origin, *Proc. Natl. Acad. Sci. USA* **69**:3360.

Kelly, D. C., and Robertson, J. S., 1973, Icosahedral cytoplasmic deoxyriboviruses, *J. Gen. Virol. Suppl.* **20**:17.

Kelly, D. C., and Tinsley, T. W., 1973, Ribonucleic acid polymerase activity associated with particles of iridescent virus types 2 and 6, *J. Invert. Pathol.* **22**:199.

Kelly, D. C., Avery, R. J., and Dimmock, N. J., 1974, Failure of an influenza virus to initiate infection in enucleate BHK cells, *J. Virol.* **13**:1155.

Kettman, R., Portetelle, D., Mammerickx, M., Cleuter, Y., Dekegel, D., Calowx, M., Ghysdael, J., Burny, A., and Chantrenne, H., 1976, Bovine leukemia virus: Exogenous RNA oncogenic virus? *Proc. Natl. Acad. Sci. USA* **73**:1014.

Kiessling, A. A., and Nieman, P. E., 1972, RNA tumor virus DNA polymerase: Activity with exogenous primers, *Biochem. Biophys. Acta* **272**:147.

Kingsbury, D. W., 1966, Newcastle disease virus RNA. II. Preferential synthesis of RNA complementary to parental viral RNA by chick embryo cells, *J. Mol. Biol.* **18**:204.

Kingsbury, D. W., 1973, Cell-free translation of paramyxovirus messenger RNA, *J. Virol.* **12**:1020.

Kingsbury, D. W., and Webster, R. G., 1969, Some properties of influenza virus nucleocapsids, *J. Virol.* **4**:219.

Kingsbury, D. W., and Webster, R. G., 1973, Cell-free translation of influenza virus messenger RNA, *Virology* **56**:654.

Kleiman, J., and Moss, B., 1973, Protein kinase activity from vaccinia virions: Solubilization and separation into heat-labile and heat-stable components, *J. Virol.* **12**:684.

Kleiman, J. H., and Moss, B., 1975a, Purification of a protein kinase and two phosphate acceptor proteins from vaccinia virions, *J. Biol. Chem.* **250**:2420.

Kleiman, J. H., and Moss, B., 1975b, Characterization of a protein kinase and two phosphate acceptor proteins from vaccinia virions, *J. Biol. Chem.* **250**:2430.

Knipe, D., Rose, J. K., and Lodish, H. F., 1975, Translation of individual species of vesicular stomatitis viral mRNA, *J. Virol.* **15**:1004.

Kolakofsky, D., de la Tour, E. B., and Bruschi, A., 1974, Self-annealing of Sendai virus RNA, *J. Virol.* **14**:33.

Kondo, M., and Weissmann, C., 1972, Polyethylene sulfonate as inhibitor of initiation by Qβ replicase, *Biochim. Biophys. Acta* **259**:41.

Kotler, M., and Becker, Y., 1971, Rifampicin and distamycin A as inhibitors of Rous sarcoma virus reverse transcriptase, *Nature (London) New Biol.* **234**:212.

Kotler, M., and Becker, Y., 1972, Effect of distamycin A and Congocidine on DNA synthesis by Rous sarcoma virus reverse transcriptase, *FEBS Lett.* **22**:222.

Kruegar, R. F., and Mayer, G. D., 1970, Tilorone hydrochloride: An orally active antiviral agent, *Science* **169**:1213.

Krug, R. M., 1972, Cytoplasmic and nucleoplasmic viral RNPs in influenza virus-infected MDCK cells, *Virology* **50**:103.

Krug, R. M., and Etkind, P. E., 1973, Cytoplasmic and nuclear virus-specific proteins in influenza virus-infected MDCK cells, *Virology* **56**:334.

Kung, H.-J., Bailey, J. M., Davidson, N., Nicolson, M. O., and McAllister, R. M., 1975a, Structure, subunit composition, and molecular weight of RD-114 RNA, *J. Virol.* **16**:397.

Kung, H. J., Bailey, J. M., Davidson, N., Vogt, P. K., Nicolson, M. O., and McAllister, R. M., 1975b, Electron microscope studies of tumor virus RNA, *Cold Spring Harbor Symp. Quant. Biol.* **39**:827.

Lai, M. M. C., Duesberg, P. H., Horst, J., and Vogt, P. K., 1973, Avian tumor virus
RNA: A comparison of three sarcoma viruses and their transformation defective
derivatives by oligonucleotide fingerprinting and DNA-RNA hybridization, *Proc.
Natl. Acad. Sci. USA* **70**:2266.

Lazarowitz, S. G., Compans, R. W., and Choppin, P. W., 1971, Influenza virus
structural and non-structural proteins in infected cells and their plasma membranes,
Virology **46**:830.

Leamnson, R. N., and Reichmann, M. E., 1974, The RNA of defective vesicular stomatitis virus particles in relation to viral cistrons, *J. Mol. Biol.* **85**:551.

Leis, J. P., and Hurwitz, J., 1972a, RNA-dependent DNA polymerase activity of RNA
tumor viruses. II. Directing influence of RNA in the reaction, *J. Virol.* **9**:130.

Leis, J. P., and Hurwitz, J., 1972b, Isolation and characterization of a protein that
stimulates DNA synthesis from avian myeloblastosis virus, *Proc. Natl. Acad. Sci.
USA* **69**:2331.

Leis, J. P., Berkower, I., and Hurwitz, J., 1973, Mechanism of action of ribonuclease
H isolated from avian myeloblastosis virus and *Escherichia coli, Proc. Natl. Acad.
Sci. USA* **70**:466.

Levin, D. H., Acs, G., and Silverstein, S. C., 1970a, Chain initiation by reovirus
transcriptase *in vitro, Nature* (*London*) **227**:603.

Levin, D. H., Mendelson, N., Schonberg, M., Klett, H., Silverstein, S. C., Kapuler, A.
M., and Acs, G., 1970b, Properties of RNA transcriptase in reovirus subviral particles, *Proc. Natl. Acad. Sci. USA* **66**:890.

Levin, D. H., Kyner, D., Acs, G., and Silverstein, S. C., 1971, Messenger activity in
mammalian cell-free extracts of reovirus single-stranded RNA prepared *in vitro,
Biochem. Biophys. Res. Commun.* **42**:454.

Levinson, W., Bishop, J. M., Quintrell, N., and Jackson, J., 1970, Presence of DNA in
Rous sarcoma virus, *Nature* (*London*) **227**:1023.

Levinson, W. E., Varmus, H. E., Garapin, A.-C., and Bishop, J. M., 1972, DNA of
Rous sarcoma virus: Its nature and significance, *Science* **175**:76.

Levinson, W. E., Faras, A. J., Woodson, B., Jackson, J., and Bishop, J. M., 1973,
Inhibition of RNA-dependent DNA polymerase of Rous sarcoma virus by thiosemicarbazones and several cations, *Proc. Natl. Acad. Sci. USA* **70**:164.

Lewandowski, L. J., Kalmakoff, J., and Tanada, Y., 1969, Characterization of a
ribonucleic acid polymerase activity associated with purified cytoplasmic polyhedrosis virus of the silkworm, *Bombyx mori, J. Virol.* **4**:857.

Lewandowski, L. J., Content, J., and Leppla, S. H., 1971, Characterization of
the subunit structure of the ribonucleic acid genome of influenza virus, *J. Virol.*
8:701.

Li, K.-K., and Seto, J. T., 1971, Electron microscope study of ribonucleic acid of
myxoviruses, *J. Virol.* **7**:524.

Lieber, M. M., Benveniste, R. E., Livingston, D. M., and Todaro, G. J., 1973, Mammalian cells in culture frequently release type C viruses, *Science* **182**:56.

Lin, F. H., and Thormar, H., 1971, Characterization of ribonucleic acid from Visna
virus, *J. Virol.* **7**:582.

Lin, F. H., and Thormar, H., 1972, Properties of Maedi nucleic acid and the presence
of ribonucleic acid and deoxyribonucleic acid-dependent deoxyribonucleic acid
polymerase in the virions, *J. Virol.* **10**:223.

Lin, F. H., Genovese, M., and Thormar, H., 1973, Multiple activities of DNA
polymerase from Visna virus, *Prep. Biochem.* **3**:525.

Linial, M., and Mason, W. S., 1973, Characterization of two conditional early mutants of Rous sarcoma virus, *Virology* **53**:258.

Livingston, D. M., and Todaro, G. J., 1973, Endogenous type C virus from a cat clone with properties distinct from previously described feline type C viruses, *Virology* **53**:142.

Livingston, D. M., Scolnick, E. M., Parks, W. P., and Todaro, G. J., 1972a, Affinity chromatography of RNA-dependent DNA polymerase from RNA tumor viruses on a solid phase immunoadsorbent, *Proc. Natl. Acad. Sci. USA* **69**:393.

Livingston, D. M., Parks, W. P., Scolnick, E. M., and Ross, J., 1972b, Affinity chromatography of avian type C viral reverse transcriptase: Studies with Rous sarcoma virus transformed rat cells, *Virology* **50**:388.

Mahy, B. W. J., Hastie, N. D., and Armstrong, S. J., 1972, Inhibition of influenza virus replication by α-amanitin: Mode of action, *Proc. Natl. Acad. Sci. USA* **69**:1421.

Mahy, B. W. J., Cox, N. J., Armstrong, S. J., and Barry, R. D., 1973, Multiplication of influenza virus in the presence of cordycepin, an inhibitor of cellular RNA synthesis, *Nature (London) New Biol.* **243**:172.

Mahy, B. W. J., Brownson, J. M. T., Carroll, A. R., Hastie, N. D., and Raper, R. H., 1975, RNA polymerase activities of nuclei from influenza virus-infected cells, in: *Negative Strand Viruses* (R. D. Barry and B. W. J. Mahy, eds.), pp. 445–467, Academic Press, London.

Manly, K., Smoler, D. F., Bromfeld, E., and Baltimore, D., 1971, Forms of deoxyribonucleic acid produced by virions of the ribonucleic acid tumor viruses, *J. Virol.* **7**:106.

Marcus, P. I., Engelhardt, D. L., Hunt, J. M., and Sekellick, M. J., 1971, Interferon action: Inhibition of vesicular stomatitis virus RNA synthesis induced by virion-bound polymerase, *Science* **174**:593.

Marinozzi, V., and Fiume, L., 1971, Effect of α-amanitin on mouse and rat liver nuclei, *Exp. Cell Res.* **67**:311.

Marshall, S., and Gillespie, D., 1972, Poly U tracts absent from viral RNA, *Nature (London) New Biol.* **243**:172.

Martin, S. A., and Moss, B., 1975, Modification of RNA by mRNA guanylyltransferase and mRNA (guanine-7-)methyltransferase from vaccinia virions, *J. Biol. Chem.* **250**:9330.

Martin, S. A., and Zweerink, H. J., 1972, Isolation and characterization of two types of bluetongue virus particles, *Virology* **50**:495.

Martin, S. A., Paoletti, E., and Moss, B., 1975, Purification of mRNA guanylyltransferase and mRNA(guanine-7-)methyltransferase from vaccinia virions, *J. Biol. Chem.* **250**:9322.

Marx, P. A., Portner, A., and Kingsbury, D. W., 1974, Sendai virion transcriptase complex: Polypeptide composition and inhibition by virion envelope proteins, *J. Virol.* **13**:298.

Mason, W. S., Friis, R. R., Linial, M., and Vogt, P. K., 1974, Determination of the defective function in two mutants of Rous sarcoma virus, *Virology* **61**:559.

Mayor, H. D., and Jordan, L. E., 1968, Preparation and properties of the internal capsid components of reovirus, *J. Gen. Virol.* **3**:233.

McAllister, R. M., Nicolson, M., Gardner, M. B., Rongey, R. W., Rasheed, S., Sarma, P. S., Huebner, R. J., Hatanaka, M., Oroszlan, S., Gilden, R. V., Kabigting,

A., and Vernon, L., 1972, C-type virus released from cultured human rhabdomyosarcoma cells, *Nature (London) New Biol.* **235**:3.

McCarthy, W. J., Neser, C. F., and Roberts, D. W., 1975, RNA polymerase activity of AMSACTA MOOREI Entomopox virions, *Intervirology* **5**:69.

McDonnell, J. P., Garapin, A. C., Levinson, W. E., Quintrell, N., Fanshier, L., and Bishop, J. M., 1970, DNA polymerase of Rous sarcoma virus: Delineation of two reactions with actinomycin, *Nature (London)* **228**:433.

McGeoch, D., and Kitron, N., 1975, Influenza virion RNA-dependent RNA polymerase: Stimulation by guanosine and related compounds, *J. Virol.* **15**:686.

McLerran, C. J., and Arlinghaus, R. B., 1973, Structural components of a virus of the California encephalitis complex: LaCrosse virus, *Virology* **53**:247.

Meager, A., and Burke, D. C., 1973, Studies on the structural basis of the RNA polymerase activity of Newcastle disease virus particles, *J. Gen. Virol.* **18**:305.

Miller, L. K., and Wells, R. D., 1971, Nucleoside diphosphokinase activity associated with DNA polymerases, *Proc. Natl. Acad. Sci. USA* **68**:2298.

Miura, K.-I., Watanabe, K., and Sugiura, M., 1974, 5′-Terminal nucleotide sequences of the double-stranded RNA of silkworm cytoplasmic polyhedrosis virus, *J. Mol. Biol.* **86**:31.

Mizutani, S., and Temin, H. M., 1971a, DNA-dependent DNA polymerase and a DNA endonuclease in virions of Rous sarcoma virus, *Nature (London)* **228**:424.

Mizutani, S., and Temin, H. M., 1971b, Enzymes and nucleotides of Rous sarcoma virus, *J. Virol.* **8**:409.

Mizutani, S., and Temin, H. M., 1973, Lack of serological relationship among DNA polymerases of avian leukosis-sarcoma viruses, reticuloendotheliosis viruses, and chicken cells, *J. Virol.* **12**:440.

Mizutani, S., and Temin, H. M., 1974, Specific serological relationships among partially purified DNA polymerases of avian leukosis-sarcoma viruses, reticuloendotheliosis viruses, and avian cells, *J. Virol.* **13**:1020.

Mizutani, S., and Temin, H. M., 1975, Endogenous RNA synthesis is required for endogenous DNA synthesis by reticuloendotheliosis virus virions, in: *Fundamental Aspects of Neoplasia* (A. A. Gottlieb, O. J. Plescia, and D. H. L. Bishop, eds.), pp. 235–242, Springer-Verlag, New York.

Mizutani, S., Boettiger, D., and Temin, H. M., 1970, DNA-dependent DNA polymerase and a DNA endonuclease in virions of Rous sarcoma virus, *Nature (London)* **228**:424.

Mizutani, S., Temin, H. M., Kodama, M., and Wells, R. D., 1971, DNA ligase and exonuclease activities in virions of Rous sarcoma virus, *Nature (London) New Biol.* **230**:232.

Moelling, K., 1974, Characterization of reverse transcriptase and RNase H from Friend-murine leukemia virus, *Virology* **62**:46.

Moelling, K., 1975, Reverse transcriptase and RNase H: Present in a murine virus and in both subunits of an avian virus, *Cold Spring Harbor Symp. Quant. Biol.* **39**:969.

Moelling, K., Bolognesi, D. P., Bauer, H., Büsen, W., Plassmann, H. W., and Hausen, P., 1971, Association of viral reverse transcriptase with an enzyme degrading the RNA moiety of RNA-DNA hybrids, *Nature (London) New Biol.* **234**:240.

Mondal, H., Gallagher, R. E., and Gallo, R. C., 1975, RNA-directed DNA polymerase from human leukemic blood cells and from primate type-C virus-producing cells: High and low molecular-weight forms with variant biochemical and immunological properties, *Proc. Natl. Acad. Sci. USA* **72**:1194.

Montagnier, L., Goldé, A., and Vigier, P., 1969, A possible subunit structure of Rous sarcoma virus RNA, *J. Gen. Virol.* **4**:499.

Moreau, M., 1974, Inhibition of a vesicular stomatitis virus mutant by rifampin, *J. Virol.* **14**:517.

Morrison, T., Stampfer, M., Baltimore, D., and Lodish, H. F., 1974, Translation of vesicular stomatitis messenger RNA by extracts from mammalian and plant cells, *J. Virol.* **13**:62.

Moss, B., and Rosenblum, E. M., 1974, Vaccinia virus polyriboadenylate polymerase: Covalent linkage of the product with polyribonucleotide and polydeoxyribonucleotide primers, *J. Virol.* **14**:86.

Moss, B., Rosenblum, E. N., and Paoletti, E., 1973, Polyadenylate polymerase from vaccinia virions, *Nature (London)* **254**:59.

Moss, B., Rosenblum, E. N., and Gershowitz, A., 1975, Characterization of a polyriboadenylate polymerase from vaccinia virions, *J. Biol. Chem.* **250**:4722.

Mountcastle, W., Harter, D., and Choppin, P., 1972, The proteins of Visna virus, *Virology* **47**:542.

Moyer, S. A., and Summers, D. F., 1974, Phosphorylation of vesicular stomatitis virus *in vivo* and *in vitro*, *J. Virol.* **13**:455.

Moyer, S. A., Abraham, G., Adler, R., and Banerjee, A. K., 1975a, Methylated and blocked 5′ termini in vesicular stomatitis virus *in vivo* mRNAs, *Cell* **5**:59.

Moyer, S. A., Grubman, M. J., Ehrenfeld, E., and Banerjee, A. K., 1975b, Studies on the *in vivo* and *in vitro* messenger RNA species of vesicular stomatitis virus, *Virology* **67**:463.

Mudd, J. A., and Summers, D. F., 1970, Polysomal ribonucleic acid of vesicular stomatitis virus-infected HeLa cells, *Virology* **42**:958.

Mueller, W. E. G., Zahn, R. K., and Seidel, H. J., 1971, Inhibitors acting on nucleic acid syntheses in an oncogenic RNA virus, *Nature (London) New Biol.* **232**:143.

Munyon, W. E., Paoletti, E., and Grace, J. T., Jr., 1967, RNA polymerase activity in purified infectious vaccinia virus, *Proc. Natl. Acad. Sci. USA* **58**:2280.

Munyon, W., Paoletti, E., Ospina, J., and Grace, J. T., Jr., 1968, Nucleotide phosphohydrolase in purified vaccinia virus, *J. Virol.* **2**:167.

Muthukrishnan, S., and Shatkin, A. J., 1975, Reovirus genome RNA segments: Resistance to S_1 nuclease, *Virology* **64**:96.

Muthukrishnan, S., Both, G. W., Furuichi, Y., and Shatkin, A. J., 1975, 5′ Terminal 7-methylguanosine in eukaryotic mRNA is required for translation, *Nature (London)* **255**:33.

Nayak, D. P., and Murray, P. R., 1973, Induction of type C viruses in cultured guinea pig cells, *J. Virol.* **12**:177.

Nichols, J. L., and Waddell, M., 1973, Comparison of free and 80 S RNA-associated RNAs of mouse L cell virions, *Nature (London) New Biol.* **243**:236.

Nichols, J. L., Hay, A. J., and Joklik, W. K., 1972, 5′-Terminal nucleotide sequence in reovirus mRNA synthesized *in vitro*, *Nature (London) New Biol.* **235**:105.

Nowinski, R. C., Watson, K. F., Yaniv, A., and Spiegelman, S., 1972, Serological analysis of the deoxyribonucleic acid polymerase of avian oncornaviruses. II. Comparison of avian deoxyribonucleic acid polymerases, *J. Virol.* **10**:959.

Obijeski, J. F., Marchendo, A. T., Bishop, D. H. L., Cann, B. W., and Murphy, F. A., 1974, Comparative electrophoretic analysis of the virus proteins of four rhabdoviruses, *J. Gen. Virol.* **22**:21.

Obijeski, J. F., Bishop, D. H. L., Murphy, F. A., and Palmer, E. L., 1976, The structural proteins of La Crosse virus, *J. Virol.* **19**:985.

Oroszlan, S., Huebner, R. J., and Gilden, R. V., 1971, Species-specific and interspecific antigenic determinants associated with the structural protein of feline C-type virus, *Proc. Natl. Acad. Sci. USA* **68**:901.

Oxford, J. S., 1973a, An inhibitor of the particle-associated RNA dependent RNA polymerase of influenza A and B viruses, *J. Gen. Virol.* **18**:11.

Oxford, J. S., 1973b, Polypeptide composition of influenza B viruses and enzymes associated with the purified virus particles, *J. Virol.* **12**:827.

Oxford, J. S., and Perrin, D. D., 1974, Inhibition of the particle-associated RNA-dependent RNA polymerase activity of influenza viruses by chelating agents, *J. Gen. Virol.* **23**:59.

Oxford, J. S., and Perrin, D. D., 1975, Inhibitors of influenza virus-associated RNA-dependent RNA polymerase, in: *Negative Strand Viruses*, Vol. 1 (B. W. J. Mahy and R. D. Barry, eds.), pp. 433–444, Academic Press, New York.

Palese, P., and Koch, G., 1972, Degradation of single and double stranded RNA by frog virus 3, *Proc. Natl. Acad. Sci. USA* **69**:698.

Palese, P., and McAuslan, B. R., 1972, Virus associated DNAse: Endonuclease in a polyhedral cytoplasmic deoxyribovirus, *Virology* **49**:319.

Panet, A., Baltimore, D., and Hanafusa, T., 1975a, Quantitation of avian RNA tumor virus reverse transcriptase by radioimmunoassay, *J. Virol.* **16**:146.

Panet, A., Verma, I. M., and Baltimore, D., 1975b, Role of the subunits of the avian RNA tumor virus reverse transcriptase, *Cold Spring Harbor Symp. Quant. Biol.* **39**:919.

Paoletti, E., and Moss, B., 1972, Protein kinase and specific phosphate acceptor proteins associated with vaccinia virus cores, *J. Virol.* **10**:417.

Paoletti, E., and Moss, B., 1974, Two nucleic acid-dependent nucleoside triphosphate phosphohydrolases from vaccinia virus: Nucleotide substrate and polynucleotide cofactor specificities, *J. Biol. Chem.* **249**:3281.

Paoletti, E., Rosemond-Hornbeak, H., and Moss, B., 1974a, Two nucleic acid-dependent nucleoside triphosphate phospholydrolases from vaccinia virus: Purification and characterization, *J. Biol. Chem.* **249**:3273.

Paoletti, E., Cooper, N., and Moss, P., 1974b, Regulation of synthesis of two immunologically distinct nucleic-acid dependent nucleoside triphosphate phosphohydrolases in vaccinia virus-injected HeLa cells, *J. Virol.* **14**:578.

Parks, W. P., Scolnick, E. M., Todaro, G. J., and Aaronson, S. A., 1971, RNA-dependent DNA polymerase in primate syncytium-forming (foamy) viruses, *Nature (London)* **229**:258.

Parks, W. P., Scolnick, E. M., Ross, J., Todaro, G. J., and Aaronson, S. A., 1972, Immunologic relationships of reverse transcriptases from ribonucleic acid tumor viruses, *J. Virol.* **9**:110.

Penhoet, E., Miller, H., Doyle, M., and Blatti, S., 1971, RNA-dependent RNA polymerase activity in influenza virions, *Proc. Natl. Acad. Sci. USA* **68**:1369.

Penit, C. A., Paraf, A., Rougeon, F., and Chapeville, F., 1974, Ribonuclease sensitive DNA polymerase activity associated with particles distinct from A type and C type viral particles in murine myeloma tumor cells, *FEBS Lett.* **38**:191.

Penman, S., Rosbash, M., and Penman, M., 1970, Messenger and heterogeneous nuclear RNA in HeLa cells: Differential inhibition by cordycepin, *Proc. Natl. Acad. Sci. USA* **67**:1878.

Perlman, S. M., and Huang, A. S., 1973, RNA synthesis of vesicular stomatitis virus. V. Interaction between transcription and replication, *J. Virol.* **12**:1395.

Peterson, D. A., Baxter-Gabbard, K. L., and Levine, A. S., 1972, Avian reticuloendotheliosis virus (strain T). V. DNA polymerase, *Virology* **47**:251.

Pettersson, R., and Kaariainen, L., 1973, The ribonucleic acids of Uukuniemi virus, a non-cubical tickborne arbovirus, *Virology* **56**:608.

Pettersson, R., and von Bonsdorff, C.-H., 1975, Ribonucleoproteins of Uukuniemi virus are circular, *J. Virol.* **15**:386.

Pogo, B. G. T., and Dales, S., 1969, Two deoxyribonuclease activities within purified vaccinia virus, *Proc. Natl. Acad. Sci. USA* **63**:820.

Pogo, B. G. T., Dales, S., Bergoin, M., and Roberts, D. W., 1971, Enzymes associated with an insect poxvirus, *Virology* **43**:306.

Poiesz, B. J., Battula, N., and Loeb, L. A., 1974, Zinc in reverse transcriptase *Biochem. Biophys. Res. Commun.* **56**:959.

Pons, M. W., 1967, Effect of actinomycin D on the replication of influenza virus RNA, *Virology* **33**:150.

Pons, M. W., 1971, Isolation of influenza virus ribonucleoprotein from infected cells: Demonstration of the presence of negative-stranded RNA in viral RNP, *Virology* **46**:149.

Pons, M. W., Schultze, I. T., and Hirst, G. K., 1969, Isolation and characterization of the ribonucleoprotein of influenza virus, *Virology* **39**:250.

Portner, A., and Kingsbury, D. W., 1970, Complementary RNAs in paramyxovirions and paramyxovirus-infected cells, *Nature (London)* **228**:1196.

Pridgen, C., and Kingsbury, D. W., 1972, Adenylated sequences in Sendai virus transcripts from infected cells, *J. Virol.* **10**:314.

Pringle, C. R., 1975, Conditional lethal mutants of vesicular stomatitis virus, *Curr. Top. Microbiol. Immunol.* **69**:85.

Quintrell, N., Fanshier, L., Evans, B., Levinson, W., and Bishop, J. M., 1971, Deoxyribonucleic acid polymerase(s) of Rous sarcoma virus: Effects of virion-associated endonuclease on the enzymatic product, *J. Virol.* **8**:17.

Ramos, B. A., Courtney, R. J., and Rawls, W. E., 1972, Structural proteins of Pichinde virus, *J. Virol.* **10**:661.

Randerath, K., Rosenthal, L. J., and Zamecnik, P. C., 1971, Base composition differences between avian myeloblastosis virus transfer RNA and transfer RNA isolated from host cells, *Proc. Natl. Acad. Sci. USA* **68**:3233.

Ranki, M., and Pettersson, 1975, Uukuniemi virus contains an RNA polymerase, *J. Virol.* **16**:1420.

Repik, P., and Bishop, D. H. L., 1973, Determination of the molecular weight of animal RNA viral genomes by nuclease digestions, *J. Virol.* **12**:969.

Repik, P., Flamand, A., and Bishop, D. H. L., 1974a, The effect of interferon upon the primary and secondary transcription of vesicular stomatitis and influenza viruses, *J. Virol.* **14**:1169.

Repik, P., Flamand, A., Clark, H. F., Obijeski, J. F., Roy, P., and Bishop, D. H. L., 1974b, The detection of homologous RNA sequences among six rhabdovirus genomes, *J. Virol.* **13**:250.

Rho, H. M., and Green, M., 1974, The homopolyadenylate and adjacent nucleotides at the 3′-terminus of 30–40 S RNA subunits in the genome of murine sarcoma-leukemia virus, *Proc. Natl. Acad. Sci. USA* **71**:2386.

Rhodes, D. P., Moyer, S. A., and Banerjee, A. K., 1974, *In vitro* synthesis of methylated messenger RNA by the virion-associated polymerase of vesicular stomatitis virus, *Cell* **3**:327.

Richert, N. J., and Balduzzi, P., 1971, Mechanism of oncogenic transformation by Rous sarcoma virus, *J. Virol.* **8**:62.

Riman, J., and Beaudreau, G. S., 1970, Viral DNA-dependent DNA polymerase and the properties of the thymidine-labelled material in virions of an oncogenic virus, *Nature (London)* **228**:427.

Robert, M. S., Smith, R. G., Gallo, R. C., Sarin, P. S., and Abrell, J. W., 1972, Viral and cellular DNA polymerases: Comparison of activities with synthetic and natural RNA templates, *Science* **176**:798.

Robinson, H. L., and Robinson, W. S., 1971, Inhibition of growth of uninfected and Rous sarcoma virus-infected chick embryo fibroblasts by rifampicin. *J. Natl. Cancer Inst.* **46**:785.

Robinson, W. S., 1970, Self-annealing of subgroup 2 myxovirus RNAs, *Nature (London)* **225**:944.

Robinson, W. S., 1971a, Ribonucleic acid polymerase activity in Sendai virions and nucleocapsid, *J. Virol.* **8**:81.

Robinson, W. S., 1971b, Intracellular structures involved in Sendai virus replication, *Virology* **43**:90.

Robinson, W. S., 1971c, Sendai virus RNA synthesis and nucleocapsid formation in the presence of cycloheximide, *Virology* **44**:494.

Robinson, W. S., and Robinson, H. L., 1971, DNA polymerase in defective Rous sarcoma virus, *Virology* **44**:457.

Robinson, W. S., Pitkanen, A., and Rubin, H., 1965, The nucleic acid of Rous sarcoma virus: Purification of the virus and isolation of the nucleic acid, *Proc. Natl. Acad. Sci. USA* **54**:137.

Robinson, W. S., Clayton, D. A., and Greenman, R. L., 1974, DNA of a human hepatitis B viris candidate, *J. Virol.* **14**:384.

Rochovansky, O., and Pons, M. W., 1975, The effects of cordycepin on influenza virus replication, in: *Negative Strand Viruses* (R. D. Barry and B. W. J. Mahy, eds.), pp. 475–483, Academic Press, London.

Rokutanda, M., Fujinaga, K., Rokutanda, H., Ray, R. K., Green, M., and Gurgo, C., 1970, Formation of viral RNA-DNA hybrid molecules by the DNA polymerase of sarcoma-leukemia viruses, *Nature (London)* **227**:1025.

Rose, J. K., 1975, Heterogeneous 5′-terminal structures occur on vesicular stomatitis virus mRNAs, *J. Viol. Chem.* **250**:8098.

Rose, J. K., and Knipe, D., 1975, Nucleotide sequence complexities, molecular weights, and poly(A) content of the vesicular stomatitis virus mRNA species. *J. Virol.* **15**:994.

Rosemond-Hornbeak, H., and Moss, B., 1974, Single-stranded deoxyribonucleic acid-specific nuclease from vaccinia virus: Endonucleolytic and exonucleolytic activities, *J. Biol. Chem.* **249**:3292.

Rosemond-Hornbeak, H., Paoletti, E., and Moss, B., 1974, Single-stranded deoxyribonucleic acid-specific nuclease from vaccinia virus: Purification and characterization, *J. Biol. Chem.* **249**:3287.

Rosenthal, L. J., and Zamecnik, P. C., 1973a, Amino acid acceptor activity of the "70 S-associated" 4 S RNA from avian myeloblastosis virus, *Proc. Natl. Acad. Sci. USA* **70**:1184.

Rosenthal, L. J., and Zamecnik, P. C., 1973*b*, Minor base composition of "70 S-associated" 4 S RNA from avian myeloblastosis virus, *Proc. Natl. Acad. Sci. USA* **70**:865.

Ross, J., Scolnick, E. M., Todaro, G. J., and Aaronson, S. A., 1971, Separation of murine cellular and murine leukemia virus DNA polymerases, *Nature (London) New Biol.* **231**:163.

Ross, J., Aviv, H., Scolnick, E. M., and Leder, P., 1972, *In vitro* synthesis of DNA complementary to purified rabbit globin mRNA, *Proc. Natl. Acad. Sci. USA* **69**:264.

Rothenberg, E., and Baltimore, D., 1976, Synthesis of long, representative DNA copies of the murine RNA tumor virus genome, *J. Virol.* **17**:168.

Rott, R., and Scholtissek, C., 1970, Specific inhibition of influenza replication by α-amanitin, *Nature (London)* **228**:56.

Rott, R., Saber, S., and Scholtissek, C., 1965, Effect on myxovirus of mitomycin C, actinomycin D, and pretreatment of the host cell with ultra-violet light, *Nature (London)* **205**:1187.

Roux, L., and Kolakofsky, D., 1975, Isolation of RNA transcripts from the entire Sendai viral genome, *J. Virol.* **16**:1426.

Roy, P., and Bishop, D. H. L., 1971, Nucleoside triphosphate phosphotransferase. A new enzyme activity of oncogenic and non-oncogenic "budding" viruses, *Biochim. Biophys. Acta* **235**:191.

Roy, P., and Bishop, D. H. L., 1972, The genome homology of vesicular stomatitis virus and defective T particles and evidence for the sequential transcription of the virion RNA, *J. Virol.* **9**:946.

Roy, P., and Bishop, D. H. L., 1973, Initiation and direction of RNA transcription by vesicular stomatitis virion transcriptase, *J. Virol.* **11**:487.

Roy, P., Clark, H. F., Madore, H. P., and Bishop, D. H. L., 1975, RNA polymerase associated with virions of pike fry rhabdovirus, *J. Virol.* **15**:338.

Roy-Burman, P., 1971, Deoxyribonucleic acid polymerase associated with feline leukemia and sarcoma viruses: Properties of the enzyme and its product, *Int. J. Cancer* **7**:409.

Ruprecht, R. M., Goodman, N. C., and Spiegelman, S., 1973, Determination of natural host taxonomy of RNA tumor viruses by molecular hybridization: Application to RD-114, a candidate human virus, *Proc. Natl. Acad. Sci. USA* **70**:1437.

Samso, A., Bouloy, M., and Hannoun, C., 1975, Presence de ribonucleoproteins circulaires dans le virus Lumbo (bunyavirus), *C. R. Acad. Sci. Ser. D* **280**:779.

Sawyer, R. C., and Dahlberg, J. E., 1973, Small RNAs of Rous sarcoma virus: Characterization by two-dimensional polyacrylamide gel electrophoresis and fingerprint analysis, *J. Virol.* **12**:1226.

Schlom, J., and Spiegelman, S., 1971, DNA polymerase activities and nucleic acid components of virions isolated from a spontaneous mammary carcinoma of a rhesus monkey, *Proc. Natl. Acad. Sci. USA* **68**:1613.

Schlom, J., Harter, D. H., Burny, A., and Spiegelman, S., 1971, DNA polymerase activities in virions of Visna virus, a causative agent of "slow" neurological disease, *Proc. Natl. Acad. Sci. USA* **68**:182.

Schochetman, G., and Millward, S., 1972, Ribonucleoside diphosphate precursors for *in vitro* reovirus RNA synthesis, *Nature (London) New Biol.* **239**:77.

Schoefl, G., 1964, The effect of actinomycin D on the fine structure of the nucleolus, *J. Ultrastruct. Res.* **10**:224.

Scholtissek, C., 1969, Synthesis *in vitro* of RNA complementary to parental viral RNA by RNA polymerase induced by influenza virus, *Biochim. Biophys. Acta* **179**:389.

Scholtissek, C., Drzeniek, R., and Rott, R., 1969, Myxoviruses, in: *The Biochemistry of Viruses* (H. B. Levy, ed.), p. 219, Dekker, New York.

Scholtissek, C., Becht, H., and Rott, R., 1971, Inhibition of influenza RNA polymerase by specific antiserum, *Virology* **43**:137.

Schulze, I. T., 1972, The structure of influenza virus. II. A model based on the morphology and composition of subviral particles, *Virology* **47**:181.

Schwartz, J., and Dales, S., 1971, Biogenesis of poxviruses: Identification of four enzyme activities within purified Yaba tumor virus, *Virology* **45**:797.

Scolnick, E. M., Aaronson, S. A., and Todaro, G. J., 1970a, DNA synthesis by RNA-containing tumor viruses, *Proc. Natl. Acad. Sci. USA* **67**:1034.

Scolnick, E. M., Rands, E., Aaronson, S. A., and Todaro, G. J., 1970b, RNA-dependent DNA polymerase activity in five RNA viruses: Divalent cation requirements, *Proc. Natl. Acad. Sci. USA* **67**:1789.

Scolnick, E. M., Parks, W. P., Todaro, G. J., and Aaronson, S. A., 1972, Immunological characterization of primate C-type virus reverse transcriptases, *Nature (London) New Biol.* **235**:35.

Shatkin, A. J., and LaFiandra, A. J., 1972, Transcription by infectious subviral particles of reovirus, *J. Virol.* **10**:698.

Shatkin, A. J., and Rada, B., 1967, Reovirus-directed ribonucleic acid synthesis in infected L cells, *J. Virol.* **1**:24.

Shatkin, A. J., and Sipe, J. D., 1968, RNA polymerase activity in purified reoviruses, *Proc. Natl. Acad. Sci. USA* **61**:1463.

Sherr, C. J., Lieber, M. M., and Todaro, G. J., 1974, Mixed splenocyte cultures and graft versus host reactions selectively induce an "S-tropic" murine type C virus, *Cell* **1**:55.

Sherr, C. J., Fedele, L. A., Benveniste, R. E., and Todaro, G. J., 1975, Interspecies antigenic determinants of the reverse transcriptases and p30 proteins of mammalian type C viruses, *J. Virol.* **15**:1440.

Shimotohno, K., and Miura, K.-I., 1973, Transcription of double-stranded RNA in cytoplasmic polyhedrosis virus *in vitro, Virology* **53**:283.

Silverstein, S. C., and Dales, S., 1968, The penetration of reovirus RNA and initiation of its genetic function in L strain fibroblasts, *J. Cell Biol.* **36**:197.

Silverstein, S. C., Levin, D. H., Schonberg, M., and Acs, G., 1970, The reovirus replicative cycle: Conservation of parental RNA and protein, *Proc. Natl. Acad. Sci. USA* **67**:275.

Silverstein, S. C., Astell, C., Levin, D. H., Schonberg, M., and Acs, G., 1972, The mechanisms of reovirus uncoating and gene activation *in vivo, Virology* **47**:797.

Simpson, R. W., and Hirst, G. K., 1961, Genetic recombination among influenza viruses. I. Cross reactivation of the plague-forming capacity as a method for selecting recombinants from the progeny of crosses between influenza A strains, *Virology* **15**:436.

Simpson, R. W., and Iinuma, M., 1975, Recovery of infectious proviral DNA from mammalian cells infected with respiratory syncytial virus, *Proc. Natl. Acad. Sci. USA* **72**:3230.

Skehel, J. J., 1971a, Estimations of the molecular weight of the influenza virus genome, *J. Gen. Virol.* **11**:103.

Skehel, J. J., 1971*b*, RNA-dependent RNA polymerase activity of the influenza virus, *Virology* **45**:793.

Skehel, J. J., and Joklik, W. K., 1969, Studies on the *in vitro* transcription of reovirus RNA catalyzed by reovirus cores, *Virology* **39**:822.

Smoler, D., Molineux, I., and Baltimore, D., 1971, Direction of polymerization of the avian myeloblastosis virus DNA polymerase, *J. Biol. Chem.* **246**:7697.

Sobell, H. M., and Jain, S. C., 1972, Stereochemistry of actinomycin binding to DNA. II. Detailed molecular model of actinomycin-DNA complex and its implications, *J. Mol. Biol.* **68**:21.

Sokol, F., and Clark, H. F., 1973, Phosphoproteins, structural components of rhabdoviruses, *J. Virol.* **52**:246.

Sokol, F., and Koprowski, H., 1975, Structure-function relationships and mode of replication of animal rhabdoviruses, *Proc. Natl. Acad. Sci. USA* **72**:933.

Sokol, F., Tan, K. B., McFalls, M. L., and Madore, P., 1974, Phosphate acceptor amino acid residues in structural proteins of rhabdoviruses, *J. Virol.* **14**:145.

Soria, M., and Huang, A. S., 1973, Association of polyadenylic acid with messenger RNA of vesicular stomatitis virus, *J. Mol. Biol.* **77**:449.

Soria, M., Little, S. P., and Huang, A. S., Characterization of vesicular stomatitis virus nucleocapsids. I. Complementary 40 S RNA molecules in nucleocapsids, *Virology* **61**:270.

Spiegelman, S., Burny, A., Das, M., R., Keydar, J., Schlom, J., Trávníček, M., and Watson, K., 1970*a*, Characterization of the products of RNA-directed DNA polymerases in oncogenic RNA viruses, *Nature (London)* **227**:563.

Spiegelman, S., Burny, A., Das, M. R., Keydar, J., Schlom, J., Trávníček, M., and Watson, K., 1970*b*, DNA-directed DNA polymerase activity in oncogenic RNA viruses, *Nature (London)* **227**:1029.

Spiegelman, S., Burny, A., Das, M. R., Keydar, J., Schlom, J., Trávníček, M., and Watson, K., 1970*c*, Synthetic DNA-RNA hybrids and RNA-RNA duplexes as templates for the polymerases of the oncogenic RNA viruses, *Nature (London)* **228**:430.

Spiegelman, S., Watson, K. F., and Kacian, D. L., 1971, Synthesis of DNA complements of natural RNAs: A general approach, *Proc. Natl. Acad. Sci. USA* **68**:2843.

Stephenson, M. L., Wirthlin, R. R. S., Scott, J. F., and Zamecnik, P., 1972, The 3′-terminal nucleosides of the high molecular weight RNA of avian myeloblastosis virus. *Proc. Natl. Acad. Sci. USA* **69**:1176.

Stephenson, M., Scott, J., and Zamecnik, 1973, Evidence that the polyadenylic acid segment of "35 S" RNA of avian myeloblastosis virus is located at the 3′ OH terminus, *Biochem. Biophys. Res. Commun.* **55**:8.

Stoltzfus, C. M., Shatkin, A. J., and Banerjee, A. K., 1973, Absence of polyadenylic acid from reovirus messenger RNA, *J. Biol. Chem.* **248**:7993.

Stone, H. O., and Kingsbury, D. W., 1973, Stimulation of Sendai virion transcriptase by polyanions, *J. Virol.* **11**:243.

Stone, H. O., Portner, A., and Kingsbury, D. W., 1971, Ribonucleic acid transcriptases in Sendai virions and infected cells, *J. Virol.* **8**:174.

Stone, H. O., Kingsbury, D. W., and Darlington, R. W., 1972, Sendai virus-induced transcriptase from infected cells: Polypeptides in the transcriptive complex, *J. Virol.* **10**:1037.

Stone, L. B., Takemoto, K. K., and Martin, M. A., 1971*a*, Physical and biochemical properties of progressive pneumonia virus, *J. Virol.* **8**:573.

Stone, L. B., Skolnick, E. A., Takemoto, K. K., and Aaronson, S. A., 1971*b*, Visna virus: A slow virus with a RNA dependent DNA polymerase, *Nature (London)* **229**:257.

Stone, M. P., Smith, R. E., and Joklik, W. K., 1975, 35 S a and b RNA subunits of avian RNA tumor virus strains cloned and passaged in chick`and duck cells, *Cold Spring Harbor Symp. Quant. Biol.* **39**:859.

Strand, M., and August, J. T., 1971, Protein kinase and phosphate acceptor proteins in Rauscher murine leukemia virus, *Nature (London)* New Biol. **233**:137.

Stromberg, K., 1972, Surface-active agents for isolation the core component of avian myeloblastosis virus, *J. Virol.* **9**:684.

Szilágyi, J. F., and Pringle, C. R., 1972, Effect of temperature-sensitive mutations on the virion-associated RNA transcriptase of vesicular stomatitis virus, *J. Mol. Biol.* **71**:281.

Szilágyi, J. F., and Uryvayer, L., 1973, Isolation of an infectious ribonucleoprotein from vesicular stomatitis virus containing an active RNA transcriptase, *J. Virol.* **11**:279.

Taylor, J. M., and Illmensee, R., 1975, Site on the RNA of an avian sarcoma virus at which primer is bound, *J. Virol.* **16**:553.

Taylor, J. M., Hampson, A. W., Layton, J. E., and White, D. O., 1970, The polypeptides of influenza virus. IV. An analysis of nuclear accumulation, *Virology* **42**:744.

Taylor, J. M., Faras, A. J., Varmus, H. E., Goodman, H. M., Levinson, W. E., and Bishop, J. M., 1973, Transcription of ribonucleic acid by the ribonucleic acid directed deoxyribonucleic acid polymerase of Rous sarcoma virus and deoxyribonucleic acid polymerase of *Escherischia coli, Biochemistry* **12**:460.

Temin, H., and Mizutani, S., 1970, RNA-dependent DNA polymerase in virions of Rous sarcoma virus, *Nature (London)* **226**:1211.

Thach, R. E., Robertson, D. L., Baenziger, N. L., and Dobbertin, D. C., 1975, Reverse transcriptase associated with A-type particles from murine myeloma cells, *Cold Spring Harbor Symp. Quant. Biol.* **39**:963.

Thacore, H., and Youngner, J. S., 1969, Cells persistently infected with Newcastle disease virus. I. Properties of mutants isolated from persistently infected L cells, *J. Virol.* **4**:244.

Thacore, H., and Youngner, J. S., 1970, Cells persistently infected with Newcastle disease virus. II. Ribonucleic acid and protein synthein cells infected with mutants isolated from persistently infected L cells, *J. Virol.* **6**:42.

Theilin, G. H., Gould, D., Fowler, M., and Dungworth, D. L., 1971, C-type virus in tumor tissue of a woolly monkey (*Lagothrix*) with fibrosarcoma, *J. Natl. Cancer Inst.* **47**:881.

Todaro, G. J., 1972, "Spontaneous" release of type C viruses from clonal lines of "spontaneously" transformed BALB/3T3 cells, *Nature (London)* New Biol. **240**:157.

Todaro, G. J., Benveniste, R. E., Lieber, M. M., and Livingston, D. M., 1973, Infectious type C viruses released by normal cat embryo cells, *Virology* **55**:505.

Todaro, G. J., Benveniste, R. E., Lieber, M. M., and Sherr, C. J., 1974*a*, Characterization of a type C virus released from the porcine cell line PK(L5), *Virology* **58**:65.

Todaro, G. J., Sherr, C. J., Benveniste, R. E., Lieber, M. M., and Melnick, J. L., 1974*b*, Type C viruses of baboons: Isolation from normal cell cultures, *Cell* **2**:55.

Todaro, G. J., Benveniste, R. E., Callahan, R., Liebar, M. M., and Sherr, C. J., 1975,

Endogenous primate and feline type C viruses. *Cold Spring Harbor Symp. Quant. Biol.* **39**:1159.

Trávníček, M., 1968, RNA with amino acid-acceptor activity isolated from an oncogenic virus, *Biochim. Biophys. Acta* **166**:757.

Trávníček, M., 1969, Some properties of amino acid-acceptor RNA isolated from avian tumor virus BA 1 strain A (avian myeloblastosis), *Biochim. Biophys. Acta* **182**:427.

Trávníček, M., and Říman, J., 1973, Occurrence of aminoacyl-tRNA synthetase in an RNA oncogenic virus, *Nature (London) New Biol.* **241**:60.

Tronick, S. R., Scolnick, E. M., and Parks, W. P., 1972, Reversible inactivation of the deoxyribonucleic acid polymerase of Rauscher leukemia virus, *J. Virol.* **10**:885.

Tronick, S. R., Stephenson, J. R., Verma, I. M., and Aaronson, S. A., 1975, Thermolabile reverse transcriptase of a mammalian leukemia virus mutant temperature sensitive in its replication and sarcoma virus helper functions, *J. Virol.* **16**:1476.

Tuominen, F. W., and Kenney, T., 1971, Inhibition of the DNA polymerase of Rauscher leukemia virus by single-stranded polyribonucleotides, *Proc. Natl. Acad. Sci. USA* **68**:2198.

Twardzik, D. R., Papas, T. S., and Portugal, F. H., 1974, DNA polymerase in virions of a reptilian type C virus, *J. Virol.* **13**:166.

Tyrrell, D. A. J., Almeida, J. D., Cunningham, C. H., Dowdle, W. R., Hofstad, M. S., McIntosh, K., Tajima, M., Zakstelskaya, L. Y., Easterday, B. C., Kapikian, A., and Bingham, R. W., 1975, Coronaviridae, *Intervirology* **5**:76.

Vaheri, A., and Hanafusa, H., 1971, Effect of rifampicin and a derivative on cells transformed by Rous sarcoma virus, *Cancer Res.* **31**:2032.

Verma, I., 1975, Studies on reverse transcriptase of RNA tumor viruses. III. Properties of purified Moloney murine leukemia virus DNA polymerase and associated RNase H, *J. Virol.* **15**:843.

Verma, I. M., and Baltimore, D., 1973, Purification of the RNA-directed DNA polymerase from avian myeloblastosis virus and its assay with polynucleotide templates, in: *Methods in Enzymology,* Vol. 29 (L. Grossman and K. Moldave, eds.), p. 135, Academic Press, New York.

Verma, I. M., Temple, G. F., Fan, H., and Baltimore, D., 1971, A covalently linked RNA-DNA molecule as the initial product of the RNA tumor virus DNA polymerase, *Nature (London) New Biol.* **233**:131.

Verma, I. M., Temple, G. F., Fan, H., and Baltimore, D., 1972, *In vitro* synthesis of DNA complementary to rabbit reticulocyte 10 S RNA, *Nature (London) New Biol.* **235**:163.

Verma, I. M., Mason, W. S., Drost, S. D., and Baltimore, D., 1974*a*, DNA polymerase activity from two temperature sensitive mutants of Rous sarcoma virus is thermolabile, *Nature (London)* **251**:27.

Verma, I. M., Meuth, N. L., Fan, H., and Baltimore, D., 1974*b*, Hamster leukemia virus DNA polymerase: Unique structure and lack of demonstrable endogenous activity, *J. Virol.* **13**:1075.

Verwoerd, D. W., and Huismans, H., 1972, Studies on the *in vitro* and *in vivo* transcription of the bluetongue virus genome, *Onderstepoort, J. Vet. Res.* **39**:185.

Vilaginès, R., and McAuslan, B. R., 1971, Proteins of polyhedral cytoplasmic deoxyvirus. II. Nucleotide phospholydrolase activity associated with frog virus 3, *J. Virol.* **7**:619.

Villareal, L. P., and Holland, J. J., 1973, Synthesis of poly(A) *in vitro* by purified virions of vesicular stomatitis virus, *Nature (London) New Biol.* **245**, 17.

Vogt, P. K., 1973, The genome of avian RNA tumor viruses: A discussion of four models, in: *Possible Episomes in Eukaryotes: Proceedings of the Fourth Lepetit Colloquium* (L. G. Silvestri, ed.), p. 35, North-Holland, Amsterdam.

Wachsman, J. T., Levin, D. H., and Acs, G., 1970, Ribonucleoside triphosphate-dependent pyrophosphate exchange of reovirus cores, *J. Virol.* **6**:563.

Wagner, R. R., Prevec, L., Brown, F., Summers, D. F., Sokol, F., and MacLeod, R., 1972, Classification of rhabdovirus proteins: A proposal, *J. Virol.* **10**:1228.

Waite, M. R. F., and Allen, P. T., 1975, RNA-directed DNA polymerase activity of reticuloendotheliosis virus: Characterization of endogenous and exogenous reactions, *J. Virol.* **16**:872.

Wang, L.-H., and Duesberg, P. H., 1973, DNA polymerase of murine sarcoma-leukemia virus: Lack of detectable RNase H and low activity with viral RNA and natural DNA templates, *J. Virol.* **12**:1512.

Wang, S., Kothari, R. M., Taylor, M., and Hung, P., 1973, Transfer RNA activities of Rous sarcoma and Rous associated viruses, *Nature (London) New Biol.* **242**:133.

Warrington, R. C., Hayward, C., and Kapuler, A. M., 1973, Conformational studies of reovirus single-stranded RNAs synthesized *in vitro, Biochim. Biophys. Acta* **331**:231.

Watanabe, Y., Kudo, H., and Graham, A. F., 1967a, Selective inhibition of reovirus ribonucleic acid synthesis by cycloheximide, *J. Virol.* **1**:36.

Watanabe, Y., Prevec., L., and Graham, A. F., 1967b, Specificity in transcription of the reovirus genome, *Proc. Natl. Acad. Sci. USA* **58**:1040.

Watanabe, Y., Gauntt, C. J., and Graham, A. F., 1968, Reovirus-induced ribonucleic acid polymerase, *J. Virol.* **2**:869.

Watson, K. F., Nowinski, C., Yaniv, A., and Spiegelman, S., 1972, Serological analysis of the deoxyribonucleic acid polymerase of avian oncornaviruses. I. Preparation and characterization of monospecific antiserum with purified deoxyribonucleic acid polymerase, *J. Virol.* **10**:95.

Watson, K. F., Moelling, K., and Bauer, H., 1973, Ribonuclease H activity present in purified DNA polymerase from AMV, *Biochem. Biophys. Res. Commun.* **51**:232.

Weber, G. H., Kiessling, A. A., and Beaudreau, G. S., 1971, Deoxyribonucleic acid polymerase activity associated with strain MC29 tumor virus, *J. Virol.* **7**:214.

Wei, C. M., and Moss, B., 1974, Methylation of newly synthesized viral messenger RNA by an enzyme in vaccinia virus, *Proc. Natl. Acad. Sci. USA* **71**:3014.

Wei, C. M., and Moss, B., 1975, Methylated nucleotides block 5′-terminus of vaccinia virus messenger RNA. *Proc. Natl. Acad. Sci. USA* **72**:318.

Weimann, B. J., Schmidt, J., and Wolfrum, D. I., 1974, RNA-dependent DNA polymerase and ribonuclease H from Friend virions, *FEBS Lett.* **43**:37.

Weiss, S. R., and Bratt, M. A., 1974, Polyadenylate sequences on Newcastle disease virus mRNA synthesized *in vivo* and *in vitro, J. Virol.* **13**:1220.

Weissbach, A., Bolden, A., Muller, R., Hanafusa, H., and Hanafusa, T., 1972, Deoxyribonucleic acid polymerase activities in normal and leukovirus-infected chicken embryo cells, *J. Virol.* **10**:321.

Wells, R. D., Fluegel, R. M., Larson, J. E., Schendel, P. F., and Sweet, R. W., 1972, Comparison of some reactions catalyzed by deoxyribonucleic acid polymerase from avian myeloblast, *Biochemistry* **11**:621.

Wertz, G. W., and Levine, M., 1973, RNA synthesis by vesicular stomatitis virus and a small plaque mutant: Effects of cycloheximide, *J. Virol.* **12**:253.

Wilson, S. H., and Kuff, E. L., 1972, A novel DNA polymerase activity found in association with intracisternal A-type particles, *Proc. Natl. Acad. Sci. USA* **69**:1531.

Wu, A. M., Sarngadharan, M. G., and Gallo, R. C., 1974, Separation of ribonuclease H and RNA-directed DNA polymerase (reverse transcriptase) of murine type-C RNA tumor viruses, *Proc. Natl. Acad. Sci. USA* **71**:1871.

Yang, S. S., and Wivel, N. A., 1973, Analysis of high molecular weight ribonucleic acid associated with intracisternal A particles, *J. Virol.* **11**:287.

Yang, S. S., Herrera, F. M., Smith, R. G., Reitz, M. S., Lancini, G., Ting, R. C., and Gallo, R. C., 1972, Rifamycin antibiotics: inhibitors of Rauscher murine leukemia virus reverse transcriptase cd of purified DNA polymerases from human normal and leukemic lymphoblasts, *J. Natl. Cancer Inst.* **49**:7.

Zaides, V. M., Selimova, L. M., Zhirnov, O. P., and Bukrinskaya, A. G., 1975, Protein synthesis in Sendai virus-infected cells, *J. Gen. Virol.* **27**:319.

Zweerink, H. J., and Joklik, W. K., 1970, Studies on the intracellular synthesis of reovirus-specified proteins, *Virology* **41**:501.

CHAPTER 4

Animal Virus–Host Genome Interactions

Walter Doerfler*

Institute of Genetics
University of Cologne
Cologne, Germany

1. INTRODUCTION

Interactions between the viral and host genomes have been studied in numerous systems and in regard to different aspects. In the course of this interaction, viral genetic material often is inserted by covalent linkage into the host chromosome. One consequence of this insertion is the fixation of the viral genome or specific parts of it in the host cell. The effects of viral integration on host genetic functions are less clear and still require intensive investigations.

Animal viruses can interact with their host cells in different ways: (1) The virus can actively replicate and finally destroy the cell in a lytic or productive infection. (2) The infection can become abortive or nonproductive and lead to the apparent or inapparent persistence or the eventual loss of the viral genome. As a consequence of viral infection, the properties of the cell can be fundamentally altered and a malignantly transformed cell can arise. In such cells, the viral genome or parts of it persist and continue to be expressed. (3) Lastly, viral infections can remain latent for an extended time period and kill the host after many years. This last type of virus–host interaction can best

---

* This chapter is dedicated to my parents.

be followed in animals and has been termed "slow virus" infection (Hotchin, 1971; ter Meulen *et al.,* 1972; Fuccillo *et al.,* 1974). Only in recent years has it been recognized that in several species viral genetic information can be transmitted from one generation to the next. These so-called endogenous viral genomes have been extensively studied in avian and murine systems. Their role in the causation of neoplasia may be a crucial one; however, this role is not understood.

This chapter will be restricted to viruses which have been shown or are presumed to insert their DNA into the genomes of the host cells. This linear insertion by covalent phosphodiester bonds between the viral and host DNAs is called "integration." It is proposed that the term "integration" in *sensu strictiori* should be used only according to this definition and should be differentiated from terms like "association" or "persistence" of viral DNA.

For several transformed cell lines, it has been demonstrated that specific segments of the viral genome persist, probably in an integrated form. In a few transformed cell lines, the viral genetic information can be localized on specific chromosomes. Even in these very carefully studied systems, the term "integration" can be used in the sense proposed above, except that one will have to specify precisely which sections of the viral genome are integrated and/or where the site of integration is located. In the near future, one can expect a great deal of information on specific sites of integration of viral genomes. Up to now, the genetic analysis of the host and viral genome interactions has not been very sophisticated.

The demonstration of covalent linkage between viral and cellular DNA does not necessarily imply that this linkage is the result of a true integration event. Such linkage may alternatively be explained by recombination between viral and cellular DNA, and conceptually one should differentiate between these possibilities, although practically it may be very difficult to unravel the biochemical mechanisms by which covalent linkage of viral and host DNAs has been effected.

The best-studied and most definitely established examples of integrated viral genomes are those of the temperate bacteriophages, e.g., λ, P2, P22, and Mu. Among animal viruses, integration of viral DNA was shown to occur in DNA and RNA tumor viruses. To account for the state of the viral genome in virus-transformed cells, the model of lysogeny was first proposed by Lwoff (1953). However, integration of animal virus DNA is not restricted to transformed cells, but is observed in productively and abortively infected cells as well. Perhaps integration is a general phenomenon in some of the animal viruses which contain DNA or replicate via a DNA intermediate. Evidence has been presented (Burger and Doerfler, 1974; Doerfler *et al.*, 1974; Hirai and Defendi, 1974; Hölzel and Sokol, 1974; Schick *et al.*, 1976) that in cells productively infected with adenovirus or simian virus 40 (SV40) a considerable number of viral genome equivalents appear in a high molecular weight form. The significance of this integration and/or recombination event between virus and host genes and its role in viral replication and in the expression of host genes are the subject of intensive investigation. In cells in which viral genomes are actively replicating, it has been technically difficult to demonstrate the presence of integrated viral genes and to distinguish unequivocally between integrated viral genes and oligomeric forms of free viral DNA. However, there is increasing evidence that at least a portion of the high molecular weight form of viral DNA is covalently linked to cellular DNA (see Section 4.3 for detailed discussion of this problem).

Since integrated viral genes have become fixed to the genome of the host, the control of viral gene expression might become subject to the regulation by host genes, or, *vice versa,* viral genes might influence the transcription of host functions. Even in the most extensively analyzed systems, *viz.* in bacteria lysogenic for temperate bacteriophages, very little is known about the effect of integrated viral genes on host functions. Taylor discovered in 1963 that infection by bacteriophage Mu caused mutations in *Escherichia coli* in those genes into which viral DNA was inserted. Similarly, phage λ which carried a deletion in its attachment site for the interaction with the host genome was shown to cause mutations in those genes where it integrated (Shimada *et al.*, 1973). It will be interesting to investigate whether the integration of viral DNA can elicit mutations in eukaryotic cells as well. The model of integrated viral genes causing mutations in eukaryotic cells might explain some of the genetically stable alterations in virus-transformed cells. However, a very detailed genetic analysis of both the host genes at the site(s) of integration and the integrated viral genes will be required to test the validity of the proposed analogy with bacteriophages.

In this chapter, the emphasis will be on the integration of the DNA of animal viruses. For obvious reasons, current research in this field has been most extensive with the oncogenic viruses. A brief summary of topics related to lysogeny and temperate bacteriophages will also be given. Many problems in relation to the integration of animal virus

DNA remain unresolved, and it is obvious that improved technology will be required to obtain definitive proof for integration *per se* and exact information on the localization of viral genetic material in eukaryotic chromosomes.

The early work on integration of animal virus DNA has been reviewed by Temin (1971), Winocour (1971), and Sambrook (1972). These authors emphasized work on integration of the DNA of oncogenic viruses. More up-to-date discussions of the subject are found in *The Molecular Biology of Tumor Viruses* (edited by Tooze, 1973) and in the 1974 *Cold Spring Harbor Symposium on Tumor Viruses*. A summary of the evidence for integration of viral genomes in prokaryotic and eukaryotic systems has been published (Doerfler, 1975), and the integration of DNA tumor virus genomes has been reviewed (Martin and Khoury, 1976).

2. METHODS USED TO DEMONSTRATE INTEGRATION OF VIRAL GENOMES

A brief summary of the methods currently employed in research on integration will first be presented. In the work with bacteriophages, genetic methods were used very effectively to establish the integrated state of the viral genome. A genetic analysis of comparable sophistication has not yet been achieved for viral genomes integrated into the genomes of eukaryotic cells. Therefore, it was necessary to develop biochemical and biophysical methods to obtain evidence for the covalent linkage between viral and host DNAs. Improved cytogenetic techniques using cell hybrids (Ruddle, 1973) and chromosome banding techniques (Caspersson *et al.*, 1970) are being developed rapidly and will permit a more refined genetic analysis. Application of these methods has led to the localization of presumably virus-specific functions (e.g., the T antigen) or the entire SV40 genome on specific chromosomes of SV40-transformed or SV40-infected cells (Croce *et al.*, 1973). Other investigators have succeeded in demonstrating chromosomal uncoiling at specific sites on one particular chromosome in human cells infected with adenoviruses (McDougall *et al.*, 1973). Although a large number of methods have already been employed to explore the state of the viral genome, improved technology will be required for the final proof of covalent linkage between viral and host sequences in eukaryotic cells.

2.1. Work with Temperate Bacteriophages and Lysogenic Cells

2.1.1. Genetic Techniques

It was established that the gene order in the prophage of bacteriophage λ was a permutation of that in the vegetative phage (Calef and Licciardello, 1960) and that bacterial and prophage markers were cotransduced (Rothman, 1965; Franklin *et al.*, 1965). Genetic analysis of deletion mutants of phage λ which originated from faulty excision (e.g., λ*dg*) and the genetic analysis of transducing phage genomes (Arber, 1958; Campbell, 1959; Hogness and Simmons, 1964) provided further evidence for integration.

2.1.2. Electron Microscopy and Heteroduplex Mapping

Martuscelli *et al.* (1971) compared the length of bacterial episomes from normal cells and from cells lysogenic for bacteriophage Mu and demonstrated the presence of the phage genome by direct electron microscopic measurements of the bacterial episomes.

More detailed information was obtained by heteroduplex mapping of phage genes integrated into bacterial episomes and heteroduplex mapping of defective phage genomes (Simon *et al.*, 1971; Fiandt *et al.*, 1971; Hsu and Davidson, 1972).

2.2. Experimental Approaches to the Analysis of Viral Genomes in Chromosomes of Eukaryotic Cells

2.2.1. Separation of Cellular and Viral DNA under Conditions Which Denature DNA

Cellular and viral DNA can be separated either by zone sedimentation in alkaline sucrose density gradients (Sambrook *et al.*, 1968; Doerfler, 1968; 1969; Burlingham and Doerfler, 1971; Hirai and Defendi, 1971; Burger and Doerfler, 1974; Doerfler *et al.*, 1974; Schick *et al.*, 1976) or by equilibrium sedimentation in alkaline CsCl density gradients (Doerfler, 1968; 1970; Burger and Doerfler, 1974). In some systems, the cellular DNA was made "heavy" by substitution with 5-bromodeoxyuridine.

Varmus *et al.* (1976) used alkaline sucrose gradients in a zonal rotor to differentiate between integrated and unintegrated avian sarcoma virus-specific DNA.

2.2.2. "Network Technique" and Hybridization Methods

Cellular DNA is denatured and subsequently reannealed without prior fragmentation. Reannealing of the highly repetitive sequences leads to network formation (Britten and Kohne, 1968). The complexes contain cellular and integrated viral DNA, but do not trap appreciable amounts of free viral DNA. These complexes can be isolated by low-speed centrifugation (Varmus *et al.*, 1973*b*).

The viral sequences integrated into cellular DNA can be detected by DNA-DNA or DNA-RNA filter hybridization using viral DNA or RNA (cRNA) synthesized *in vitro* on a viral DNA template. The use of the DNA-RNA hybridization technique employing cRNA has frequently led to an overestimation of the number of viral genomes integrated (Haas *et al.*, 1972). For a more reliable quantitation of the number of integrated viral genome equivalents per cell, the analysis by DNA reassociation kinetics (Britten and Kohne, 1968; Wetmur and Davidson, 1968) has proved useful (Gelb *et al.*, 1971). A major shortcoming of this approach became apparent when it was applied to the analysis of cells in which only a part of the viral genome was present. This disadvantage can be avoided when specific fragments of the viral DNA generated by restriction endonucleases are used for the analysis by DNA reassociation kinetics.

Khoury *et al.* (1974*a*) used the separate strands of SV40 DNA (Westphal, 1970) to quantitate viral genetic material in transformed cells.

2.2.3. Cotranscription of Integrated Viral Genomes

Integrated viral genomes can be cotranscribed with cellular genes. Such "mixed transcripts" are isolated on polyacrylamide gels which contain formamide to avoid the formation of unspecific RNA aggregates. High molecular weight RNA molecules of a size exceeding that of unit-length viral RNA can be shown to contain viral and cellular sequences. This RNA is first hybridized to viral DNA on filters and is subsequently eluted from the filters and hybridized to cellular DNA in the second step of the reaction (Lindberg and Darnell, 1970;

Tonegawa *et al.*, 1970; Acheson *et al.*, 1971; Wall and Darnell, 1971; Rozenblatt and Winocour, 1972; Weinberg *et al.*, 1974). It has also been shown that a large proportion of the high molecular weight RNA hybridized to SV40 DNA on a filter remains sensitive to RNase. This result has been interpreted to indicate that the SV40-specific, high molecular weight RNA contains covalently linked cellular sequences (Jaenisch, 1972).

2.2.4. Excision of Integrated Genomes

Excision of integrated viral genomes from cellular DNA with the aid of restriction enzymes (Botchan and McKenna, 1973) was followed by analysis of the excised viral genomes by DNA-DNA hybridization (Schick *et al.*, 1976). Analysis of substituted viral genomes (compare Section 2.2.5.) was also possible by cleavage with restriction enzymes and separation of the fragments on agarose or polyacrylamide gels (Brockman *et al.*, 1973; Rozenblatt *et al.*, 1973).

2.2.5. Substituted Viral Genomes

Demonstration of "substituted viral genomes" was achieved in SV40 and polyoma virions; these presumably originate from faulty excision of integrated viral DNA (compare defective phage genomes). Substituted SV40 genomes are usually shorter than normal viral DNA and contain increasing lengths of cellular DNA depending on the number of high-multiplicity passages of the virus (Lavi and Winocour, 1972; Brockman *et al.*, 1973; Lavi *et al.*, 1973; Rozenblatt *et al.*, 1973; Martin *et al.*, 1973; Lavi and Winocour, 1974).

2.2.6. Heteroduplex Mapping

Heteroduplex mapping has been used mainly in the analysis of substituted viral genomes of SV40 and the SV40–adenovirus hybrids.

2.2.7. *In Situ* Hybridization

Preparations of metaphase chromosomes from virus-transformed or virus-infected cells are treated with alkali to "denature" the DNA, and virus-specific sequences are visualized by hybridization with virus-specific cRNA of high specific radioactivity and subsequent autoradiography (Gall and Pardue, 1969). The cRNA is synthesized *in vitro* on a viral DNA template with DNA-dependent RNA polymerase from *Escherichia coli*.

This method does not permit the unequivocal differentiation between true integration and association of viral DNA with chromosomes. Another limitation of this method lies in the fact that not all viral genomes are transcribed faithfully by the *E. coli* polymerase.

2.2.8. Work with Cell Hybrids

Fusion of virus-transformed cells with permissive cells allowing viral replication and rescue of the viral genome has been successful with SV40 (Gerber, 1966; Koprowski *et al.*, 1967; Tournier *et al.*, 1967; Watkins and Dulbecco, 1967) but only in one case with polyoma virus (Fogel and Sachs, 1969). So far, viral genomes have not been rescued from adenovirus-transformed cells. Fusion of different cell types is effected by UV-inactivated Sendai virus (Harris and Watkins, 1965) or by treatment with polyethyleneglycol. This technique represents one of the most powerful tools to prove that the entire viral genome persists in some strains of virus-transformed cells. However, rescue of the viral genome does not shed light on the physical state of the viral genome in transformed cells.

2.2.9. Chemical Induction of Virus

Similar reservations apply to results of experiments in which virus was chemically induced from transformed cells (Burns and Black, 1968; Rothschild and Black, 1970) or apparently normal cells (Teich *et al.*, 1973).

2.2.10. Mendelian Genetics

Mendelian genetics has proved to be a very successful approach to the localization of the genome of murine viruses in high incidence leukemia strains of mice (Rowe, 1973; Lilly and Pincus, 1973; Chattopadhyay *et al.*, 1975).

3. BACTERIOPHAGE MODELS

Temperate bacteriophage and lysogenic bacteria have frequently been viewed as models for those DNA animal viruses whose genomes can be integrated into host DNA. Therefore, it might be useful to present by way of introduction a brief summary of the most important features of lysogeny. There are clearly parallels between lysogeny and the integrated state of viral genomes in eukaryotic cells. Although existing analogies should not be overemphasized, it may be useful to present in context the known gamut of interactions between viral and host genomes in prokaryotic and eukaryotic cells.

3.1. Bacteriophage λ

A number of outstanding reviews on different aspects of bacteriophage λ have been published (Dove, 1968; Signer, 1968; Echols, 1972; Herskowitz, 1973). The most comprehensive survey on λ is found in *The Bacteriophage Lambda* (1971). A review on lysogeny has also been included in Volume 8 of *Comprehensive Virology*.

3.1.1. Lytic vs. Lysogenic Infection

After infection of the host cells by bacteriophage λ, the phage can either start replicating and eventually destroy the host cell or the phage DNA can become integrated into the cellular DNA with a concomitant repression of most of the viral gene functions: The cell becomes lysogenized. In the lysogenic state, the viral genome replicates in the integrated form synchronously with the host chromosome. Repression of viral gene functions is effected by a phage-coded protein, the repressor (Ptashne, 1967*a,b*), which interacts, probably in a dimeric form (Chadwick *et al.,* 1970), with two sites on the viral genome (O_l and O_r; see Fig. 1), and thus presumably prevents the host RNA polymerase from transcribing the early viral genes. The λ repressor is coded for by gene *c*I (Kaiser, 1957). Transcription of the *c*I gene continues in the lysogenic cell and is essential for the maintenance of lysogeny. A certain intracellular titer of repressor bestows upon the cell immunity toward superinfection with the same phage. Since the interaction between the repressor and the O_l and O_r sites on the viral genome is highly specific, the lambdoid phages, e.g., phages 434, 424,

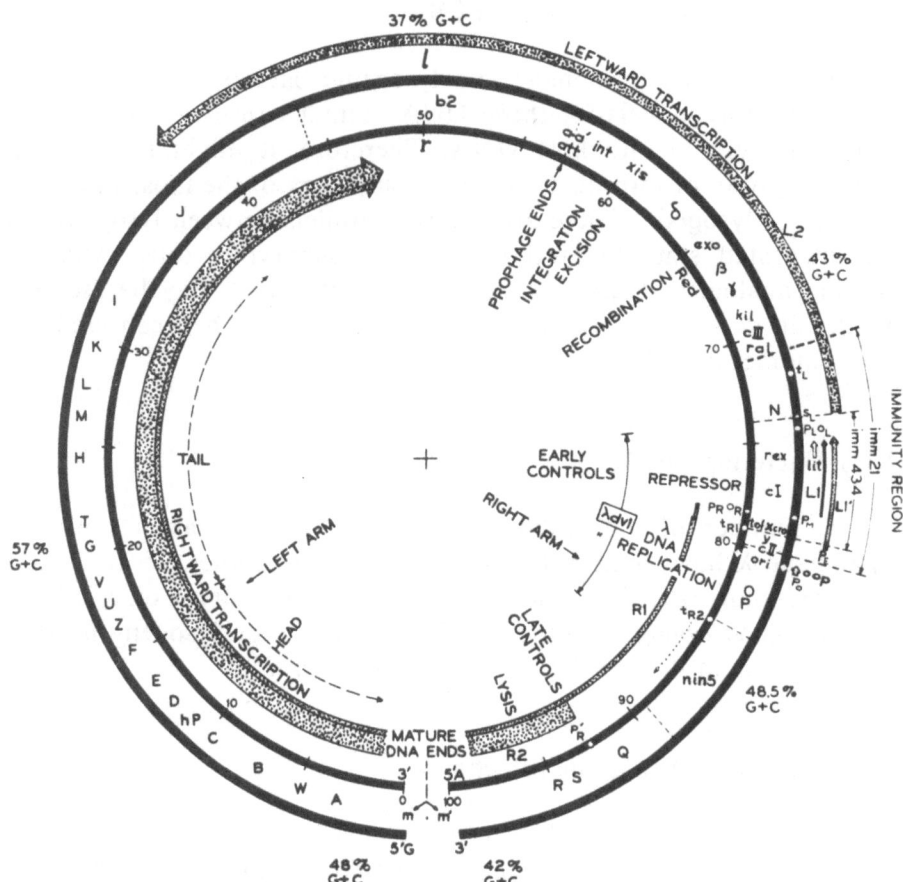

Fig. 1. The λ DNA map. Modified from Szybalski *et al.* (1970) and from Szybalski (1974). *A, W, B, C, Nu3, D, E, F*$_I$, *F*$_{II}$: Genes involved in formation of phage head (Kaiser and Masuda, 1973). Gene *A* is responsible for the formation of the cohesion of the DNA and the cleavage of polymeric head precursor DNA (Wang and Kaiser, 1973). Gene *E* protein makes up 95% of the mass of the head proteins (Casjens *et al.*, 1970; Buchwald *et al.*, 1970). The phage head is a $T = 7$ structure, containing 420 subunits each of proteins E and D, which are distributed over the entire surface of the phage head (Casjens and Hendrix, 1974). Gene *F*$_{II}$ probably codes for the attachment site on heads for tails (Casjens, 1974). *Z, U, V, G, T, H, M, L, K, I, J*: Genes involved in formation of phage tail. Gene *V*: Major tail protein which is distributed along the length of the phage tail. Gene *J*: Determines host range. *int*: Responsible for prophage insertion (and excision). *xis*: Responsible for prophage excision. *exo* or *redα*: λ exonuclease; promotes general recombination in *red*A$^-$ and *pol*A$^-$ host cells. *β* or *redβ*: β protein; promotes general recombination and facilitates growth in *rec*A$^-$ host cells. *c*III: Establishment of immunity. *N*: Positive regulator, regulates transcription of genes to the left of *N* and to the right of *cro*. *rex*: Inhibits the growth of T4*r*II mutants. *c*I: λ repressor; maintenance of immunity (Kaiser, 1957). *cro, tof*, or *fed*: Negative control of immunity; responsible for turnoff of transcription of genes *N* through *int*. *y*: Interacts with the *c*II/

and 21, which differ from λ only in their immunity regions (Fig. 1), are not susceptible to λ repression and can replicate in a cell lysogenic for phage λ.

The regulation of the synthesis of λ repressor has been investigated by Reichardt (1975a,b), who studied in detail the role that the establishment promoter (*pre*) and the maintenance promoter (*prm*) and the *N*, *c*II, *c*III, and *cro* gene products play in lysogenization. Recently, Ptashne *et al.* (1976) summarized our present understanding of the autoregulation and function of the λ repressor.

In the series of events involved in lysogenization of a cell, three steps are essential: (1) establishment of lysogeny, (2) maintenance of lysogeny, and (3) capacity for induction. Establishment of lysogeny is a very complex process requiring, at the same time, expression of the early viral genes necessary for integration, repression of the late genes responsible for lytic development, and a highly specific recombination event between specific sites on the phage (*POP'*)* and the host (*BOB'*)* genomes (Fig. 2). The phage-specific *int* function is necessary for the integration of the phage genome. The products of genes *c*II and

---

* According to Shimada *et al.* (1975), the bacterial attachment site carries two sequences *B* and *B'* which flank a core sequence *O*. The phage attachment site exhibits two different sequences *P* and *P'* which flank the same core sequence *O*. The designation *O* for this core sequence indicates that recombination within the attachment sites involves a region (Shulman and Gottesman, 1973).

---

*c*III oligomer. *c*II: Establishment of immunity. *O,P*: DNA replication. *Q*: Positive regulator of late gene transcription. *S,R*: Cellular lysis; *R* is the gene for the endolysin, an endopeptidase. *m,m'*: Left and right cohesive ends of λ DNA; the nucleotide sequence of the cohesive ends was determined by Wu and Taylor (1971):

$$
\begin{array}{ll}
5' & 3' \\
\text{GGCGGCGACCT}\!\!-\!\!-\!\!-\ \ -\!\!-\!\!-\text{OH} & \\
\quad\quad\text{HO}\!\!-\!\!-\!\!-\ \ -\!\!-\!\!-\text{CCCGCCGCTGGA} & \\
3' & 5'
\end{array}
$$

att: Site of interaction ("attachment") of λ DNA with host DNA during integration step. $P \cdot P'$ $(a \cdot a')$ designates the phage DNA site and $B \cdot B'$ $(b \cdot b')$ the host bacterial DNA site (*cf.* footnote on p. 289). P_L: Early leftward promoter (mutants *sex* and *t27* define this promoter and are deficient in transcription of genes *N* through *int*). O_L: Operator controlling transcription of genes *N* through *int*; left binding site of the λ repressor; site of V2 mutation (*vir*L). P_R: Early rightward promoter situated in the *x* region; λ x^- mutations are deficient in transcription of genes *cro* through *Q*. $P_{R'}$: Late rightward promoter; required for late gene expression. Site of action of gene *Q* product. O_R: Operator controlling transcription of genes *cro* through *Q*; right binding site of the λ repressor; site of V3 and V1 mutations. *ori*: Origin of DNA replication. *t*: Termination for transcription. i^{21}, i^{434}: Immunity regions which are altered in lambdoid phages 21 and 434.

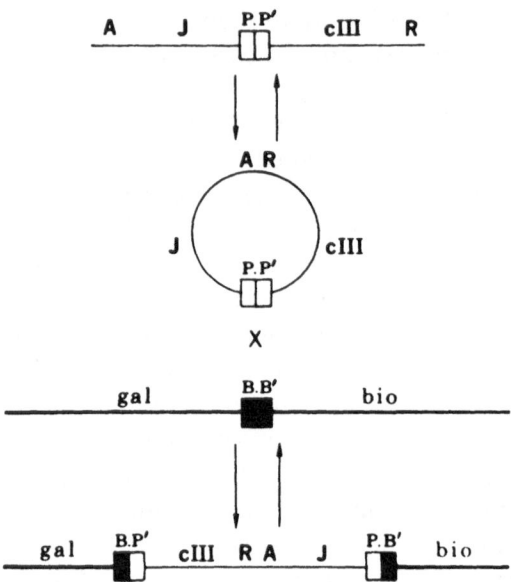

Fig. 2. Circularization of the λ chromosome and prophage insertion and excision according to the Campbell (1962) model. *A, J, c*III, and *R* are λ genes whose sequence is permuted in the process of insertion. *B · B′* and *P · P′* are the bacterial and phage attachment sites, respectively. Upon insertion of the prophage, the distance between the bacterial genes *gal* (utilization of galactose) and *bio* (synthesis of biotin) is increased (Rothman, 1965). From Gottesman and Weisberg (1971).

*c*III, perhaps in the form of an oligomeric complex, are essential for the establishment of lysogeny (Echols and Green, 1971), since mutants in these genes exhibit a marked deficiency in the frequency of lysogenic conversion. Recent evidence suggests that the *c*II and *c*III proteins act at a single site in the *y* region (Fig. 1) of λ DNA to provide both positive and negative regulation (Court *et al.,* 1975). Lysogeny is maintained through the action of the repressor.

For the induction of lytic development in a lysogenic host, it is necessary to release repression and to excise the viral genome. Several agents known to inhibit cellular DNA synthesis (irradiation with ultraviolet light, mitomycin C, thymine deprivation, etc.) inactivate the repressor by a mechanism that is not understood. It has been demonstrated that the inactivation of the λ repressor by ultraviolet light or mitomycin C is accompanied by proteolytic cleavage of the repressor (Roberts and Roberts, 1975). In the mutant λ*c*I857 the repressor is temperature sensitive and can be inactivated by high temperatures. Upon inactivation of the repressor, viral genes can be expressed, the viral genome is excised through the action of the *int* and *xis* functions, and late transcription starts leading to lytic development.

The factors which determine the outcome of an infection with phage λ, and lead to either lysis or lysogeny, are poorly understood. At high multiplicity of infection, lysogenization is favored over the lytic cycle. Cells in "poor nutritional conditions" are lysogenized at higher frequency than those actively growing. At temperatures higher than 37°C, the frequency of lysogenization of the cells decreases. In cells deficient in the enzyme adenylyl cyclase or in the catabolite activator protein (CAP), the frequency of lysogenization is also reduced (Grodzicker *et al.*, 1972). Echols *et al.* (1975) postulated a compensatory regulatory mechanism. Supposedly, a delay in the late functions of lytic development is coupled with a lower production of λ repressor.

Phages P22 (Levine, 1972) and P2 (Bertani and Bertani, 1971) are very similar to λ in many respects. On the other hand, phage P1 apparently does not have to integrate its genome and is able to replicate as a free episome synchronously with the host chromosome (Ikeda and Tomizawa, 1968).

3.1.2. On the Mechanism of Integration

The molecular mechanism of integration of the λ genome is only partly understood (Gottesman and Weisberg, 1971; Gottesman, 1974). In 1962, Campbell suggested a model for integration in phage λ (Fig. 2). This model was later proven to be correct in its essential features (Rothman, 1965; Franklin *et al.*, 1965; Young and Sinsheimer, 1964; Bode and Kaiser, 1965; Dove and Weigle, 1965; Gellert, 1967; Tomizawa and Ogawa, 1968). Upon entering the cell, the infecting phage DNA molecules are circularized, and, by reciprocal recombination between the *POP'* site on the phage DNA and a specific site (called *att* or *BOB'*) (Signer and Beckwith, 1966; Signer *et al.*, 1969) on the host chromosome, the viral genome is inserted between the *gal* and *bio* sites at 17 min on the genetic map of *E. coli*. This recombination step can be formalized as follows:

$$\text{---}POP'\text{---} x\text{---}BOB'\text{---}\rightleftharpoons\text{---}BOP'\text{---}POB'\text{---}$$

(*BOP'* and *POB'* represent the junctions between the bacterial ——— and phage ~ ~~ chromosomes.)

Rothman (1965) has been able to show that insertion of the λ chromosome leads to a reduction in the frequency of cotransduction of the *gal* and *bio* markers by phage P1. The insertion of the λ chromosome into the bacterial DNA is alkali stable and is therefore presumed to be effected by covalent bonds (Freifelder and Meselon, 1970; Folkmanis and Freifelder, 1972). In the process of integration, the gene order of the phage is permuted (Calef and Licciardello, 1960; Rothman, 1965; Franklin *et al.*, 1965).

In 1967, Weisberg and Gallant demonstrated that integration of the phage genome (the prophage) was dependent on protein synthesis. At the same time, mutants deficient in the integration function (*int⁻* mutants) were isolated in several laboratories. Thus it appeared likely that a phage gene was responsible for the synthesis of the "integrase" function (Zissler, 1967; Gingery and Echols, 1967; Gottesman and Yarmolinsky, 1968*a*). The *int* protein with a molecular weight of approximately 42,000 (Hendrix, 1971; Ausubel *et al.*, 1971) has been characterized to some extent.

The recombination step catalyzed by the *int* function is highly specific for the viral and host sites. This has been clearly demonstrated in experiments crossing *int⁺ red⁻* λ mutants in a *rec⁻* host. In this cross, the general recombination genes of the phage (*red*) (Kayajanian, 1972) and the recombination system of the host (*rec*) are inoperative. The phage recombinants found in these experiments were restricted to the region between *b2* and *int* (Weil and Signer, 1968; Echols *et al.*, 1968). It is not understood how the *int* protein (perhaps in cooperation with other functions) recognizes the *POP′* and *BOB′* sites.

Lambda *int⁻* mutants are able to lysogenize at a frequency of approximately 10⁻⁶ per infected cell. Such lysogens apparently can be formed by the action of the general *rec* system of the host cell. Gottesman and Yarmolinsky (1968*b*) demonstrated that in these lysogens the prophage was not inserted at the *BOB′* site but rather at random locations on the host chromosome. Thus the high frequency and specificity of the prophage insertion depend on the very efficient integration system that is coded for, at least in an essential part, by the *int* gene of phage λ.

When the primary λ attachment site (*BOB′*) between the *gal* and *bio* markers on the map of *E. coli* was deleted, λ DNA integrated at secondary sites. Some of these lysogens proved to be mutants, and this mutagenesis was not random. The inserted prophage reduced the expression of distal genes in the operon into which λ DNA was inserted (Shimada *et al.*, 1973).

Integration at secondary sites in hosts that lack the bacterial *att* site is due to *int*-promoted recombination and occurs at approximately

0.5% the normal frequency. These secondary sites are different from the normal prophage attachment and the phage *att* sites (Shimada *et al.*, 1975).

How does the *int* function recognize specific nucleotide sequences at the site of integration? The bipartite structure of the attachment site (POP′) on the λ chromosome has been elucidated by Parkinson and collaborators (Parkinson and Huskey, 1971; Parkinson, 1971; Davis and Parkinson, 1971). Presently, the attachment region (*POP′*) on the λ genome is being precisely mapped with the aid of restriction endonucleases in several laboratories. The nucleotide sequence of this site may be known before long.

Integration resembles genetic recombination in that it proceeds via breakage and rejoining, but differs from general recombination by its requirement for the *int* function. DNA synthesis is not required for insertion or excision. However, λ mutants deficient in DNA replication lysogenize less efficiently than wild-type λ (Brooks, 1965).

Excision is the reversal of integration and entails recombination of the *P* and *P′* sites of the prophage. For efficient excision, the products of both the *int* and *xis* genes are necessary. In rare cases, faulty excision occurs as a result of illegitimate recombination between sites in the prophage and the bacterial chromosome (Campbell, 1962), giving rise to phage particles which are defective in phage genes but carry bacterial markers (e.g., the *gal* operon in λ*dg*). Such phage particles can transduce bacterial markers, e.g., various parts of or the entire *gal* operon.

The expression of the *int* and *xis* genes is under the control of the *N* gene product and therefore regulated by the leftward operator and promoter. After infection of *E. coli* with a λ*N*⁻ mutant, integration cannot occur, but the λ*N*⁻ genome replicates once with each round of cellular replication, and persists in the cell as a nonintegrated plasmid (Signer, 1969). Many details of the differential control of genes *int* and *xis* are still unknown. One of the paradoxes of regulation in λ centers around the question of how an integrated phage genome escapes excision, since both integration and excision are highly efficient processes and both the *int* and *xis* genes belong to the *N* operon. Two explanations have been proposed, but neither is entirely satisfactory:

1. Weisberg and Gottesman (1971) provided evidence that the *xis* gene product is relatively unstable.
2. Shimada and Campbell (1974*a*) have discovered a weak leftward promoter, *p*I, which is probably located in or near the *xis*

gene and may be independent of immunity. This secondary promoter may uncouple the otherwise obligatory cotranscription of the *int* and *xis* functions and drive the equilibrium established between the insertion–excision reactions in the direction of integration.

Mutations in the vicinity of *xis* have been isolated which may define a third gene, called *hen,* which also appears to be involved in the integration process (Chung, Green, and Echols, cited by Shimada and Campbell, 1974*b*). There are probably a number of additional factors required for integration, some of which have recently been analyzed more precisely. Guarneros and Echols (1973) have shown that a thermolabile element is required for integration but not for excision. Freifelder *et al.* (1975), Gottesman and Gottesman (1975*a*), and Nash (1975*a*) have carried out detailed studies on the requirements for integration.

Nash (1975*b*) reported on the development of an *in vitro* system for the study of integrative recombination of bacteriophage λ DNA. This *in vitro* recombination was dependent on *int* gene product, the presence of a thermolabile component, ATP, Mg^{2+}, spermidine, and a monovalent cation. The DNA from λ *att* B–*att* P, a phage variant with two attachment sites, served as substrate for the reaction. The enzymatic activity was extracted from *E. coli* containing phage λ gene products and was inhibited by the *xis* gene product. An *in vitro* system allowing the study of site-specific recombination between *att* P was described by Syvanen (1974).

Similarly, Gottesman and Gottesman (1975*b*) demonstrated the excision of prophage λ DNA in a cell-free system. The substrate for this reaction was λ DNA carrying two attachment sites. The *in vitro* system was very efficient, converting 25–35% of the substrate within 30 min. For the reaction, ATP and Mg^{2+} were required; spermidine stimulated the reaction. There did not seem to be a requirement for RNA or DNA synthesis, and polynucleotide ligase was probably not involved in the reaction.

Upon infection of *E. coli* with bacteriophage λ, polylysogens are formed at a frequency of approximately 40%. Most of the polylysogens are dilysogens which are produced by the *int* system or the *int* and *red* systems together. The *int*-promoted polylysogens are generated either by sequential insertion of monomers or by integration of λ polymers (Freifelder and Levine, 1975).

3.2. Bacteriophage Mu

In 1963, Taylor discovered phage Mu in K12 strains of *E. coli*. The striking feature of this phage was that it caused polar mutations in a large number of host genes (Taylor, 1963). In prototrophic *E. coli* lysogenized by phage Mu, 2% of the survivors were auxotrophic. The site of phage-induced mutation could be correlated with the location of insertion of the phage genome into the host chromosome (Jordan *et al.*, 1968; Toussaint, 1969; Martuscelli *et al.*, 1971; Boram and Abelson, 1971; Bukhari and Zipser, 1972). Thus it became likely that phage Mu could integrate its genome into the *E. coli* chromosome practically at random. In this respect, the temperate *E. coli* phage Mu differed markedly from other temperate phages, such as phage λ and phage P22 (Levine, 1972), which usually have one specific site of integration, and phage P2, with a limited number of attachment sites (Calendar and Lindahl, 1969).

3.2.1. The DNA of Phage Mu

The morphology of phage Mu was described by To *et al.* (1966) and by Martuscelli *et al.* (1971). The DNA of the phage is a double-stranded molecule of 28×10^6 daltons (Torti *et al.*, 1970). The length of the linear DNA molecule as measured in electron micrographs ranged between 12.9 ± 0.1 μm (Martuscelli *et al.*, 1971) and 14.5 ± 0.7 μm (Torti *et al.*, 1970). More detailed information about the anatomy of the DNA of phage Mu was obtained by denaturation and renaturation of the DNA and subsequent electron microscopy: In addition to regular duplex molecules, more complex structures were observed. Most of the renatured molecules carried split ends on one terminus, and about 50% of the renatured molecules had a split end and a loop (the so-called G loop) on the same terminus (Bade, 1972; Daniell *et al.*, 1973*a,b*). The split ends of Mu DNA are probably due to host DNA which remains after the excision of the prophage (Daniell *et al.*, 1973*a,b*). The split ends are not detectable in the prophage form of Mu DNA (Hsu and Davidson, 1974). The F loop is explained by an inversion of some of the phage DNA sequences (Hsu and Davidson, 1974). This inversion was also found in the prophage (Hsu and Davidson, 1972). In the region of the G loop, the sequence occurs in either the direct or the inverted order and is bracketed by inverted repeat sequences of approximately 50 nucleotides

(Hsu and Davidson, 1974). The inversion is thought to be generated by reciprocal recombination between the inverted repeated sequences (Hsu and Davidson, 1974). The G loop was also observed in the DNA of defective λ phage which carried 14% of Mu DNA on one terminus (Daniell *et al.*, 1973*b*). The G loop was also found when phage Mu was replicated in *rec*A⁻ and *rec*BC⁻ hosts (Wijffelman *et al.*, 1972; Daniell *et al.*, 1973*b*). The arrangement of the DNA of phage Mu is schematically depicted in Fig. 3.

The restriction endonuclease *Eco*RI cuts Mu DNA twice, producing fragments A, B, and C of molecular weights 11.6×10^6, 9.5×10^6, and 3.1×10^6, respectively. Fragment C corresponds to the immunity end and fragment B carries the G segment, the variable end, and host sequences (Bade *et al.*, 1975; Bukhari and Taylor, 1975).

Except for a minor fraction (1.45% of the late RNA) of the phage genome, the RNA is transcribed from that strand of Mu DNA which binds the larger amount of poly(U,G) (Bade, 1972).

3.2.2. The Genetics of Phage Mu

A large number of mutants of phage Mu have been isolated, and 21 cistrons have been identified (Abelson *et al.*, 1973). Only two of the phage functions are known; *c* is responsible for immunity and *lys* is required for the lysis of the host. There is agreement among several laboratories that the gene orders in the prophage and the vegetative phage are identical and that there is a specific site on the phage genome which undergoes recombination with an unlimited number of sites in the host chromosome (Bukhari and Metlay, 1973; Faelen and Toussaint, 1973; Howe, 1973*a*; Wijffelman *et al.*, 1973). Moreover, it is apparent from a genetic analysis of the prophage that it can be inserted

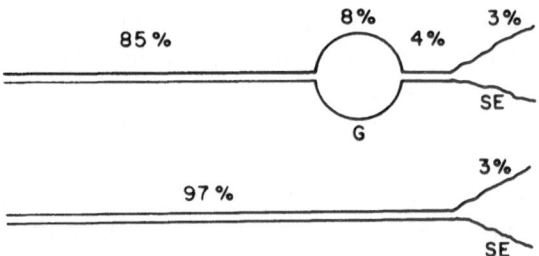

Fig. 3.   The two types of molecules observed after denaturation and self-renaturation of Mu DNA. The variable end, branched in renatured molecules, is designated SE, and the internal bubble is designated G. From Bukhari and Allet (1975).

in either of two orientations relative to the orientation of the host gene in which the prophage integrates (Boram and Abelson, 1973; Bukhari and Metlay, 1973; Howe, 1973*a*; Hsu and Davidson, 1972; Wijffelman *et al.*, 1973; Zeldis *et al.*, 1973). Phage Mu has been recognized to be capable of generalized transduction of host markers at a frequency of 10^{-7} to 10^{-9} transductants per plaque-forming unit (Howe, 1973*b*).

3.2.3. Integration of Phage Mu

Phage Mu induces strong polar mutations in approximately 2% of the lysogens (Taylor, 1963; Taylor and Trotter, 1967; Jordan *et al.*, 1968). These mutants have a very low reversion frequency of 10^{-10} or less. Linear insertion of the DNA of phage Mu1 was directly demonstrated by Martuscelli *et al.* (1971), who isolated covalently closed circular DNA molecules from *E. coli* strains carrying the episome *F′ lac* or *F′ lac* Mu$^+$ and compared the lengths of these molecules. The mean length of the *F′ lac* episome of *E. coli* was 37.6 ± 0.4 μm and that of the *F′ lac* Mu$^+$ lysogen 53.2 ± 0.4 μm. The difference of 15.6 μm was close to the values of 12.9 μm and 14.5 μm for the DNA of vegetative phage Mu determined by Martuscelli *et al.* (1971) and Torti *et al.* (1970), respectively.

Insertion of the phage genome was also documented genetically. Boram and Abelson (1971) analyzed a Mu-induced *chl*D mutant located between the insertion site of λ and the *gal* operon, and demonstrated that λ did not transduce *gal* genes from this mutant and that transducing particles of λ isolated from this strain contained parts of the Mu genome. Bukhari and Zipser (1972) investigated 76 different Mu-induced mutations in the gene of the *lac* operon of *E. coli* and provided powerful evidence that the insertion of the Mu genome in the Z gene was random. It could also be demonstrated that there were no major deletions of host DNA at the site of integration.

Little is known about the enzymatic mechanism that is responsible for the insertion of Mu DNA, except that it is phage encoded, recognizes a specific site on the phage DNA, and has apparently no specific requirements as to recognition of host sequences. The integration mechanism of Mu can catalyze the insertion of the DNA of a λ N^- mutant in a *E. coli* *rec*A$^-$ host, but this mechanism is not operative in a Mu lysogen (Faelen *et al.*, 1971). This artificial insertion of λ into the host chromosome is not site specific. Phage Mu does not have a recombination system of its own, since in a *rec*A$^-$ host recombination frequencies are extremely low.

However, phage Mu can utilize the recombination system provided by coinfection with phage lambda (Wijffelman *et al.,* 19.72). In a *rec*A⁻ host, phage Mu can also promote the integration of an F^+ episome into the host chromosome (van de Putte and Gruijthuijsen, 1972). Lambda particles have been isolated which carried different segments of Mu DNA, while parts of the λ genome were deleted. The deleted fractions of the λ genome did not necessarily correspond in length to the insertions of Mu DNA (Bukhari and Allet, 1975).

Hsu and Davidson (1972) have demonstrated by electron microscopy of heteroduplex DNA molecules that the Mu genome is linearly inserted into the F' episome of *E. coli* and that the prophage contains the same sequences in identical order as in the vegetative phage.

After induction of cells lysogenic for phage Mu, covalently closed circular DNA molecules were observed (Hsu and Davidson, 1972; Waggoner *et al.,* 1974). The length of these molecules ranged from 36.5 to 156.7 kilobases. These heterogeneous molecules were not simply mono- or oligomers of the phage chromosome, but contained phage genomes linked to cellular DNA (Waggoner *et al.,* 1974).

Genetic evidence for asymmetrical excision of the prophage Mu was presented by Schroeder and van de Putte (1974). According to this study, the prophage DNA is excised very precisely at a specific site on one end and unspecifically on the other end of the phage genome. This finding explains the occurrence of the split ends of the vegetative phage DNA.

4. ADENOVIRUSES

The mechanism of replication of adenoviruses in human cells in tissue culture has been analyzed in great detail, and we understand to some extent the intricacies of viral macromolecular syntheses. In recent years, the genetic characterization of the adenovirus genome has. progressed, and genetic maps based on recombination data (Ginsberg and Young, Vol. 9, this series) or on the analysis of the transcription of the viral genome (Pettersson and Philipson, 1974; Philipson *et al.,* 1974; Tal *et al.,*1974; Sharp *et al.,* 1974; Söderlund *et al.,*1976; Craig *et al.,*1975; Ortin *et al.,*1976) or its translation (Lewis *et al.,*1975; Atkins *et al.,*1975; Öberg *et al.,* 1975; Saborio and Öberg, 1976) have been constructed. With the advent of restriction endonucleases, a more refined dissection of the viral genome has become possible, and, as a consequence, studies on transcription and translation have profited from this improved technology.

The adenovirus system offers the additional advantage that the three principal types of virus–host cell interactions can be investigated: (1) Adenoviruses infect human cells productively and replicate to extremely high titers. (2) Certain types of adenoviruses cannot replicate in hamster cells (adenovirus type 12) and undergo an abortive infection (Strohl, 1969a,b; Doerfler, 1969). Rat cells, on the other hand, are semipermissive for adenovirus types 2 and 5, and the virus multiplies to a very limited extent or possibly only in a small number of the infected cells (Gallimore, 1974). (3) Adenoviruses transform hamster, rat, and rabbit cells in tissue culture, and most of the established transformed cell lines are oncogenic when injected into animals. Transformation of cells can be achieved also with isolated viral DNA (Graham and van der Eb, 1973a,b) and even with viral DNA fragments (Graham et al., 1974a,b) generated by restriction endonucleases.

Thus it has become possible to apply the information gained on the molecular biology of adenoviruses in the productive system to investigations on the molecular mechanisms underlying the abortive cycle(s) and/or malignant transformation of cells by viruses. There is mounting evidence that an as yet unknown number of host factors play an important role in viral replication (Nishimoto et al., 1975) and an even more decisive part in the complex events involved in viral transformation. With the dearth of information on the genetics of animal cells, progress in the elucidation of these host factors has been very slow.

It has been recognized that one of the key events in the interaction of oncogenic animal viruses with their host cells is the integration of the viral into the cellular genome. Since the initial demonstration of integration of animal virus DNA in the adenovirus and simian virus 40 systems (Doerfler, 1968, 1970; Sambrook et al., 1968), it has become apparent that the DNA of most—if not all—DNA and RNA tumor viruses can be inserted into the host chromosome. It is well documented that the genetic information encoded in the genome of RNA tumor viruses is transcribed by the reverse transcriptase into DNA (Temin and Baltimore, 1972) prior to integration into the host genome (Varmus et al., 1974). Evidence for integration of the viral genome has also been presented for Visna virus (Haase and Varmus, 1973) and even for respiratory syncytial virus (Simpson and Iinuma, 1975) and measles virus (Zhdanov, 1975). As far as is known, respiratory syncytial virus is not a tumor virus and has previously been classified among the negative-strand viruses. In its replication cycle, respiratory syncytial virus apparently can traverse the regular pathway and replicate also via the DNA route. There is evidence that the DNA intermediate may become integrated (Simpson and Iinuma, 1975).

Recently, it has been claimed that the genetic information encoded in measles virus RNA can also be transcribed into DNA and that this DNA can persist in an integrated state (Zhdanov, 1975).

Considering all the evidence on integrated viral genomes, it is conceivable that insertion of viral DNA is a general phenomenon and that, in some cases, integration may even be required for viral replication. At present, it is unknown if and, if so, how the event of integration is linked to the transformed state of the cell. Investigations of this "missing link" have now come to the fore of research on oncogenic viruses.

In principle, three different approaches have been chosen to examine the integrated state of viral genetic material: (1) At early times after infection of cells with oncogenic DNA or RNA viruses, the fate of the parental or newly synthesized viral DNA was followed and viral DNA was shown to be integrated at least in part into the host chromosome (Doerfler, 1968, 1970; zur Hausen and Sokol, 1969; Burlingham and Doerfler, 1971; Hirai *et al.*, 1971; Collins and Sauer, 1972; Hirai and Defendi, 1972; Ralph and Colter, 1972; Manor *et al.*, 1973; Varmus *et al.*, 1973*b*; 1974; Burger and Doerfler, 1974; Hölzel and Sokol, 1974; Doerfler *et al.*, 1974; Schick *et al.*, 1976; Fanning and Doerfler, 1977). This type of study approaches the problem from a general point of view and aims at an investigation of the mechanism of integration. (2) The integrated state of the viral DNA in cells transformed by oncogenic viruses was the subject of equally intensive research (Sambrook *et al.*, 1968, 1974; Dulbecco, 1968; Green, 1970*b*; Gelb *et al.*, 1971; Hirai and Defendi, 1971; Smith *et al.*, 1972; Gelb and Martin, 1973; Markham and Baluda, 1973; Pettersson and Sambrook; 1973; Varmus *et al.*, 1973*b*, 1974; Hill *et al.*, 1974; Sharp *et al.*, 1974*a*; Gallimore *et al.*, 1974; Bellett, 1975; Fanning and Doerfler, 1976; Groneberg *et al.*, 1977). This approach focuses on the virus-transformed cells, which, in a sense, represent a terminal state and possibly a special case in the gamut of interactions between susceptible cells and oncogenic viruses. A great deal of work has been devoted to studies on the expression of the viral genome in virus-transformed cells (see below). (3) In simian virus 40 and polyoma virus, populations of substituted viral genomes have been detected which carry covalently linked cellular sequences (Lavi and Winocour, 1972, 1974; Lavi *et al.*, 1973; Rozenblatt *et al.*, 1973; Brockman *et al.*, 1973; Martin *et al.*, 1973; Winocour *et al.*,

1974). Similar structures have also been found with adenoviruses (Tjia *et al.*, 1977). Such substituted genomes are thought to arise by faulty excision of viral DNA and subsequent amplification.

4.1. The Adenovirus System

Adenoviruses were discovered in 1953 by Rowe and colleagues in human adenoids. There are 34 human types of adenoviruses (Béládi, 1972), and some 50 types have been isolated from other species. Human adenoviruses can be propagated to very high titers in human cells growing in suspension culture (Green and Piña, 1963) and cause characteristic cytopathic effects (Boyer *et al.*, 1959). Some types of human adenoviruses (types 2 and 5) can also replicate in hamster cells. In 1962, Trentin and colleagues and Huebner and colleagues discovered that human adenovirus types 12 and 18 caused tumors in newborn hamsters. Pope and Rowe (1964) and McBride and Wiener (1964) demonstrated that hamster cells could be transformed in tissue culture by adenovirus types 12 and 18. In subsequent years, a large number of cell types were transformed by human adenoviruses, e.g., rat embryo cells by adenovirus type 2 (Freeman *et al.*, 1967; Gallimore, 1974), baby hamster kidney (BHK-21) cells by adenovirus type 12 (Strohl *et al.*, 1967), rabbit cells by adenovirus type 12 (Levinthal and Petersen, 1965), and hamster cells by adenovirus type 2 (Lewis *et al.*, 1974*a,b*).

Piña and Green (1965) and Huebner (1967) have divided the human adenoviruses into three classes according to their oncogenic potential: The highly oncogenic group is comprised of adenovirus types 12, 18, and 31; the group of intermediate oncogenicity is comprised of adenovirus types 3, 7, 8, 11, 14, 16, and 21; and the nononcogenic or rather weakly oncogenic group consists of the remaining adenovirus types (McAllister *et al.*, 1969). Adenovirus types 2 and 5, initially thought to be nononcogenic, were subsequently shown to transform cells in tissue culture (Freeman *et al.*, 1967; Lewis *et al.*, 1974*a,b*; Gallimore, 1974), although these adenovirus types do not lead to tumor formation upon injection into animals (Lewis *et al.*, 1974*a*). Piña and Green (1965) detected an interesting correlation between the degree of oncogenicity and the G+C content of the DNA of human-type adenoviruses. The highly oncogenic group had G+C contents of 48–49%, the intermediate group G+C contents of 50–53%, and the weakly oncogenic group G+C contents of 56–60%. The biological significance of this discovery remains unknown.

Adenovirus-transformed cells do not produce virions or viral structural proteins; they do, however, contain the T antigen (Huebner *et al.*, 1963; Huebner, 1967), which is presumably virus coded. Viral DNA sequences are efficiently transcribed in transformed cells and the transcripts become associated with polysomes in the cytoplasm of transformed cells (Fujinaga and Green, 1966, 1967*a,b,* 1968, 1970; Fujinaga *et al.*, 1969; Green *et al.*, 1970). The genes transcribed in transformed cells correspond to genes transcribed early in productive infection (Green *et al.*, 1970). Late viral genes are not expressed in transformed cells. Green *et al.* (1970) examined cell line 8617, a rat cell line transformed by adenovirus type 2 (Freeman *et al.*, 1967) and found RNA sequences homologous to 4-10% of both strands of the viral DNA. After restriction endonucleases became available, more detailed analyses of the viral sequences transcribed in adenovirus-transformed rat and hamster cells were carried out in several laboratories.

Gallimore (1974) isolated and characterized nine independent lines of rat cells transformed by adenovirus type 2. Sambrook *et al.* (1974), Gallimore *et al.* (1974), Sharp *et al.* (1974*a*), and Flint *et al.* (1975) determined the fragments of the viral genome which persisted in these transformed cell lines (Section 4.5). A transcriptional pattern common to most transformed cell lines was recognized in that the left terminus of the viral genome was found to be consistently expressed. In accordance with such a general pattern, Sharp *et al.* (1974*b*) and Flint *et al.* (1975) observed that the viral messenger RNA sequences isolated from transformed rat cell lines corresponded to those viral DNA sequences shown to persist in these cells (Sambrook *et al.*, 1974; Gallimore *et al.*, 1974; Sharp *et al.*, 1974*a*). One class of Ad2-transformed rat cells contained RNA complementary to a segment from 0.03 to 0.10 on the physical map of Ad2 DNA. Two other classes of Ad2-transformed rat cells carried RNA sequences complementary to two or three regions transcribed early in lytic infection (Sharp *et al.*, 1974*b*).

Taube *et al.* (1974) and Hoffmann and Darnell (1975) demonstrated that presumably integrated adenovirus sequences in transformed cells were preferentially transcribed in certain phases of the cell cycle. Both nuclear and cytoplasmic adenovirus-specific RNA were synthesized more actively early in S phase in adenovirus type 2 transformed rat embryo cells (Hoffmann and Darnell, 1975).

Adenoviruses are able to infect cells abortively; the best-studied system is that of BHK-21 cells infected with adenovirus type 12 (Doerfler, 1968, 1969, 1970; Strohl, 1969*a,b*; Doerfler and Lundholm, 1970; Doerfler *et al.*, 1972*b*; Raška and Strohl, 1972; Weber and Mak,

1972; Ortin and Doerfler, 1975; Fanning and Doerfler, 1976; Ortin *et al.*, 1976). In this system, the block in viral replication is early; i.e., viral DNA replication is not detectable (Doerfler and Lundholm, 1970; Raška and Strohl, 1972). An exceedingly small fraction (2×10^{-5}) of the BHK-21 cells infected with adenovirus type 12 are transformed (Strohl *et al.*, 1967, 1970; Strohl, 1969*a*).

Two hamster cell lines transformed by adenovirus type 12 have been analyzed for the persistence of adenovirus type 12 DNA (Fanning and Doerfler, 1976) and for the type of RNA sequences transcribed in these cells (Ortin *et al.*, 1976). T637 cells are a line of Ad12-transformed BHK-21 cells (Strohl *et al.*, 1970), and HA12/7 cells were derived from Syrian hamster kidney cells by transformation with Ad12 (zur Hausen, 1973). It will be detailed in Section 4.5 which fragments of the Ad12 genome persist in these cell lines (Fanning and Doerfler, 1976). The size-class analysis of Ad12-specific mRNA in abortively (Ortin and Doerfler, 1975) and productively (Scheidtmann *et al.*, 1975) infected cells, as well as in Ad12-transformed cells (Ortin and Doerfler, 1975), revealed mRNA populations of practically identical size in cells early in productive infection and in abortively infected and transformed cells. These RNA molecules exhibited molecular weights of 2.1×10^6, 1.9×10^6, 1.5×10^6, 0.88×10^6, and 0.67×10^6. Using the restriction endonucleases *Eco*RI (from *E. coli*) and *Bam*HI (from *Bacillus amyloliquefaciens*) the Ad12-specific mRNA and nuclear RNA from productively infected KB cells (early), from abortively infected BHK-21 cells, and from transformed lines T637 and HA12/7 were mapped on the Ad12 genome (Ortin *et al.*, 1976). The results of these studies are summarized in Fig. 4. The data indicate that in Ad12-transformed hamster cells as well as in abortively infected BHK-21 cells predominantly the *Eco*RI A, C, and F fragments (T637 cells, and BHK-21·Ad12 system), and the A and C fragments (HA12/7 cells) are transcribed and become associated with polysomes. These fragments are also expressed early (8 h postinfection) in productively infected KB cells. It is striking that the nuclear RNA isolated from these systems contains additional sequences which do not become associated with polysomes. In the case of HA12/7 cells, for example, in addition to sequences complementary to *Eco*RI fragments A and C, RNA transcribed from fragments D and F is present in nuclear RNA (Ortin *et al.*, 1976).

Human adenoviruses infect simian cells also abortively (Rapp *et al.*, 1966), unless helper viruses are used (Feldman *et al.*, 1966; Friedman *et al.*, 1970). In this type of abortive adenovirus infection, viral DNA replication is permitted, but very little, if any, infectious

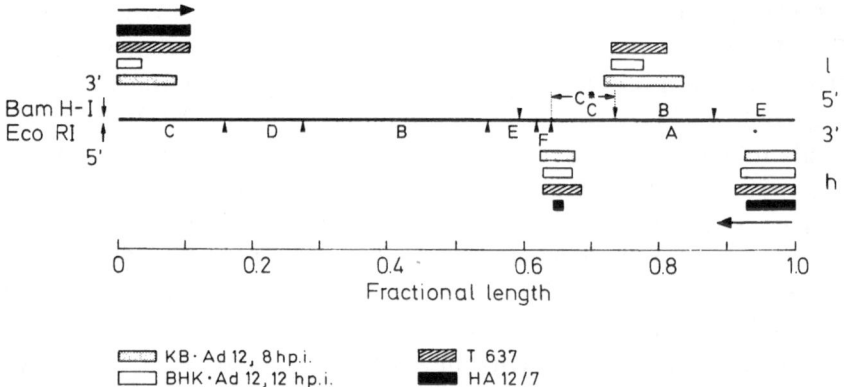

Fig. 4. Maps of messenger RNA of adenovirus type 12 DNA early in productively infected KB cells, in abortively infected BHK-21 cells, and in two lines of Ad12 transformed hamster cells. The top line represents the Ad12 DNA molecule with the l and h strands. The letters A–F refer to the *Eco*RI fragments of Ad12 DNA. Fragmentation of the *Eco*RIA fragment by the *Bam*HI endonuclease is also shown. The different bars designate the regions of the Ad12 genome in which the l or h strand is transcribed in cell lines as indicated. Arrows indicate direction of transcription. Modified from Ortin *et al.* (1976).

virus is produced. However, when African green monkey kidney cells are pretreated with 5-iododeoxyuridine, human adenoviruses can replicate even in the absence of a helper virus (Jerkofsky and Rapp, 1975; Staal and Rowe, 1975).

An endonuclease was reported to be associated with purified preparations of adenovirus type 2 (Burlingham and Doerfler, 1972) and could be isolated in association with the penton capsomers isolated from extracts of infected cells (Burlingham *et al.*, 1971). More recently, it was shown that this endonuclease could be separated from the penton (Doerfler and Philipson, 1973; Reif *et al.*, 1977a) and could be purified and characterized as a separate entity (Reif *et al.*, 1977a,b).

Adenoviruses and cells infected or transformed by adenoviruses have proven to be extremely useful systems in the study of the interaction between oncogenic viruses and host cells.

As it is impossible to give a complete summary on this interesting virus system, I should like to refer the reader to several reviews of the adenovirus field covering various aspects of the interaction of adenoviruses with their host cells: Schlesinger (1969), Green (1970a), Philipson and Pettersson (1973), Strohl (1973), Tooze (1973), zur Hausen (1973), Philipson and Lindberg (Vol. 3, this series), Philipson *et al.* (1975), Doerfler (1975), and McDougall (1976).

4.2. The Viral DNA

Adenovirus DNA can be isolated in pure form only by use of proteolytic enzymes (van der Eb *et al.,* 1969; Green *et al.,* 1967; Doerfler, 1969). Adenovirus DNA obtained in this way is a linear, double-stranded molecule with molecular weights between 20×10^6 and 25×10^6, depending on the serotype (Green and Piña, 1964).

The DNAs of adenovirus types 2 (Doerfler and Kleinschmidt, 1970), 12 (Doerfler *et al.,* 1972*a*), and 5 (Ellens *et al.,* 1974) were shown to have unique denaturation patterns; therefore, the gene orders in these DNA molecules could not be circularly permuted. The DNAs of all adenoviruses examined carry inverted terminal repetitions (Garon *et al.,* 1972; Wolfson and Dressler, 1972) in which both termini have complementary sequences of varying length, depending on the serotype. Upon denaturation, adenovirus DNA is therefore capable of giving rise to single-stranded circular molecules. Adenovirus DNA could be isolated as a circular molecule by treating purified virions with guanidinium hydrochloride (Robinson *et al.,* 1973) or sarkosyl (Doerfler *et al.,* 1974; Brown *et al.,* 1975) without the use of proteolytic enzymes. The circles of adenovirus DNA appear to be stabilized by a protein linker, but not by a covalent phosphodiester bond. The anatomy of the adenovirus genome has been further clarified by the use of several restriction endonucleases which recognize palindromic sequences in the viral genome. The restriction enzyme *Eco*RI from *E. coli* cleaves the DNAs of adenovirus types 2 (Pettersson *et al.,* 1973) and 12 (Mulder *et al.,* 1974*a*) into six specific fragments. A restriction nuclease from *Haemophilus parainfluenzae* (*Hpa*I) generates seven specific fragments (Gallimore *et al.,* 1974). The location of these fragments on the viral DNA has been determined by denaturation mapping in the electron microscope (Mulder *et al.,* 1974*b*).

For a series of investigations, it was very useful to have available a technique with which the infectivity of isolated viral DNA could be tested. Graham and van der Eb (1973*a,b*) developed such methods and could also demonstrate the transformation of cells by intact DNA or by fragments of viral DNA (Graham and van der Eb, 1973*b*; Graham *et al.,* 1974*a,b*).

4.3. Integration of Adenovirus DNA in Productively Infected Cells

Human KB cells (Eagle, 1955) are productively infected by adenovirus type 2, and each cell produces in excess of 10^5 virus parti-

cles. Considerable efforts have been made by many laboratories to elucidate the mechanism of replication of the viral DNA.

Several laboratories have reported the isolation of partly single-
stranded structures as possible intermediates in viral DNA replication
(van der Vliet and Sussenbach, 1972; Sussenbach et al., 1973; van der
Eb, 1973; Pettersson, 1973; Robin et al., 1973; Doerfler et al., 1973;
Ellens et al., 1974; Pearson, 1975). Lavelle et al. (1975) reported that
approximately 20% of the intracellular viral DNA was single stranded
and that both strands were equally represented. There is at present no
evidence for the involvement of covalently closed supercoiled circular
forms of viral DNA in the replication process (Horwitz, 1971; Doerfler
et al., 1973; Pearson, 1975). It is conceivable that circular viral DNA
molecules stabilized by a protein linker (Robinson et al., 1973) play a
role in viral DNA replication.

Adenovirus DNA replication is discontinuous and involves the synthesis of viral DNA sedimenting at 10 S in alkaline gradients (Horwitz,
1971; Winnacker, 1975; Vlak et al., 1975), and there seem to be origins
of replication at either end of the DNA molecule (Winnacker, 1974;
Schilling et al., 1975; Tolun and Pettersson, 1975; Weingärtner et al.,
1976). The data on the replication of Ad2ND$_1$ DNA published by
Horwitz (1974) suggest that replication of this DNA starts at both ends
of the molecule. Sussenbach et al. (1972) have suggested a model of
adenovirus DNA replication according to which replication starts at the
AT-rich right molecular end of adenovirus DNA and proceeds by displacement of that parental strand which has the higher buoyant density
in alkaline CsCl density gradients (Ellens et al., 1974; Sussenbach et al.,
1972, 1973). Although this model has stimulated a great deal of experimentation, definite answers to the problem of adenovirus DNA replication are not yet available and recent results do not entirely support this
early model.

Investigations on the molecular mechanism of adenovirus DNA
replication cannot be complete without an account of the different intracellular forms of viral DNA in productively infected cells. KB
cells productively infected with adenovirus type 2 were investigated. The analysis revealed at least four size classes of newly
synthesized DNA (> 100 S, 40–100 S, 34 S, and < 20 S) which were
resolved by velocity sedimentation in alkaline sucrose density gradients (Burlingham and Doerfler, 1971; Doerfler et al., 1971, 1974;
Burger and Doerfler, 1974; Schick et al., 1976). It could be shown
by DNA-DNA hybridization that all four size classes contained
viral DNA sequences. The time course of synthesis of the different
forms of viral DNA suggested that the high molecular weight forms of

adenovirus DNA appeared starting 2–8 h postinfection, at a time when unit-size viral DNA sedimenting at 34 S under alkaline conditions was not apparent (Schick *et al.*, 1976) (Fig. 5). Extensive control experiments demonstrated that the presence of high molecular weight forms of viral DNA could not be explained by unspecific inclusion of viral DNA into cellular DNA, by the association of viral DNA with protein and/or RNA, or by the occurrence of a covalently closed supercoiled circular form of adenovirus DNA (Burger and Doerfler, 1974; Schick *et al.*, 1976; Fanning and Doerfler, 1977). The high molecular weight viral DNA could then represent (1) an oligomeric form or (2) viral DNA or fragments of adenovirus DNA covalently integrated into cellular DNA (Fig. 7 and Table 1). It is also conceivable that a mixture of either type of molecules accounts for the fast-sedimenting viral DNA. In order to distinguish between these possibilities, two sets of experiments were performed. First, the high molecular weight DNA from adenovirus type 2 infected KB cells was resedimented in alkaline CsCl density gradients and was found to band in a density stratum intermediate between that of the viral and cellular marker DNAs (Burger and Doerfler, 1974; Schick *et al.*, 1976). Upon ultrasonic treat-

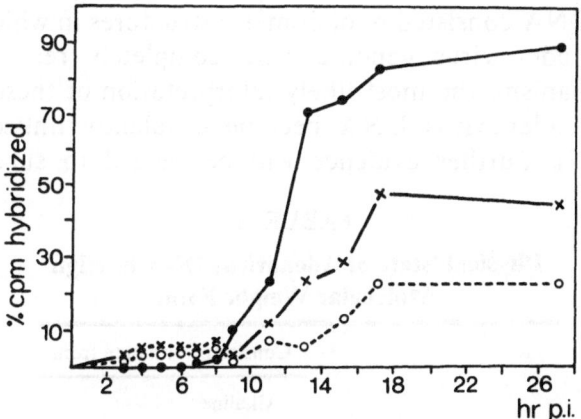

Fig. 5. Analysis by DNA-DNA hybridization of the different size classes of newly synthesized DNA in adenovirus type 2 infected KB cells. KB cells were infected with adenovirus type 2 and were labeled with [³H]thymidine at time intervals starting 2–4 h until 26–28 h postinfection. At the end of the labeling period, cells were lysed in alkali and the different size classes of DNA were separated on alkaline sucrose gradients. For each time period, representative fractions were pooled and analyzed for viral sequences by DNA-DNA hybridization to adenovirus type 2 DNA. The values of hybridization obtained for the 2–4 h postinfection period are plotted at the 3 h time point, the ones for the 3–5 h postinfection period at the 4 h time point, etc. ○, > 100 S DNA; ×, 40–100 S DNA; ●, 34 S (viral) DNA. From Schick *et al.* (1976).

ment of the labeled, fast-sedimenting DNA prior to resedimentation in alkaline CsCl density gradients, the labeled DNA shifted to the density positions of viral and cellular DNAs (Burger and Doerfler, 1974). These results are consistent with the model of viral DNA covalently linked to cellular DNA (Table 1). In a second type of experiment, the viral DNA sequences were excised from cellular DNA with restriction endonucleases (for review of restriction enzymes, see Nathans and Smith, 1975). The DNA in adenovirus type 2 infected cells was labeled with [³H]thymidine between 6 and 12 h postinfection and was purified in the native condition. The high molecular weight DNA was isolated by velocity sedimentation in neutral sucrose gradients (Fig. 6a), was mixed with ¹⁴C-labeled adenovirus type 2 marker DNA, and was cleaved with the *Eco*RI restriction endonuclease (Pettersson *et al.*, 1973; Green *et al.*, 1974). The fragmented DNA was analyzed by gel electrophoresis (Fig. 6b), and the virus-specific sequences were identified by DNA-DNA hybridization (Fig. 6c). There were virus-specific sequences corresponding in size to most of the marker DNA fragments, and, in addition, virus-specific sequences were found which were intermediate in size between the *Eco*RI A and B and B and C fragments (Schick *et al.*, 1976). Similar results were obtained using the *Bam*HI endonuclease (Baczko *et al.*, manuscript in preparation). Unless one chose to postulate that the high molecular weight viral DNA consisted of oligomeric structures in which the *Eco*RI sites on the adenovirus genome were completely rearranged by an obscure mechanism, the most likely interpretation of these data is that fragments of adenovirus DNA became covalently linked to cellular DNA (Fig. 7). Further evidence will be needed to substantiate this

TABLE 1

Physical State of Adenovirus DNA in High Molecular Weight Form

	Consistent with data from	
Model	Alkaline CsCl density gradients	*Eco*RI excision analysis
1. Oligomeric		
a. True	No	No
b. Rearranged	No	Perhaps
2. Integration		
a. Intact genome	Yes	No
b. Fragmented genome	Yes	Yes
3. Recombination	Yes	Yes

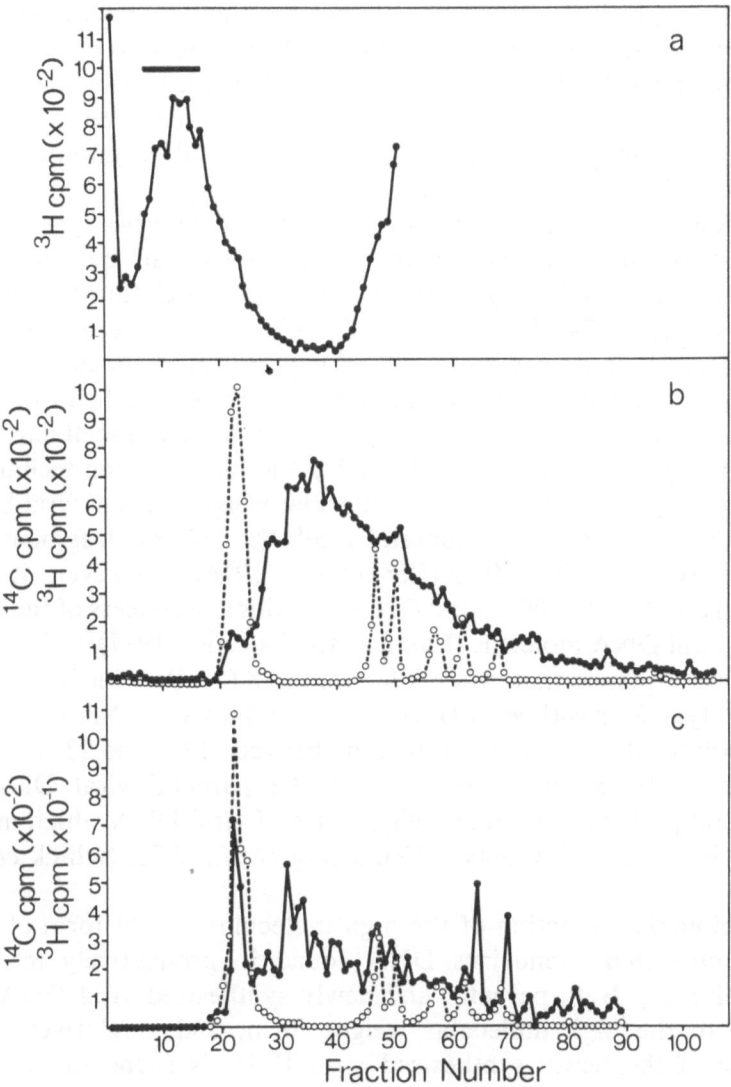

Fig. 6. Analysis by gel electrophoresis of the *Eco*RI restriction enzyme fragments of the cellular DNA peak synthesized 6–12 h postinfection in Ad2-infected KB cells. (a) Distribution of the ³H-labeled DNA synthesized 6–12 h postinfection in a neutral sucrose gradient. (b) Fractions 7–17 indicated by the horizontal bar (a) were pooled and mixed with 2–5 μg of ¹⁴C-labeled Ad2 DNA. The mixture was incubated with the *Eco*RI restriction endonuclease. The fragmented DNA was analyzed by electrophoresis on a polyacrylamide-agarose gel. ●, ³H-labeled DNA; ○, ¹⁴C-labeled adenovirus type 2 marker DNA representing the six specific *Eco*RI fragments (Pettersson *et al.*, 1973). (c) The DNA from individual gel slices was hybridized to adenovirus type 2 DNA on filters. The ³H and ¹⁴C activity curves (symbols as in b) represent the position of adenovirus type 2 specific sequences. From Schick *et al.* (1976).

interpretation, particularly in view of the fact that the high molecular weight forms of simian virus 40 (SV40) DNA described in productively infected monkey cells (Hirai and Defendi, 1972, 1974; Hölzel and Sokol, 1974) have been shown to consist, at least partly, of circular (Martin *et al.*, 1976) or linear (Rigby and Berg, personal communication) oligomeric molecules.

For KB cells productively infected with adenovirus type 2, the newly synthesized, high molecular weight DNA can be shown to contain both viral and cellular sequences by reassociation techniques (Doerfler *et al.*, 1974; Fanning and Doerfler, 1977). A quantitative analysis of results obtained from several experiments indicated that, at various times after infection, the high molecular weight form of viral DNA constituted between 5% and 26% of the total intracellular viral DNA (Fanning and Doerfler, 1977). Using the reassociation technique, evidence was obtained that the high molecular weight form of viral DNA isolated 18 h postinfection contained all the *Eco*RI fragments of adenovirus type 2 DNA (Doerfler *et al.*, 1974). However, it was established that the 40–100 S viral DNA contains a net excess of the right end of the viral DNA molecule (Fanning and Doerfler, 1977).

There is evidence that, after infection of KB cells with ³H-labeled adenovirus type 2 or with viral DNA, the parental viral DNA is cleaved endonucleolytically to pieces sedimenting between 19 S and 23 S. Starting 2–6 h postinfection, a fraction of the parental viral DNA is converted into a high molecular weight form of viral DNA which might constitute integrated viral genes (Groneberg *et al.*, 1975; Schick *et al.*, 1976).

The biological function of the high molecular weight form of presumably integrated adenovirus DNA found in productively infected cells is unknown. Both parental and newly synthesized viral DNA are converted to the high molecular weight form, and a relatively high percentage of the newly synthesized viral DNA is found in this size class. Moreover, it is striking that the appearance of this size class of viral DNA in productively infected cells precedes that of unit-size adenovirus DNA. Thus it is conceivable that the high molecular weight viral DNA plays a role in replication or transcription of the viral genome early in the infection cycle. In this context, it is worth noting that the 40–100 S class of high molecular weight DNA was found also in uninfected cells, and that pulse-chase experiments suggested that this form of cellular DNA was an intermediate in the synthesis of cellular DNA (Schick *et al.*, 1976). The DNA in the 40–100 S size class has a molecular weight of 32×10^6 to 250×10^6 or of 16–125 μm length. Newly synthesized DNA of 30–180 μm length has been detected in Chinese hamster ovary cells (Hozier and Taylor, 1975).

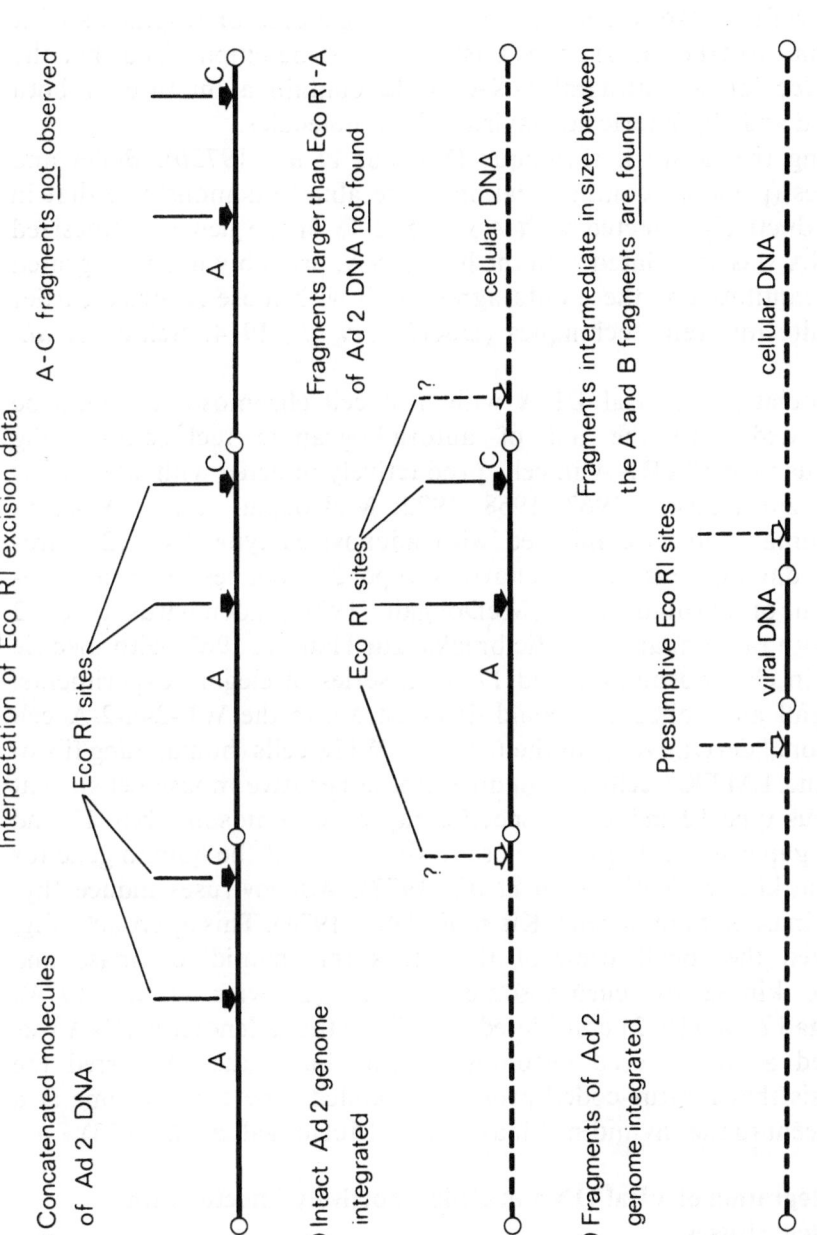

Fig. 7. Possible interpretations of data presented in Fig. 6. For clarity, only the A and C *Eco*RI sites on the adenovirus type 2 DNA are shown (filled arrows). The location of the *Eco*RI sites on the cellular DNA is unknown (open arrows). The data observed best fit alternative 3; i.e., fragments of viral DNA are integrated into cellular DNA (*cf.* Table 1, possibility 2b).

If the high molecular weight viral DNA does indeed constitute adenovirus DNA covalently linked to cellular DNA, the possibility has to be considered that this linkage is due to a recombination event between cellular DNA and the intact viral genome or fragments of it rather than to true integration. Lastly, it must be emphasized that the high molecular weight viral DNA could contain a mixture of both integrated and oligomeric adenovirus DNA molecules.

Using the network technique (Varmus *et al.*, 1973*b*), Bellet and colleagues (personal communication) were able to demonstrate that in cells productively infected with Ad5, 15–20% of the newly synthesized viral DNA was associated with the host DNA, probably in an integrated form. Quantitatively, these data agreed well with those reported earlier using quite different techniques (Doerfler *et al.*, 1974; Schick *et al.*, 1976).

Association of viral DNA with host cell chromosomes could be demonstrated with the aid of autoradiographic techniques using adenovirus type 12 cRNA in cells productively infected with adenovirus type 12 (zur Hausen, 1967, 1968, 1972; McDougall *et al.*, 1974*a,b*). When human cells are infected with adenovirus type 2 or 12, chromosome damage ensues. Adenovirus type 2 produces random fragmentation of chromosomes (McDougall, 1971); adenovirus type 12 causes both random and specific breaks (zur Hausen, 1967) with specific damage in chromosomes 1 and 17. In a series of elegant experiments, McDougall and colleagues could demonstrate in the WL-24a-2-A cell line, a clonal derivative from the fusion of WI38 cells (human lung fibroblasts) and LMTK⁻ cells (thymidine kinase negative mouse cells), that adenovirus type 12 induced a specific gap on chromosome No. 17 and that this gap was located in or close to the region of the human gene for thymidine kinase (McDougall *et al.*, 1973). Adenoviruses induce thymidine kinase in human cells (Kit *et al.*, 1970, 1974). This approach (Fig. 8) allowed the localization of the genes for thymidine kinase and galactose kinase on chromosome No. 17 (Elsevier *et al.*, 1974). McDougall *et al.* (1972) considered it unlikely that adenovirus DNA was integrated at the altered chromosome site but rather preferred the hypothesis that a virus-coded protein "uncoiled" the chromosome at a site adjacent to the thymidine kinase locus (McDougall *et al.*, 1973).

4.4. Integration of Viral DNA in Cells Abortively Infected with Adenoviruses

Adenovirus type 12 infects BHK-21 cells abortively (Doerfler, 1968, 1969; Strohl, 1969*a,b*). Viral particles can be detected electron

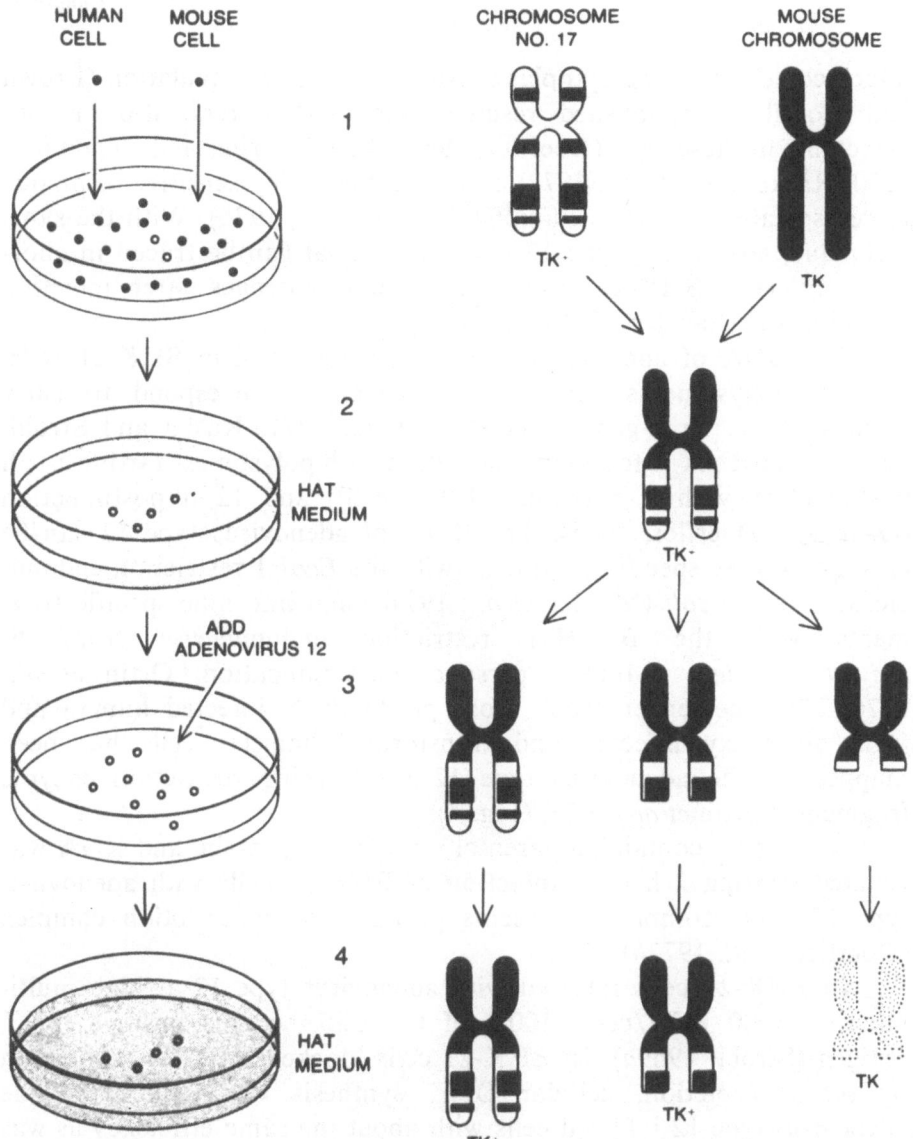

Fig. 8.   Induced chromosome breakage defined the location of the gene for the enzyme thymidine kinase (TK), which had previously been mapped to chromosome No. 17. The investigators fused mouse cells deficient in TK with human cells (1). A hybrid line in which there was a translocation of the long arm of human No. 17 to a mouse chromosome was viable in HAT medium, showing that the TK gene was on the long arm of No. 17 (2). Adenovirus 12 was added to induce chromosome breakage at several different points, and cells with different translocation products were grown in nonselective mediums (3). When they were tested in HAT medium (4), it was found that the cells with translocation products retaining the whole long arm of No. 17 or most of the arm still had the TK gene; the cells that had lost more of the arm did not have the gene. Comparison of the three break points located the TK gene. This figure was taken from Ruddle and Kucherlapati (1974).

microscopically in the cytoplasm 10–20 min after inoculation (Brown and Doerfler, unpublished results). Viral DNA replication is not detectable in these cells (Doerfler, 1969, 1970; Doerfler and Lundholm, 1970; Raška and Strohl, 1972) even with the highly sensitive technique of reassociation kinetics (Fanning and Doerfler, 1976). With the same technique, adenovirus type 12 genetic material can be traced in abortively infected BHK-21 cells even several passages after infection (Fanning and Doerfler, 1976).

The DNA of adenovirus type 12 is transcribed in BHK-21 cells, and the polysome-associated viral sequences correspond to early regions of the viral genome (Raška *et al.*, 1971; Raška and Strohl, 1972). Viral RNA is found in association with polysomes starting 5–7 h postinfection with a maximum between 10 and 12 h postinfection (Ortin and Doerfler, 1975). The DNA of adenovirus type 12 can be cleaved into six specific fragments with the *Eco*RI restriction endonuclease from *E. coli* (Mulder *et al.*, 1974*a*) and into nine specific fragments with the *Bam*H-I restriction endonuclease from *B. amyloliquefaciens* (Mulder, personal communication; Ortin *et al.*, 1976). The messenger RNA from productively infected human and from abortively infected and transformed hamster cells has been mapped on the adenovirus type 12 DNA using restriction enzyme fragments (Ortin *et al.*, 1976) (Fig. 4).

A complex containing parental viral DNA, protein, and RNA was isolated starting 2 h after infection of BHK-21 cells with adenovirus type 12. This complex represents probably a transcription complex (Doerfler *et al.*, 1972*b*).

In BHK-21 cells infected with adenovirus type 12 at high multiplicities (1000 PFU/cell), 100% of the cells became positive for T antigen (Strohl, 1969*a*). In BHK-21 cells blocked in DNA replication by serum depletion, cellular DNA synthesis was induced in the adenovirus type 12 infected cells with about the same efficiency as was T-antigen synthesis (Strohl, 1969*b*). The newly synthesized DNA in these cells had a molecular weight between 5×10^6 and 10×10^6 (Strohl, 1969*b*; Doerfler, 1969).

The DNAs of BHK-21 cells and adenovirus type 12 can be separated either by size in alkaline sucrose density gradients (Burlingham and Doerfler, 1971) or by equilibrium centrifugation in neutral and alkaline CsCl density gradients, provided that the cellular DNA is substituted with the thymidine analogue 5-bromodeoxyuridine (Doerfler, 1968, 1970). When BHK-21 cells are grown for 4 days in the presence of 5 μg of 5-bromodeoxyuridine per milliliter, approximately 50% of the thymidine residues in the cellular DNA are replaced by 5-

bromodeoxyuridine (Doerfler, 1968). Such cells were infected with high multiplicities of [³H]thymidine-labeled adenovirus type 12, and the distribution of the parental label between the viral and cellular density positions was determined in neutral and alkaline CsCl density gradients at various times after infection. Starting approximately 16 h postinfection, parental viral label was detected in the cellular density position (Doerfler, 1968). Upon ultrasonic treatment of the DNA prior to equilibrium sedimentation, the labeled DNA banded in a buoyant density position intermediate between that of viral and cellular DNA in both neutral and alkaline CsCl density gradients, indicating that the parental viral label in the cellular DNA position was due to viral DNA integrated into cellular DNA by covalent linkage (Doerfler, 1968, 1970). These results represented the first evidence for integration of viral DNA in mammalian cell DNA.

These interpretations were confirmed by the results of experiments in which the parental viral ³H label isolated from the cellular density position in alkaline CsCl density gradients could be shown to hybridize to both viral and cellular DNA fixed to membrane filters (Doerfler, 1970). Similar experiments were performed by zur Hausen and Sokol (1969) with adenovirus type 12 infected Nil-2 cells (a hamster cell line). These authors showed that parental label isolated from the cellular density position hybridized to viral DNA.

In experiments in which the fate of parental viral DNA was followed in infected cells, it is critical to ascertain that the ³H label incorporated into cellular DNA was not due to reincorporation of solubilized mononucleotides. This possibility was ruled out in the experiments described above (1) by the finding that the parental label incorporated into cellular DNA hybridized with viral and cellular DNA and (2) by the observation that the same relative amounts of integrated viral genomes were found when DNA replication was chemically inhibited to >96% by cytosine arabinoside from 2 h before to 44 h after infection (Doerfler, 1970). Unlike in the case of λ phage, inhibition of protein synthesis by cycloheximide affected integration only slightly. After fragmentation of the intracellular DNA to pieces of $1-2 \times 10^6$ daltons by ultrasonic treatment, 75% of the parental viral label in the density stratum of cellular DNA shifted to an intermediate density position. This result indicated that fragments of viral DNA were integrated, rather than the entire adenovirus type 12 genome (Doerfler, 1970). In several lines of adenovirus-transformed cells it could be shown that specific fragments of viral DNA persisted (Sambrook et al., 1974) (compare Section 4.5), whereas the entire genome was found to persist only in rare cases.

In BHK-21 cells abortively infected with adenovirus type 12, approximately 30% of the cell-associated, parental viral DNA is linked covalently to cellular DNA. Seventy percent remains as free viral DNA. The viral DNA is fragmented prior to integration. The time course of appearance of viral DNA fragments argues in favor of a precursor role of viral DNA fragments for the integrated form of viral DNA (Burlingham and Doerfler, 1971). It is conceivable that the fragments of the parental viral DNA are generated by the adenovirus-associated endonuclease (Burlingham and Doerfler, 1972). This endo-nuclease was found in association with the penton capsomer of the adenovirion (Burlingham et al., 1971). Marusyk et al. (1975) and Cajean-Feroldi et al. (1977) have reported on the association of endonucleases with the pentons of several serotypes of adenoviruses. The activity described by Marusyk et al. (1975) was presumably single-strand specific. The actual role of the adenovirus-associated endonuclease for the integration process remains uncertain. There is suggestive evidence that this endonuclease may be responsible for the fragmentation of cellular DNA (Doerfler, 1969; Strohl, 1969b; Burlingham and Doerfler, 1971).

Evidence has been published that the DNA of adenovirus type 12 can become associated with chromosomes of human embryonic kidney cells (zur Hausen, 1967) and of BHK-21 and Nil-2 hamster cells (zur Hausen, 1968). There was, however, no indication that this association was specific for one particular chromosome (zur Hausen, 1973). Stich and Yohn (1967) reported pulverization of chromosomes in hamster cells infected with adenovirus type 12. Specific chromosome changes have not been detected. There is a high degree of chromosome damage (Stich and Yohn, 1970; McDougall et al., 1974b), and these alterations are dependent on virus dose (zur Hausen, 1968).

Yamamoto et al. (1972) isolated less tumorigenic mutants of adenovirus type 12 and tested these mutants for the capability to integrate into the genome of nonproductively infected rat brain cells. Using gradient techniques (Doerfler, 1968, 1970), Yamamoto et al. (1972) found that the less tumorigenic mutants (particularly the mutant lt3) were able to integrate, although they did not transform cells.

4.5. Integration of Viral DNA in Adenovirus-Transformed Cells

Adenoviruses are oncogenic in newborn hamsters and can transform a variety of rodent cells in tissue culture (Freeman et al., 1967; Gallimore, 1974). Virus DNA or fragments of viral DNA (Sambrook et al., 1974; Fanning and Doerfler, 1976) and virus-specific RNA (Green et al., 1970; Sharp et al., 1974a,b; Ortin et al., 1976) have been detected in adenovirus-transformed cells. When a large number of different human tumors were screened for the persistence of adenovirus DNA or for the presence of viral RNA or proteins, no evidence for the involvement of adenoviruses in human oncogenesis was obtained (McAllister *et al.*, 1972). It was reported that in 34 human gastrointestinal and 22 lung tumors, adenovirus type 12 DNA was not detectable (< 0.1 genome copies per cell) (Mackey *et al.,* 1976).

There were several reports from Green's laboratory (Green *et al.*, 1970; Green, 1970*b*, 1972) and from Fujinaga's laboratory (Fujinaga *et al.*, 1974) adducing evidence for the integrated state of adenovirus DNA in adenovirus-transformed cells. Using the network technique of reassociation, Bellet (1975) provided further evidence for integrated viral DNA in hamster cells transformed by CELO virus (chicken embryo lethal orphan virus), an avian adenovirus. Using different techniques, Groneberg *et al.* (1977) could show that adenovirus type 12 DNA was integrated in the DNA of four virus-transformed hamster cell lines.

The most accurate measurements of the number of viral gene copies per transformed cell came from determining the enhancement of the rate of reassociation of radioactively labeled viral DNA in the presence of the DNA from transformed cells (Gelb *et al.*, 1971). The first results of Pettersson and Sambrook (1973) using adenovirus-transformed cells estimated that there was close to one copy of adenovirus 2 DNA per transformed rat cell. More precise data have been obtained by Sambrook *et al.* (1974), Gallimore *et al.* (1974), and Sharp *et al.* (1974*a,b*). These authors used specific fragments of adenovirus type 2 DNA generated by the restriction enzymes *Eco*RI from *E. coli* and *Hpa*I from *H. parainfluenzae.* Each of these fragments was reassociated in separate experiments in the presence and absence of the DNA from ten different lines of adenovirus type 2 transformed rat cells. The scheme presented in Fig. 9 summarizes the data obtained by Sambrook *et al.* (1974), Sharp *et al.* (1974*a*), and Gallimore *et al.* (1974). In each adenovirus type 2 transformed cell line, a specific set of viral DNA segments is present. It is common to all of the cell lines examined that the left molecular end of the viral DNA persists. Only in one adenovirus type 2 transformed rat cell line (T_2C_4) does the entire viral genome seem to be represented. At this point, it is interesting to recall the data obtained with hamster cells abortively infected with adenovirus type 12 where fragments of viral DNA have been shown to be integrated (Doerfler, 1970). Moreover, the pattern of persisting fragments of viral genetic material explains the observations that infectious virus could not be rescued so far from adenovirus transformed cells by treatment with ultraviolet light or chemicals (Landau *et al.*, 1966) or by fusion with permissive cells (cited in Tooze, 1973, p. 451).

	HpaI							
	4.0		24.2	26.5	58.8	86.4		98.6
CELL LINE	E	C		F	A	B	D	G
F 17	3.5	3.5		0.0	0.0	0.0	0.0	0.0	0.0
F 18	2.9	2.9		0.0	0.0	0.0	0.0	0.0	0.0
F 19	4.9	4.9		0.0	0.0	0.0	0.0	0.0	0.0
B1	6.0	6.0		0.0	0.0	0.0	0.0	0.0	0.0
2T8	4.5	4.5		0.0	0.0	0.0	0.0	0.0	0.0
2T4	6.9	6.9		0.0	0.0	0.0	0.0	0.0	0.0
REM	6.37	1.8		1.6	0.0	0.3	1.6	3.0
F4	ND	15.1		ND	16.5	5.1	2.2	20.4
T_2C_4	6.9	1.9		0.4	0.9	2.5	1.5	2.3
8617	13.0	1.1		0.0	0.0	0.8	2.0	3.4

	EcoRI			76.5		
		59.7	71.9	83.9	89.8	
CELL LINE	A	B	F	D	E	C
F 17	<~14%> 0.0	0.0	0.0	0.0	0.0	0.0
F 18	<~14%> 0.0	0.0	0.0	0.0	0.0	0.0
F 19	<~14%> 0.0	0.0	0.0	0.0	0.0	0.0
B1	<~14%> 0.0	0.0	0.0	0.0	0.0	0.0
2T8	<~14%> 0.0	0.0	0.0	0.0	0.0	0.0
2T4	<~14%> 0.0	0.0	0.0	0.0	0.0	0.0
REM	0.7	0.3	0.0	0.3	1.9	2.6
F4	16.3	12.3	0.0	0.0	0.0	3.0
T_2C_4	1.1	1.8	4.2	4.3	2.6	1.4
8617	0.4	0.0	0.0	1.6	1.7	1.7

Fig. 9. Persistence of adenovirus type 2 sequences in adenovirus 2 transformed rat cells. Radioactively labeled adenovirus type 2 DNA was cleaved with the restriction endonucleases from *Haemophilus parainfluenzae* (*Hpa*I) and RI from *Escherichia coli* (*Eco*RI), respectively. The sites of cleavage by these enzymes are indicated on the adenovirus type 2 genome by vertical arrows, and the fractional lengths of the fragments are shown by percentage figures. Ten different rat cell lines transformed by adenovirus type 2 (Gallimore, 1974) were examined by the reassociation technique for the presence of adenovirus type 2 genetic material. Each of the fragments indicated was annealed with itself and in the presence of DNA from the different rat cell lines. From the enhancement of the reassociation rate, the number of gene copies per cell was calculated for each of the different lines. The numbers of copies for each of the cell lines are listed under the respective fragments. The number of copies of the various segments of adenovirus type 2 DNA that are present in diploid quantities of DNA extracted from each of the transformed cell lines are shown. The figures given for REM, F4, T_2C_4, and 8617 cells are calculated assuming that the total sequence of each of the fragments is present in the cells. The number of copies of *Hpa*I fragment C present in each diploid quantity of DNA extracted from F17, F18, F19, B1, 2T8, and 2T4 cells was calculated on the assumption that only 50% of the sequences of the fragment are present in the cells. From Sambrook *et al.* (1974).

Four oncogenic hamster cell lines transformed by adenovirus type 12 were analyzed for the presence of adenovirus type 12 DNA sequences (Fanning and Doerfler, 1976). The lines T637 (Strohl *et al.*, 1970), HA12/7 (zur Hausen, 1973), and A2497-2 and A2497-3 (A. M. Lewis, personal communication) were derived from BHK-21 cells, primary Syrian hamster kidney cells, and inbred hamster embyro cells, respectively, and all contained viral sequences. The results of a series of reassociation experiments in which the rate of reassociation of each of the *Eco*RI fragments of adenovirus type 12 DNA was measured in the presence and in the absence of DNA from the transformed hamster cells (Fanning and Doerfler, 1976) indicated that all *Eco*RI fragments of adenovirus type 12 DNA were present, although in nonstoichiometric amounts. A similar study was carried out by Green *et al.* (1976), who found by reassociation kinetics and network techniques that in two adenovirus type 12 transformed hamster embryo cell lines eight to ten copies of all *Eco*RI fragments of the viral DNA persisted.

Graham *et al.* (1974*a,b*) were able to demonstrate that a fragment on the left terminus of the adenovirus 2 or 5 DNA molecule comprising the segment between 1% and 6% on the fractional length scale was sufficient to transform rat embryo cells when these cells were directly infected with specific fragments of viral DNA (Graham and van der Eb, 1973*a,b*). These data are consistent with those of Sambrook *et al.* (1974), Sharp *et al.* (1974*a*), and Gallimore *et al.* (1974). Furthermore, the data of Graham *et al.* (1974*a,b*) confirm the notion that even fragments of adenovirus DNA might be integrated as was outlined in Section 4.4. Lewis *et al.* (1976) analyzed the gene products of this segment by selecting messenger RNA from transformed cells on the transforming segment of adenovirus type 5 DNA and translating this mRNA in a wheat germ protein-synthesizing system. The products of this translation experiment were polypeptides of molecular weights of 40,000 and 15,000. The function of these polypeptides is as yet unknown.

There is yet another line of evidence which supports the model of integrated viral genomes in adenovirus-transformed cells: High molecular weight nuclear RNA isolated from adenovirus-transformed cells was shown to contain both viral and cellular sequences in the same molecule (Wall and Darnell, 1971; Tseui *et al.*, 1972; Wall *et al.*, 1973; Georgieff *et al.*, 1974). These molecules are presumed to be precursors for messenger RNA, since they contained sequences of polyadenylic acid which were possibly added during the processing of adenovirus messenger RNA molecules (Philipson *et al.*, 1971; Wall *et al.*, 1972;

Georgieff *et al.*, 1974). The presence of viral and cellular sequences in the same RNA molecule suggests that these molecules were derived from viral sequences integrated into the host genome.

Sharp *et al.* (1974*b*) mapped viral RNA sequences in five independently cloned adenovirus type 2 transformed rat cell lines. One class of these cells carries RNA sequences which are complementary to only that segment of adenovirus type 2 DNA which lies between 0.03 and 0.10 fractional unit on the physical map. This region is one of the segments which are transcribed early in a productive infection. The two other classes of adenovirus type 2 transformed rat cells contain RNA sequences complementary to two or three regions of the genome which are transcribed early.

The data describing the transcription map of adenovirus type 12 DNA in the adenovirus type 12 transformed hamster lines T637 and HA12/7 have been presented in Fig. 4 in Section 4.1 (Ortin *et al.*, 1976). It has yet to be determined whether the general pattern is that in adenovirus type 12 transformed hamster cells the entire genome persists, in contrast to the adenovirus type 2 genome in transformed rat cells. To my knowledge, there are no reports to date about successful attempts to induce the adenovirus type 12 genome by fusion of transformed with permissive cells. Thus it is conceivable that different sections of the adenovirus type 12 genome are integrated in different parts of the host genome. It is also possible that intact viral genomes persist in some of the cell lines. There is indeed evidence that the adenovirus type 12 genome is cleaved endonucleolytically in hamster cells (Burlingham and Doerfler, 1971) and that fragments of the viral DNA become integrated (Doerfler, 1970). Lastly, McDougall *et al.* (1976) reported that in adenovirus-transformed cells, viral genetic material was associated with a limited number of chromosomes.

With the technique of *in situ* hybridization and autoradiography, Loni and Green (1973) detected virus-specific sequences over the nuclei of cells transformed by adenovirus types 2, 7, and 12. Karyotype analysis of adenovirus-transformed rat cells revealed a high incidence of polyploidy and a variety of abnormal forms of chromosomes, such as rings, dicentrics, gaps, breaks, and fragments (McDougall *et al.*, 1974*b*). More recently, McDougall *et al.* (1976) attempted to localize the virus genetic material on specific chromosomes of cells transformed by adenovirus types 2, 5, and 12. The data obtained to date would indicate that there is not one specific chromosome which carries adenovirus DNA, but rather several different chromosomes.

Adenovirus type 2 transformed rat cell lines differ in their oncogenic potential, depending on the multiplicity of infection that was used originally to transform the cells. Lines derived from cells which were infected at 50 PFU/cell produce tumors in animals without immunosuppression, transformed cells which originated from an infection using 5–40 PFU/cell generate tumors only under immunosuppression, and cells transformed with less than 5 PFU/cell are unable to induce tumors in animals even under immunosuppression (Gallimore, 1974; McDougall *et al.*, 1974*a*; McDougall, 1976). The explanations for this phenomenon are somewhat tentative. It has been suggested that in the interaction of the viral and host genomes several sites in the cellular genome have to be altered to transform a cell to a fully oncogenic cell, whereas the requirements for a cell to acquire some of the properties of malignant transformation might be less stringent.

Although the molecular biology of adenoviruses has been studied in great detail and although there is a large amount of information about the interaction of the viral genome with the host DNA, it is obvious that the decisive questions about the specificity of this interaction can only now be approached.

5. SIMIAN VIRUS 40 (SV40)

Among the DNA tumor viruses, simian virus 40 (SV40) together with adenoviruses and cells infected or transformed by these viruses are the most extensively investigated systems. There is good evidence that the viral genome can be inserted into the host genome of infected and transformed cells. In SV40, viral particles are known to contain substituted genomes in which various parts of the viral genome have been replaced by host sequences. The occurrence of these substitutions is best explained by insertion of the viral DNA into the host genome and faulty excision (Lavi and Winocour, 1972).

5.1. The SV40 System

Simian virus 40 was discovered in 1960 by Sweet and Hilleman as a contaminant in monkey kidney cells. Eddy *et al.* (1961, 1962) demonstrated that SV40 was oncogenic in newborn hamsters, and Shein and Enders (1962) were able to transform cells in tissue culture. SV40 infects monkey cells productively, whereas mouse cells are nonpermissive for this virus, and human and hamster cells are semipermissive for SV40, i.e., some cells in semipermissive cultures produce virus, while others are infected abortively. Rat cells, guinea pig cells, rabbit cells, and bovine cells are infected non- or semipermissively. These different cell lines can all be transformed in tissue culture by the virus.

The parameters influencing viral transformation were studied most extensively in the 3T3 mouse cell line infected with SV40. When very high multiplicities of infection were used (10^6 PFU/cell), up to 40% of the cells were transformed (Todaro and Green, 1966a). For fixation of the transformed state of the cell, it was necessary that the SV40-infected 3T3 cells replicate at least once (Todaro and Green, 1966b). The genetic properties of the host cell seemed to play a decisive role in the frequency of transformation. Cells obtained from patients with Fanconi's anemia, Down's syndrome, or Klinefelter's syndrome had considerably increased transformation frequencies (Todaro et al., 1966; Potter et al., 1970). A careful analysis of the properties of SV40-transformed cells has been published by Risser and Pollack (1974).

There is mounting evidence that the SV40 gene *A* function is required for the maintenance of transformation (Martin and Chou, 1975; Tegtmeyer, 1975; Brugge and Butel, 1975; Osborn and Weber, 1975). It has been proposed that the *A* gene of SV40 codes for T antigen and that T antigen regulates its own synthesis. T antigen is a 100,000 dalton protein which is probably phosphorylated (Tegtmeyer et al., 1975). Many laboratories have suggested that the T antigen can bind to DNA (Carroll et al., 1974; Tegtmeyer et al., 1975). The site of interaction of T antigen on the SV40 DNA is probably specific and coincides with the origin of DNA synthesis (Reed et al., 1975a).

Shin et al. (1975) have provided evidence for the notion that transformation of cells by SV40 involves a series of cellular events. This series can be experimentally analyzed with the aid of variants of transformed cells that exhibit only a subset of the parameters which characterize a fully transformed cell that would produce tumors upon injection into animals. This operational definition of a series of events does not yet shed light on the mechanism underlying the conversion of normal cells to any of the states defined by the subsets.

The DNA of SV40 is a supercoiled, covalently closed circular molecule of molecular weight $3.0-3.6 \times 10^6$ (Crawford and Black, 1964; Tai et al., 1972; Finch and Crawford, Vol. 5, this series). The sedimentation behavior of SV40 DNA has been analyzed in detail (for review, see Crawford, 1969). Form I DNA, the supercoiled, covalently closed circular molecule, sediments at 20 S in neutral gradients. Form II, the uncoiled circular molecule in which one of the strands has been nicked, sediments at 16 S. Component III, sedimenting more slowly than form II, is present in some preparations of SV40 and represents linear DNA molecules of host origin. It has been very elegantly demonstrated that form I of SV40 DNA contains 24 ± 2 superhelical turns per molecule (Keller and Wendel, 1974; Keller, 1975b).

The interactions of a nicking–closing enzyme (Pulleyblank et al., 1975) DNA-ligase and DNA-relaxing enzyme (Keller, 1975b) with PM2 DNA and SV40 DNA, respectively, have been studied in great detail. It has been recognized that the DNA of SV40 can be isolated as a minichromosome both from lytically infected cells (Griffith et al., 1975; Griffith, 1975) and from the virion (Germond et al., 1975).

The DNA of SV40 has been extensively mapped with the aid of restriction enzymes (Fig. 10). Danna and Nathans (1971) discovered that the restriction enzyme from *H. influenzae* (Smith and Wilcox, 1970; Kelly and Smith, 1970) cut both strands of SV40 DNA at 11 specific sites which were mapped on the DNA relative to each other (Danna et al., 1973) and relative to the unique cleavage site of the restriction enzyme $R \cdot R_1$ from *E. coli* (Mulder and Delius, 1972; Morrow and Berg, 1972). The cleavage sites on the SV40 DNA of the restriction endonuclease from *H. parainfluenzae* have also been mapped (Sack and Nathans, 1973) (Fig. 11). The SV40 DNA fragments in six adenovirus–SV40 (nondefective ND_1–ND_5 and the defective E46[+]) hybrids (Lebowitz et al., 1974) have been related to the SV40 map. A map of the SV40 genome has been published at the Cold Spring Harbor Tumor Virus Meeting 1975 and is the result of the work from several laboratories (see the caption of Fig. 10) using 12 different restriction enzymes.

Several functions and the regions of early and late transcription have been localized on the SV40 genome (Danna et al., 1973; Khoury et al., 1973b; Lebowitz et al., 1974) (Fig. 11). Detailed analyses of the SV40 genome are presently being continued in several laboratories using restriction endonucleases from many different microorganisms (Fig. 10). This analysis is also aided by the isolation of SV40 variants with specifically altered genomes (Brockman and Nathans, 1974; Mertz et al., 1974). The laboratories of Fiers (Fiers et al., 1974) and of Weissman (Dhar et al., 1974) are in the process of sequencing the entire SV40 genome.

Tegtmeyer and Ozer (1971) have isolated temperature-sensitive mutants of SV40 which fall into four complementation groups. The defects in five of these mutants have been mapped on the SV40 genome by marker rescue experiments using *H. influenzae* and *H. parainfluenzae* restriction enzyme fragments of wild-type SV40 DNA (Lai

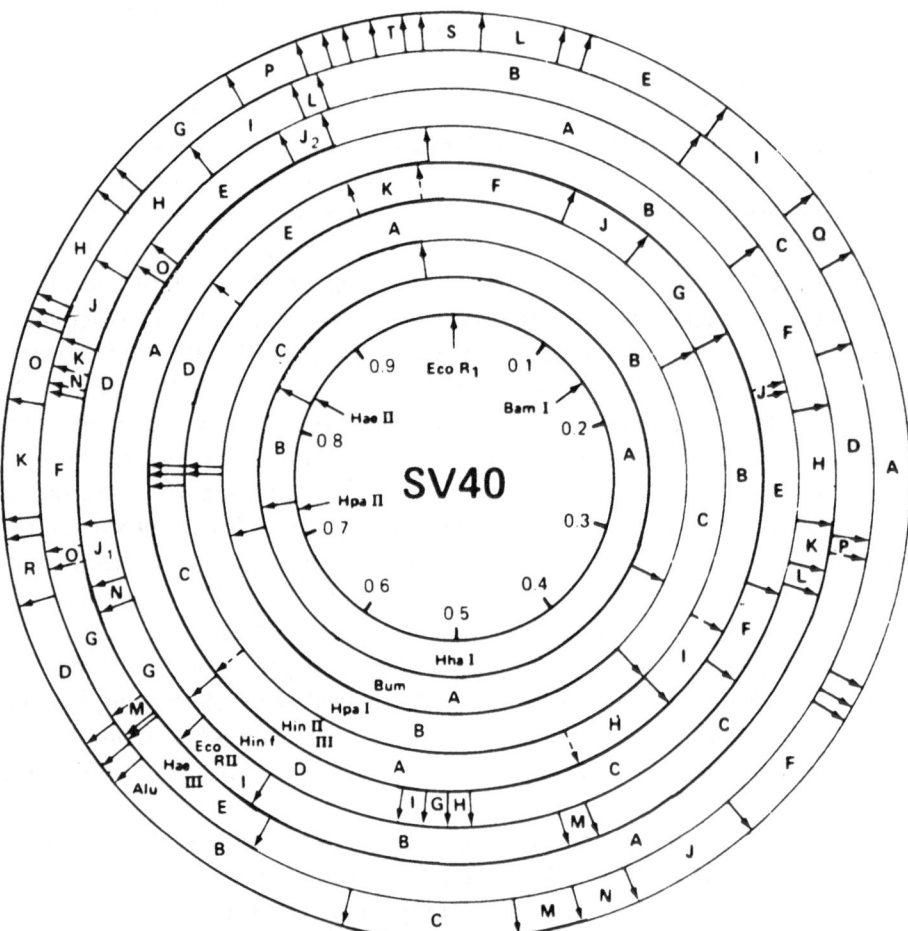

Fig. 10.   Map of the positions at which SV40 DNA is cleaved by several restriction enzymes. The cleavage site of *E. coli* R·R$_1$ is at zero, and map distances are given as fractional lengths of the SV40 genome clockwise from the *Eco*RI cleavage site. The positions of the cleavage sites were taken from the following sources: *Eco*RI, Morrow and Berg (1972) and Mulder and Delius (1972); *Bam*I, J. Sambrook and M. Mathews (unpublished); *Hpa*II, Sharp *et al.* (1973); *Hae*II, Roberts *et al.* (1975); *Hha*I, K. N. Subramanian, S. Zain, R. J. Roberts, and S. Weissman (unpublished); *Bum*I, M. Mathews (unpublished); *Hpa*I, Danna *et al.* (1973) and Sharp *et al.* (1973); *Hind*II/III, Danna *et al.* (1973) and Yang *et al.* (1975); *Hin*f, K. N. Subramanian, S. Zain, R. J. Roberts, and S. Weissman (unpublished); *Eco*RII, and *Hae*III, Subramanian *et al.* (1974); *Alu*I, Yang *et al.* (1975).

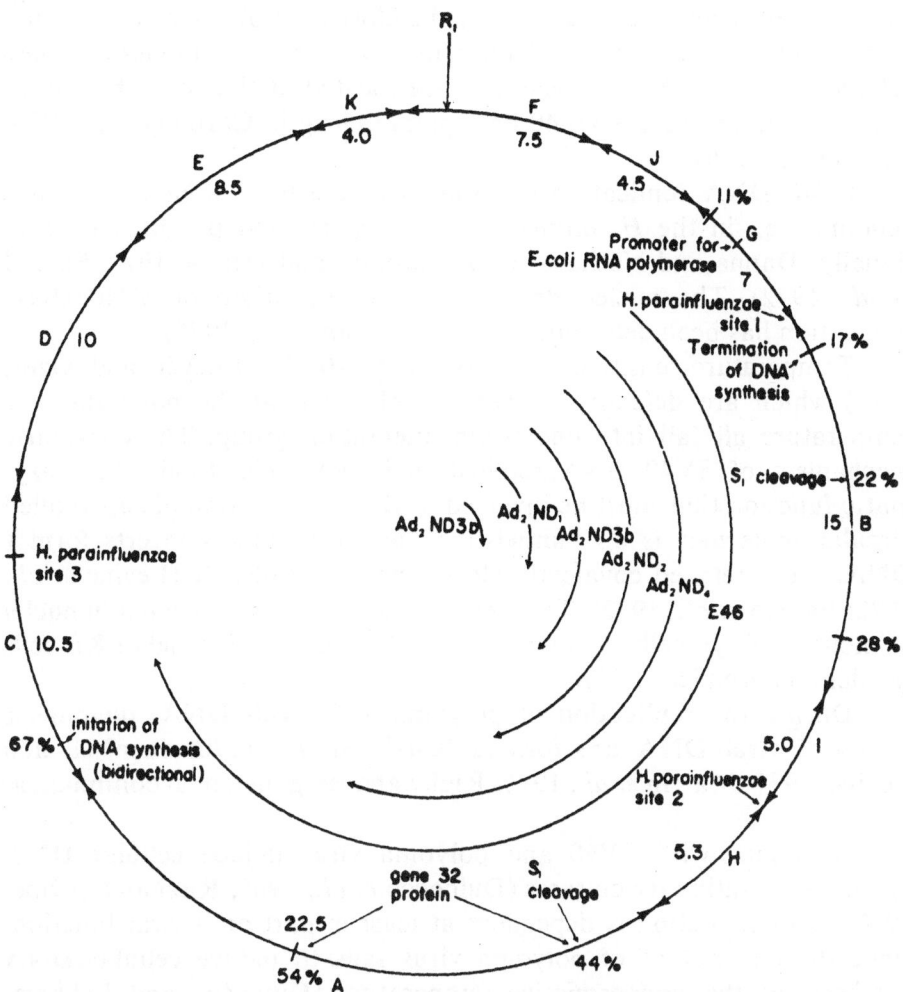

Fig. 11. Physical map of the SV40 genome. The sites at which bacterial restriction endonucleases cleave the viral genome are indicated by "R₁," "*H. parainfluenzae*," and double arrowheads for the *Eco*RI endonuclease, the endonuclease from *Haemophilus parainfluenzae,* and the endonuclease from *Haemophilus influenzae,* respectively. The sites at which SV40 DNA replication is initiated and terminated, the site of a promoter for the *E. coli* RNA polymerase, and the site of binding to the gene 32 protein of phage T4 are also indicated. The segments of the SV40 genome integrated into defective and nondefective adenovirus–SV40 hybrids are also shown. From Tooze (1973).

and Nathans, 1974). A new class of viable deletion mutants of SV40 has been constructed by cleavage of the closed circular DNA molecule, 5′-exonuclease digestion, and infection of permissive monkey kidney cells with this DNA. The deletions are located at the sites of endonuclease digestion and are 15–40 base pairs in length (Carbon *et al.*, 1975; Scott *et al.*, 1976).

SV40 DNA replication is initiated at the 67% point of the viral genome, i.e., in the *H. influenzae* C fragment, and proceeds bidirectionally (Danna and Nathans, 1972; Nathans and Danna, 1972; Fareed *et al.*, 1972). The nucleotide sequence of the origin of SV40 DNA replication has been determined (Subramanian *et al.*, 1975).

Temperature-sensitive mutants of SV40 (Tegtmeyer and Ozer, 1971) which are defective in DNA replication at the nonpermissive temperature all fall into one complementation group. The enzymatic mechanism of SV40 DNA replication is not understood. One enzymatic function that must be involved in DNA replication of supercoiled circular molecules is an "untwisting" activity which converts form I DNA into a relaxed covalently closed circular molecule (Levine *et al.*, 1970; Fareed *et al.*, 1972). This activity has been demonstrated in nuclei of mouse embryo cells (Champoux and Dulbecco, 1972) and of KB cells (Keller and Wendel, 1974).

During the replication of polyoma and SV40 DNA, oligomeric forms of viral DNA are formed (Cuzin *et al.*, 1970; Jaenisch and Levine, 1971; Martin *et al.*, 1976; Rigby and Berg, personal communication).

Infections with SV40 and polyoma virus induce cellular DNA synthesis in stationary cultures (Dulbecco *et al.*, 1965; Ritzi and Levine, 1970). This induction is dependent at least in part on a viral function, since the *ts*3 mutant of polyoma virus fails to induce cellular DNA synthesis at the nonpermissive temperature (Dulbecco and Eckhart, 1970; Eckhart and Dulbecco, 1974). To a certain extent, the temperature sensitivity of the *ts*3 mutant can be diminished by the addition of fresh medium and serum to the culture (Eckhart and Dulbecco, 1974). The importance of the induction of cellular DNA synthesis for viral DNA replication cannot be assessed with certainty. Concomitant with or following the induction of cellular DNA synthesis, SV40 or polyoma infection leads to an increase in the activity of numerous enzymes in the infected cells (for review, see Tooze, 1973). Most of these enzymes are involved in DNA replication. Weil *et al.* (1974) have postulated that SV40 and polyoma infection exert a pleiotropic effect on the host genome, thus leading to induction of cellular DNA synthesis, induction of host enzymes, and the complex events that manifest themselves in the transformed state of the cell.

5.2. Integration in the Productive System

As in the productive adenovirus system, it is very difficult to prove integration of SV40 DNA in productively infected cells. The main problem arises from the finding that high molecular weight viral DNA which is synthesized *de novo* late in lytically infected cells contains linear (Rigby and Berg, personal communication) and circular forms (Martin *et al.*, 1976) of oligomeric SV40 DNA (dimers to hexamers). By digestion with restriction endonucleases, oligomeric circles of SV40 DNA have been shown to consist exclusively of SV40 sequences (Martin *et al.*, 1976). In contrast, it will be recalled that high molecular weight viral DNA which is synthesized in KB cells early after productive infection with adenovirus yields cleavage patterns after digestion with *Eco*RI or *Bam*H–I which are different from those of virion DNA (Schick *et al.*, 1976; Baczko *et al.*, manuscript in preparation).

In SV40-infected cells, integrated sequences are probably also present. However, almost all the evidence which has been presented for integration so far does not rigorously differentiate between oligomeric viral DNA and viral sequences covalently linked to cellular DNA. With this reservation, which does not apply to the observation of substituted SV40 genomes (Lavi and Winocour, 1972), we enter into the discussion of the evidence available for integrated SV40 sequences.

Evidence for the integration of SV40 DNA into the host genome in permissively infected African green monkey kidney (CV-1) cells was first presented by Hirai and Defendi (1972). According to their studies, insertion of viral DNA began 20 h postinfection and remained at a constant level up to 48 h postinfection. In this study, viral and cellular DNAs isolated from nuclei of infected cells were separated by precipitation of the high molecular weight DNA (Hirt, 1967) and by zone velocity sedimentation in alkaline sucrose density gradients. Viral sequences were detected by nucleic acid hybridization using ^{3}H-labeled cRNA. Inhibition of DNA synthesis by D-arabinosylcytosine did not interfere with integration.

In productively infected CV-1 cells, Waldeck *et al.* (1973) found SV40 DNA to be preferentially integrated into cellular DNA of 1.5–2 times the contour length of SV40 DNA.

Hirai *et al.* (1974a) proposed that induction of cellular DNA synthesis might be a consequence of viral DNA integration. When CV-1 cells were infected with the temperature-sensitive early SV40 mutant *ts*101 (Robb and Martin, 1972), induction of cellular DNA synthesis and synthesis of the T, V, and U antigens and of viral DNA were blocked at the nonpermissive temperature. Under these conditions, integration of viral DNA was absent in productively infected CV-1 cells (Hirai *et al.*, 1974*b*). However, the *ts*101 mutant behaved very differently in nonproductive hamster cells: The *ts*101 mutant induced T antigen and integrated into the host genome but did not induce cellular DNA synthesis. Thus there must be a block in this mutant which leads to an uncoupling of the integration event and the induction of cellular DNA synthesis in hamster cells.

The number of integrated viral genomes in productively infected cells was estimated by Hölzel and Sokol (1974), who devised an elaborate procedure to ascertain that CV-1 cellular DNA was not contaminated with free viral DNA: Cellular DNA was first precipitated by the Hirt method (1967) and subsequently subjected to velocity sedimentation in alkaline sucrose gradients, equilibrium centrifugation in ethidium bromide/CsCl density gradients, and an additional velocity sedimentation in an alkaline sucrose density gradient. From the results of reconstitution experiments in which CV-1 DNA and free viral DNA had been artificially mixed, the authors of this report calculated that cellular DNA extensively purified by this procedure contained as little as 0.006% free viral DNA. The cellular DNA isolated by the same procedure from CV-1 cells productively infected with SV40 contained more than 20,000 integrated viral genome equivalents per cell. In view of the reservations explained above, these findings will have to be reinvestigated, particularly with respect to a more precise quantitation.

The mechanism by which viral genomes are integrated and the significance of the integration process for viral DNA replication are under intensive investigation in a number of laboratories. Little definitive information is as yet available. The observation that in *ts*101-infected monkey cells viral DNA replication and integration are absent (Hirai *et al.*, 1974*b*) could still be coincidental and does not necessarily prove that these events are causally related. On the other hand, even in the well-characterized system of bacteriophage λ the enzymatic mechanism required for the integration and excision events is only now beginning to be unraveled (Syvanen, 1974; Nash, 1975*b*; Gottesman and Gottesman, 1975*a,b*).

An adenosine triphosphate-dependent polynucleotide ligase was detected in uninfected chicken, hamster, mouse, monkey, and human cells, as well as in mouse embryo, monkey kidney, and HeLa cells infected with polyoma virus, SV40, and vaccinia virus, respectively (Sambrook and Shatkin, 1969). Presumably, a ligase could be involved in the integration event. However, the presence of a ligase in uninfected and infected cells obviously does not mean that this ligase is necessarily involved in the process of integration of viral DNA.

5.3. Substituted SV40 Genomes

When SV40 was serially passaged at high multiplicities of infection, noninfectious virus particles were formed which contained supercoiled circular DNA. This DNA was heterogeneous in size and shorter than the DNA from SV40 passaged at low multiplicities (Yoshiike, 1968). Shortly after Yoshiike's 1968 report, it was demonstrated that SV40 grown in the monkey BSC-1 cell line contained closed circular viral DNA which hybridized also to cellular DNA (Aloni *et al.*, 1969). Apparently, host sequences had been covalently linked to SV40 DNA. Virus particles containing DNA with both viral and cellular sequences covalently linked have to be distinguished from the "pseudovirions" which occur in SV40 preparations grown on African green monkey kidney cells (Levine and Teresky, 1970) and in polyoma virus (Michel *et al.*, 1967; Winocour, 1967). In pseudovirions, linear host cell DNA is encapsidated into virus particles (see Volume 5 of *Comprehensive Virology* for a detailed discussion of pseudovirions).

The first systematic investigation of SV40 particles carrying host sequences in the viral DNA was presented by Lavi and Winocour (1972). When BSC-1 monkey cells were infected with SV40 at low multiplicities of infection (0.032 PFU/cell), hardly any hybridization of this SV40 DNA to host cell DNA was observed. However, with increasing multiplicities of infection (0.16–3000 PFU/cell), an increasing amount of the newly synthesized SV40 DNA hybridized to cell DNA. This result was obtained regardless of whether the SV40 DNA was extracted from purified virions or directly from the infected cells. Plaque-purified virus did not contain cellular sequences; however, upon passage at high multiplicity, the viral DNA acquired host sequences. To explain the occurrence of such "substituted" SV40 genomes, Lavi and Winocour (1972) postulated that this recombination event between the SV40 and host genomes was the consequence of integration and excision of the viral DNA during lytic infection. The substituted viral genomes which arose in this way might have an advantage during DNA replication, probably because they contain multiple sites for the initiation of viral DNA. Thus the substituted genomes would be amplified after their generation by faulty excision. An alternate mechanism envisaged was an increase in the frequency of recombination between viral and host genomes with increasing amounts of cellular DNA incorporated into the viral DNA.

The SV40 DNAs from virions obtained after infection at low and high multiplicities were also analyzed for sequence homology by the technique of heteroduplex mapping (Tai *et al.*, 1972). The DNA from virus passaged at low multiplicity carried deletions in 2% of the molecules, but substitutions could not be detected. The DNA derived from virions passaged seven times undiluted or at high multiplicities carried substitutions in 12% or 7%, and deletions in 13% or 11% of the molecules, respectively. More than 80% of the molecules carrying substitutions were shorter than SV40 DNA. The substitutions comprised about 20–30% of the wild-type SV40 sequences.

Substituted SV40 DNA could be isolated starting with four different single-plaque isolates from two different SV40 strains (777 and 776), when these were serially passaged undiluted (Lavi *et al.*, 1973). In these substituted genomes, the host DNA was predominantly of the reiterated type. The majority of the substituted genomes were noninfectious and were again shorter than wild-type SV40 DNA without homology to host DNA. Evidence was presented that the substituted SV40 genomes were able to replicate and incorporate [³H]thymidine into host sequences.

Substituted SV40 DNA in which 66% of the molecules contained host DNA was digested with the restriction endonuclease from *H. influenzae* and subsequently analyzed by electrophoresis on polyacrylamide gels. Nine fragments could be resolved which had molecular weights different from those of the fragments obtained by digestion of wild-type SV40 DNA. Two of the nine fragments contained highly reiterated cellular sequences. The majority of the fragments were thought to carry unique host sequences (Rozenblatt *et al.*, 1973).

More recent results from Winocour's laboratory (Frenkel *et al.*, 1974; Winocour *et al.*, 1974) indicated that the integrated host sequences were not a random selection of all host sequences and that the cellular sequences were predominantly of the nonreiterated type. Starting from a single-plaque isolate, the host sequences were similar in substituted genomes arising from a particular set of serial passages, but they were different for passages originating from different single-plaque

isolates. There was no homology between the cellular sequences in substituted genomes of polyoma virus (Lavi and Winocour, 1974) and
SV40.

The sites of recombination between the cellular and viral genomes
are nonrandom but probably multiple and are preferentially located in
the nonreiterated regions of the cellular genome (Winocour *et al.*,
1974). According to these authors, the occasional inclusion of
reiterated cellular DNA sequences into substituted SV40 genomes may
be due to excision at a site where reiterated and unique cellular DNA
sequences overlap. Obviously, the biological significance of this insertion–excision mechanism is unknown. It will be very interesting to
investigate whether substituted SV40 genomes have the capacity to
transduce cellular genetic markers, in a manner similar to transducing
phage.

In certain instances, substituted SV40 genomes contained a
reiteration of the sequence in the *Hin*C and D fragments of the SV40
genome. An extreme example of this type was observed by Brockman
et al. (1973) in the sV particles from late-passage virus. These particles
carried DNA in which nonreiterated host DNA and an SV40 segment
equivalent to part of the *Hin*C fragment (site of initiation of DNA
replication) were tandemly repeated. SV40 variants with specifically
altered genomes were isolated by Brockman and Nathans (1974). The
locations of rearrangements in the DNA of defective SV40 particles
have also been mapped using the *Eco*RI restriction endonuclease and
heteroduplex formation with wild-type SV40 DNA (Risser and Mulder,
1974). Triplications of specific viral sequences in SV40-like DNA have
been reported by Fareed *et al.* (1974) and Khoury *et al.* (1974*b*).

Lee *et al.* (1975) have cloned variant clones of SV40 which
evolved from high-multiplicity serial passage of the virus. The structure
of several cloned variant genomes was determined by analysis with
restriction endonucleases. The structures of a series of such evolutionary variants of SV40 are depicted in Fig. 12. It is striking that the
initiation site for DNA replication is preserved even in the case of
variant *ev*1103, where only a small segment of SV40 DNA persists in
tandem replication with cellular DNA sequences.

Of all substitutions in the SV40 genome, 75–80% were found to
occur at specific sites, i.e., at the 25%, 29%, and 31% points relative to
the *Eco*RI cleavage site. These data suggested that there was some
specificity with respect to the site of integration on the viral genome
(Chow *et al.*, 1974).

Yoshiike *et al.* (1974) were able to rescue a defective SV40 genome

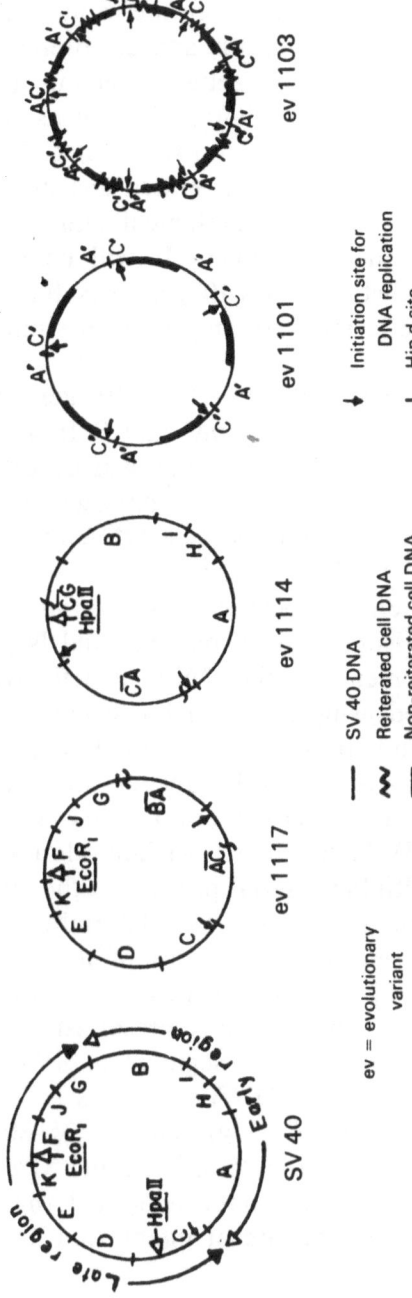

Fig. 12.   Cleavage maps of SV40 DNA and the DNA of four evolutionary variants of SV40. From Brockman *et al.* (1975).

from a transformed mouse cell line. The rescued genome had a deletion and an insertion at two different sites. Oren *et al.* (1976) have analyzed the unique cellular sequences in defective SV40 genomes and compared these sequences in the DNA which was derived from several independently isolated single plaques upon high multiplicity serial passage. The data of Oren *et al.* are consistent with the interpretation that the SV40-host recombination events which generate substituted SV40 variants occur at preferred sites in the cellular genome.

Considering all the evidence available at present, the most likely mechanism for the generation of substituted SV40 genomes is that of integration and faulty excision of the viral genome during lytic infection (see also preceding section). Integration may occur at only a limited number of sites in the host chromosome. There is evidence that the site of recombination on the viral genome is specific. It is considered likely that, upon repeated passage, the substituted viral DNA molecules have a selective advantage during replication due to the presence of multiple sites of initiation for DNA replication (Winocour *et al.*, 1974). Substituted genomes may accumulate because of an enhanced capacity for recombination with the host, once host sequences have been incorporated into the SV40 genome.

5.4. Mixed Transcripts Containing Viral and Host Sequences

Further evidence for the notion of integrated SV40 DNA in productively infected and transformed (see also Section 5.5) cells comes from the analysis of the virus-specific RNA in these cells. Lindberg and Darnell (1970) reported that the heterogeneous nuclear RNA from SV40-transformed cells contained sequences homologous to SV40 DNA and was considerably longer than the virus-specific RNA isolated from polysomes. Presumably, heterogeneous nuclear RNA was processed to the mRNA associated with polysomes and was possibly transcribed from integrated viral genomes. The RNA isolated from SV40-transformed 3T3 cells (mouse cells) was 0.75–3 times the length of single-stranded SV40 DNA. Similar results and conclusions were presented for the SV40-specific RNA isolated from SV40-transformed and lytically infected cells by Darnell *et al.* (1970), Tonegawa *et al.* 1970), and Weinberg *et al.* (1972, 1974). Reviewing this evidence, the possibility should be considered that at least some of the "mixed transcripts" might have been due to aggregation.

The conclusion that this high molecular weight RNA contained both viral and cellular sequences covalently linked was further supported by the results of DNA-RNA hybridization experiments. The high molecular weight nuclear RNA was hybridized to viral DNA, eluted from the filters, and subsequently hybridized to cellular DNA in a second reaction. Very similar results were described for SV40-transformed cells (Wall and Darnell, 1971) and for lytically infected cells (Rozenblatt and Winocour, 1972). Jaenisch (1972) chose a slightly different approach and demonstrated that a large portion of the SV40-specific RNA of longer than genome length was sensitive to digestion by pancreatic RNase after annealing to viral DNA.

Thus it appears possible that integrated SV40 genomes are cotranscribed with adjacent cellular sequences both in lytically infected and in transformed cells. We do not understand at present what role cellular and/or viral regulatory functions play in the transcription process. Recent data by Laub and Aloni (1975) and Khoury *et al.* (1975*b*) suggest that in cells productively infected with SV40 two control mechanisms regulate viral gene expression, one at the transcriptional level and the other at the posttranscriptional level.

5.5.   Integration of SV40 DNA in Transformed and Abortively Infected Cells

Cell lines transformed by SV40 do not produce virus. However, the virus can be rescued from some of these lines by cocultivation with permissive cells or by fusion of the transformed cells with permissive cells (Gerber, 1966; Koprowski *et al.*, 1967; Watkins and Dulbecco, 1967; Tournier *et al.*, 1967). Fusion is usually induced by UV-inactivated Sendai virus (Harris and Watkins, 1965). Croce and Koprowski (1973) and Poste *et al.* (1974) have demonstrated that rescue of SV40 from SV40-transformed cells is possible by fusion with anucleate monkey cells and that nondividing confluent cultures of monkey cells are most efficient in the rescue process.

In an attempt to determine the number of viral genomes in SV40-transformed cells, SV40 complementary ^{3}H-labeled RNA was synthesized *in vitro* with the DNA-dependent RNA polymerase from *E. coli,* and this RNA was used in DNA-RNA hybridization experiments for the quantitation of the number of viral gene copies in transformed cells (Westphal and Dulbecco, 1968; Sambrook *et al.*, 1968). It should be mentioned that Westphal and Kiehn (1970) demonstrated that the *E. coli* polymerase transcribed the SV40 genome faithfully; thus an exact SV40-specific probe was available for these studies. Although hybridization methods gave qualitatively satisfactory results, this procedure proved to be problematic as to the exact quantitation of the number of viral gene copies per cell, since Haas *et al.* (1972) found that DNA-RNA hybrids failed to be retained on filters when the DNA was saturated with cRNA. Thus the calibration curves on which the enumeration of viral gene copies was initially based were not precise enough.

The state of the viral DNA in SV40-transformed cells was elucidated by Sambrook *et al.* (1968). High molecular weight cellular DNA was isolated in alkaline sucrose density gradients and was shown to contain viral DNA covalently linked to cellular DNA. Free supercoiled circular SV40 DNA could not be found in these cells. This study, however, could not rigorously rule out the possibility that circular oligomers of free viral DNA cosedimented with high molecular weight cellular DNA. This report by Sambrook *et al.* (1968) presented one of the first lines of evidence for the integrated state of SV40 DNA in transformed cells.

The presence of the SV40 surface (S) and tumor (T) antigens could be correlated with the persistence of SV40 genomes in SV40-transformed cells (A. S. Levine *et al.*, 1970). Furthermore, these workers used improved hybridization techniques for the quantitation of SV40 genomes and arrived at values considerably lower than those reported by Westphal and Dulbecco (1968) and Sambrook *et al.* (1968). Benjamin (1968) found no homology between the DNA of SV40 or polyoma virus and the mitochondrial DNA of cells transformed by these viruses.

Further improvements in the techniques to quantitate viral genomes in transformed cells came with the application of DNA reassociation kinetics (Britten and Kohne, 1968; Wetmur and Davidson, 1968) to the measurement of viral gene equivalents in transformed cells (Gelb *et al.*, 1971). In four out of five SV40-transformed 3T3 cell lines, Gelb *et al.* (1971) detected only one SV40 genome equivalent, and three SV40 equivalents were measured in the fifth cell line. In African green monkey kidney cells transformed by SV40, less than one viral genome equivalent was observed. Similar quantitative data were obtained by Ozanne *et al.* (1973). In SV40-transformed cells, the viral DNA appeared to be linked to the nonreiterated sequences of cellular DNA (Gelb and Martin, 1973).

In SV40-transformed cells, different segments of the SV40 genome were found at different frequencies by the technique of reassociation kinetics using SV40 DNA fragments produced by sequential cleavage with the restriction endonucleases *Eco*RI and *Hpa*I (Sambrook *et al.*, 1974). Thus transformed cells may contain one or several entire viral genome equivalents in addition to various fragments at different frequencies. The SVT2 line of SV40-transformed mouse cells contains six copies of a segment of SV40 DNA which carries the early region of the SV40 genome and one copy of the late region per cell. The early genes were found to be transcribed (Botchan *et al.*, 1974).

Botchan and McKenna (1973) investigated the integrated SV40 genome in high molecular weight DNA from SV40-transformed mouse cells (SVT2 cells) with the restriction endonuclease *Eco*RI. Since this enzyme cleaves SV40 DNA only at one site, one would expect two fragments containing viral sequences, if there were one *Eco*RI cleavage site in the cellular DNA on either side of the viral genome. SVT2 DNA was cleaved with the *Eco*RI endonuclease, the fragments were resolved by electrophoresis on agarose gels, and the SV40-specific sequences were localized by hybridization with ^{3}H-labeled cRNA. Two peaks of SV40 DNA were found. The SV40-containing fragments had molecular weights of 3.1×10^6 and 1.8×10^6. The sum of these molecular weights is higher than the molecular weight of SV40 DNA. This finding indicates that there is covalent linkage between viral and host DNA. There are probably a limited number of integration sites on the host chromosome. This finding does not rule out the presence of oligomeric forms of SV40 with altered restriction sites. The DNA from five different lines of SV40 transformed mouse cells was cut by restriction endonucleases, transferred to membrane filters, and viral DNA sequences were detected by hybridization with ^{32}P-labeled SV40 cRNA and autoradiography of the filters. Each of the lines showed a different pattern in the distribution of SV40 specific DNA sequences. This finding indicated that the structure and the location of the integrated SV40 DNA in the host DNA were different in each of the lines (Ketner and Kelly, 1976). Similar results were obtained for 11 lines of SV40 transformed rat cells (Botchan *et al.*, 1976).

The integration of SV40 DNA was also demonstrated in SV40-transformed Chinese hamster embryo cells (Hirai and Defendi, 1971). Hybridization of SV40 cRNA with the nucleolus-associated DNA and the total nuclear DNA from SV40-transformed Chinese hamster cells revealed SV40-specific sequences in both DNA fractions. These results are consistent with multiple integration sites of the SV40 genome (Hirai *et al.*, 1974c).

Moreover, it was shown that DNA became associated in an alkaline-stable form with the DNA of abortively infected Chinese hamster embryo cells 15–20 h postinfection and that integration was not inhibited by D-arabinosylcytosine and only partly inhibited by cycloheximide (Hirai *et al.*, 1971). Collins and Sauer (1972) investigated the fate of SV40 DNA infecting nonpermissive 3T3 cells. At 48 h postinfection with viral DNA, a large number of genome equivalents had become associated with cellular DNA in an alkali-stable form. Cellular DNA replication did not appear to be a prerequisite for the integration process. Some SV40-transformed cells, the so-called flat transformants or abortively transformed cells, differ from regular transformants in that they do not grow in multiple layers of cells, and upon repeated passage lose the characteristics of transformed cells altogether. Smith *et al.* (1972) reported that some of these abortively transformed cell lines contained as many as five SV40 genome equivalents per diploid cell. Boyd and Butel (1972) actually succeeded in isolating infectious SV40 DNA from SV40-transformed cells. More recently, Fried (1975) measured precisely the midpoint of transition in strand separation of DNA from the SV40-transformed 11A 8 cell line and concluded that at least 50% of the SV40 DNA was integrated at a single site in the cellular genome.

The pattern of transcription of the SV40 genome in SV40-transformed lines (Reich *et al.*, 1966) differs from that in productively infected cells (Martin, 1970; Khoury *et al.*, 1973*a,b*; Ozanne *et al.*, 1973; Martin and Khoury, 1973). In 11 transformed lines, the early viral DNA was predominantly transcribed. In only three lines were minor portions of the late strand expressed. The stable virus-specific RNA detected in transformed cells includes those RNA species transcribed early in the productive cycle (Aloni *et al.*, 1968; Oda and Dulbecco, 1968; Sauer and Kidwai, 1968; Sambrook *et al.*, 1972; Khoury *et al.*, 1973*a,b*). The SV40-specific RNA present in several SV40-transformed lines has been precisely mapped on the SV40 genome using the *Hind*III restriction enzyme fragments of SV40 DNA (Khoury *et al.*, 1975*a*). Khoury *et al.* (1975*b*) provided evidence for posttranscriptional selection of SV40 specific RNA.

In human cells transformed by SV40, the chromosome was identified which carried the SV40 genome (Croce *et al.*, 1973, 1974*a,b*). These authors used techniques developed by Weiss *et al.* (1968) and by Weiss (1970) and fused mouse cells deficient in thymidine kinase with SV40-transformed human cell lines deficient in hypoxanthine phosphoribosyltransferase. A concordant segregation of the gene(s) for SV40 T antigen and human chromosome C-7 was observed. By fusing hybrid clones with the CV-1 cell line, which is permissive for SV40, it could be shown that the viral capsid proteins were present only in those cells which were T-antigen positive. Furthermore, it was possible to rescue defective SV40 virions from hybrid subclones containing human chromosome 7, but not from those subclones which had lost that chromosome (Croce *et al.*, 1974*a*). These data indicate linkage between human chromosome C-7, the genes for the SV40 T-antigen, and the integrated SV40 genome. Furthermore, Croce *et al.* (1974*b*) concluded that the SV40 genome in human chromosome C-7 introduced gene(s) which were mandatory for continued cell division and might code for "transforming factors" (Croce and Koprowski, 1975). Moreover, it was shown that tumors induced in "nude" mice after injection of hybrids between mouse peritoneal macrophages and SV40-transformed human cells contained cells which retained the human chromosome 7. This chromosome still carried the SV40 genome, and the T antigen of SV40 was expressed in the tumor cells (Croce *et al.*, 1975). Monkey cells can be transformed by the adenovirus 7-SV40 hybrid virus (see Section 7) (Jensen and Defendi, 1968). In such transformants, the expression of the T antigen of SV40 is linked to the monkey chromosome which corresponds to human chromosome C-7 (Croce *et al.*, 1974*c*).

These data represented conclusive evidence for the chromosomal location of the integrated SV40 genomes in at least two SV40-transformed cell lines and correlated the presence of the viral genome on a specific chromosome with the characteristics of virus-transformed cells.

Shani *et al.* (1975) demonstrated that metaphase chromosomes which were isolated from an inducible line of SV40-transformed Chinese hamster cells could be transferred into permissive BSC-1 cells. This transfer of chromosomes led to activation of infectious SV40 in one cell in 4×10^5 treated cells. Chromosomes were fractionated according to size; the capacity to transfer the inducibility for infectious SV40 to cells rested with two size classes of chromosomes (Shani *et al.*, 1975).

Attempts to identify the gene(s) of SV40 which are responsible for transformation were made by Graham *et al.* (1974*b*) using the technique of transforming cells with fragments of viral DNA (Graham and van der Eb, 1973*b*). It was found that specific SV40 DNA fragments comprising 59% of the genome were sufficient for transformation. These fragments contained the early region of the SV40 genome. Cells transformed by the early region of the SV40 genome (59% of the genome) were all T-antigen positive, but virus could not be rescued from these cells by fusion (Abrahams *et al.*, 1975).

Before concluding this section, the experiments of Jaenisch and Mintz (1974) should be mentioned. These authors infected isolated mouse embryos at the early blastula stage with SV40 DNA, reimplanted these embryos into pregnant animals, and demonstrated the persistence of SV40 genomes in several organs of the newborn mice originating from the infected blastulas. So far, it is not known whether the persisting viral genomes have any effect on the development of the first or future generations of these persistently infected mice or on the occurrence of tumors in these animals.

6. POLYOMA VIRUS

In the history of the work on DNA tumor viruses, polyoma virus has played a major role, and many important contributions have been made and continue to be made with this virus. Since relatively little work has been done on the integration of polyoma virus, I shall not discuss this virus system in great detail, but rather mention briefly the experimental evidence available for the interaction of polyoma virus DNA and host DNA. The first evidence for the persistence of the genome of polyoma virus (and of SV40) in virus-transformed cells came from experiments in which a small portion of the RNA from transformed cells was found to hybridize specifically to viral DNA (Benjamin, 1966). In mouse kidney cells productively infected with polyoma virus, Acheson *et al.* (1971) detected, late after infection, giant RNA molecules which were larger than the viral genome. Beard *et al.* (1976) have reported that the giant RNA hybridized almost exclusively with that strand of polyoma DNA which was transcribed late in the lytic infection cycle. It could not be decided whether these RNA molecules stemmed from multiple cycles of transcription or represented molecules transcribed jointly from cellular DNA and adjacent integrated polyoma genomes.

By gradient centrifugation, Ralph and Colter (1972) isolated cellular DNA from mouse cells productively infected with polyoma virus. The cellular DNA, presumably uncontaminated by free polyoma, was hybridized with ^{3}H-labeled polyoma cRNA. The experimental evidence suggested that polyoma genomes were present in the integrated form. Following infection of mouse and hamster (BHK-21) cells, both the parental and the newly synthesized polyoma DNAs were found associated with cellular DNA in an alkali-stable form (Babiuk and Hudson, 1972). In human cells, little if any polyoma DNA could be detected in association with the cellular DNA.

Fogel (1975) succeeded in inducing the synthesis of viral capsid proteins in polyoma-transformed cell lines. Polyoma-transformed cells can revert to apparently untransformed cells. However, they retain the T antigen and the same number of viral DNA molecules per cell as transformed cells (Shani *et al.*, 1972). Thus evidence for integration of the polyoma genome in lytically infected and transformed cells is rather scanty at the present time and work on this system has been by far less extensive than with SV40 or adenoviruses.

7. THE ADENO-SV40 HYBRID VIRUSES

As has been discussed in the sections on adenoviruses and SV40, there is evidence that these viruses are able to integrate their genomes into the host chromosome both in productive and in abortive infections, and that the viral genomes persist in transformed cells probably in the integrated state as well. The adeno-SV40 hybrid viruses represent a group of viruses in which SV40 genes are integrated into the adenovirus genome. The study of these hybrid viruses in which the DNAs of two oncogenic viruses are covalently linked (Baum *et al.*, 1966) has yielded a great deal of information about the anatomy of the hybrid DNA and the function of the SV40 genes incorporated into the adenovirus DNA. However, very little if any information is available as yet on the mechanism of the integration event. Obviously, two different mechanisms can be envisaged:

1. Direct recombination between the viral genomes.
2. Integration of the two viral genomes into the host chromosomes at the same site or a closely linked site followed by faulty excision of the viral genomes.

The first adeno-SV40 hybrid viruses discovered (Huebner *et al.*, 1964*b*; Rowe and Baum, 1964; Rapp *et al.*, 1964) were defective viruses which could not be propagated without a nonhybrid adenovirus as a helper. The defectiveness hampered the detailed genetic analysis of the hybrids. In 1969, Lewis and colleagues isolated the first nondefective adenovirus 2–SV40 hybrid ($Ad2^+ND_1$, nondefective hybrid 1) which was able to replicate with one-hit kinetics both in human embryonic kidney cells and in African green monkey kidney cells (Lewis *et al.*, 1969). In cytolytic infections, the $Ad2^+ND_1$ hybrid virus induced a new SV40-specific antigen, the U antigen, but not the SV40-specific T antigen (Lewis and Rowe, 1971). The U antigen was detected by using sera from SV40 tumor-bearing hamsters and was stable to heating at 50°C for 30 min.

The DNA molecules from the $Ad2^+ND_1$ hybrids contained Ad2 and SV40 sequences covalently linked (Crumpacker et al., 1970; Levin et al., 1971), as in the adenovirus 7–SV40 defective hybrid virus (Baum et al., 1966). The hybrid virions did not contain supercoiled circular SV40 DNA (Crumpacker et al., 1970). Only a fraction of the $Ad2^+ND_1$ genome consisted of SV40 sequences, which corresponded to an equivalent of 17% of the SV40 genome (Figs. 13 and 14). The biophysical characterization of the $Ad2^+ND_1$ DNA revealed that it was practically indistinguishable from Ad2 DNA (Crumpacker et al., 1971). An analysis of the SV40 genes transcribed in the $Ad2^+ND_1$ genome in lytically infected cells demonstrated that the SV40 segment present in the $Ad2^+ND_1$ DNA corresponded to some, but not all, of that part of the SV40 genome which was transcribed early (Oxman et al., 1971).

In 1973, Lewis and colleagues reported the isolation of four new nondefective adenovirus 2–simian virus 40 hybrid viruses, designated $Ad2^+ND_2$, $Ad2^+ND_3$, $Ad2^+ND_4$, and $Ad2^+ND_5$, which were all derived from the same Ad2-SV40 hybrid population as $Ad2^+ND_1$ was. These hybrids differed significantly in their biological properties. The hybrids $Ad2^+ND_1$, $Ad2^+ND_2$, and $Ad2^+ND_4$ grew in both human embryonic kidney (HEK) and primary African green monkey kidney cells, whereas $Ad2^+ND_3$ and $Ad2^+ND_5$ could replicate only in HEK cells. Moreover, the nondefective hybrids differed in the spectrum of the SV40-specific antigens which they induced in productively infected cells (Lewis et al., 1973; Kelly and Lewis, 1973) (Fig. 14).

The biophysical properties of the DNAs of the nondefective hybrids $Ad2^+ND_1$–$Ad2^+ND_5$ were very similar to those of Ad2 DNA. Ad2 and SV40 genes were linked covalently in all of the nondefective hybrid DNAs. The fraction of SV40 genes in each hybrid DNA was determined by hybridization with SV40 complementary RNA (Henry et al., 1973), by competition hybridization of the RNAs which were synthesized in cells lytically infected with each of the hybrids (Levine et al., 1973), and by heteroduplex mapping of the hybrid DNA molecules with the RI restriction endonuclease product of SV40 DNA (Kelly and Lewis, 1973; Morrow et al., 1973). The schemes in Figs. 11 and 13 summarize the results of these experiments and indicate precisely which sections of the SV40 genome were inserted into the adenovirus genome. The SV40 genes integrated into the Ad2 genome corresponded exclusively to early SV40 genes (Levine et al., 1973; Morrow et al., 1973;

Lebowitz *et al.*, 1974) (Figs. 11 and 13). Depending on the size of the inserted SV40 fragment, the deletions in the Ad2 genome varied in size (Fig. 14). All the fragments of SV40 DNA occurring in the hybrid viruses had a common left end in the *Hin*G fragment of SV40 and represented an overlapping series (Kelly and Lewis, 1973; Lebowitz *et al.*, 1974) (Figs. 11 and 13). It is interesting to note that the common end points of all the nondefective hybrids and of the E46⁺ hybrid lie in the *Hin*G fragment, which contains also the site of termination of SV40

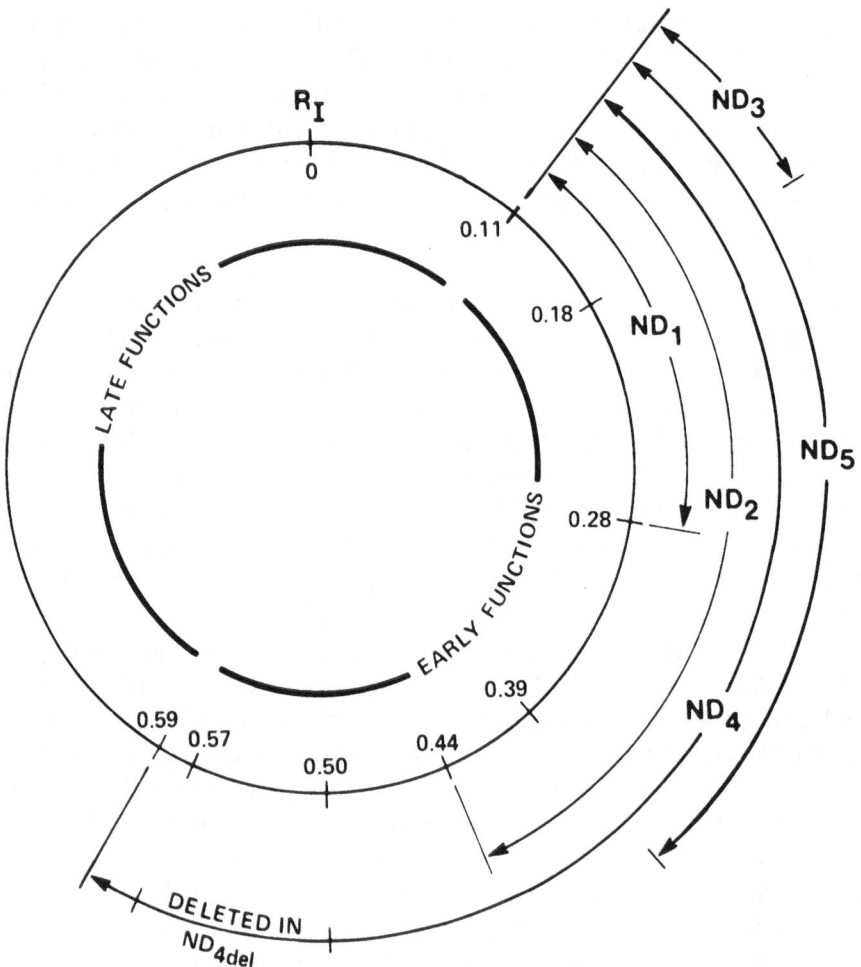

Fig. 13. Map positions of SV40 segments in nondefective hybrids on the SV40 genome. On the circular map, the *Eco*RI cleavage point is the origin. The fragments of the SV40 genome inserted in the nondefective adenovirus–SV40 hybrids are designated by arrows. The common point of origin of the known hybrids lies at 11%, in the *Hin*G fragment. From Morrow *et al.* (1973).

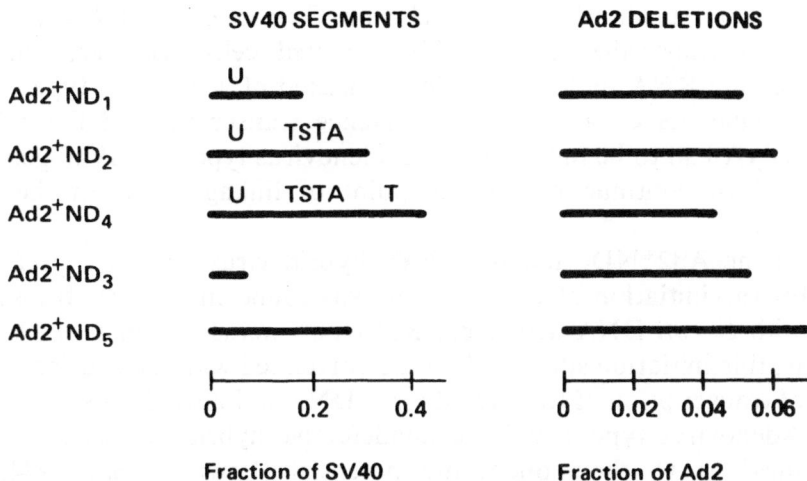

Fig. 14. SV40 segments form an overlapping series. The letters above each hybrid segment designate the SV40-specific antigens induced by each hybrid. The deletions in the adenovirus type 2 genome for each of the hybrids are also shown. The common end point for the deletion is 0.14 fractional adenovirus type 2 length unit from one end of each hybrid. From Kelly and Lewis (1973).

DNA replication (Danna and Nathans, 1972). These and other findings raise the possibility that all SV40-Ad2 hybrids originated from a single recombinational event.

Although the nondefective hybrid $Ad2^+ND_5$ apparently contains the SV40 genes responsible for the U and TSTA antigens, as can be seen in Fig. 14, cells infected with the $Ad2^+ND_5$ hybrid do not produce these antigens (Levine et al., 1973; Kelly and Lewis, 1973). There is no definitive explanation for this phenomenon at the moment. Lewis and Rowe (1973) suggested that a frameshift caused by the insertion of the SV40 genes might be responsible for the lack of antigen expression. The same authors calculated the genes for the SV40-specific transplantation antigen (TSTA) to be located between 0.17 and 0.43 fractional map unit on the SV40 map.

The synthesis of the SV40-specific T antigen was found to be very sensitive to interferon inhibition in cells infected with SV40, whereas the adenovirus 2 specific T antigen is expressed in interferon-treated cells infected with adenovirus type 2. When cells were infected with the nondefective hybrid $Ad2^+ND_4$, the synthesis of the T antigens of both SV40 and adenovirus type 2 became interferon resistant. This result was interpreted to indicate that the expression of the SV40 T antigen in the $Ad2^+ND_4$ was under the control of adenovirus type 2 genes (Oxman et *al.*, 1974). To substantiate this conclusion, Oxman *et al.* (1974) examined the RNA transcribed in Ad2$^+$ND$_4$-infected cells and were able to demonstrate RNA molecules which consisted of adenovirus type 2 and SV40 sequences covalently linked. Hence transcription of the hybrid virus DNA had to be initiated in the adenovirus type 2 or SV40 genome and had to continue across the point of linkage between the two genomes.

In the Ad2$^+$ND$_1$ and Ad2$^+$ND$_3$ hybrid virus DNAs, a preferred site for the initiation of transcription was found in *in vitro* transcription with *E. coli* DNA-dependent RNA polymerase (Zain *et al.*, 1973). So far, this initiation site could not be correlated with the site for initiation of transcription of the hybrid virus DNA in infected cells.

Adenovirus type 2 and the nondefective hybrid viruses were also examined as to their oncogenic potential (Lewis *et al.*, 1974*a,b*). Adenovirus type 2 and the Ad2$^+$ND$_1$–Ad2$^+$ND$_5$ hybrid viruses failed to produce tumors when injected into newborn hamsters. They did, however, transform hamster kidney cells in tissue culture. The transformed cells contained the adenovirus type 2 specific T antigen and adenovirus type 2 specific RNA. Upon injection into hamsters, the transformed cells produced in some cases tumors with the histopathology characteristic of adenovirus-induced tumors, in others tumors characteristic of SV40. Thus adenovirus type 2 was able to transform both rat cells (Freeman *et al.*, 1967; Gallimore, 1974) and hamster cells (Lewis *et al.*, 1974*a,b*), and it therefore appears clear that adenovirus type 2 has oncogenic potential whose expression might depend on the host and/or the experimental conditions employed.

8. HERPESVIRUS

8.1. Epstein-Barr Virus

The problem of integration of viral DNA in herpesvirus-infected cells has been investigated with a number of different herpesviruses. The most extensively studied system is that of the Epstein-Barr virus (EBV) DNA persisting in human lymphoblastoid cell lines (for review, see Klein, 1973; zur Hausen, 1975). This virus has been implicated in the causation of human diseases, such as Burkitt's lymphoma (Burkitt, 1962, 1963), nasopharyngeal carcinoma (Old *et al.*, 1966), and infectious mononucleosis (Henle *et al.*, 1968). Burkitt's lymphoma is a proliferative disease of the lymphoreticular system and occurs mainly in children. Burkitt's lymphoma is endemic in certain zones of Equatorial Africa (Burkitt, 1963; Burkitt and O'Connor, 1961) and New Guinea.

The nasopharyngeal carcinoma is observed in certain regions of Southern China (Shanmugaratnam, 1967) and Singapore (Muir and Shanmugaratnam, 1967), where it represents the most frequent malignancy. It is also observed in East Equatorial Africa and Tunesia.

From patients with Burkitt's lymphoma, lymphoblastoid cell lines have been established in tissue culture. Four different Epstein-Barr virus (EBV) specific antigens have been detected in these cells:

1. Structural antigens of the EBV (Henle and Henle, 1966).
2. Early virus-specific antigens (Henle *et al.*, 1970).
3. Membrane antigens (Klein *et al.*, 1966).
4. EBV nuclear antigen (EBNA), which has been demonstrated by complement fixation and immunofluorescence (Reedman and Klein, 1973). This antigen may correspond to the T antigen in papovavirus- or adenovirus-infected and -transformed cells.

The synthesis of EBV capsid antigens and of virus particles could be induced by 5-bromodeoxyuridine treatment of the Raji line and the NC37 line (Gerber, 1972; Hampar *et al.*, 1972). The Raji line (Pulvertaft, 1965) and the NC37 line were derived from patients with Burkitt's lymphoma (Burkitt, 1962, 1963), and these cells did not detectably produce EBV or viral antigens prior to induction. An enhanced induction of EBV was observed when cells in the early S phase of the cell cycle were treated with an inhibitor of DNA synthesis such as 1-β-D-arabinofuranosylcytosine, hydroxyurea, or excess thymidine (Hampar *et al.*, 1974).

In 1970, zur Hausen and Schulte-Holthausen demonstrated by DNA-DNA hybridization that EBV DNA persisted in the "nonproducing" Raji line. With the same technique, the EBV genome could be detected in cells derived from other cases of Burkitt's lymphoma and nasopharyngeal carcinoma, whereas cells from control tumors were free of EBV DNA (zur Hausen *et al.*, 1970). Improved estimates of the number of viral DNA copies came from studies employing DNA-DNA hybridization techniques using Epstein-Barr virus DNA-complementary RNA (Nonoyama and Pagano, 1971; zur Hausen, 1972). According to Nonoyama and Pagano (1971), the Raji line contained 65 EBV genome equivalents, line F265 contained 100 equivalents, and line NC37 contained 80 genome equivalents. Zur Hausen (1972) detected

hybridizable material in cells derived from 27 cases of Burkitt's lymphoma and from six cases of nasopharyngeal carcinoma and came to
comparable quantitative estimates. Nonoyama and Pagano (1973) and
Kawai *et al.* (1973) further refined the analysis by using the technique
of reassociation kinetics for the analysis of EBV DNA in cells from
Burkitt's lymphomas and nasopharyngeal carcinomas and could confirm their earlier estimates. Klein *et al.* (1974) have described human
lymphoblastoid cell lines which did not contain detectable amounts of
EBV DNA or virus-specific antigens.

The physical state of the EBV genome was investigated by techniques described in previous sections. The results of *in situ* cRNA-
DNA hybridization experiments led zur Hausen (1972) to conclude that
the EBV DNA was associated with the chromosomes of Raji cells. In
the nonproducing Raji line, viral DNA was found not to be covalently
linked to the host genome, and it was rather suggested that at least a
portion of the viral genome was linked to the host genome by alkali-
labile bonds (Nonoyama and Pagano, 1972; Adams *et al.*, 1973;
Tanaka and Nonoyama, 1974).

The DNA of EBV (ρ = 1.718 g/cm³) (Schulte-Holthausen and zur
Hausen, 1970) and of human cells (ρ = 1.700 g/cm³) can be separated by
equilibrium centrifugation in CsCl density gradients. When the DNA
from Raji cells which contain 50–60 copies of EBV DNA per cell is
analyzed by equilibrium sedimentation in neutral CsCl density gradients,
a large proportion of the intracellular EBV DNA remains associated
with cellular DNA. This association is stable toward pronase and phenol
treatment. Part of the viral DNA linked to cell DNA shifts to intermediate density positions upon shear breakage of the cellular DNA
(Adams *et al.*, 1973). However, after alkali denaturation, the viral DNA
separates from the host chromosome (Jehn *et al.*, 1972; Adams *et al.*,
1973; Tanaka and Nonoyama, 1974). On the basis of these results, it was
concluded at the time that the Epstein-Barr virus DNA was linearly
integrated into the host chromosome by alkali-labile bonds.

Human lymphoid cell lines which carry the EBV genomes contain
the viral DNA as integrated DNA and as covalently closed circular
DNA molecules which have been visualized in the electron microscope
(Fig. 15). Both forms of viral DNA seem to have been isolated from Burkitt's tumor biopsies and from cells of nasopharyngeal carcinomas which
were grown in nude mice. Thus, during latency, the EBV genome
appears to have the properties of an episome (Lindahl *et al.*, 1976;
Adams and Lindahl, 1975). Human lymphoid cell lines devoid of EBV
genomes did not contain the circular DNA molecules.

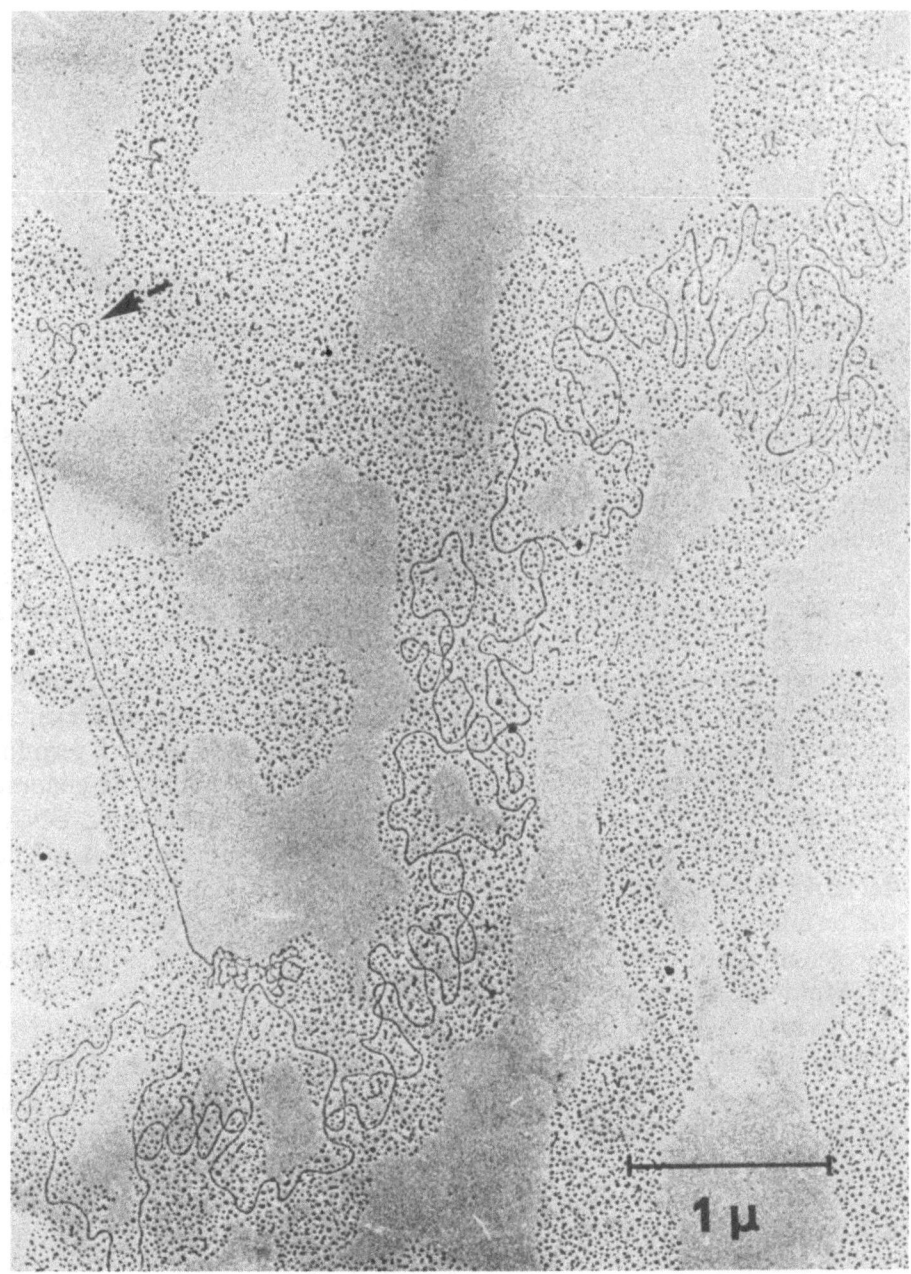

Fig. 15.   Electron micrograph of circular Epstein-Barr virus DNA molecule which was isolated from cells taken from a Burkitt's tumor biopsy. From Lindahl *et al.* (1975).

When Epstein-Barr virus was inoculated into cottontop marmosets, one out of three animals developed lymphoproliferative disease. In tumor and infiltrated spleen cells isolated from these animals, between one and two viral genome equivalents per cell were detected (Wolf *et al.*, 1974).

8.2. *Herpesvirus saimiri* and *ateles*

Two other interesting herpesviruses have been isolated from primates: *herpesvirus saimiri* and *herpesvirus ateles* (Melendez *et al.*, 1972a) were derived from squirrel monkeys (*Saimiri scuireus*) and spider monkeys, respectively. In the natural hosts, these viruses persist in the cells of the peripheral blood. When other species of marmosets, owl monkeys, cinamon ring tail monkeys (Melendez *et al.*, 1970), or New Zealand rabbits (Melendez *et al.*, 1972b) are infected by these viruses, malignant lymphomas or lymphatic leukemias develop.

These virus systems offer an additional advantage in that productive systems are available which permit one to produce larger amounts of virus and viral DNA. The DNA of *Herpesvirus saimiri* is genetically heterogeneous as revealed by equilibrium centrifugation in CsCl density gradients (Fleckenstein and Wolf, 1974). Electron microscopic examination and partial denaturation of the DNA revealed G+C-rich termini of the DNA molecule (Fleckenstein *et al.*, 1975). The M-genome molecule (M-DNA) is composed of 70% of light sequences (L-DNA, 36% G+C) and 30% of heavy sequences (H-DNA, 71% G+C). The defective H genome consists exclusively of H-DNA which is located on the termini of the molecule (Bornkamm *et al.*, 1976). The anatomy of the genome has been further analyzed with restriction endonucleases (C. Mulder *et al.*, personal communication).

Lymphoblastoid cells and tumor cells induced by *Herpesvirus saimiri* can be shown to contain viral DNA (Fleckenstein, personal communication) in multiple copies. The physical state of the DNA has not yet been investigated.

8.3. Herpes Simplex Virus Type 2

A role of herpes simplex virus type 2 in human cancers, particularly in human cervical carcinoma, was suggested on the basis of epidemiological studies as early as 1842, by Rigoni-Stern (cited in zur Hausen, 1975), and the issue is still far from being settled (Naib *et al.*,

1966; Rawls *et al.*, 1969). Experimental tumors have been produced by human herpes simplex virus in hamsters and in mice (Nahmias *et al.*, 1970, 1971). Duff and Rapp (1971*a,b*) demonstrated that hamster cells could be transformed *in vitro* by UV-inactivated herpesvirus type 2 and the transformed cells did not contain leukosis virus markers (Rapp *et al.*, 1972). Transformation of rodent cells by UV-inactivated herpes simplex virus types 1 and 2 was shown by Garfinkle and McAuslan (1974), whereas Darai and Munk (1973) reported that human embryonic lung cells exhibited properties of transformed cells after infection with UV-inactivated herpesvirus type 2. Frenkel *et al.* (1972) reported that in a single case of a human cervical tumor, which did not contain infectious herpes simplex virus type 2, the tumor cells contained 39% of the viral DNA covalently linked to highly repetitive sequences of host DNA and this DNA was transcribed into RNA. This finding has so far not been repeated in other laboratories. It has been reported that approximately 50 cervical carcinomas have been tested by DNA-cRNA hybridization and by reassociation kinetics for the presence of herpesvirus simplex type 2 DNA. None was found (zur Hausen, 1975). The technique of using cRNA or nick-translated DNA to probe for viral sequences in tumor cells is potentially limited by the lack of complete transcription or replication of the large viral DNA molecule. To my knowledge, this problem has not yet been thoroughly investigated for herpesvirus DNA.

Frenkel *et al.* (1975) have examined five hamster embryo cell lines transformed with UV-irradiated herpes simplex virus type 2 for the presence of viral DNA sequences. All five lines carried viral sequences with variations in different lines from 5% to 38% of the DNA to approximately three copies of viral DNA per cell. The physical state of the herpes simplex virus type 2 DNA in these cells is not known and is currently being investigated with the aid of restriction enzymes.

8.4. Marek's Disease

Before concluding this part of the chapter, another tumor disease caused by a herpesvirus has to be mentioned, Marek's disease (Marek, 1907; Biggs *et al.*, 1968; Biggs, 1973). Marek's disease is a lymphoproliferative illness in chicken which is caused by a herpesvirus (Calnek *et al.*, 1970; Nazerian and Witter, 1970). This disease can be prevented by inoculation of chickens with the related, non-oncogenic herpesvirus of turkey (Purchase *et al.*, 1971). Nazerian *et al.* (1973) established by DNA-RNA hybridization experiments using RNA complementary to

Marek's disease virus DNA that in cells isolated from five different Marek's disease tumors (derived from ovary, liver, and testis) 3–15 viral genome equivalents per cell were present. The state of these viral genomes in the cells has not yet been determined.

In summary, one can conclude that in herpesvirus-transformed cells and in cells from tumors in the etiology of which herpesviruses have been shown to play a role, viral genetic material does persist. The best-investigated system is that of the Epstein-Barr virus. The DNA of this virus appears to be present in an integrated form and as a free supercoiled circular plasmid. To determine the role, if any, of herpesviruses in the causation of human malignancies, much further work will be required.

Herpesvirus infections in humans and in experimental animals are very interesting from another point of view. These infections can lead to a latent state in the organism (see Rapp and Jerkofsky, 1973, for review). There is recent experimental evidence using the *in situ* hybridization technique to demonstrate viral DNA sequences in the nuclei of neuronal cells of trigeminal ganglia isolated from latently herpesvirus-infected rabbits and guinea pigs. In freshly infected animals, viral DNA was not observed in neurons (Schulte-Holthausen, 1977). This system should lend itself to a study of the molecular aspects of viral latency. Stevens (1975) has published a comprehensive review of herpesvirus latency.

9. INTEGRATION OF THE GENOME OF RNA TUMOR VIRUSES

Within the scope of this chapter, it is impossible to even attempt to summarize the entire RNA tumor virus field. Many extensive reviews on RNA tumor viruses have been published (Green, 1970*a*; Vigier, 1970; Temin, 1971; Temin and Baltimore, 1972; Gallo, 1972; Hill and Hillova, 1974; Wyke, 1975; see also Volumes 4 and 9 of *Comprehensive Virology* and Chapter 5 in this volume). Therefore, this chapter will be limited to a survey of the evidence for integration of the RNA tumor virus genomes.

9.1. The Provirus Hypothesis

In 1964, Temin proposed the provirus hypothesis to explain his findings on the replication of the RNA tumor virus genome. In its simplest form, the provirus hypothesis states that after infection the genome of RNA tumor viruses is transcribed into a DNA copy, that this DNA is integrated into the host chromosome, and that the progeny RNA is synthesized on the DNA template. This at the time daring hypothesis was based on the observation that the replication of and transformation by Rous sarcoma virus (RSV) could be blocked by actinomycin D (Temin, 1963) and by inhibitors of DNA replication (Temin, 1964). Further evidence for the involvement of DNA in the replication of the RSV genome was published by Bader (1964), Bader and Bader (1970), Balduzzi and Morgan (1970), and Boettiger and Temin (1970). The latter authors demonstrated that transformation of chicken embryo fibroblasts by RSV was blocked 50–90% when RSV-infected cells were maintained for 18–24 h in medium containing 5-bromodeoxyuridine and were then exposed to visible light.

Temin's provirus hypothesis met skepticism for many years, mainly because it postulated a direct violation of the central dogma of molecular biology. This dilemma was resolved when Temin and Mizutani (1970), and independently Baltimore (1970), discovered the enzyme reverse transcriptase in Rous sarcoma and Rauscher mouse leukemia virions, respectively. The reverse transcriptase was able to catalyze the synthesis of DNA on a natural or synthetic RNA template and required all four deoxyribonucleoside triphosphates. The properties of the reverse transcriptase and the details of the reaction have been extensively studied, and have been reviewed by Temin and Baltimore (1972). Shortly after the discovery of the reverse transcriptase, it became clear that several enzymes were associated with RNA tumor viruses (Mizutani et al., 1970) and were apparently involved in the replication of the RNA tumor virus genome. It was also recognized that a DNA-RNA hybrid was formed as an intermediate in the reverse transcription process (Rokutanda et al., 1970; Spiegelman et al., 1970). In the meantime, a huge amount of information has accumulated on reverse transcriptase and its role in RNA tumor virus-infected and RNA tumor virus-transformed cells.

9.2. Virus-Specific DNA in RSV-Transformed Cells

Physical evidence for the occurrence of Rous sarcoma virus-specific DNA in RSV-transformed cells came from the work of Varmus et al. (1972, 1973a). Using the reassociation technique, these authors could demonstrate that Rous sarcoma virus-specific DNA was present in RSV-transformed rat and mouse cells but not in normal cells. Two populations of double-stranded DNA were identified, corresponding to approximately 5% and 30% of the 70 S RSV genome. The RSV-specific DNA was detected in the nonrepetitive fraction of the DNA from mouse and rat cells transformed by B77 avian sarcoma virus or the Schmidt-Ruppin strain of RSV. The results bridged the gap at that time between the abundance of information on the mechanism of the reverse transcriptase reaction and Temin's provirus hypothesis (Temin, 1964) postulating a DNA intermediate in cells infected and/or transformed by RSV.

9.3. Integrated RSV DNA

Britten and Kohne (1968) had demonstrated that when unsheared cellular DNA was incubated to C_0t values which allowed the reassociation of repetitive but not unique sequences, "networks" of DNA were formed which precipitated and could be separated from the remainder of the DNA by low speed centrifugation. Using this technique, Varmus et al. (1973b) were able to show that RSV-transformed permissive (duck) and nonpermissive (mammalian) cells contained RSV-specific DNA covalently integrated. In XC cells, derived from a tumor induced in a rat by the Prague strain of RSV, viral DNA was found to be integrated into cell DNA containing repeated sequences. The apparent discrepancy with respect to the results described in Section 9.2 on RSV DNA integrated into nonrepetitive sequences may be due to differences in the strains of transformed cells or the transforming virus.

In mouse 3T3 cells, 0.8 copy of RSV-specific DNA per cell was found in the integrated state 12 h postinfection; in duck cells, four to six integrated copies of RSV-specific DNA per cell were detected between 6 and 24 h postinfection. The time course of appearance of integrated RSV-specific DNA is very similar to that reported in cells infected with adenovirus type 12 (Doerfler, 1968) and SV40 (Hirai and Defendi, 1971). Varmus et al. (1976) also used the technique of alkali lysis of cells and zone velocity sedimentation of cellular DNA in alkaline sucrose gradients to substantiate their findings on integrated viral DNA sequences.

In another report, Varmus et al. (1974) described the sequence of steps early after infection of permissive cells by Rous sarcoma virus. Within the first 3 h postinfection, double-stranded, RSV-specific DNA of molecular weight 6×10^6 was synthesized in the cytoplasm of infected cells by the RSV-associated DNA polymerase. The RSV-specific DNA was presumably converted into a supercoiled, covalently closed circular molecule. This conclusion was still tentative, as it rested solely on the results of equilibrium sedimentation experiments in CsCl–ethidium bromide density gradients and of velocity sedimentation in alkaline sucrose gradients. Integration into the nuclear host DNA began early after infection. Integration of the RSV-specific DNA could be inhibited to 83% by ethidium bromide (1.0 μg/ml). The effect of ethidium bromide on cell transformation by RSV was not yet clear; preliminary results suggested that transformation was suppressed by ethidium bromide.

Bishop *et al.* (1975) proposed the following scheme for the replication of RNA tumor viruses:

$$
\begin{array}{c}
\text{Viral RNA} \\
\text{DNA-RNA hybrid*} \\
\text{Open circular duplex DNA*} \\
\text{Closed circular duplex DNA} \\
\text{Recombination with chromosomal DNA} \\
\text{Transcription} \\
\swarrow \qquad \searrow \\
\text{Genome RNA} \qquad \text{Viral messenger}
\end{array}
$$

The structures identified by asterisks have not yet been found experimentally.

The reverse transcriptase is an enzyme which is encoded in the viral genome, since deletions or temperature-sensitive mutants (Linial and Mason, 1973) in the polymerase function are unable to replicate or transform at the nonpermissive temperature (Hanafusa and Hanafusa, 1971). The synthesis of virus-specific DNA on an RNA template by the RNA-dependent DNA polymerase is initiated by a tRNA molecule, probably by cellular Trp-RNA. The viral DNA is first detected in the cytoplasm. Among the viral DNA molecules, supercoiled circular structures are found with a molecular weight of 6×10^6. The viral RNA is synthesized in the nucleus by a host polymerase, and it is the integrated form of viral DNA which presumably serves as template. The major species of virus-specific RNA are 10–15% longer than the 35 S viral genome; thus cellular sequences are possibly cotranscribed.

Evidence for the covalent linkage of avian myeloblastosis virus (AMV) specific DNA to the DNA of chicken embryo fibroblasts and to the DNA of AMV-transformed cells was presented by Markham and Baluda (1973). Free AMV-specific DNA was not found in these cells.

An elegant way of demonstrating the association of RSV-specific DNA with the DNA of RSV-transformed cells was worked out by Hill and Hillova (1972, reviewed in 1974) and by Hill *et al.* (1974). These authors isolated the cellular DNA from XC rat cells transformed by the Prague strain of RSV, purified the DNA after alkali denaturation and neutralization by equilibrium centrifugation in neutral CsCl density gradients, and demonstrated that this DNA, in both the native and denatured forms, had the capacity to transform chicken embryo fibroblasts in culture. These results provided further support for the replication scheme of RNA tumor viruses via a DNA intermediate and demonstrated that the entire genetic equivalent of RSV required for the replication and transformation events was associated with the genome of RSV-transformed mammalian cells.

Certain strains of avian RNA tumor viruses can transform fibroblasts (Temin, 1971). Only specific viral genes are responsible for the transformation event (Martin, 1970; Kawai and Hanafusa, 1971; Bader, 1972). Deletion mutants of RNA tumor viruses exist which are unable to transform cells in culture and cannot induce sarcomas in animals. The RNA of these deletion mutants lacks 10–20% of the genetic information (Duesberg and Vogt, 1973; Lai *et al.*, 1973; Neiman *et al.*, 1974). Stehelin *et al.* (1976a) isolated DNA specific for nucleotide sequences missing in the deletion mutants by reverse transcription and exhaustive hybridization with RNA from the deletion mutant and the wild-type sarcoma-inducing strain. This DNA has been designated "$cDNA_{sarc}$," since it must contain the genetic information required for transformation and sarcoma induction. The $cDNA_{sarc}$ offers a probe to search for the transforming nucleotide sequence in normal and transformed cells.

The DNA of uninfected cells from several species (chicken, quail, duck, turkey, emu) contains sequences homologous to DNA_{sarc} from avian sarcoma virus. These sequences appear to be transcribed even in uninfected cells and in quail cells transformed by methylcholanthrene (Stehelin *et al.*, 1976b). The quail cells transformed by methylcholanthrene do not contain DNA or RNA complementary to other parts of the avian sarcoma genome.

Stehelin *et al.* (1976a) analyzed the degree of genetic relatedness among the DNA_{sarc} genes in different avian species by the thermal denaturation of duplexes between cellular DNA and $cDNA_{sarc}$. A divergence in the $cDNA_{sarc}$ sequences was found corresponding to the phylogenetic relatedness of the avian species.

Hamster cells transformed by avian sarcoma virus do not produce virus. Certain clones of the avian sarcoma virus-transformed hamster cells revert to phenotypically normal cells. Like the transformed cells, the "revertants" contain both integrated sarcoma-specific DNA and virus-specific mRNA. Hence neither integration nor transcription of the *src* genes alone can be sufficient for phenotypic transformation of the cell (Bishop *et al.*, 1975).

10. ENDOGENOUS VIRAL DNA

Several lines of evidence support the conclusion that probably all avian and murine cell lines contain the genome of RNA tumor viruses either in part or *in toto*. In many instances, this genetic information can be activated by various physical and chemical agents to produce infectious virions. It is likely, although not proven in all cases, that the virus-specific information is transmitted vertically from generation to generation in the form of the provirus, i.e., virus-specific DNA integrated into the host genome. The postulate of vertically transmitted virus-specific DNA is part of the so-called oncogene hypothesis proposed by Huebner and Todaro (1969). The evidence for endogenous viruses and their activation was extensively reviewed by Rowe *et al.*, 1972*a,b*), by Rowe (1973), by Lilly and Pincus (1973), and by Hanafusa *et al.* (1974).

It had been observed by several investigators that supposedly virus-negative strains of mice or cells derived from these animals contained virus particles of the C type (Bernhard, 1960). Furthermore, it became apparent that a large number of cell lines established from presumably normal animals contained the group-specific (*gs*) antigen (Huebner *et al.*, 1964*a*) or virion antigens (Geering *et al.*, 1966; Stockert *et al.*, 1971) of known RNA tumor viruses. The widespread presence of the *gs* antigen in normal murine tissues could also be demonstrated by the very sensitive radioimmune assay (Parks *et al.*, 1973).

10.1. Induction of RNA Tumor Viruses in "Normal" Cells

When "normal" BALB/c mouse embryo cells were maintained in culture by frequent passage at high cell density (3T12 line), some of the established cell lines started to release murine leukemia viruses (Aaronson *et al.*, 1969). This observation, among others, led to the hypothesis that the virus-specific genetic information was transmitted vertically as part of the host genome (Huebner and Todaro, 1969). Very similar results were reported for noninfectious AKR mouse embryo cell lines after repeated transfers in culture (Rowe *et al.*, 1971). The detection of virus induction became possible after Rowe *et al.* (1970) developed a quantitative assay for murine leukemia viruses. The induction of virus replication could be enhanced in several clonally derived sublines of AKR cells by irradiation with X-rays or ultraviolet light, or by transformation with SV40. Virus could also be induced by X-irradiation in BALB/3T3 cells (Pollack *et al.*, 1970). Rowe *et al.* (1971) concluded from a quantitative study that probably all the cells in the AKR lines contained the entire viral genome. At the same time, Weiss *et al.* (1971) were able to demonstrate that avian tumor viruses could be induced in normal, *gs* antigen-negative chicken cells after exposure to ionizing radiation, chemical carcinogens, or mutagens. These authors established the identity of the induced virions with the known avian tumor viruses by comparing several physical, chemical, and genetic parameters. Todaro (1972) found a striking correlation between the transformed state of "spontaneously" occurring transformed clones of BALB/c 3T6 and 3T12 cells and the release of high titers of type C virions and concluded that the control of expression of the endogenous type C viruses was affected by the transformed state.

More quantitative investigations became possible after the discovery that treatment of AKR and BALB/c cells with 5-iododeoxyuridine (5-IUdR) or 5-bromodeoxyuridine (5-BUdR) induced murine leukemia viruses in 0.1–0.5% of the cells (Lowy *et al.*, 1971; Aaronson *et al.*, 1971). This induction occurred as early as 3 days after the addition of 5-IUdR or 5-BUdR. The halogenated pyrimidines had to be incorporated into DNA in order to exert the virus-inducing effect (Teich *et al.*, 1973).

10.2. Virus-Specific DNA and RNA in "Uninfected" Cells

With the availability of *in vitro* synthesized, radioactively labeled DNA copies of the RNA tumor virus genomes, it became possible to screen the DNA from "normal" (i.e., nonproducing) avian and murine cell lines for the presence of virus-specific nucleotide sequences. Using a radioactively labeled viral DNA probe, it was determined with the aid of reassociation techniques that chicken cells which did not synthesize RSV-specific *gs* antigen or the chickhelper factor (Hanafusa *et al.*, 1970) contained up to 16 viral gene copies per diploid cell (Varmus *et al.*, 1972). A similar number of viral gene equivalents was measured in *gs* antigen-positive cells which were also RAV-I (Rous-associated virus) positive. Similar results were reported by Baluda (1972) and by Neiman (1973), who found with the DNA-RNA hybridization technique that normal chicken embryos and adult chickens contained 1.7–4.6 viral genome equivalents per cell. Leukemia cells and cells infected with RSV or AMV carried 4 and 13 viral DNA equivalents, respectively.

The number of murine leukemia virus DNA equivalents in "normal" AKR, C58, and NIH/3T3 cells was measured by reassociation experiments using ^{3}H-labeled Kirsten murine leukemia virus DNA (Gelb *et al.*, 1973) and was found to range from 9.5 to 13.8 equivalents per cell. The number of viral DNA copies could not be correlated with differences in leukemia incidence among different mouse strains. Gelb *et al.* (1973) concluded that the murine leukemia virus-specific DNA was probably covalently linked to cellular DNA, since the viral DNA cosedimented with high molecular weight chicken cell DNA. This conclusion was premature, as the cellular DNA had been isolated from neutral sucrose gradients.

In addition to RNA tumor virus-specific DNA, viral RNA was also found in uninfected chicken (Hayward and Hanafusa, 1973) and mouse (BALB/c 3T3) cell lines (Benveniste *et al.*, 1973). Cells producing the avian tumor virus group-specific antigen and the virus-related helper factor carried 3–40 copies of viral RNA, whereas RSV-infected cells contained 3000–4000 viral RNA molecules per cell (Hayward and Hanafusa, 1973). Similarly, a low level of virus-related RNA was present in presumably normal BALB/c 3T3 cells and the amount of RNA was markedly increased in murine leukemia virus-transformed cells (Benveniste *et al.*, 1973).

More detailed studies, in which the kinetics of reassociation of a virus-specific, ^{3}H-labeled DNA probe were analyzed in the presence of DNA from "normal" AKR mouse cells and from "normal" NIH Swiss mouse cells, revealed that the AKR cells contained two sets of viral DNA sequences, one represented ten times and the other four times per haploid AKR cell genome. The NIH Swiss cells carried 15 copies of only one set of viral DNA sequences (Chattopadhyay *et al.*, 1974).

In "normal" cells, the integrated form of the endogenous avian myeloblastosis virus DNA was associated with cell sequences which were reiterated 1200 times (Evans *et al.*, 1974). Each DNA equivalent corresponded in size to the 35 S RNA subunit of the virion. In AMV-infected cells, however, additional viral sequences might be integrated adjacent to unique cellular sequences, in tandem with endogenous viral

DNA. Chattopadhyay and colleagues (1976) presented evidence that the presence of viral sequences in mouse and rat cells influences the efficiency of integration of exogenous viral DNA.

10.3. Genetic Analysis

Several laboratories have been using classical Mendelian genetics to investigate host control factors which influence the susceptibility to infection with murine leukemia viruses (for review, see Rowe, 1973). The 23 strains of naturally occurring murine leukemia viruses which were tested in this respect fell into two classes: They were able to infect mouse embryo cells derived either from NIH Swiss mice (N-tropic viruses) or from BALB/c mice (B-tropic viruses). When cells from the F_1 hybrid mouse generation from an N-type $\times$ B-type cross were tested, they were resistant to both N-type and B-type viruses. Backcross studies suggested that a single genetic locus was responsible for the resistance toward virus infection (Pincus *et al.,* 1971*a*). The results of further experiments indicated that this N-B locus was identical to the *Fv-1* locus (Lilly, 1970) which had been found to determine sensitivity of cells to infection with Friend leukemia virus. Sensitivity of certain mouse strains to infection with Friend leukemia virus is also determined by a second genetic locus, *Fv-2* (Lilly, 1970), which does not seem to be effective in tissue culture (Pincus *et al.,* 1971*b*). All mouse strains found to be sensitive at the *Fv-1* locus were susceptible to N-type virus.

The analysis of genetic crosses between different mouse strains for the expression of infectious murine leukemia virus became feasible after a reliable plaque assay had been developed (Rowe *et al.,* 1970), and after the discovery that virus was present in rather large amounts in the tail tissue of mice starting at 2–6 weeks of age (Rowe and Pincus, 1972). These studies led to the conclusion that AKR mice had two unlinked, autosomal loci either of which sufficed to induce infectious virus when crossed into $Fv-1^n$ strains which did not produce virus. One of these loci (V_1) was located on linkage group I, 25–30 map units from the locus for albino (Rowe, 1972). The *Fv-1* locus suppressed the expression of infectious virus. This suppression could be overcome by induction with 5-iododeoxyuridine. The host range of the virus acquired by the hybrid generation from crosses with the AKR strain was always of the AKR type. These results were interpreted as evidence that the virus-inducing loci contained the murine leukemia virus genome (Rowe and Hartley, 1972). The virus-inducing locus, designated *Akv-1*, which presumably carries the integrated viral DNA, was located on the genetic map of AKR mice. This locus was found to map on linkage group I, about 12 map units from *Gpi-1*, the locus for the isozymes of glucosephosphate isomerase (Rowe *et al.*, 1972*b*). Later studies seemed to suggest that there were at least four different sites for virus induction in the mouse genome (Rowe, 1973; Lilly and Pincus, 1973).

Chattopadhyay *et al.* (1975) could demonstrate by reassociation kinetics that the DNA of the AKR mouse contained sequences of murine leukemia viruses which were absent in the DNA of NIH Swiss mice. On the other hand, NIH mice which contained the AKR locus for murine leukemia virus induction, *Akv-1*, did carry the same sequences. In three-point crosses in which the *Akv-1* locus was segregated, the murine leukemia virus sequences were also lost. Chattopadhyay and collaborators concluded that the *Akv-1* locus contained viral sequences (for further discussion of RNA tumor virus genetics, see Vogt, Vol. 9, this series).

11. OUTLOOK

11.1. Integration

In many virus systems, the interaction between the viral and host genomes results in the covalent integration of the entire virus genome or of fragments of viral DNA. This phenomenon has been well documented genetically and biochemically for several phage systems, most notably for λ, μ, P22, and P2. There is mounting evidence from several different experimental approaches that integration of viral DNA is also occurring with animal viruses. Adenovirus, SV40, polyoma virus, and herpesvirus—i.e., the oncogenic DNA viruses—have been studied intensely in this respect. Herpesvirus DNA seems to be similar to a plasmid in that it can probably exist in both the integrated and the free supercoiled circular state, although additional evidence is needed in this system. Equally well analyzed are the oncogenic RNA viruses. In this case, the reverse transcripts of the viral genomes are circularized and subsequently inserted into the host DNA. There is evidence that these steps are essential for viral replication.

Integration has been shown for apparently nononcogenic viruses as well: Visna virus replicates via a DNA intermediate; thus occurrence of an integration event would not be surprising, and, in fact, there is evidence for integration of Visna DNA. On the other hand, even RNA viruses previously not known to replicate through a DNA intermediate, such as respiratory syncytial virus (Simpson and Iinuma, 1975) and measles virus (Zhdanov, 1975), are now thought to be able to integrate. The consequences of these findings for virus latency and slow virus disease remain to be investigated. First results in this respect are promising (Hall and ter Meulen, 1975).

But not only has viral genetic information been shown to be integrated in experimental systems or in tumor cells, but also complete copies of C-type virus DNA (endogenous virus) have been shown to be present in the DNA of somatic cells of the chicken, Chinese hamster, Syrian hamster, mouse, rat, pig, and baboon (Todaro *et al.,* 1974). These genomes are presumably transmitted vertically from one generation of animals to the next. On the basis of comparative studies on the presence of endogenous viral DNA in different species and the phylogenetic relationships between these species, it has been suggested that these endogenous genomes may also have been transmitted horizontally from one species to another at some time during evolution (Todaro *et al.,* 1974). Endogenous viral genes, and perhaps all viral genes, therefore can be assigned a dualistic nature, since it will prove difficult in some cases to designate a gene as viral or as belonging to the host.

11.2. Need for More Refined Technology

The evidence for integration of viral genomes into host DNA adduced from work with different experimental systems has been presented. It is obvious that in several cases rigorous proof for integration is still lacking and that there is a definite need for improved techniques. More refined methods will be especially important for the analysis of the cellular sequences adjacent to the integrated viral genes. In addition to approaches which proved useful such as sequential hybridization (Lavi and Winocour, 1972), equilibrium centrifugation in alkaline CsCl density gradients (Doerfler, 1968; 1970), and application of restriction endonucleases (Botchan and McKenna, 1973; Schick *et al.,* 1976; Ketner and Kelly, 1976; Botchan *et al.,* 1976; Groneberg *et al.,* 1977), nucleic acid labeling techniques can be exploited: "nicked translation" (Kelly *et al.,* 1969; Rigby and Berg, personal communication) or labeling with ^{125}I (Anderson and Folk, 1976) or Hg^{2+} (Dale and Ward, 1975).

11.3.   Site of Integration

One of the most interesting questions to pursue is that concerning the site or sites of integration of viral DNA into the eukaryotic chromosome. This problem can be approached at the level of individual chromosomes or of specific nucleotide sequences. Information on the site of integration may be essential to understanding the interrelationship of integration and transformation.

Only in a few instances can the site of integration of viral genes be localized on a specific chromosome of eukaryotic cells: SV40 DNA can become associated with human chromosome C-7 (Croce *et al.,* 1973). In human cells infected with adenovirus type 12, a specific break on chromosome No. 17 close to the thymidine kinase locus has been reported (McDougall *et al.,* 1973). However, there is no evidence to suggest that the site of this break is identical with the site of integration.

With the advanced state of bacterial genetics, detailed information is available on the site(s) of integration of the DNA of bacteriophage λ or μ. In eukaryotic cells, a comparable level of genetic sophistication has not yet been accomplished. However, recent developments in chromosome banding techniques (Caspersson *et al.,* 1970) and chromosomal analysis (Ruddle *et al.,* 1971; Boone *et al.,* 1972; Ruddle, 1973) point out new approaches for investigations on the sites of interaction with viral genomes. Using classical genetic techniques, progress in the localization of the integrated viral genome on the host chromosome has been made for murine leukemia virus DNA (Rowe, 1973; Lilly and Pincus, 1973).

Viral genomes integrated into host DNA have been detected in unique and in reiterated sequences. The question arises of whether the viral genome can be represented singly or in tandem at multiple sites, e.g., in reiterated sequences.

11.4.   Pattern of Integration

Viral DNA can become integrated as an intact molecule in single or multiple copies or in fragments in single or multiple copies. In the preceding discussion, various examples of these possibilities have been described. The factors which determine the establishment of a certain pattern of integration are unknown; however, it is reasonable to propose that the specific virus–host cell combination may play a decisive role. Other parameters—the phase of the cell cycle in which cells are infected, experimental conditions, etc.—may also be important. Lastly, when transformed cells are analyzed, it must be considered that transformed clones have usually been selected under stringent experimental conditions (e.g., for the ability to grow in agar). In this way, highly specific patterns of integration may be selected which are not necessarily representative of a wider gamut of possibilities in individual virus–host cell pairs.

In those cases where integration of a large number of viral DNA copies—intact genomes or fragments—has been found, it will still have to be investigated whether integration occurs in tandem or in a more complicated arrangement, e.g., according to the onion skin model (Smithies, 1973).

In mouse 3T3 cells, there is evidence that the number of stable SV40 integration sites is limited (Prasad *et al.,* 1975). Mouse 3T3 cells were transformed at high multiplicity by a temperature-sensitive (*ts*) mutant of SV40. Subsequently, the transformed cells were superinfected with wild-type SV40 at high multiplicity. Isolated clones from these colonies were fused with permissive monkey cells. The rescued virus was exclusively of the *ts* type. On the other hand, when 3T3 cells were transformed by the *ts* mutant at low multiplicity and superinfected with wild-type SV40 at high multiplicity, with the transformed cells then being fused again with permissive cells, most of the clones released both wild-type and *ts* SV40 virions (Prasad *et al.,* 1975).

11.5. Integration—A General Phenomenon

There is evidence for integration of viral genetic information in many different systems, not only in cells transformed by DNA and RNA tumor viruses but also in cells abortively and productively infected by these viruses. The question of whether integration is required for replication of the virus cannot be decided at the moment. The observation that DNA virus genomes of measles or respiratory syncytial virus, i.e., of RNA viruses usually not replicating via a DNA intermediate, exist and presumably can also become inserted lends support to the idea that integration of viral genomes is a general phenomenon, although these latter examples may reflect very special circumstances.

11.6. Consequences for the Host

The consequences of integration of viral DNA into the host chromosome have hardly been studied in eukaryotic systems, although the intriguing model of a mutator phage in *E. coli* (Taylor, 1963) has been known for many years. The conceptual link between the integration event and the transformed state of the cell is missing. At this point, one can only speculate that the site of integration may be decisive for triggering the events leading to transformation. It is also conceivable that viral DNA has to be integrated in a specific combination of sites. In any event, the number of sites at which viral DNA can integrate appears to be limited. Apart from the phenomenon of integration, it will be interesting to investigate whether integration of viral DNA enhances the frequency of mutation in eukaryotic cells.

Endogenous viral genomes are present in all cells of many species and are transmitted vertically. Again, it is left to conjecture what the presence of the viral genes means to the host, e.g., during development or evolution.

11.7. The Mechanism of Integration

Even in such well-analyzed bacteriophage systems as λ and μ, the enzymatic mechanism of integration is only partly understood, and *in vitro* systems to study this mechanism have only recently become available (Syvanen, 1974; Nash, 1975*a,b*; Gottesman and Gottesman, 1975*a,b*). At least part of the enzymatic apparatus required for integration of λ DNA is encoded in the viral genome. The molecular basis of the highly specific recognition between unique sites on the viral and host genomes in the case of λ is also unknown.

The mechanism of integration of animal virus genomes has not yet been investigated. The circular genomes of the papovaviruses would fulfill one of the requirements of the Campbell (1962) model. The adenovirus genome appears to exist in a protein-stabilized circular form in the virion (Robinson *et al.*, 1973), and intracellular circular forms of adenovirus DNA have been observed. In Epstein-Barr virus-transformed cells, circular forms of the viral DNA can be detected (Lindahl *et al.*, 1975), and circular viral DNA is also present in cells at early times after infection with Rous sarcoma virus (Varmus *et al.*, 1974). Thus, in many systems where integration of viral DNA has been described, circular viral DNA, covalently linked or linker stabilized, has been found.

In several instances, specific fragments of viral DNA were shown to persist in virus-transformed cells. Is it conceivable that viral DNA is fragmented after infection, that fragments are circularized and subsequently integrated? Such a mechanism has been proposed in adenovirus infection (Doerfler, 1970; Burlingham and Doerfler, 1971; Doerfler *et al.*, 1974; Groneberg *et al.*, 1975; Fanning and Doerfler, 1976).

With regard to the recognition between specific sites on the host and viral genomes, it is tempting to speculate that certain palindromic and/or repetitive sequences known to abound in mammalian cell DNA (Pyeritz and Thomas, 1973; Wilson and Thomas, 1974) could function as recognition sites in the host chromosome for the integration event and that palindromic sequences which exist in the viral genomes might serve as the viral counterparts in the reciprocal recombination event.

In this context, it would be worth investigating whether any type of foreign DNA introduced into a eukaryotic cell can be bound covalently to the host DNA. This problem is also pertinent to chromosome transfer studies (McBride and Ozer, 1973) and in particular to the fate of the genetic information transmitted to another cell.

11.8. Repressorlike Mechanism in Virus-Transformed Eukaryotic Cells

The question of whether a virus-specific, "repressor"-like substance can be detected in virus-transformed cells remains unresolved. Virus-free extracts of SV40-transformed (Cassingena and Tournier, 1968; Cassingena *et al.*, 1969) and adenovirus type 12 transformed (Champe *et al.*, 1972; 1975) cells were shown to contain a factor which inhibited viral replication in permissive cells infected with SV40 and adenovirus type 12, respectively. Weber (1974), on the other hand, could not find a repressor in adenovirus-transformed cells. Similarly, extracts of Burkitt's lymphoma cells were shown to block the replication of *Herpesvirus hominis* in permissive cells (Rabson *et al.*, 1971). However, several laboratories have reported that SV40 was able to replicate in SV40-transformed monkey cells (Jensen and Koprowski, 1969; Barbanti-Brodano *et al.*, 1970; Rapp and Trulock, 1970; Sauer and Hahn, 1970). This last result is not consistent with the presence of a "repressor" in the classical sense, but neither does it conclusively rule out this possibility.

11.9. Integration and Transformed State

Several hypotheses have been proposed to explain how tumor viruses cause malignant transformation of cells (Huebner and Todaro, 1969; Levine and Burger, 1972; Comings, 1973).

There is no proof for a causal relationship between the integrated state of viral genetic material in the host DNA and the altered metabolic conditions of cells malignantly transformed by oncogenic viruses. For a mutational theory of cancer, it would be appealing to postulate that the integration of viral DNA could permanently alter host genes which have crucial regualtory functions (Mu-effect). Perhaps integration of viral DNA leads to a "reorganization" of cellular genes, to translocations or transpositions. With this model, several mechanisms are possible which relate integration to transformation:

1. Integration alters host regulatory genes directly, because of somatic mutation. Transformation would then be the direct consequence of the loss of essential functions or the acquisition of new functions.
2. Integration serves mainly to fix the viral genome in the cell, and thus allows the continued synthesis of virus-specific functions which elicit processes leading to the transformation of the cell.
3. Obviously, these two mechanisms are not mutually exclusive, but may both be operative.

A mutational theory of viral transformation would offer a unifying concept for malignant transformation of cells by chemical, physical, and viral agents. In this context, it is interesting to mention that a wide variety of carcinogens have been shown to be mutagens (see Kier *et al.,* 1974). And yet a mutational theory in its simplest form will not suffice to explain all the phenomena in viral transformation. It is clear that "normal" (i.e., nontransformed) cells of many, perhaps all species carry viral genetic material in an integrated form. In some cases, the viral genes are transcribed and translated.

Even in cells transformed by carcinogenic hydrocarbons, e.g., methylcholanthrene, virus genetic information is expressed. This has been demonstrated for the *src* gene in chick embryo fibroblasts (Stehelin *et al.,* 1976). Furthermore, it has been shown in many instances that viral genetic material can be induced chemically and that the cells will be transformed only after induction of viral genes. Perhaps such endogenous viral genomes play a role also in cells latently infected with viruses and in certain cases of genetic disease.

In this context, one should remember that in prokaryotes strong polar mutations can be caused by the insertion of host DNA sequences (800 and 1400 nucleotide pairs long) into the control region of certain operons (for review, see Starlinger and Saedler, 1972; 1976). These authors have pointed out that such transmissible genetic elements could play an important role in many different systems. In a similar fashion, integration of viral genomes might affect control regions in the eukaryotic genome.

The possibility has to be considered that in the course of repeated exits and entrances of viral DNA, host genetic material may be transferred from one cell to another, as has been demonstrated for transducing bacteriophage. It is not known whether such a mechanism is operative in eukaryotes, although substituted genomes of SV40 and adenovirus contain cellular sequences (Lavi and Winocour, 1972; Tjia *et al.*, 1977), and thus it is quite likely that such transfers can occur.

In plants, the crown gall tumors have been known for many decades (see Braun, 1972, for review). Many different families of plants are affected by this disease. It was shown that these tumors are caused by a bacterium, *Agrobacterium tumefaciens*. A recent comparative study of oncogenic and non-oncogenic strains of *A. tumefaciens* revealed that from the oncogenic strains large plasmids could be isolated. In nonpathogenic strains, the plasmids were not found. The plasmids measured 54.1 μm in length, which corresponded to a molecular weight of 112×10^6 (Zaenen *et al.*, 1974). These authors postulate that the plasmids represent the "tumor-inducing" principle in the bacteria. So far, there is no evidence that the bacterial plasmid DNA can be integrated into the plant DNA. In any event, this system offers a very interesting model for the expression of prokaryotic genes in eukaryotic cells and for oncogenicity. The tumorigenic effects of *A. tumefaciens* have been reviewed by Drlica and Kado (1975) and by Lippincott and Lippincott (1975).

For the viral genome, insertion into the host chromosome may offer a number of advantages. For viral transcription, particularly of early genes, it might be essential to link viral genes to cellular sites at which the cellular polymerase(s) can bind. Even if this linkage to cellular genes were not an essential step, it might render viral transcription much more efficient. Under conditions unfavorable to viral replication, integration might salvage the viral genome from degradation. Lastly, one has to ask whether the site(s) of viral insertion could have any relation to the origin of the viral genetic material in evolution.

On the cellular side, integration of viral genes could bestow upon the transformed cell properties which would make it fit to survive selective pressure exerted either by the specific culture conditions employed *in vitro* (e.g., agar technique) or by the immunological defense systems in an intact organism.

ACKNOWLEDGMENTS

I wish to thank all members of the Virology Section at the Institute of Genetics in Cologne for their contributions and discussions. I am very grateful to Ms. R. Koenig and Ms. R. I. Prüfer for typing the manuscript. Research performed in the author's laboratory was supported in different periods by the Deutsche Forschungsgemeinschaft (SFB74), by NATO Research Grant 639, and by the American Cancer Society (Grants No. E565 and VC-14A).

12. REFERENCES

Aaronson, S. A., Hartley, J. W., and Todaro, G. J., 1969, Mouse leukemia virus: "Spontaneous" release by mouse embryo cells after long term *in vitro* cultivation, *Proc. Natl. Acad. Sci. USA* **64**:87.

Aaronson, S. A., Todaro, G. J., and Scolnick, E. M., 1971, Induction of murine C-type viruses from clonal lines of virus-free BALB/3T3 cells, *Science* **174**:157.

Abelson, J., Boram, W., Bukhari, A. I., Faelen, M., Howe, M., Metlay, M., Taylor, A. L., Toussaint, A., van de Putte, P., Westmaas, G. C., and Wijffelman, C. A., 1973, Summary of the genetic mapping of prophage mu, *Virology* **54**:90.

Abrahams, P. J., Mulder, C., van de Voorde, A., Warnaar, S. O., and van der Eb, A. J., 1975, Transformation of primary rat kidney cells by fragments of simian virus 40 DNA, *J. Virol.* **16**:818.

Acheson, N. H., Buetti, E., Scherrer, K., and Weil, R., 1971, Transcription of the polyoma virus genome: Synthesis and cleavage of giant late polyoma-specific RNA, *Proc. Natl. Acad. Sci. USA* **68**:2231.

Adams, A., and Lindahl, T., 1975, Epstein-Barr virus genomes with properties of circular DNA molecules in carrier cells, *Proc. Natl. Acad. Sci. USA* **72**:1477.

Adams, A., Lindahl, T., and Klein, G., 1973, Linear association between cellular DNA and Epstein-Barr virus DNA in a human lymphoblastoid cell line, *Proc. Natl. Acad. Sci. USA* **70**:2888.

Aloni, Y., Winocour, E., and Sachs, L., 1968, Characterization of the simian virus 40-specific RNA in virus-yielding and transformed cells, *J. Mol. Biol.* **31**:415.

Aloni, Y., Winocour, E., Sachs, L., and Torten, J., 1969, Hybridization between SV40 DNA and cellular DNA's, *J. Mol. Biol.* **44**:333.

Anderson, D. M., and Folk, R. W., 1976, Iodination of DNA. Studies of the reaction and iodination of papovavirus DNA, *Biochemistry* **15**:1022.

Arber, W., 1958, Transduction des charactères gal par le bacteriophage lambda, *Arch. Sci.* **11**:261.

Atkins, J. F., Lewis, J. B., Anderson, C. W., Baum, P. R., and Gesteland, R. F., 1975, Mapping of adenovirus 2 genes by translation of RNA selected by hybridization, in: *In Vitro Transcription and Translation of Viral Genomes* (A. L. Haenni and G. Beaud, eds.), pp. 293–298, INSERM, Paris.

Ausubel, F., Voynow, P., Signer, E., and Mistry, J., 1971, Purification of proteins determined by two nonessential genes in lambda, in: *The Bacteriophage Lambda* (A. D. Hershey, ed.), pp. 395–405, Cold Spring Harbor Laboratory, Cold Spring Harbor, N.Y.

Babiuk, L. A., and Hudson, J. B., 1972, Integration of polyoma virus DNA into mammalian genomes, *Biochem. Biophys. Res. Commun.* **47**:111.

Bader, J. P., 1964, The role of deoxyribonucleic acid in the synthesis of Rous sarcoma virus, *Virology* **22**:462.

Bader, J. P., 1972, Temperature-dependent transformation of cells infected with a mutant of Bryan Rous sarcoma virus, *J. Virol.* **10**:267.

Bader, J. P., and Bader, A. V., 1970, Evidence for a DNA replicative genome for RNA-containing tumor viruses, *Proc. Natl. Acad. Sci. USA* **67**:843.

Balduzzi, P., and Morgan, H. R., 1970, Mechanism of oncogenic transformation by Rous sarcoma virus. I. Intracellular inactivation of cell-transforming ability of Rous sarcoma virus by 5-bromodeoxyuridine and light, *J. Virol.* **5**:470.

Baltimore, D., 1970, Viral RNA-dependent DNA polymerase, *Nature (London)* **226**:1209.

Baluda, M. A., 1972, Widespread presence, in chickens, of DNA complementary to the RNA genome of avian leukosis viruses, *Proc. Natl. Acad. Sci. USA* **69**:576.

Barbanti-Brodano, G., Swetley, P., and Koprowski, H., 1970, Superinfection of simian virus 40-transformed permissive cells with simian virus 40, *J. Virol.* **6**:644.

Baum, S. G., Reich, P. R., Hybner, C. J., Rowe, W. P., and Weissman, S. M., 1966, Biophysical evidence for linkage of adenovirus and SV40 DNA's in adenovirus 7–SV40 hybrid particles, *Proc. Natl. Acad. Sci. USA* **56**:1509.

Beard, P., Acheson, N. H., and Maxwell, I. H., 1976, Strand-specific transcription of polyoma virus DNA early in productive infection and in transformed cells, *J. Virol.* **17**:20.

Béládi, I., 1972, Adenoviruses, in: *Strains of Human Viruses* (M. Majer and S. A. Plotkin, eds.), pp. 1–19, Karger, Basel.

Bellett, A. J. D., 1975, Covalent integration of viral DNA into cell DNA in hamster cells transformed by an avian adenovirus, *Virology* **65**:427.

Benjamin, T. L., 1966, Virus-specific RNA in cells productively infected or transformed by polyoma virus, *J. Mol. Biol.* **16**:359.

Benjamin, T. L., 1968, Absence of homology between polyoma or SV40 viral DNA and mitochondrial DNA from virus-induced tumors, *Virology* **36**:685.

Benveniste, R. E., Todaro, G. J., Scolnick, E. M., and Parks, W. P., 1973, Partial transcription of murine type C viral genomes in BALB/c cell lines, *J. Virol.* **12**:711.

Bernhard, W., 1960, The detection and study of tumor viruses with the electron microscope, *Cancer Res.* **20**:712.

Bertani, L. E., and Bertani, G., 1971, Genetics of P2 and related phages, *Adv. Genet.* **16**:199.

Biggs, P. M., 1973, Marek's disease, in: *The Herpesviruses* (A. S. Kaplan, ed.), pp. 557–594, Academic Press, New York.

Biggs, P. M., Churchill, A. E., Rootes, D. G., and Chubb, R. C., 1968, The etiology of Marek's disease—An oncogenic herpes-type virus, *Perspect. Virol.* **6**:211.

Bishop, J. M., Deng, C. T., Moscovici, C., Stavnezer, E., Stehelin, D., Varmus, H. E., and Vogt, P., 1975, Avian RNA tumor viruses: Viral gene expression and cellular transformation. *Int. Virol.* **3**:32.

Bode, V. C., and Kaiser, A. D., 1965, Changes in the structure and activity of λ DNA in a superinfected immune bacterium, *J. Mol. Biol.* **14**:399.

Boettiger, D., and Temin, H. M., 1970, Light inactivation of focus formation by chicken embryo fibroblasts infected with avian sarcoma virus in the presence of 5-bromodeoxyuridine, *Nature (London)* **228**:622.

Boone, C., Chen, T.-R., and Ruddle, F. H., 1972, Assignment of three human genes to chromosomes (LDH-A to 11, TK to 17, and IDH to 20) and evidence for translocation between human and mouse chromosomes in somatic cell hybrids, *Proc. Natl. Acad. Sci. USA* **69**:510.

Boram, W., and Abelson, J., 1971, Bacteriophage mu integration: On the mechanism of mu-induced mutations, *J. Mol. Biol.* **62**:171.

Boram, W., and Abelson, J., 1973, Bacteriophage mu integration: On the orientation of the prophage, *Virology* **54**:102.

Bornkamm, G. W., Delius, H., Fleckenstein, B., Werner, F.-J., and Mulder, C., 1976, Structure of *Herpesvirus saimiri* genomes: Arrangement of heavy and light sequences in the M genome, *J. Virol.* **19**:154.

Botchan, M., and McKenna, G., 1973, Cleavage of integrated SV40 by RI restriction endonuclease, *Cold Spring Harbor Symp. Quant. Biol.* **38**:391.

Botchan, M., Ozanne, B., Sugden, B., Sharp, P. A., and Sambrook, J., 1974, Viral DNA in transformed cells. III. The amounts of different regions of the SV40 genome present in a line of transformed mouse cells, *Proc. Natl. Acad. Sci. USA* **71**:4183.

Botchan, M. Topp, W., and Sambrook, J., 1976, The arrangement of simian virus 40 sequences in the DNA of transformed cells, *Cell* **9**:269.

Boyd, V. A. L., and Butel, J. S., 1972, Demonstration of infectious deoxyribonucleic acid in transformed cells. I. Recovery of simian virus 40 from yielder and nonyielder transformed cells, *J. Virol.* **10**:399.

Boyer, G. S., Denny, F. W., Jr., and Ginsberg, H. S., 1959, Sequential cellular changes produced by types 5 and 7 adenoviruses in HeLa cells and in human amniotic cells: Cytological studies aided by fluorescein-labeled antibody, *J. Exp. Med.* **110**:827.

Braun, A. C., 1972, The relevance of plant tumor systems to an understanding of the basic cellular mechanism underlying tumorigenesis, *Progr. Exp. Tumor Res.* **15**:165.

Britten, R. J., and Kohne, D. E., 1968, Repeated sequences in DNA, *Science* **161**:529.

Brockman, W. W., and Nathans, D., 1974, The isolation of simian virus 40 variants with specifically altered genomes, *Proc. Natl. Acad. Sci. USA* **71**:942.

Brockman, W. W., Lee, T. N. H., and Nathans, D., 1973, The evolution of new species of viral DNA during serial passage of simian virus 40 at high multiplicity, *Virology* **54**:384.

Brooks, K., 1965, Studies in the physiological genetics of some suppressor-sensitive mutants of bacteriophage λ, *Virology* **26**:489.

Brown, D. T., Westphal, M., Burlingham, B. T., Winterhoff, U., and Doerfler, W., 1975, Structure and composition of the adenovirus type 2 core, *J. Virol.* **16**:366.

Brugge, J. S., and Butel, J. S., 1975, Role of simian virus 40 gene A function in maintenance of transformation, *J. Virol.* **15**:619.

Buchwald, M., Murialdo, H., and Siminovitch, L., 1970, The morphogenesis of bacteriophage lambda. II. Identification of the principal structural proteins, *Virology* **42**:390.

Bukhari, A. I., and Allet, B., 1975, Plaque-forming λ-mu hybrids, *Virology* **63**:30.

Bukhari, A. I., and Metlay, M., 1973, Genetic mapping of prophage mu, *Virology* **54**:109.

Bukhari, A. I., and Taylor, A. L., 1975, Influence of insertions on packaging of host sequences covalently linked to bacteriophage Mu DNA, *Proc. Natl. Acad. Sci. USA* **72**:4399.

Bukhari, A. I., and Zipser, D., 1972, Random insertion of mu-1 DNA within a single gene, *Nature (London) New Biol.* **236**:240.

Burger, H., and Doerfler, W., 1974, Intracellular forms of adenovirus DNA. III. Integration of the DNA of adenovirus type 2 into host DNA in productively infected cells, *J. Virol.* **13**:975.

Burkitt, D. P., 1962, A children's cancer dependent on climatic factors, *Nature (London)* **194**:232.

Burkitt, D. P., 1963, A lymphoma syndrome in tropical Africa, *Int. Exp. Pathol.* **2**:67.

Burkitt, D. P., and O'Connor, G. T., 1961, Malignant lymphoma in African children, *Cancer* **14**:258.

Burlingham, B. T., and Doerfler, W., 1971, Three size-classes of intracellular adenovirus deoxyribonucleic acid, *J. Virol.* **7**:707.

Burlingham, B. T., and Doerfler, W., 1972, An endonuclease in cells infected with adenovirus and associated with adenovirions, *Virology* **48**:1.

Burlingham, B. T., Doerfler, W., Pettersson, U., and Philipson, L., 1971, Adenovirus endonuclease: Association with the penton of adenovirus type 2, *J. Mol. Biol.* **60**:45.

Burns, W. H., and Black, P. H., 1968, Analysis of simian virus 40-induced transformation of hamster kidney tissue *in vitro*. V. Variability of virus recovery from cell clones inducible with mitomycin C and cell fusion, *J. Virol.* **2**:606.

Cajean-Feroldi, C., Chardonnet, Y., and Auray, J., 1977, Deoxyribonuclease activity associated with adenovirus 5 and 7, *Eur. J. Biochem.*, in press.

Calef, E., and Licciardello, G., 1960, Recombination experiments on prophage host relationships, *Virology* **12**:81.

Calendar, R., and Lindahl, G., 1969, Attachment of prophage P2: Gene order at different host chromosomal sites, *Virology* **39**:867.

Calnek, B. W., Aldinger, H. K., and Kohne, D. E., 1970, Feather follicle epithelium: A source of enveloped and infectious cell-free herpes virus from Marek's disease, *Avian Dis.* **14**:219.

Campbell, A., 1959, Ordering of genetic sites in bacteriophage λ by the use of galactose-transducing defective phages, *Virology* **9**:293.

Campbell, A. M., 1962, Episomes, *Adv. Genet.* **11**:101.

Carbon, J., Shenk, T. E., and Berg, P., 1975, Biochemical procedure for production of small deletions in simian virus 40 DNA, *Proc. Natl. Acad. Sci. USA* **72**:1392.

Carroll, R. B., Hager, L., and Dulbecco, R., 1974, Simian virus 40 T antigen binds to DNA, *Proc. Natl. Acad. Sci. USA* **71**:3754.

Casjens, S. H., 1974, Bacteriophage lambda F$_{II}$ gene protein: Role in head assembly, *J. Mol. Biol.* **90**:1.

Casjens, S. H., and Hendrix, R. W., 1974, Locations and amounts of the major structural proteins in bacteriophage lambda, *J. Mol. Biol.* **88**:535.

Casjens, S. H., Hohn, T. H., and Kaiser, A. D., 1970, Morphological proteins of phage lambda: Identification of the major head protein as the product of gene E, *Virology* **42**:496.

Caspersson, T., Zech, L., Johansson, C., and Modest, E. J., 1970, Identification of human chromosomes by DNA-binding fluorescent agents, *Chromosoma* **30**:215.

Cassingena, R., and Tournier, P., 1968, Mise en évidence d'un "répresseur" spécifique dans des cellules d'espèces différentes transformées par le virus SV40, *C. R. Acad. Sci. Ser. D* **267**:2251.

Cassingena, R., Tournier, P., May, E., Estrade, S., and Bourali, M., 1969, Synthèse du "répresseur" du virus SV40 dans l'infection productive et abortive, *C. R. Acad. Sci. Ser. D* **268**:2834.

Chadwick, P., Pirrotta, V., Steinberg, R., Hopkins, N., and Ptashne, M., 1970, The λ and 434 phage repressors, *Cold Spring Harbor Symp. Quant. Biol.* **35**:283.

Champe, P. C., Strohl, W. A., and Schlesinger, R. W., 1972, Demonstration of an adenovirus-inhibitory factor in adenovirus-induced hamster tumor cells, *Virology* **50**:482.

Champe, P. C., Strohl, W. A., and Schlesinger, R. W., 1975, Characterization of the viral inhibitory factor (VIF) demonstrable in adenovirus-induced tumor and transformed cells. I. Effect on DNA and viral capsid protein synthesis, *Virology* **68**:317.

Champoux, J. J., and Dulbecco, R., 1972, An activity from mammalian cells that untwists superhelical DNA—A possible swivel for DNA replication, *Proc. Natl. Acad. Sci. USA* **69**:143.

Chattopadhyay, S. K., Lowy, D. R., Teich, N. M., Levine, A. S., and Rowe, W. P., 1974, Evidence that the AKR murine-leukemia-virus genome is complete in DNA of the high-virus AKR mouse and imcomplete in the DNA of the "virus-negative" NIH mouse, *Proc. Natl. Acad. Sci. USA* **71**:167.

Chattopadhyay, S. K., Rowe, W. P., Teich, N. M., and Lowy, D. R., 1975, Definitive evidence that the murine C-type virus inducing locus *AKv-1* is viral genetic material, *Proc. Natl. Acad. Sci. USA* **72**:906.

Chattopadhyay, S. K., Rowe, W. P., and Levine, A. S., 1976, Quantitative studies of integration of murine leukemia virus after exogenous infection, *Proc. Natl. Acad. Sci. USA* **73**:4095.

Chow, L. T., Boyer, H. W., Tischer, E. G., and Goodman, H. M., 1974, Electron microscopic mapping of the attachment sites on SV40 DNA during lytic infection, *Cold Spring Harbor Symp. Quant. Biol.* **39**:109.

The Cold Spring Harbor Symposium on Quantitative Biology, 1975, Vol. 39: *Tumor Viruses,* Cold Spring Harbor Laboratory, Cold Spring Harbor, N.Y.

Collins, C. J., and Sauer, G., 1972, Fate of infecting simian virus 40 deoxyribonucleic acid in nonpermissive cells: Integration into host deoxyribonucleic acid, *J. Virol.* **10**:425.

Comings, D. E., 1973, A general theory of carcinogenesis, *Proc. Natl. Acad. Sci. USA* **70**:3324.

Court, D., Green, L., and Echols, H., 1975, Positive and negative regulation by the *c*II and *c*III gene products of bacteriophage λ, *Virology* **64**:484.

Craig, E. A., McGrogan, M., Mulder, C., and Raskas, H. J., 1975, Identification of early adenovirus type 2 RNA species transcribed from the left-hand end of the genome, *J. Virol.* **16**:905.

Crawford, L. V., 1969, Nucleic acids of tumor viruses, *Adv. Virus Res.* **14**:89.

Crawford, L. V., and Black, P. H., 1964, The nucleic acid of simian virus 40, *Virology* **24**:388.

Croce, C. M., and Koprowski, H., 1973, Enucleation of cells made simple and rescue of SV40 by enucleated cells made even simpler, *Virology* **51**:227.

Croce, C. M., and Koprowski, H., 1975, Assignment of gene(s) for cell transformation to human chromosome 7 carrying the simian virus 40 genome, *Proc. Natl. Acad. Sci. USA* **72**:1658.

Croce, C. M., Girardi, A. J., and Koprowski, H., 1973, Assignment of the T-antigen gene of simian virus 40 to human chromosome C-7, *Proc. Natl. Acad. Sci. USA* **70**:3617.

Croce, C. M., Huebner, K., Girardi, A. J., and Koprowski, H., 1974a, Rescue of defective SV40 from mouse–human hybrid cells containing human chromosome 7, *Virology* **60**:276.

Croce, C. M., Huebner, K., Girardi, A. J., and Koprowski, H., 1974b, Genetics of cell transformation by simian virus 40, *Cold Spring Harbor Symp. Quant. Biol.* **39**:335.

Croce, C. M., Huebner, K., and Koprowski, H., 1974c, Chromosome assignment of the T-antigen gene of simian virus 40 in African green monkey kidney cells transformed by adeno 7–SV40 hybrid, *Proc. Natl. Acad. Sci. USA* **71**:4116.

Croce, C. M., Aden, D., and Koprowski, H., 1975, Somatic cell hybrids between mouse peritoneal macrophages and simian virus 40–transformed human cells: II. Presence of human chromosome 7 carrying simian virus 40 genome in cells of tumors induced by hybrid cells, *Proc. Natl. Acad. Sci. USA* **72**:1397.

Crumpacker, C. S., Levin, M. J., Wiese, W. H., Lewis, A. M., Jr., and Rowe, W. P., 1970, Adenovirus type 2–simian virus 40 hybrid population: Evidence for a hybrid deoxyribonucleic acid molecule and the absence of adenovirus-encapsidated circular simian virus 40 deoxyribonucleic acid, *J. Virol.* **6**:788.

Crumpacker, C. S., Henry, P. H., Kakefuda, T., Rowe, W. P., Levin, M. J., and Lewis, A. M., Jr., 1971, Studies of nondefective adenovirus 2–simian virus 40 hybrid viruses. III. Base composition, molecular weight and conformation of the Ad2$^+$ND$_1$ genome, *J. Virol.* **7**:352.

Cuzin, F., Dieckmann, M., Vogt, M., and Berg, P., 1970, Induction of virus multiplication in 3T3 cells transformed by a thermosensitive mutant of polyoma virus, *J. Mol. Biol.* **47**:317.

Dale, R. M. K., and Ward, D. C., 1975, Mercurated polynucleotides: New probes for hybridization and selective polymer fractionation, *Biochemistry* **14**:2458.

Daniell, E., Abelson, J., Kim, J. S., and Davidson, N., 1973a, Heteroduplex structures of bacteriophage mu DNA, *Virology* **51**:237.

Daniell, E., Boram, W., and Abelson, J., 1973b, Genetic mapping of the inversion loop in bacteriophage mu DNA, *Proc. Natl. Acad. Sci. USA* **70**:2153.

Danna, K. J., and Nathans, D., 1971, Specific cleavage of simian virus 40 DNA by restriction endonuclease of *Hemophilus influenzae*, *Proc. Natl. Acad. Sci. USA* **68**:2913.

Danna, K. J., and Nathans, D., 1972, Bidirectional replication of simian virus 40 DNA, *Proc. Natl. Acad. Sci. USA* **69**:3097.

Danna, K. J., Sack, G. H., Jr., and Nathans, D., 1973, Studies of simian virus 40 DNA: A cleavage map of the SV40 genome, *J. Mol. Biol.* **78**:363.

Darai, G., and Munk, K., 1973, Human embryonic lung cells abortively infected with herpesvirus hominis type 2 show some properties of cell transformation, *Nature (London) New Biol.* **241**:268.

Darnell, J. E., Pagoulatos, G. N., Lindberg, U., and Balint, R., 1970, Studies on the relationship of mRNA to heterogeneous nuclear RNA in mammalian cells, *Cold Spring Harbor Symp. Quant. Biol.* **35**:555.

Davis, R. W., and Parkinson, J. S., 1971, Deletion mutants of bacteriophage lambda. III. Physical structure of *att* φ, *J. Mol. Biol.* **56**:403.

Dhar, R., Subramanian, K., Zain, B. S., Pan, J., and Weissman, S. M., 1974, Nucleotide sequence about the 3′ terminus of SV40 DNA transcripts and the region where DNA synthesis is initiated, *Cold Spring Harbor Symp. Quant. Biol.* **39**:153.

Doerfler, W., 1968, The fate of the DNA of adenovirus type 12 in baby hamster kidney cells, *Proc. Natl. Acad. Sci. USA* **60**:636.

Doerfler, W., 1969, Nonproductive infection of baby hamster kidney cells (BHK21) with adenovirus type 12, *Virology* **38**:587.

Doerfler, W., 1970, Integration of the DNA of adenovirus type 12 into the DNA of baby hamster kidney cells, *J. Virol.* **6**:652.

Doerfler, W., 1975, Integration of viral DNA into the host genome, *Curr. Top. Microbiol. Immunol.* **71**:1.

Doerfler, W., and Kleinschmidt, A. K., 1970, Denaturation pattern of the DNA of adenovirus type 2 as determined by electron microscopy, *J. Mol. Biol.* **50**:579.

Doerfler, W., and Lundholm, U., 1970, Absence of replication of the DNA of adenovirus type 12 in BHK21 cells, *Virology* **40**:754.

Doerfler, W., and Philipson, L., 1973, Isolation and characterization of the endonuclease from KB cells infected with adenovirus type 2, *Cold Spring Harbor Tumor Virus Meeting Abstracts,* p. 43.

Doerfler, W., Hirsch-Kauffmann, M., and Lundholm, U., 1971, How is the replication of adenovirus DNA regulated? *First European Biophysics Congress,* pp. 495–501.

Doerfler, W., Hellmann, W., and Kleinschmidt, A. K., 1972a, The DNA of adenovirus type 12 and its denaturation pattern, *Virology* **47**:507.

Doerfler, W., Lundholm, U., and Hirsch-Kauffmann, M., 1972b, Intracellular forms of adenovirus DNA. I. Evidence for a DNA–protein complex in baby hamster kidney cells infected with adenovirus type 12, *J. Virol.* **9**:297.

Doerfler, W., Lundholm, U., Rensing, U., and Philipson, L., 1973, Intracellular forms of adenovirus DNA. II. Isolation in dye-buoyant density gradients of a DNA-RNA complex from KB cells infected with adenovirus type 2, *J. Virol.* **12**:793.

Doerfler, W., Burger, H., Ortin, J., Fanning, E., Brown, D. T., Westphal, M., Winterhoff, U., Weiser, B., and Schick, J., 1974, Integration of adenovirus DNA into the cellular genome, *Cold Spring Harbor Symp. Quant. Biol.* **39**:505.

Dove, W. F., 1968, The genetics of the lambdoid phages, *Annu. Rev. Genet.* **2**:305.

Dove, W. F., and Weigle, J. J., 1965, Intracellular state of the chromosome of bacteriophage lambda. I. The eclipse of infectivity of the bacteriophage DNA. *J. Mol. Biol.* **12**:620.

Drlica, K. A., and Kado, C. I., 1975, Crown gall tumors: Are bacterial nucleic acids involved? *Bacteriol. Rev.* **39**:186.

Duesberg, P. H., and Vogt, P. K., 1973, RNA species obtained from clonal lines of avian sarcoma and from avian leukosis virus, *Virology* **54**:207.

Duff, R., and Rapp, F., 1971*a*, Oncogenic transformation of hamster cells after exposure to herpes simplex virus type 2, *Nature (London) New Biol.* **233**:48.

Duff, R., and Rapp, F., 1971*b*, Properties of hamster embryo fibroblasts transformed *in vitro* after exposure to ultraviolet-irradiated herpes simplex virus type 2, *J. Virol.* **8**:469.

Dulbecco, R., 1968, The state of the DNA of polyoma virus and SV40 in transformed cells, *Cold Spring Harbor Symp. Quant. Biol.* **33**:777.

Dulbecco, R., and Eckhart, W., 1970, Temperature-dependent properties of cells transformed by a thermosensitive mutant of polyoma virus, *Proc. Natl. Acad. Sci. USA* **67**:1775.

Dulbecco, R., Hartwell, L. H., and Vogt, M., 1965, Induction of cellular DNA synthesis by polyoma virus, *Proc. Natl. Acad. Sci. USA* **53**:403.

Eagle, H., 1955, Propagation in a fluid medium of human epidermoid carcinoma, strain KB, *Proc. Soc. Exp. Biol. Med.* **89**:362.

Echols, H., 1972, Developmental pathways for the temperate phage: Lysis vs. lysogeny, *Annu. Rev. Genet.* **6**:157.

Echols, H., and Green, L., 1971, Establishment and maintenance of repression by bacteriophage lambda: The role of the *c*I, *c*II, and *c*III proteins, *Proc. Natl. Acad. Sci. USA* **68**:2190.

Echols, H., Gingery, R., and Moore, L., 1968, Integrative recombination function of bacteriophage λ: Evidence for a site-specific recombination enzyme, *J. Mol. Biol.* **34**:251.

Echols, H., Green, L., Kudrna, R., and Edlin, G., 1975, Regulation of phage λ development with the growth rate of host cells: A homeostatic mechanism, *Virology* **66**:344.

Eckhart, W., and Dulbecco, R., 1974, Properties of the *ts*3 mutant of polyoma virus during lytic infection, *Virology* **60**:359.

Eddy, B. E., Borman, G. S., Berkeley, W. H., and Young, R. D., 1961, Tumors induced in hamsters by injection of rhesus monkey kidney cell extracts, *Proc. Soc. Exp. Biol. Med.* **107**:191.

Eddy, B. E., Borman, G. S., Grubbs, G. E., and Young, R. D., 1962, Identification of the oncogenic substance in rhesus monkey kidney cell cultures as simian virus 40, *Virology* **17**:65.

Ellens, D. J., Sussenbach, J. S., and Jansz, H. S., 1974, Studies on the mechanism of replication of adenovirus DNA. III. Electron microscopy of replicating DNA, *Virology* **61**:427.

Elsevier, S. M., Kucherlapati, R. S., Nichols, E. A., Creagan, R. P., Giles, R. E., Ruddle, F. H., Willecke, K., and McDougall, J. K., 1974, Assignment of the gene for galactokinase to human chromosome 17 and its regional localisation to band q 21–22, *Nature (London)* **251**:633.

Evans, R. M., Baluda, M. A., and Shoyab, M., 1974, Differences between the integration of avian myeloblastosis virus DNA in leukemic cells and of endogenous viral DNA in normal chicken cells, *Proc. Natl. Acad. Sci. USA* **71**:3152.

Faelen, M., and Toussaint, A., 1973, Isolation of conditional defective mutants of temperate phage mu-1 and deletion mapping of the mu-1 prophage, *Virology* **54**:117.

Faelen, M., Toussaint, A., and Couturier, M., 1971, Mu-1 promoted integration of a λ-gal phage in the chromosome of *E. coli, Mol. Gen. Genet.* **113**:367.

Fanning, E., and Doerfler, W., 1976, Intracellular forms of adenovirus DNA. V. Viral DNA sequences in hamster cells abortively infected and transformed with human adenovirus type 12, *J. Virol.* **20**:373.

Fanning, E., and Doerfler, W., 1977, Intracellular forms of adenovirus DNA. VI. Quantitation and characterization of the four size-classes of adenovirus type 2 DNA in human KB cells, *Virology,* submitted.

Fareed, G. C., Garon, C. F., and Salzman, N. P., 1972, Origin and direction of SV40 DNA replication, *J. Virol.* **10**:484.

Fareed, G. C., Byrne, J. C., and Martin, M. A., 1974, Triplication of a unique genetic segment in a simian virus 40-like virus of human origin and evolution of new viral genomes, *J. Mol. Biol.* **87**:275.

Feldman, L. A., Butel, J. S., and Rapp, F., 1966, Interaction of a simian papovavirus and adenoviruses. I. Induction of adenovirus tumor antigen during abortive infection of simian cells, *J. Bacteriol.* **91**:813.

Fiandt, M., Hradecna, Z., Lozeron, H. A., and Szybalski, W., 1971, Electron micrographic mapping of deletions, insertions, inversions, and homologies in the DNAs of coliphages lambda and phi 80, in: *The Bacteriophage Lambda* (A. D. Hershey, ed.), pp. 329–354, Cold Spring Harbor Laboratory, Cold Spring Harbor, N.Y.

Fiers, W., Danna, K., Rogiers, R., Vandevoorde, A., van Herreweghe, J., van Heuverswyn, H., Volckaer, G., and Yang, R., 1974, Approaches to the sequence determination of SV40 DNA, *Cold Spring Harbor Symp. Quant. Biol.* **39**:179.

Fleckenstein, B., and Wolf, H., 1974, Purification and properties of *Herpesvirus saimiri* DNA, *Virology* **58**:55.

Fleckenstein, B., Bornkamm, G. W., and Ludwig, H., 1975, Repetitive sequences in complete and defective genomes of *Herpesvirus saimiri, J. Virol.* **15**:398.

Flint, S. J., Gallimore, P. H., and Sharp, P. A., 1975, Adenovirus transcription. I. Comparison of viral RNA sequences in adenovirus 2-transformed and lytically infected cells, *J. Mol. Biol.* **96**:47.

Fogel, M., 1975, Polyoma virus-transformed rat cell lines inducible for viral capsid antigen synthesis, *Virology* **65**:446.

Fogel, M., and Sachs, L., 1969, The activation of virus synthesis in polyoma-transformed cells, *Virology* **37**:327.

Folkmanis, A., and Freifelder, D., 1972, Studies on lysogeny in *Escherichia coli* with bacteriophage λ: Physical observation of the insertion process, *J. Mol. Biol.* **65**:63.

Franklin, N. C., Dove, W. F., and Yanofsky, C., 1965, The linear insertion of a prophage into the chromosome of *E. coli* shown by deletion mapping, *Biochem. Biophys. Res. Commun.* **18**:910.

Freeman, A. E., Black, P. H., Vanderpool, E. A., Henry, P. H., Austin, J. B., and Huebner, R. J., 1967, Transformation of primary rat embryo cells by adenovirus type 2, *Proc. Natl. Acad. Sci. USA* **58**:1205.

Freifelder, D., and Levine, E. E., 1975, The formation of polylysogens during infection of *Escherichia coli* with bacteriophage λ, *Virology* **63**:428.

Freifelder, D., and Meselson, M., 1970, Topological relationship of prophage λ to the bacterial chromosome in lysogenic cells, *Proc. Natl. Acad. Sci. USA* **65**:200.

Freifelder, D., Baran, N., Chud, L., Folkmanis, A., and Levine, E. E., 1975, Requirements for insertion of bacteriophage λ DNA into the DNA of *Escherichia coli, J. Mol. Biol.* **91**:401.

Frenkel, N., Roizman, B., Cassai, E., and Nahmias, A., 1972, A DNA fragment of herpes simplex 2 and its transcription in human cervical cancer tissue, *Proc. Natl. Acad. Sci. USA* **69**:3784.

Frenkel, N., Lavi, S., and Winocour, E., 1974, The host DNA sequences in different populations of serially passaged SV40, *Virology* **60**:9.

Frenkel, N., Locker, H., Cox, B., Roizmann, B., and Rapp, F., 1976, Herpes simplex virus DNA in transformed cells: Sequence complexity in five hamster cell lines and one derived hamster tumor, *J. Virol.* **18**:885.

Fried, A. H., 1975, Temperature dependence of strand separation of the DNA molecules containing integrated SV40 DNA in transformed cells, *Nucl. Acids Res.* **2**:1591.

Friedman, M. P., Lyons, M. J., and Ginsberg, H. S., 1970, Biochemical consequences of type 2 adenovirus and simian virus 40 double infections of African green monkey kidney cells, *J. Virol.* **5**:586.

Fuccillo, D. A., Kurent, J. E., and Sever, J. L., 1974, Slow virus diseases, *Annu. Rev. Microbiol.* **28**:231.

Fujinaga, K., and Green, M., 1966, The mechanism of viral carcinogenesis by DNA mammalian viruses: I. Viral-specific RNA in polyribosomes of adenovirus tumor and transformed cells, *Proc. Natl. Acad. Sci. USA* **55**:1567.

Fujinaga, K., and Green, M., 1967a, Mechanism of viral carcinogenesis by DNA mammalian viruses, II. Viral-specific RNA in tumor cells induced by "weakly" oncogenic human adenoviruses, *Proc. Natl. Acad. Sci. USA* **57**:806.

Fujinaga, K., and Green, M., 1967b, Mechanism of viral carcinogenesis by DNA mammalian viruses. IV. Related virus-specific ribonucleic acids in tumor cells induced by "highly" oncogenic adenovirus types 12, 18, and 31, *J. Virol.* **1**:576.

Fujinaga, K., and Green, M., 1968, Mechanism of viral carcinogenesis by DNA mammalian viruses. V. Properties of purified viral-specific RNA from human adenovirus-induced tumor cells, *J. Mol. Biol.* **31**:63.

Fujinaga, K., and Green, M., 1970, Mechanism of viral carcinogenesis by DNA mammalian viruses. VII. Viral genes transcribed in adenovirus type 2 infected and transformed cells, *Proc. Natl. Acad. Sci. USA* **65**:375.

Fujinaga, K., Piña, M., and Green, M., 1969, The mechanism of viral carcinogenesis by DNA mammalian viruses. VI. A new class of virus-specific RNA molecules in cells transformed by group C human adenoviruses, *Proc. Natl. Acad. Sci. USA* **64**:255.

Fujinaga, K., Sekikawa, K., Yamazaki, H., and Green, M., 1974, Analysis of multiple viral genome fragments in adenovirus 7-transformed hamster cells, *Cold Spring Harbor Symp. Quant. Biol.* **39**:633.

Gall, J. G., and Pardue, M. L., 1969, Formation and detection of RNA-DNA hybrid molecules in cytological preparations, *Proc. Natl. Acad. Sci. USA* **63**:378.

Gallimore, P. H., 1974, Interactions of adenovirus type 2 with rat embryo cells: Permissiveness, transformation and *in vitro* characteristics of adenovirus transformed rat embryo cells, *J. Gen. Virol.* **25**:263.

Gallimore, P. H., Sharp, P. A., and Sambrook, J., 1974, Viral DNA in transformed cells. II. A study of the sequences of adenovirus 2 DNA in nine lines of transformed rat cells using specific fragments of the viral genome, *J. Mol. Biol.* **89**:49.

Gallo, R. C., 1972, RNA-dependent DNA polymerase in viruses and cells: Views on the current state, *Blood* **39**:117.

Garfinkle, B., and McAuslan, B. R., 1974, Transformation of cultured mammalian cells by viable herpes simplex virus subtypes 1 and 2, *Proc. Natl. Acad. Sci. USA* **71**:220.

Garon, C. F., Berry, K. W., and Rose, J. A., 1972, A unique form of terminal redundancy in adenovirus DNA molecules, *Proc. Natl. Acad. Sci. USA* **69**:2391.

Geering, G., Old, L. J., and Boyse, E. A., 1966, Antigens of leukemias induced by naturally occurring murine leukemia virus: Their relation to the antigens of Gross virus and other murine leukemia viruses, *J. Exp. Med.* **124**:753.

Gelb, L. D., and Martin, M. A., 1973, Simian virus 40 DNA integration within the genome of virus-transformed mammalian cells, *Virology* **51**:351.

Gelb, L. D., Kohne, D. E., and Martin, M. A., 1971, Quantitation of simian virus 40 sequences in African green monkey, mouse and virus-transformed cell genomes, *J. Mol. Biol.* **57**:129.

Gelb, L. D., Milstein, J. B., Martin, M. A., and Aaronson, S. A., 1973, Characterization of murine leukemia virus-specific DNA present in normal mouse cells, *Nature (London) New Biol.* **244**:76.

Gellert, M., 1967, Formation of covalent circles of lambda DNA by *E. coli* extracts. *Proc. Natl. Acad. Sci. USA* **57**:148.

Georgieff, M., Bachenheimer, S., and Darnell, J., E., 1974, An examination of the nuclear RNA of adenovirus-transformed cells, *Cold Spring Harbor Symp. Quant. Biol.* **39**:475.

Gerber, P., 1964, Virogenic hamster tumor cells: Induction of virus synthesis, *Science* **145**:833.

Gerber, P., 1966, Studies on the transfer of subviral infectivity from SV40-induced hamster tumor cells to indicator cells, *Virology* **28**:501.

Gerber, P., 1972, Activation of Epstein-Barr virus by 5-bromodeoxyuridine in "virus-free" human cells, *Proc. Natl. Acad. Sci. USA* **69**:83.

Germond, J. E., Hirt, B., Oudet, P., Gross-Bellard, M., and Chambon, P., 1975, Folding of the DNA double helix in chromatin-like structures from simian virus 40, *Proc. Natl. Acad. Sci. USA* **72**:1843.

Gingery, R., and Echols, H., 1967, Mutants of bacteriophage λ unable to integrate into the host chromosome, *Proc. Natl. Acad. Sci. USA* **58**:1507.

Gottesman, M. E., 1974, The integration and excision of bacteriophage lambda, *Cell* **1**:69.

Gottesman, M. E., and Weisberg, R. A., 1971, Prophage insertion and excision, in: *The Bacteriophage Lambda* (A. D. Hershey, ed.), pp. 113–138, Cold Spring Harbor Laboratory, Cold Spring Harbor, N.Y.

Gottesman, M. E., and Yarmolinsky, M. B., 1968a, Integration-negative mutants of bacteriophage lambda, *J. Mol. Biol.* **31**:487.

Gottesman, M. E., and Yarmolinsky, M. B., 1968b, The integration and excision of the bacteriophage lambda genome, *Cold Spring Harbor Symp. Quant. Biol.* **33**:735.

Gottesman, S., and Gottesman, M., 1975a, Elements involved in site-specific recombination in bacteriophage lambda, *J. Mol. Biol.* **91**:489.

Gottesman, S., and Gottesman, M., 1975b, Excision of prophage λ in a cell-free system, *Proc. Natl. Acad. Sci. USA* **72**:2188.

Graham, F. L., and van der Eb, A. J., 1973a, A new technique for the assay of infectivity of human adenovirus 5 DNA, *Virology* **52**:456.

Graham, F. L., and van der Eb, A. J., 1973b, Transformation of rat cells by DNA of human adenovirus 5, *Virology* **54**:536.

Graham, F. L., van der Eb, A. J., and Heijneker, H. L., 1974a, Size and location of the transforming region in human adenovirus type 5 DNA, *Nature (London)* **251**:687.

Graham, F. L., Abrahams, P. J., Mulder, C., Heijneker, H. L., Warnaar, S. O., de Vries, F. A. J., Fiers, W., and van der Eb, A. J., 1974b, Studies on *in vitro* transformation by DNA and DNA fragments of human adenoviruses and simian virus 40, *Cold Spring Harbor Symp. Quant. Biol.* **39**:637.

Green, M., 1970a, Oncogenic viruses, *Annu. Rev. Biochem.* **39**:701.

Green, M., 1970b, Effect of oncogenic DNA viruses on regulatory mechanisms of cells, *Fed. Proc.* **29**:1265.

Green, M., 1972, Molecular basis for the attack on cancer, *Proc. Natl. Acad. Sci. USA* **69**:1036.

Green, M., and Piña, M., 1963, Biochemical studies on adenovirus multiplication. IV. Isolation, purification, and chemical analysis of adenovirus, *Virology* **20**:199.

Green, M., and Piña, M., 1964, Biochemical studies on adenovirus multiplication. VI. Properties of highly purified tumorigenic human adenoviruses and their DNAs, *Proc. Natl. Acad. Sci. USA* **51**:1251.

Green, M., Piña, M., Kimes, R., Wensink, P. C., MacHattie, L. A., and Thomas, C. A., Jr., 1967, Adenovirus DNA. I. Molecular weight and conformation, *Proc. Natl. Acad. Sci. USA* **57**:1302.

Green, M., Parsons, J. T., Piña, M., Fujinaga, K., Caffier, H., and Landgraf-Leurs, I., 1970, Transcription of adenovirus genes in productively infected and in transformed cells, *Cold Spring Harbor Symp. Quant. Biol.* **35**:803.

Green M. R., Chinnadurai, G., Mackey, J. K., and Green, M., 1976, A unique pattern of integrated viral genes in hamster cells transformed by highly oncogenic human adenovirus 12, *Cell* **7**:419.

Green, P. S., Bellack, M. C., Goodman, H. M., and Boyer, H. W., 1974, The *Eco*RI restriction endonuclease, in: *Methods in Molecular Biology 6*, (R. Wickner, ed.), pp. 413–428, Academic Press, New York.

Griffith, J. D., 1975, Chromatin structure: Deduced from a minichromosome, *Science* **187**:1202.

Griffith, J., Dieckmann, M., and Berg, P., 1975, Electron microscope localization of a protein bound near the origin of simian virus 40 DNA replication, *J. Virol.* **15**:167.

Grodzicker, T., Arditti, R. R., and Eisen, H., 1972, Establishment of repression by lambdoid phage in catabolite activator protein and adenylate cyclase mutants of *Escherichia coli*, *Proc. Natl. Acad. Sci. USA* **69**:366.

Groneberg, J., Brown, D. T., and Doerfler, W., 1975, Uptake and fate of the DNA of adenovirus type 2 in KB cells, *Virology* **64**:115.

Groneberg, J., Chardonnet, Y., and Doerfler, W., 1977, Integrated viral sequences in adenovirus type 12-transformed hamster cells, *Cell* **10**:101.

Guarneros, G., and Echols, H., 1973, Thermal assymetry of site-specific recombination by bacteriophage λ, *Virology* **52**:30.

Haas, M., Vogt, M., and Dulbecco, R., 1972, Loss of simian virus 40 DNA-RNA hybrids from nitrocellulose membranes: Implications for the study of virus–host DNA interactions, *Proc. Natl. Acad. Sci. USA* **69**:2160.

Haase, A. T., and Varmus, H. E., 1973, Demonstration of a DNA provirus in the lytic growth of Visna virus, *Nature (London) New Biol.* **245**:237.

Hall, W., and ter Meulen, V., 1976, RNA homology between SSPE and measles viruses, *Nature (London)* **264**:474.

Hampar, B., Derge, J. G., Martos, L. M., and Walker, J. L., 1972, Synthesis of Epstein-Barr virus after activation of the viral genome in a "virus-negative" human lymphoblastoid cell (Raji) made resistant to 5-bromodeoxyuridine, *Proc. Natl. Acad. Sci. USA* **69**:78.

Hampar, B., Derge, J. G., and Showalter, S. D., 1974, Enhanced activation of the repressed Epstein-Barr viral genome by inhibitors of DNA synthesis, *Virology* **58**:298.

Hanafusa, H., and Hanafusa, T., 1971, Noninfectious RSV deficient in DNA polymerase, *Virology* **43**:313.

Hanafusa, H., Hanafusa, T., Kawai, S., and Luginbuhl, R. E., 1974, Genetic control of expression of endogenous virus genes in chicken cells, *Virology* **58**:439.

Hanafusa, T., Hanafusa, H., and Miyamoto, T., 1970, Recovery of a new virus from apparently normal chick cells by infection with avian tumor viruses, *Proc. Natl. Acad. Sci. USA* **67**:1797.

Harris, H., and Watkins, J. F., 1965, Hybrid cells derived from mouse and man: Artificial heterokaryons of mammalian cells from different species, *Nature (London)* **205**:640.

Hayward, W. S., and Hanafusa, H., 1973, Detection of avian tumor virus RNA in uninfected chicken embryo cells, *J. Virol.* **11**:157.

Hendrix, R. W., 1971, Identification of proteins coded in phage lambda, in: *The Bacteriophage Lambda* (A. D. Hershey, ed.), pp. 355–370, Cold Spring Harbor Laboratory, Cold Spring Harbor, N.Y.

Henle, G., and Henle, W., 1966, Immunofluorescence in cells derived from Burkitt's lymphoma, *J. Bacteriol.* **91**:1248.

Henle, G., Henle, W., and Diehl, V., 1968, Relation of Burkitt's tumor-associated herpes-type virus to infectious mononucleosis, *Proc. Natl. Acad. Sci. USA* **59**:94.

Henle, W., Henle, G., Zajac, B. A., Pearson, G., Waubke, R., and Scriba, M., 1970, Differential reactivity of human serums with early antigens induced by Epstein-Barr virus, *Science* **169**:188.

Henry, P. H., Schnipper, L. E., Samaha, R. J., Crumpacker, C. S., Lewis, A. M., Jr., and Levine, A. S., 1973, Studies of non-defective adenovirus 2–simian virus 40 hybrid viruses. VI. Characterization of the DNA from five nondefective hybrid viruses, *J. Virol.* **11**:665.

Herskowitz, I., 1973, Control of gene expression in bacteriophage lambda, *Annu. Rev. Genet.* **7**:289.

Hill, M., and Hillova, J., 1972, Virus recovery in chicken cells treated with Rous sarcoma cell DNA, *Nature (London) New Biol.* **237**:35.

Hill, M., and Hillova, J., 1974, RNA and DNA forms of the genetic material of C-type viruses and the integrated state of the DNA form in the cellular chromosome, *Biochim. Biophys. Acta* **355**:7.

Hill, M., Hillova, J., Dantchev, D., Mariage, R., and Gonbin, G., 1974, Infectious viral DNA in Rous sarcoma virus transformed nonproducer and producer animal cells, *Cold Spring Harbor Symp. Quant. Biol.* **39**:1015.

Hirai, K., and Defendi, V., 1971, Homology between SV40 DNA and DNA of normal and SV40-transformed Chinese hamster cells, *Biochem. Biophys, Res. Commun.* **42**:714.

Hirai, K., and Defendi, V., 1972, Integration of simian virus 40 DNA into the DNA of permissive monkey kidney cells, *J. Virol.* **9**:705.

Hirai, K., and Defendi, V., 1974, Factors affecting the process and extent of integration of the viral genome, *Cold Spring Harbor Symp. Quant. Biol.* **39**:325.

Hirai, K., Lehmann, J., and Defendi, V., 1971, Integration of SV40 DNA into the DNA of primary infected Chinese hamster cells, *J. Virol.* **8**:708.

Hirai, K., Campbell, G., and Defendi, V., 1974a, Changes of regulation of host DNA synthesis and viral DNA integration in SV40-infected cells, in: *Control of Proliferation in Animal Cells* (B. Clarkson and R. Baserga, eds.), pp. 151–161, Cold Spring Harbor Laboratory, Cold Spring Harbor, N.Y.

Hirai, K., Robb, J. A., and Defendi, V., 1974b, Integration of SV40 DNA and induction of cellular DNA synthesis after a *ts* SV40 infection, *Virology* **59**:266.

Hirai, K., Henner, D., and Defendi, V., 1974c, Hybridization of simian virus 40 complementary RNA with nucleolus-associated DNA isolated from simian virus 40-transformed Chinese hamster cells, *Virology* **60**:588.

Hirt, B., 1967, Selective extraction of polyoma DNA from infected mouse cell cultures, *J. Mol. Biol.* **26**:365.

Hoffmann, P. R., and Darnell, J. E., 1975, Differential accumulation of virus-specific RNA during the cell cycle of adenovirus-transformed rat embryo cells, *J. Virol.* **15**:806.

Hogness, D. S., and Simmons, J. R., 1964, Breakage of λ dg DNA: Chemical and genetic characterization of each isolated half-molecule, *J. Mol. Biol.* **9**:411.

Hölzel, F., and Sokol, F., 1974, Integration of progeny simian virus 40 DNA into the host cell genome, *J. Mol. Biol.* **84**:423.

Horwitz, M. S., 1971, Intermediates in the synthesis of type 2 adenovirus DNA, *J. Virol.* **8**:675.

Horwitz, M. S., 1974, Location of the origin of DNA replication in adenovirus type 2, *J. Virol.* **13**:1046.

Hotchin, J., 1971, *Persistent and Slow Virus Infections*, Karger, Basel.

Howe, M. M., 1973a, Prophage deletion mapping of bacteriophage mu-1, *Virology* **54**:93.

Howe, M. M., 1973b, Transduction by bacteriophage mu-1, *Virology* **55**:103.

Hozier, J. C., and Taylor, J. H., 1975, Length distribution of single-stranded DNA in Chinese hamster ovary cells, *J. Mol. Biol.* **93**:181.

Hsu, M.-T., and Davidson, N., 1972, Structure of inserted bacteriophage mu-1 DNA and physical mapping of bacterial genes by mu-1 DNA insertion, *Proc. Natl. Acad. Sci. USA* **69**:2823.

Hsu, M.-T., and Davidson, N., 1974, Electron microscope heteroduplex study of the heterogeneity of mu phage and prophage DNA, *Virology* **58**:229.

Huebner, R. J., 1967, Adenovirus-directed tumor and T-antigens, in: *Perspectives in Virology*, Vol. V (M. Pollard, ed.), pp. 147–166, Academic Press, New York.

Huebner, R. J., and Todaro, G. J., 1969, Oncogenes of RNA tumor viruses as determinants of cancer, *Proc. Natl. Acad. Sci. USA* **64**:1087.

Huebner, R. J., Rowe, W. P., and Lane, W. T., 1962, Oncogenic effects in hamsters of human adenovirus types 12 and 18, *Proc. Natl. Acad. Sci. USA* **48**:2051.

Huebner, R. J., Rowe, W. P., Turner, H. C., and Lane, W. T., 1963, Specific adenovirus complement-fixing antigens in virus-free hamster and rat tumors, *Proc. Natl. Acad. Sci. USA* **50**:379.

Huebner, R. J., Armstrong, D., Okuyan, M., Sarma, P. S., and Turner, H. C., 1964a, Specific complement-fixing viral antigens in hamster and guinea pig tumors induced by the Schmidt-Ruppin strain of avian sarcoma, *Proc. Natl. Acad. Sci. USA* **51**:742.

Huebner, R. J., Chanock, R. M., Rubin, B. A., and Casey, M. J., 1964b, Induction by adenovirus type 7 of tumors in hamsters having the antigenic characteristics of SV40 virus, *Proc. Natl. Acad. Sci. USA* **52**:1333.

Ikeda, H., and Tomizawa, J., 1968, Prophage P1, an extrachromosomal replication unit, *Cold Spring Harbor Symp. Quant. Biol.* **33**:791.

Jaenisch, R., 1972, Evidence for SV40 specific RNA containing virus and host specific sequences, *Nature (London) New Biol.* **235**:46.

Jaenisch, R., and Levine, A., 1971, DNA replication in SV40-infected cells. V. Circular and catenated oligomers of SV40 DNA, *Virology* **44**:480.

Jaenisch, R., and Mintz, B., 1974, Simian virus 40 DNA sequences in DNA of healthy adult mice derived from preimplantation blastocysts injected with viral DNA, *Proc. Natl. Acad. Sci. USA* **71**:1250.

Jehn, U., Lindahl, T., and Klein, G., 1972, Fate of virus DNA in the abortive infection of human lymphoid cell lines by Epstein-Barr virus, *J. Gen. Virol.* **16**:409.

Jensen, F., and Defendi, V., 1968, Transformation of African green monkey kidney cells by irradiated adenovirus 7–simian virus 40 hybrid, *J. Virol.* **2**:173.

Jensen, F. C., and Koprowski, H., 1969, Absence of repressor in SV40-transformed cells, *Virology* **37**:687.

Jerkofsky, M., and Rapp, F., 1975, Stimulation of adenovirus replication in simian cells in the absence of a helper virus by pretreatment of the cells with iododeoxyuridine, *J. Virol.* **15**:253.

Jordan, E., Saedler, H., and Starlinger, P., 1968, 0° and strong-polar mutations in the *gal* operon are insertions, *Mol. Gen. Genet.* **102**:353.

Kaiser, A. D., 1957, Mutations in a temperate bacteriophage affecting its ability to lysogenize *Escherichia coli, Virology* **3**:42.

Kaiser, A. D., and Masuda, T., 1973, *In vitro* assembly of bacteriophage lambda heads, *Proc. Natl. Acad. Sci. USA* **70**:260.

Kathmann, P., Schick, J., Winnacker, E.-L., and Doerfler, W., 1976, Isolation and preliminary characterization of temperature-sensitive mutants of adenovirus type 2, *J. Virol.* **19**:43.

Kawai, S., and Hanafusa, H., 1971, The effects of reciprocal changes in temperature on the transformed state of cells infected with a Rous sarcoma virus mutant, *Virology* **46**:470.

Kawai, Y., Nonoyama, M., and Pagano, J. S., 1973, Reassociation kinetics for Epstein-Barr virus DNA: Nonhomology to mammalian DNA and homology of viral DNA in various diseases, *J. Virol.* **12**:1006.

Kayajanian, C., 1972, A reanalysis of *Red*-mediated reciprocal recombination in bacteriophage λ, *Virology* **49**:599.

Keller, W., 1975a, Characterization of purified DNA-relaxing enzyme from human tissue culture cells, *Proc. Natl. Acad. Sci. USA* **72**:2550.

Keller, W., 1975b, Determination of the number of superhelical turns in simian virus 40 DNA by gel electrophoresis, *Proc. Natl. Acad. Sci. USA* **72**:4876.

Keller, W., and Wendel, I., 1974, Stepwise relaxation of supercoiled SV40 DNA, *Cold Spring Harbor Symp. Quant. Biol.* **39**:199.

Kelly, R. B., Atkinson, M. R., Huberman, J. A., and Kornberg, A., 1969, Excision of thymine dimers and other mismatched sequences by DNA polymerase of *E. coli, Nature (London)* **224**:495.

Kelly, T. J., Jr., and Lewis, A. M., Jr., 1973, Use of nondefective adenovirus-simian virus 40 hybrids for mapping the simian virus 40 genome, *J. Virol.* **12**:643.

Kelly, T. J., Jr., and Smith, H. O., 1970, A restriction enzyme from *Hemophilus influenzae*. II. Base sequence of the recognition site, *J. Mol. Biol.* **51**:393.

Ketner, G., and Kelly, T. J., Jr., 1976, Integrated simian virus 40 sequences in transformed cell DNA: Analysis using restriction endonucleases, *Proc. Natl. Acad. Sci. USA* **73**:1102.

Khoury, G., Byrne, J. C., Takemoto, K. K., and Martin, M. A., 1973a, Patterns of simian virus 40 deoxyribonucleic acid transcription. II. In transformed cells, *J. Virol.* **11**:54.

Khoury, G., Martin, M. A., Lee, T. N. H., Danna, K. J., and Nathans, D., 1973b, A map of simian virus 40 transcription sites expressed in productively infected cells, *J. Mol. Biol.* **78**:377.

Khoury, G., Howley, P., Brown, M., and Martin, M., 1974a, The detection and quantitation of SV40 nucleic acid sequences using single-stranded SV40 DNA probes, *Cold Spring Harbor Symp. Quant. Biol.* **39**:147.

Khoury, G., Fareed, G. C., Berry, K., Martin, M. A., Lee, T. N. H., and Nathans, D., 1974b, Characterization of a rearrangement in viral DNA: Mapping of the circular simian virus 40-like DNA containing a triplication of a specific one-third of the viral genome, *J. Mol. Biol.* **87**:289.

Khoury, G., Martin, M. A., Lee, T. N. H., and Nathans, D., 1975a, A transcriptional map of the SV40 genome in transformed cell lines, *Virology* **63**:263.

Khoury, G., Howley, P., Nathans, D., and Martin, M. 1975b, Posttranscriptional selection of SV40 specific RNA, *J. Virol.* **15**:433.

Kier, L. D., Yamasaki, E., and Ames, B. N., 1974, Detection of mutagenic activity in cigarette smoke condensates, *Proc. Natl. Acad. Sci. USA* **71**:4159.

Kit, S., Nakajima, K., and Dubbs, D. R., 1970, Origin of thymidine kinase in adenovirus-infected human cell lines, *J. Virol.* **5**:446.

Kit, S., Leung, W. C., Jorgensen, G., Trkula, D., and Dubbs, D. R., 1974, Subcellular localization and properties of thymidine kinase from adenovirus-infected cells, *J. Gen. Virol.* **24**:281.

Klein, G., 1973, The Epstein-Barr virus, in: *The Herpes Viruses* (A. S. Kaplan, ed.), pp. 521–555, Academic Press, New York.

Klein, G., Clifford, P., Klein, E., and Stjernswärd, J., 1966, Search for tumor-specific immune reactions in Burkitt lymphoma patients by the membrane immunofluorescence reaction, *Proc. Natl. Acad. Sci. USA* **55**:1628.

Klein, G., Lindahl, T., Jondal, M., Leibold, W., Menézes, J., Nilsson, K., and Sundström, C., 1974, Continuous lymphoid cell lines with characteristics of B cells (bone-marrow-derived) lacking the Epstein-Barr virus genome and derived from three human lymphomas, *Proc. Natl. Acad. Sci. USA* **71**:3283.

Koprowski, H., Jensen, F. C., and Steplewski, Z., 1967, Activation of production of infectious tumor virus SV40 in heterokaryon cultures, *Proc. Natl. Acad. Sci. USA* **58**:127.

Lai, C.-J., and Nathans, D., 1974, Mapping temperature-sensitive mutants of simian virus 40: Rescue of mutants by fragments of viral DNA, *Virology* **60**:466.

Lai, M. M. C., Duesberg, P. H., Horst, J., and Vogt, P. K., 1973, Avian tumor virus RNA: A comparison of three sarcoma viruses and their transformation-defective derivatives by oligonucleotide fingerprinting and DNA-RNA hybridization, *Proc. Natl. Acad. Sci. USA* **70**:2266.

Landau, B. J., Larson, V. M., Devers, G. A., and Hilleman, M. R., 1966, Studies on induction of virus from adenovirus and SV40 hamster tumors. 1. Chemical and physical agents, *Proc. Soc. Exp. Biol. Med.* **122**:1174.

Laub, O., and Aloni, Y., 1975, Transcription of simian virus 40. V. Regulation of simian virus 40 gene expression, *J. Virol.* **16**:1171.

Lavelle, G., Patch, C., Khoury, G., and Rose, J., 1975, Isolation and partial characterization of single-stranded adenoviral DNA produced during synthesis of adenovirus type 2 DNA, *J. Virol.* **16**:775.

Lavi, S., and Winocour, E., 1972, Acquisition of sequences homologous to host DNA by closed circular SV40 DNA, *J. Virol.* **9**:309.

Lavi, S., and Winocour, E., 1974, Accumulation of closed-circular polyoma DNA molecules containing host DNA during serial passage of the virus, *Virology* **57**:296.

Lavi, S., Rozenblatt, S., Singer, M. F., and Winocour, E., 1973, Acquisition of sequences homologous to host DNA by closed circular simian virus 40 DNA. II. Further studies on the serial passage of virus clones, *J. Virol.* **12**:492.

Lebowitz, P., Kelly, T. J., Jr., Nathans, D., Lee, T. N. H., and Lewis, A., Jr., 1974, A colinear map relating the SV40 DNA segments of six adenovirus–SV40 hybrids to the DNA fragments produced by restriction endonuclease cleavage of SV40 DNA, *Proc. Natl. Acad. Sci. USA* **71**:441.

Lee, T. N. H., Brockman, W. W., and Nathans, D., 1975, Evolutionary variants of simian virus 40: Cloned substituted variants containing multiple initiation sites for DNA replication, *Virology* **66**:53.

Levin, M. J., Crumpacker, C. S., Lewis, A. M., Jr., Oxman, M. N., Henry, P. H., and Rowe, W. P., 1971, Studies of nondefective adenovirus 2–simian virus 40 hybrid viruses. II. Relationship of Ad2 DNA and SV40 DNA in the Ad2$^+$ND$_1$genome, *J. Virol.* **7**:343.

Levine, A. J., and Burger, M. M., 1972, A working hypothesis explaining the maintenance of the transformed state by SV40 and polyoma, *J. Theor. Biol.* **37**:435.

Levine, A. J., and Teresky, A. K., 1970, Deoxyribonucleic acid replication in simian virus 40-infected cells. II. Detection and characterization of simian virus 40 pseudovirions, *J. Virol.* **5**:451.

Levine, A. J., Kang, H. S., and Billheimer, F. E., 1970, DNA replication in SV40 infected cells. I. Analysis of replicating SV40 DNA, *J. Mol. Biol.* **50**:549.

Levine, A. S., Oxman, M. N., Henry, P. H., Levin, M. J., Diamandopoulos, G. T., and Enders, J. F., 1970, Virus-specific DNA in SV40-exposed hamster cells: Correlation with S and T antigens, *J. Virol.* **6**:199.

Levine, A. S., Levin, M. J., Oxman, M. N., and Lewis, A. M., Jr., 1973, Studies of nondefective adenovirus 2–simian virus 40 hybrid viruses. VII. Characterization of the SV40 RNA species induced by five nondefective hybrid viruses, *J. Virol.* **11**:672.

Levine, M., 1972, Replication and lysogeny with phage P22 in *Salmonella typhimurium, Curr. Top. Microbiol. Immunol.* **58**:135.

Levinthal, J. D., and Petersen, W., 1965, *In vitro* transformation and immunofluorescence with human adenovirus type 12 in rat and rabbit kidney cells, *Fed. Proc.* **24**:174.

Lewis, A. M., Jr., and Rowe, W. P., 1971, Studies on nondefective adenovirus-simian virus 40 hybrid viruses. I. A newly characterized simian virus 40 antigen induced by the Ad2$^+$ND$_1$ virus, *J. Virol.* **7**:189.

Lewis, A. M., Jr., and Rowe, W. P., 1973, Studies of nondefective adenovirus 2–simian virus 40 hybrid viruses. VIII. Association of simian virus 40 transplantation antigen with a specific region of the early viral genome, *J. Virol.* **12**:836.

Lewis, A. M., Jr., Levin, M. J., Wiese, W. H., Crumpacker, C. S., and Henry, P. H., 1969, A nondefective (competent) adenovirus–SV40 hybrid isolated from the Ad2-SV40 hybrid population, *Proc. Natl. Acad. Sci. USA* **63**:1128.

Lewis, A. M., Jr., Levine, A. S., Crumpacker, C. S., Levin, M. J., Samaha, R. J., and Henry, P. H., 1973, Studies of nondefective adenovirus 2–simian virus 40 hybrid viruses. V. Isolation of additional hybrids which differ in their simian virus 40-specific biological properties, *J. Virol.* **11**:655.

Lewis, A. M., Jr., Rabson, A. S., and Levine, A. S., 1974a, Studies of nondefective adenovirus 2–simian virus 40 hybrid viruses. X. Transformation of hamster kidney cells by adenovirus 2 and the nondefective hybrid viruses, *J. Virol.* **13**:1291.

Lewis, A. M., Jr., Breeden, J. H., Wewerka, Y. L., Schnipper, L. E., and Levine, A. S., 1974b, Studies of hamster cells transformed by adenovirus 2 and the nondefective Ad2-SV40 hybrids, *Cold Spring Harbor Symp. Quant. Biol.* **39**:651.

Lewis, J. B., Atkins, J. F., Anderson, C. W., Baum, P. R., and Gesteland, R. F., 1975, Mapping of late adenovirus genes by cell-free translation of RNA selected by hybridization to specific DNA fragments, *Proc. Natl. Acad. Sci. USA* **72**:1344.

Lewis, J. B., Atkins, J. F., Baum, P. R., Solem, R., Gesteland, R. F., and Anderson, C. W., 1976, Location and identification of the genes for adenovirus type 2 early polypeptides, *Cell* **7**:141.

Lilly, F., 1970, *Fv-2*: Identification and location of a second gene governing the spleen focus response to Friend leukemia virus in mice, *J. Natl. Cancer Inst.* **45**:163.

Lilly, F., and Pincus, T. H., 1973, Genetic control of murine viral leukemogenesis, *Adv. Cancer Res.* **17**:231.

Lindahl, T., Adams, A., Bjursell, G., Bornkamm, G. W., Kaschka-Dierich, C., and Jehn, U., 1976, Covalently closed circular duplex DNA of Epstein-Barr virus in human lymphoid cell line. *J. Mol. Biol.* **102**:511.

Lindberg, U., and Darnell, J. E., 1970, SV40-specific RNA in the nucleus and polyribosomes of transformed cells, *Proc. Natl. Acad. Sci. USA* **65**:1089.

Linial, M., and Mason, W. S., 1973, Characterization of two conditional early mutants of Rous sarcoma virus, *Virology* **53**:258.

Lippincott, J. A., and Lippincott, B. B., 1975, The genus *Acrobacterium* and plant tumorigenesis, *Annu. Rev. Microbiol.* **29**:377.

Loni, M. C., and Green, M., 1973, Detection of viral DNA sequences in adenovirus-transformed cells by *in situ* hybridization, *J. Virol.* **12**:1288.

Lowy, D. R., Rowe, W. P., Teich, N., and Hartley, J. W., 1971, Murine leukemia virus: High-frequency activation *in vitro* by 5-iododeoxyuridine and 5-bromodeoxyuridine, *Science* **174**:155.

Lwoff, A., 1953, Lysogeny, *Bacteriol. Rev.* **17**:269.

Mackey, J. K., Rigden, P. M., and Green, M., 1976, Do highly oncogenic group A human adenoviruses cause human cancer? Analysis of human tumors for adenovirus 12 transforming DNA sequences, *Proc. Natl. Acad. Sci. USA* **73**:4657.

Manor, H., Fogel, M., and Sachs, L., 1973, Integration of viral into chromosomal deoxyribonucleic acid in an inducible line of polyoma transformed cells, *Virology* **53**:174.

Marek, J., 1907, Multiple Nervenentzündung (Polyneuritis) bei Hühnern, *Deutsch. Tierärztl. Wochenschr.* **15**:417.

Markham, P. D., and Baluda, M. A., 1973, Integrated state of oncornavirus DNA in normal chicken cells and in cells transformed by avian myeloblastosis virus, *J. Virol.* **12**:721.

Martin, G. S., 1970, Rous sarcoma virus: A function required for the maintenance of the transformed state, *Nature (London)* **227**:1021.

Martin, M. A., 1970, Characteristics of SV40 DNA transcription during lytic infection, abortive infection, and in transformed mouse cells, *Cold Spring Harbor Symp. Quant. Biol.* **35**:833.

Martin, M. A., and Khoury, G., 1973, Transcription of SV40 DNA in lytically infected and transformed cells, in: *Virus Research* (C. F. Fox and W. S. Robinson, eds.), pp. 33–50, Academic Press, New York.

Martin, M. A., and Khoury, G., 1976, Integration of DNA tumor virus genomes, *Curr. Top. Microbiol. Immunol.* **73**:35.

Martin, M. A., Gelb, L. D., Fareed, G. C., and Milstein, J. B., 1973, Reassortment of simian virus 40 DNA during serial undiluted passage, *J. Virol.* **12**:748.

Martin, M. A., Howley, P. M., Byrne, J. C., and Garon, C. F., 1976, Characterization of supercoiled, oligomeric SV40 DNA molecules in productively infected cells, *Virology* **71**:28.

Martin, R. G., and Chou, J. Y., 1975, Simian virus 40 functions required for the establishment and maintenance of malignant transformation, *J. Virol.* **15**:599.

Martuscelli, J., Taylor, A. L., Cummings, D. J., Chapman, V. A., DeLong, S. S., and Cañedo, L., 1971, Electron microscopic evidence for linear insertion of bacteriophage mu-1 in lysogenic bacteria, *J. Virol.* **8**:551.

Marusyk, R. G., Morgan, A. R., and Wadell, G., 1975, Association of endonuclease activity with serotypes belonging to the three subgroups of human adenoviruses, *J. Virol.* **16**:456.

McAllister, R. M., Nicolson, M. O., Reed, G., Kern, J., Gilden, R. V., and Huebner, R. J., 1969, Transformation of rodent cells by adenovirus 19 and other group D adenoviruses, *J. Natl. Cancer Inst.* **43**:917.

McAllister, R. M., Gilden, R. V., and Green, M., 1972, Adenoviruses in human cancer, *Lancet* **i**:831.

McBride, O. W., and Ozer, H. L., 1973, Transfer of genetic information by purified metaphase chromosomes, *Proc. Natl. Acad. Sci. USA* **70**:1258.

McBride, W. D., and Wiener, A., 1964, *In vitro* transformation of hamster kidney cells by human adenovirus type 12, *Proc. Soc. Exp. Biol. Med.* **115**:870.

McDougall, J. K., 1971, Adenovirus-induced chromosome aberrations in human cells, *J. Gen. Virol.* **12**:43.

McDougall, J. K., 1976, Adenoviruses—Interaction with the host cell genome, *Prog. Med. Virol.* **21**:118.

McDougall, J. K., Dunn, A. R., and Jones, K. W., 1972, *In situ* hybridization of adenovirus RNA and DNA, *Nature (London) New Biol.* **236**:346.

McDougall, J. K., Kucherlapati, R., and Ruddle, F. H., 1973, Localization and induction of the human thymidine kinase gene by adenovirus 12, *Nature (London) New Biol.* **245**:172.

McDougall, J. K., Dunn, A. R., and Gallimore, P. H., 1974a, Recent studies on the characteristics of adenovirus-infected and -transformed cells, *Cold Spring Harbor Symp. Quant. Biol.* **39**:591.

McDougall, J. K., Vause, K. E., Gallimore, P. H., and Dunn, A. R., 1974b, Cytogenetic studies in permissive and abortive infections by adenovirus type 12, *Int. J. Cancer* **14**:236.

Melendez, L. V., Hunt, R. D., Daniel, M. D., Fraser, C. E. O., Garcia, F. G., and

Williamson, M. E., 1970, Lethal reticuloproliferative disease induced in *Cebus albifrons* monkeys by *Herpesvirus saimiri, Int. J. Cancer* **6**:431.

Melendez, L. V., Castellanos, H., Barahona, H. H., Daniel, M. D., Hunt, R. D., Fraser, C. E. O., Garcia, F. G., and King, N. W., 1972a, Two new herpesviruses from spider monkeys (*Ateles geoffroyi*), *J. Natl. Cancer Inst.* **49**:233.

Melendez, L. V., Hunt, R. D., Daniel, M. D., Fraser, C. E. O., Barahona, H. H., King, N. W., and Garcia, F. G., 1972b, *Herpesviruses saimiri* and *ateles*—Their role in malignant lymphomas of monkeys, *Fed. Proc.* **31**:1643.

Mertz, J. E., Carbon, J., Herzberg, M., Davis, R. W., and Berg, P., 1974, Isolation and characterization of individual clones of simian virus 40 mutants containing deletions, duplications and insertions in their DNA, *Cold Spring Harbor Symp. Quant. Biol.* **39**:69.

Michel, M. R., Hirt, B., and Weil, R., 1967, Mouse cellular DNA enclosed in polyoma viral capsids (pseudovirions), *Proc. Natl. Acad. Sci. USA* **58**:1381.

Mizutani, S., Boettiger, D., and Temin, H. M., 1970, A DNA-dependent DNA polymerase and a DNA endonuclease in virions of Rous sarcoma virus, *Nature (London)* **228**:424.

Morrow, J. F., and Berg, P., 1972, Cleavage of simian virus 40 DNA at a unique site by a bacterial restriction enzyme, *Proc. Natl. Acad. Sci. USA* **69**:3365.

Morrow, J. F., Berg, P., Kelly, T. J., Jr., and Lewis, A. M., Jr., 1973, Mapping of simian virus 40 early functions on the viral chromosome, *J. Virol.* **12**:653.

Muir, C. S., and Shanmugaratnam, K., 1967, in: *Cancer of the Nasopharynx* (C. S. Muir and K. Shanmugaratnam, eds.), p. 47, UICC Monograph Series, No. 1.

Mulder, C., and Delius, H., 1972, Specificity of the break produced by restriction endonuclease R1 in simian virus 40, as revealed by partial denaturation mapping, *Proc. Natl. Acad. Sci. USA* **69**:3215.

Mulder, C., Sharp, P. A., Delius, H., and Pettersson, U., 1974a, Specific fragmentation of DNA of adenovirus serotypes 3, 5, 7, and 12, and adeno-SV40 hybrid virus Ad2⁺ND₁ by restriction endonuclease R · EcoRI, *J. Virol.* **14**:68.

Mulder, C., Arrand, J. R., Delius, H., Keller, W., Pettersson, U., Roberts, R. J., and Sharp, P. A., 1974b, Cleavage maps of DNA from adenovirus types 2 and 5 by restriction endonucleases EcoRI and HpaI, *Cold Spring Harbor Symp. Quant. Biol.* **39**:397.

Nahmias, A. J., Naib, Z. M., Josey, W. E., Murphey, F. A., and Luce, C. F., 1970, Sarcomas after inoculation of newborn hamsters with herpes virus hominis type 2 strains, *Proc. Soc. Exp. Biol. Med.* **134**:1065.

Nahmias, A. J., Naib, Z. M., and Josey, W. E., 1971, Herpes virus hominis type 2 infection—Association with cervical cancer and perinatal disease, in *Perspectives in Virology,* Vol. VII (M. Pollard, ed.), pp. 73–89, Academic Press, New York.

Naib, Z. M., Nahmias, A. J., and Josey, W. E., 1966, Cytology and histopathology of cervical herpes simplex infection, *Cancer* **19**:1026.

Nash, H. A., 1975a, Integrative recombination in bacteriophage lambda: Analysis of recombinant DNA, *J. Mol. Biol.* **19**:501.

Nash, H. A., 1975b, Integrative recombination of bacteriophage lambda DNA *in vitro, Proc. Natl. Acad. Sci. USA* **72**:1072–1076.

Nathans, D., and Danna, K. J., 1972, Specific origin in SV40 DNA replication, *Nature (London) New Biol.* **236**:200–202.

Nathans, D., and Smith, H. O., 1975, Restriction endonucleases in the analysis and restructuring of DNA molecules, *Annu. Rev. Biochem.* **44**:273.

Nazerian, K., and Witter, R. L., 1970, Cell-free transmission and in vivo replication of Marek's disease virus, *J. Virol.* **5**:388.

Nazerian, K., Lindahl, T., Klein, G., and Lee, L. F., 1973, Deoxyribonucleic acid of Marek's disease virus in virus-induced tumors, *J. Virol.* **12**:841.

Neiman, P. E., 1973, Measurement of endogenous leukosis virus nucleotide sequences in the DNA of normal avian embryos by RNA-DNA hybridization, *Virology* **53**:196.

Neiman, P. E., Wright, S. E., McMillin, C., and MacDonnell, D., 1974, Nucleotide sequence relationships of avian RNA tumor viruses: Measurement of the deletion in a transformation-defective mutant of Rous sarcoma virus, *J. Virol.* **13**:837.

Nishimoto, T., Raskas, H. J., and Basilico, C., 1975, Temperature-sensitive cell mutations that inhibit adenovirus 2 replication, *Proc. Natl. Acad. Sci. USA* **72**:328.

Nonoyama, M., and Pagano, J. S., 1971, Detection of Epstein-Barr viral genome in nonproductive cells, *Nature (London) New Biol.* **233**:103.

Nonoyama, M., and Pagano, J. S., 1972, Separation of Epstein-Barr virus DNA from large chromosomal DNA in non-virus-producing cells, *Nature (London) New Biol.* **238**:169.

Nonoyama, M., and Pagano, J. S., 1973, Homology between Epstein-Barr virus DNA and viral DNA from Burkitt lymphoma and nasopharyngeal carcinoma determined by DNA-DNA reassociation kinetics, *Nature (London)* **242**:44.

Öberg, B., Saborio, J., Persson, T., Everitt, E., and Philipson, L., 1975, Identification of the *in vitro* translation products of adenovirus m-RNA by immunoprecipitation, *J. Virol.* **15**:199.

Oda, K., and Dulbecco, R., 1968, Regulation of transcription of the SV40 DNA in productively infected and in transformed cells, *Proc. Natl. Acad. Sci. USA* **60**:525.

Old, L. J., Boyse, E. A., Oettgen, H. F., de Harven, E., Geering, G., Williamson, B., and Clifford, P., 1966, Precipitating antibody in human serum to an antigen present in cultured Burkitt's lymphoma cells, *Proc. Natl. Acad. Sci. USA* **56**:1699.

Oren, M., Kuff, E. L., and Winocour, E., 1976, The presence of common host sequences in different populations of substituted SV40 DNA, *Virology* **73**:419.

Ortin, J., and Doerfler, W., 1975, Transcription of the genome of adenovirus type 12. I. Viral m-RNA in abortively infected and transformed cells, *J. Virol.* **15**:27.

Ortin, J., Scheidtmann, K. H., Greenberg, R., Westphal, M., and Doerfler, W., 1976, Transcription of the genome of adenovirus type 12. III. Maps of stable RNA from productively infected human cells and abortively infected and transformed hamster cells, *J. Virol.* **20**:355.

Osborn, M., and Weber, K., 1975, Simian virus 40 gene A function and maintenance of transformation, *J. Virol.* **15**:636.

Oxman, M. N., Levine, A. S., Crumpacker, C. S., Levin, M. J., Henry, P. H., and Lewis, A. M., Jr., 1971, Studies of nondefective adenovirus 2–simian virus 40 hybrid viruses. IV. Characterization of the simian virus 40 ribonucleic acid species induced by wild-type simian virus 40 and by the nondefective hybrid virus, Ad2$^+$ND$_1$, *J. Virol.* **8**:215.

Oxman, M. N., Levin, M. J., and Lewis, A. M., Jr., 1974, Control of simian virus 40 gene expression in adenovirus–simian virus 40 hybrid viruses: Synthesis of hybrid adenovirus 2–simian virus 40 RNA molecules in cells infected with a nondefective adenovirus 2–simian virus 40 hybrid virus, *J. Virol.* **13**:322.

Ozanne, B., Sharp, P. A., and Sambrook, J., 1973, Transcription of simian virus 40. II. Hybridization of RNA extracted from different lines of transformed cells to the separated strands of simian virus 40 DNA, *J. Virol.* **12**:90.

Parkinson, J. S., 1971, Deletion mutants of bacteriophage lambda. II. Genetic properties of *att*-defective mutants, *J. Mol. Biol.* **56**:385.

Parkinson, J. S., and Huskey, R. J., 1971, Deletion mutants of bacteriophage lambda. I. Isolation and initial characterization, *J. Mol. Biol.* **56**:369.

Parks, W. P., Livingston, D. M., Todaro, G. J., Benveniste, R. J., and Scolnick, E. M., 1973, Radioimmunoassay of mammalian type C viral protein. III. Detection of viral antigen in normal murine cells and tissues, *J. Exp. Med.* **137**:622.

Pearson, G. D., 1975, Intermediate in adenovirus type 2 replication, *J. Virol.* **16**:17.

Pettersson, U., 1973, Some unusual properties of replicating adenovirus type 2 DNA, *J. Mol. Biol.* **81**:521.

Pettersson, U., and Philipson, L., 1974, Synthesis of complementary RNA sequences during productive adenovirus infection, *Proc. Natl. Acad. Sci. USA* **71**:4887.

Pettersson, U., and Sambrook, J., 1973, Amount of viral DNA in the genome of cells transformed by adenovirus type 2, *J. Mol. Biol.* **73**:125.

Pettersson, U., Mulder, C., Delius, H., and Sharp, P. A., 1973, Cleavage of adenovirus type 2 DNA into six unique fragments by endonuclease R · RI, *Proc. Natl. Acad. Sci. USA* **70**:200.

Philipson, L., and Pettersson, U., 1973, Structure and function of virion proteins of adenoviruses, *Prog. Exp. Tumor. Res.* **18**:1.

Philipson, L., Wall, R., Glickman, G., and Darnell, J. E., 1971, Addition of poly-adenylate sequences to virus-specific RNA during adenovirus replication, *Proc. Natl. Acad. Sci. USA* **68**:2806.

Philipson, L., Pettersson, U., Lindberg, U., Tibbetts, C., Vennström, B., and Persson, T., 1974, RNA synthesis and processing in adenovirus-infected cells, *Cold Spring Harbor Symp. Quant. Biol.* **39**:447.

Philipson, L., Pettersson, U., and Lindberg, U., 1975, Molecular biology of adenoviruses, in: *Virology Monographs,* Vol. 14 (S. Gard and G. Hallauer, eds.), pp. 1–115, Springer-Verlag, Vienna, New York.

Piña, M., and Green, M., 1965, Biochemical studies on adenovirus multiplication. IX. Chemical and base composition and analysis of 28 human adenoviruses, *Proc. Natl. Acad. Sci. USA* **54**:547.

Pincus, T. H., Hartley, J. W., and Rowe, W. P., 1971*a*, A major genetic locus affecting resistance to infection with murine leukemia viruses. I. Tissue culture studies of naturally occurring viruses, *J. Exp. Med.* **133**:1219.

Pincus, T. H., Rowe, W. P., and Lilly, F., 1971*b*, A major genetic locus affecting resistance to infection with murine leukemia viruses. II. Apparent identity to a major locus described for resistance to Friend murine leukemia virus, *J. Exp. Med.* **133**:1234.

Pollack, E. J., Aaronson, S. A., and Todaro, G. J., 1970, X-irradiation of Balb 3T3: Sarcoma-forming ability and virus induction, *Int. J. Radiat. Biol.* **17**:97.

Pope, J. H., and Rowe, W. P., 1964, Immunofluorescent studies of adenovirus 12 tumors and of cells transformed or infected by adenoviruses, *J. Exp. Med.* **120**:577.

Poste, G., Schaeffer, B., Reeve, P., and Alexander, D. J., 1974, Rescue of simian virus 40 (SV40) from SV40-transformed cells by fusion with anucleate monkey cells and variation in the yield of virus rescued by fusion with replicating or nonreplicating monkey cells, *Virology* **60**:85.

Potter, C. W., Potter, A. M., and Oxford, J. S., 1970, Comparison of transformation and T antigen induction in human cell lines, *J. Virol.* **5**:293.

Prasad, I., Zouzias, D., and Basilico, C., 1975, Simian virus 40 integration sites in the genome of virus-transformed mouse cells, *J. Virol.* **16**:897.

Ptashne, M., 1967*a*, Isolation of the λ phage repressor, *Proc. Natl. Acad. Sci. USA* **57**:306.

Ptashne, M., 1967*b*, Specific binding of the lambda phage repressor to lambda DNA, *Nature (London)* **214**:232.

Ptashne, M., Backman, K., Humayun, H. Z., Jeffrey, A., Maurer, R., Meyer, B., and Sauer, R. T., 1976, Autoregulation and function of a repressor in bacteriophage lambda, *Science* **194**:156.

Pulvertaft, R. J. V., 1965, A study of malignant tumours in Nigeria by short-term tissue culture, *J. Clin. Pathol.* **18**:261.

Purchase, H. G., Witter, R. L., Okazaki, W., and Burmester, B. R., 1971, Vaccination against Marek's disease, *Perspect. Virol.* **7**:91.

Pyeritz, R. E., and Thomas, C. A., Jr., 1973, Regional organization of eukaryotic DNA sequences as studied by the formation of folded rings, *J. Mol. Biol.* **77**:57.

Rabson, A. S., Tyrrell, S. A., and Legallais, F. Y., 1971, An inhibitor of *Herpesvirus hominis* in extracts of cultures of Burkitt's lymphoma cells, *Proc. Soc. Exp. Biol. Med.* **137**:264.

Ralph, R. K., and Colter, J. S., 1972, Evidence for the integration of polyoma virus DNA in a lytic system, *Virology* **48**:49.

Rapp, F., and Jerkofsky, M. A., 1973, Persistent and latent infections, in: *The Herpesviruses* (A. S. Kaplan, ed.), pp. 271–289, Academic Press, New York.

Rapp, F., and Trulock, S. C. 1970, Susceptibility to superinfection of simian cells transformed by SV40, *Virology* **40**:961.

Rapp, F., Melnick, J. L., Butel, J. S., and Kitahara, T., 1964, The incorporation of SV40 genetic material into adenovirus 7 as measured by intranuclear synthesis of SV40 tumor antigen, *Proc. Natl. Acad. Sci. USA* **52**:1348.

Rapp, F., Feldman, L. A., and Mandel, M., 1966, Synthesis of virus deoxyribonucleic acid during abortive infection of simian cells by human adenoviruses, *J. Bacteriol.* **92**:931.

Rapp, F., Conner, R., Glaser, R., and Duff, R., 1972, Absence of leukosis virus markers in hamster cells transformed by herpes simplex virus type 2, *J. Virol.* **9**:1059.

Raška, K., Jr., and Strohl, W. A., 1972, The response of BHK21 cells to infection with type 12 adenovirus. VI. Synthesis of virus-specific RNA, *Virology* **47**:734.

Raška, K., Strohl, W. A., Holowczak, J. A., and Zimmermann, J., 1971, The response of BHK21 cells to infection with type 12 adenovirus. V. Stimulation of cellular RNA synthesis and evidence for transcription of the viral genome, *Virology* **44**:296.

Rawls, W. E., Tompkins, W., and Melnick, J. L., 1969, The association of herpes type 2 and carcinoma of the cervix, *Am. J. Epidemiol.* **89**:547.

Reed, S. I., Ferguson, J., Davis, R. W., and Stark, G. R., 1975, T-antigen binds to simian virus 40 DNA at the origin of replication, *Proc. Natl. Acad. Sci. USA* **72**:1605.

Reedman, B. M., and Klein, G., 1973, Cellular localization of an Epstein-Barr virus (EBV)-associated complement-fixing antigen in producer and non-producer lymphoblastoid cell lines, *Int. J. Cancer* **11**:499.

Reich, P. R., Black, P. H., and Weissman, S. M., 1966, Nucleic acid homology studies of SV40 virus-transformed and normal hamster cells, *Proc. Natl. Acad. Sci. USA* **56**:78.

Reichardt, L. F., 1975a, Control of bacteriophage lambda repressor synthesis after phage infection: The role of the N, cII, cIII and cro products, *J. Mol. Biol.* **93**:267.

Reichardt, L. F., 1975b, Control of bacteriophage lambda repressor synthesis: Regulation of the maintenance pathway by the cro and cI products, *J. Mol. Biol.* **93**:289.

Reif, U., Winterhoff, U., Lundholm, U., Philipson, L., and Doerfler, W., 1977a, Purification of an endonuclease from adenovirus-infected KB cells, *Eur. J. Biochem.* **73**:313.

Reif, U., Winterhoff, U., and Doerfler, W., 1977b, Characterization of the pH 4.0 endonuclease from adenovirus type 2-infected KB cells, *Eur. J. Biochem.* **73**:327.

Risser, R., and Mulder, C., 1974, Relative locations of rearrangements in the DNA of defective simian virus 40 (SV40), *Virology* **58**:424.

Risser, R., and Pollack, R., 1974, A nonselective analysis of SV40 transformation of mouse 3T3 cells, *Virology* **59**:477.

Ritzi, E., and Levine, A. J., 1970, Deoxyribonucleic acid replication in simian virus 40-infected cells. III. Comparison of SV40 lytic infection in three different monkey kidney cell lines, *J. Virol.* **5**:686.

Robb, J. A., and Martin, R. G., 1972, Genetic analysis of simian virus 40. III. Characterization of a temperature-sensitive mutant blocked at an early stage of productive infection in monkey cells, *J. Virol.* **9**:956.

Roberts, J. W., and Roberts, C. W., 1975, Proteolytic cleavage of bacteriophage lambda repressor in induction, *Proc. Natl. Acad. Sci. USA* **72**:147.

Roberts, R. J., Breitmeyer, J. B., Tabachnik, N. F., and Myers, P. A., 1975, A second specific endonuclease from *Haemophilus aegyptius, J. Mol. Biol.* **91**:121.

Robin, J., Bourgaux-Ramoisy, D., and Bourgaux, P., 1973, Single-stranded regions in replicating DNA of adenovirus type 2, *J. Gen. Virol.* **20**:233.

Robinson, A. J., Younghusband, H. B., and Bellett, A. J. D., 1973, A circular DNA–protein complex from adenoviruses, *Virology* **56**:54.

Rokutanda, M., Rokutanda, H., Green, M., Fujinaga, K., Ray, R. K., and Gurgo, C., 1970, Formation of viral RNA-DNA hybrid molecules by the DNA polymerase of sarcoma-leukemia viruses, *Nature (London)* **227**:1026.

Rothman, J. S., 1965, Transduction studies on the relation between prophage and host chromosome, *J. Mol. Biol.* **12**:892.

Rothschild, H., and Black, P. H., 1970, Analysis of SV40-induced transformation of hamster kidney tissue *in vitro*. VII. Induction of SV40 virus from transformed hamster cell clones by various agents, *Virology* **42**:251.

Rowe, W. P., 1972, Studies of genetic transmission of murine leukemia virus by AKR mice. I. Crosses with $Fv-I^n$ strains of mice, *J. Exp. Med.* **136**:1272.

Rowe, W. P., 1973, Genetic factors in the natural history of murine leukemia virus infection: G. H. A. Clowes Memorial Lecture, *Cancer Res.* **33**:3061.

Rowe, W. P., and Baum, S. G., 1964, Evidence for a possible genetic hybrid between adenovirus type 7 and SV40 viruses, *Proc. Natl. Acad. Sci. USA* **52**:1340.

Rowe, W. P., and Hartley, J. W., 1972, Studies of genetic transmission of murine leukemia virus by AKR mice. II. Crosses with $Fv-I^b$ strains of mice, *J. Exp. Med.* **136**:1286.

Rowe, W. P., and Pincus, T., 1972, Quantitative studies of naturally occurring murine leukemia virus infection of AKR mice, *J. Exp. Med.* **135**:429.

Rowe, W. P., Huebner, R. J., Gillmore, L. K., Parrott, R. H., and Ward, T. G., 1953, Isolation of a cytopathogenic agent from human adenoids undergoing spontaneous degeneration in tissue culture, *Proc. Soc. Exp. Biol. Med.* **84**:570.

Rowe, W. P., Pugh, W. E., and Hartley, J. W., 1970, Plaque assay techniques for murine leukemia viruses, *Virology* **42**:1136.

Rowe, W. P., Hartley, J. W., Lander, M. R., Pugh, W. E., and Teich, N., 1971, Noninfectious AKR mouse embryo cell lines in which each cell has the capacity to be activated to produce infectious murine leukemia virus, *Virology* **46**:866.

Rowe, W. P., Lowy, D. R., Teich, N., and Hartley, J. W., 1972a, Some implications of the activation of murine leukemia virus by halogenated pyrimidines, *Proc. Natl. Acad. Sci. USA* **69**:1033.

Rowe, W. P., Hartley, J. W., and Bremner, T. H., 1972b, Genetic mapping of a murine leukemia virus-inducing locus of AKR mice, *Science* **178**:860.

Rozenblatt, S. H., and Winocour, E., 1972, Covalently linked cell and SV40-specific sequences in an RNA from productively infected cells, *Virology* **50**:558.

Rozenblatt, S. H., Lavi, S., Singer, M. F., and Winocour, E., 1973, Acquisition of sequences homologous to host DNA by closed circular simian virus 40 DNA. III. Host sequences, *J. Virol.* **12**:501.

Ruddle, F. H., 1973, Linkage analysis in man by somatic cell genetics, *Nature (London)* **242**:165.

Ruddle, F. H., and Kucherlapati, R. S., 1974, Hybrid cells and human genes, *Sci. Am.* **231**:36.

Ruddle, F. H., Chapman, V. M., Ricciuti, F., Murnane, M., Klebe, R., and Meera Khan, P., 1971, Linkage relationships of seventeen human gene loci as determined by man–mouse somatic cell hybrids, *Nature (London) New Biol.* **232**:69.

Saborio, J. L., and Öberg, B., 1976, *In vivo* and *in vitro* synthesis of adenovirus type 2 early proteins, *J. Virol.* **17**:865.

Sack, G. H., Jr., and Nathans, D., 1973, Studies of SV40 DNA. VI. Cleavage of SV40 DNA by restriction endonuclease from *Hemophilus parainfluenzae*, *Virology* **51**:517.

Sambrook, J., 1972, Transformation by polyoma virus and simian virus 40, *Adv. Cancer Res.* **16**:141.

Sambrook, J., and Shatkin, A. J., 1969, Polynucleotide ligase activity in cells infected with simian virus 40, polyoma virus or vaccinia virus, *J. Virol.* **4**:719.

Sambrook, J., Westphal, H., Srinivasan, P. R., and Dulbecco, R., 1968, The integrated state of viral DNA in SV40-transformed cells, *Proc. Natl. Acad. Sci.* **60**:1288.

Sambrook, J., Sharp, P. A., and Keller, W., 1972, Transcription of simian virus 40. I. Separation of the strands of SV40 DNA and hybridization of the separated strands to RNA extracted from lytically infected and transformed cells, *J. Mol. Biol.* **70**:57.

Sambrook, J., Sugden, B., Keller, W., and Sharp, P. A., 1973, Transcription of SV40. III. Mapping of "early" and "late" species of RNA, *Proc. Natl. Acad. Sci. USA* **70**:3711.

Sambrook, J., Botchan, M., Gallimore, P., Ozanne, B., Pettersson, U., Williams, J., and Sharp, P. A., 1974, Viral DNA sequences in cells transformed by simian virus 40, adenovirus type 2 and adenovirus type 5, *Cold Spring Harbor Symp. Quant. Biol.* **39**:615.

Sauer, G., and Hahn, E. C., 1970, The interaction of SV40 with SV40-transformed and non-transformed monkey kidney cells, *Z. Krebsforsch.* **7**:40.

Sauer, G., and Kidwai, J. R., 1968, The transcription of the SV40 genome in productively infected and transformed cells, *Proc. Natl. Acad. Sci. USA* **61**:1256.

Scheidtmann, K. H., Ortin, J., and Doerfler, W., 1975, Transcription of the genome of adenovirus type 12. II. Viral mRNA in productively infected KB cells, *Eur. J. Biochem.* **58**:283.

Schick, J., Baczko, K., Fanning, E., Groneberg, J., Burger, H., and Doerfler, W., 1976, Intracellular forms of adenovirus DNA. IV. The integrated form of adenovirus DNA appears early in productive infection, *Proc. Natl. Acad. Sci. USA* **73**:1043.

Schilling, R., Weingärtner, B., and Winnacker, E.-L., 1975, Adenovirus type 2 DNA replication. II. Termini of DNA replication, *J. Virol.* **16**:767.

Schlesinger, R. W., 1969, Adenoviruses: The nature of the virion and of controlling factors in productive or abortive infection and tumorigenesis, *Adv. Virus Res.* **14**:1.

Schroeder, W., and van de Putte, P., 1974, Genetic study of prophage excision with a temperature inducible mutant of mu-1, *Mol. Gen. Genet.* **130**:99.

Schroeder, W. H., Bade, E. G., and Delius, H., 1974, Participation of *Escherichia coli* DNA in the replication of temperate bacteriophage mu 1, *Virology* **60**:534.

Schulte-Holthausen, H., 1977, personal communication.

Schulte-Holthausen, H., and zur Hausen, H., 1970, Partial purification of the Epstein-Barr virus and some properties of its DNA, *Virology* **40**:776.

Scott, W. A., Brockman, W. W., and Nathans, D., 1976, Biological activities of deletion mutants of simian virus 40, *Virology* **75**:319.

Shani, M., Rabinowitz, Z., and Sachs, L., 1972, Virus deoxyribonucleic acid sequences in subdiploid and subtetraploid revertants of polyoma-transformed cells, *J. Virol.* **10**:456.

Shani, M., Aloni, Y., Huberman, E., and Sachs, L., 1976, Gene activation by transfer of isolated mammalian chromosomes. II. Activation of simian virus 40 by transfer of fractionated chromosomes from transformed cells, *Virology* **70**:201.

Shanmugaratnam, K., 1967, in: *Racial and Geographical Factors in Tumour Incidence* (A. A. Shiras, ed.), p. 169, Edinburgh University Press, Edinburgh.

Sharp, P. A., Sugden, B., and Sambrook, J., 1973, Detection of two restriction endonuclease activities in *H. parainfluenzae* using analytical agarose–ethidium bromide gel electrophoresis, *Biochemistry* **12**:3055.

Sharp, P. A., Pettersson, U., and Sambrook, J., 1974a, Viral DNA in transformed cells. I. A study of the sequences of adenovirus 2 DNA in a line of transformed rat cells using specific fragments of the viral genome, *J. Mol. Biol.* **86**:709.

Sharp, P. A., Gallimore, P. H., and Flint, S. J., 1974b, Mapping of adenovirus 2 RNA sequences in lytically infected cells and transformed cell lines, *Cold Spring Harbor Symp. Quant. Biol.* **39**:457.

Shein, H. M., and Enders, J. F., 1962, Transformation induced by simian virus 40 in human renal cell cultures. I. Morphology and growth characteristics, *Proc. Natl. Acad. Sci. USA* **48**:1164.

Shimada, K., and Campbell, A., 1974a, *Int*-constitutive mutants of bacteriophage lambda, *Proc. Natl. Acad. Sci. USA* **71**:237.

Shimada, K., and Campbell, A., 1974b, Lysogenization and curing by *int*-constitutive mutants of phage λ, *Virology* **60**:157.

Shimada, K., Weisberg, R. A., and Gottesman, M. E., 1973, Prophage lambda at unusual chromosomal locations, *J. Mol. Biol.* **80**:297.

Shimada, K., Weisberg, R. A., and Gottesman, M. E., 1975, Prophage lambda at unusual chromosomal locations. III. The components of the secondary attachment sites, *J. Mol. Biol.* **93**:415.

Shin, S., Freedman, V. H., Risser, R., and Pollack, R., 1975, Tumorigenicity of virus-transformed cells in nude mice is correlated specifically with anchorage independent growth *in vitro*, *Proc. Natl. Acad. Sci. USA* **72**:4435.

Shulman, M., and Gottesman, M., 1973, Attachment site mutants of bacteriophage lambda, *J. Mol. Biol.* **81**:461.

Signer, E. R., 1968, Lysogeny: The integration problem, *Annu. Rev. Microbiol.* **22**:451.

Signer, E. R., 1969, Plasmid formation: A new mode of lysogeny by phage lambda, *Nature (London)* **223**:158.

Signer, E. R., and Beckwith, J. R., 1966, Transcription of the *lac* region of *Escherichia coli.* III. The mechanism of attachment of bacteriophage ϕ80 to the bacterial chromosome, *J. Mol. Biol.* **22**:33.

Signer, E. R., Weil, J., and Kimball, P. C., 1969, Recombination in bacteriophage λ. III: Studies on the nature of the prophage attachment region, *J. Mol. Biol.* **46**:543.

Simon, M. N., Davis, R. W., and Davidson, N., 1971, Heteroduplexes of DNA molecules of lambdoid phages: Physical mapping of their base sequence relationships by electron microscopy, in: *The Bacteriophage Lambda* (A. D. Hershey ed.), pp. 313–328, Cold Spring Harbor Laboratory, Cold Spring Harbor, N.Y.

Simpson, R. W., and Iinuma, M., 1975, Recovery of infectious proviral DNA from mammalian cells infected with respiratory syncytial virus, *Proc. Natl. Acad. Sci. USA* **72**:3230.

Smith, H. O., and Wilcox, K. W., 1970, A restriction enzyme from *Hemophilus influenzae.* I. Purification and general properties, *J. Mol. Biol.* **51**:379.

Smith, H. S., Gelb, L. D., and Martin, M. A., 1972, Detection and quantitation of simian virus 40 genetic material in abortively transformed BALB/3T3 clones, *Proc. Natl. Acad. Sci. USA* **69**:152.

Smithies, O., 1973, Immunoglobulin genes: Arranged in tandem or in parallel? *Cold Spring Harbor Symp. Quant. Biol.* **38**:725.

Söderlund, H., Pettersson, U., Vennström, B., Philipson, L., and Mathews, M. B., 1976, A new species of virus-coded low moleuclar weight RNA from cells infected with adenovirus type 2, *Cell* **7**:585.

Spiegelman, S., Burny, A., Das, M. R., Keydar, J., Schlom, J., Travnicek, M., and Watson, K., 1970, Synthetic DNA-RNA hybrids and RNA-RNA duplexes as templates for the polymerases of the oncogenic RNA viruses, *Nature (London)* **228**:430.

Staal, S. P., and Rowe, W. P., 1975, Enhancement of adenovirus infection in WI-38 and AGMK cells by pretreatment of cells with 5-iododeoxyuridine, *Virology* **64**:513.

Starlinger, P., and Saedler, H., 1972, Insertion mutations in microorganisms, *Biochimie* **54**:177.

Starlinger, P., and Saedler, H., 1976, IS-elements in microorganisms, *Curr. Top. Microbiol. Immunol.* **75**:111.

Stehelin, D., Guntaka, R. V., Varmus, H. E., and Bishop, J. M., 1976a, Purification of DNA complementary to nucleotide sequences required for neoplastic transformation of fibroblasts by avian sarcoma viruses, *J. Mol. Biol.* **101**:349.

Stehelin, D., Varmus, H. E., Bishop, J. M., Moscovici, C., and Vogt, P. K., 1976, DNA related to the transforming genes of avian sarcoma viruses is present in normal avian DNA, *Nature (London)* **260**:170.

Stevens, J. G., 1975, Latent herpes simplex virus and the nervous system, *Curr. Top. Microbiol. Immunol.* **70**:31.

Stich, H. F., and Yohn, D. S., 1967, Mutagenic capacity of adenoviruses for mammalian cells, *Nature (London)* **216**:1292.

Stich, H. F., and Yohn, D. S., 1970, Viruses and chromosomes, *Prog. Med. Virol.* **12**:78.

Stockert, E., Old, L. J., and Boyse, E. A., 1971, The G$_{IX}$ system: A cell surface allo-antigen associated with murine leukemia virus; implications regarding chromosomal integration of the viral genome, *J. Exp. Med.* **133:**1334.

Strohl, W. A., 1969*a*, The response of BHK21 cells to infection with type 12 adenoviruses. I. Cell killing and T antigen synthesis as correlated viral genome functions, *Virology* **39:**642.

Strohl, W. A., 1969*b*, The response of BHK21 cells to infection with type 12 adenovirus. II. Relationship of virus-stimulated DNA synthesis to other viral functions, *Virology* **39:**653.

Strohl, W. A., 1973, Alterations in hamster cell regulatory mechanism resulting from abortive infection with an oncogenic adenovirus, *Progr. Exp. Tumor Res.* **18:**199.

Strohl, W. A., Rabson, A. S., and Rouse, H., 1967, Adenovirus tumorigenesis: Role of the viral genome in determining tumor morphology, *Science* **156:**1631.

Strohl, W. A., Rouse, H., Teets, K., and Schlesinger, R. W., 1970, The response of BHK21 cells to infection with type 12 adenovirus. III. Transformation and restricted replication of superinfecting type 2 adenovirus, *Arch. Ges. Virusforsch.* **31:**93.

Subramanian, K. N., Pan, J., Zain, S., and Weissman, S., 1974, The mapping and ordering of fragments of SV40 DNA produced by restriction endonucleases, *Nucleic Acid Res.* **1:**727.

Subramanian, K. N., Dhar, R., and Weissman, S. M., 1975, Nucleotide sequences of a fragment of SV40 DNA spanning the origin of DNA replication, personal communication.

Sussenbach, J. S., van der Vliet, P. C., Ellens, D. J., and Jansz, H. S., 1972, Linear intermediates in the replication of adenovirus DNA, *Nature* (*London*) *New Biol.* **239:**47.

Sussenbach, J. S., Ellens, D. J., and Jansz, H. S., 1973, Studies on the mechanism of replication of adenovirus DNA. II. The nature of single-stranded DNA in replicative intermediates, *J. Virol.* **12:**1131.

Sweet, B. H., and Hilleman, M. R., 1960, The vacuolating virus, SV40, *Proc. Soc. Exp. Biol. Med.* **105:**420.

Syvanen, M., 1974, *In vitro* genetic recombination of bacteriophage λ, *Proc. Natl. Acad. Sci. USA* **71:**2496.

Szybalski, W., 1974, Genetic and molecular map of *Escherichia coli* bacteriophage λ, in: *CRC Handbook of Microbiology,* Vol. 4 (A. I. Laskin and H. A. Lechevalier, eds.), pp. 611–618, CRC Press, Cleveland.

Szybalski, W., Bøvre, K., Fiandt, M., Hayes, S., Hradecna, Z., Kumar, S., Lozeron, H. A., Nijkamp, H. J. J., and Stevens, W. F., 1970, Transcriptional units and their controls in *Escherichia coli* phage λ: Operons and scriptons, *Cold Spring Harbor Symp. Quant. Biol.* **35:**341.

Tai, H. T., and O'Brien, R. L., 1969, Multiplicity of viral genomes in a SV40-transformed hamster cell line, *Virology* **38:**698.

Tai, H. T., Smith, C. A., Sharp, P. A., and Vinograd, J., 1972, Sequence heterogeneity in closed SV40 DNA, *J. Virol.* **9:**317.

Tal, J., Craig, E. A., Zimmer, S., and Raskas, H. J., 1974, Localization of adenovirus 2 m-RNA to segments of the viral genome defined by endonuclease R·RI, *Proc. Natl. Acad. Sci. USA* **71:**4057.

Tanaka, A., and Nonoyama, M., 1974, Latent DNA of Epstein-Barr virus: Separation from high molecular weight cell DNA in a neutral glycerol gradient, *Proc. Natl. Acad. Sci. USA* **71:**4658.

Taube, S. E., McGuire, P. M., and Hodge, L. D., 1974, RNA synthesis specific for an integrated adenovirus genome during the cell cycle, *Nature (London)* **250**:416.

Taylor, A. L., 1963, Bacteriophage-induced mutation in *Escherichia coli, Proc. Natl. Acad. Sci. USA* **50**:1043.

Taylor, A. L., and Trotter, C. D., 1967, Revised linkage map of *Escherichia coli, Bacteriol. Rev.* **31**:332.

Tegtmeyer, P., 1975, Function of simian virus 40 gene A in transforming infection, *J. Virol.* **15**:613.

Tegtmeyer, P., and Ozer, H. L., 1971, Temperature-sensitive mutants of SV40: Infection of permissive cells, *J. Virol.* **8**:516.

Tegtmeyer, P., Schwartz, M., Collins, J. K., and Rundell, K., 1975, Regulation of tumor antigen synthesis by simian virus 40 gene A, *J. Virol.* **16**:168.

Teich, N., Lowy, D. R., Hartley, J. W., and Rowe, W. P., 1973, Studies of the mechanism of induction of infectious murine leukemia virus from AKR mouse embryo cell lines by 5-iododeoxyuridine and 5-bromodeoxyuridine, *Virology* **51**:163.

Temin, H. M., 1963, The effects of actinomycin D on growth of Rous sarcoma virus *in vitro, Virology* **20**:577.

Temin, H. M., 1964, The participation of DNA in Rous sarcoma virus production, *Virology* **23**:486.

Temin, H. M., 1971, Mechanism of cell transformation by RNA tumor viruses, *Annu. Rev. Microbiol.* **25**:609.

Temin, H. M., and Baltimore, D., 1972, RNA-directed DNA synthesis and RNA tumor viruses, *Adv. Virus Res.* **17**:129.

Temin, H. M., and Mizutani, S., 1970, RNA-dependent DNA polymerase in virions of Rous sarcoma virus, *Nature (London)* **226**:1211.

ter Meulen, V., Katz, M., and Müller, D., 1972, Subacute sclerosing panencephalitis: A review, *Curr. Top. Microbiol. Immunol.* **57**:1.

Tibbetts, C., Pettersson, U., Johansson, K., and Philipson, L., 1974, Relation of m-RNA from productively infected cells to the complementary strands of adenovirus type 2 DNA, *J. Virol.* **13**:370.

Tjia, S., Fanning, E., Schick, J., and Doerfler, W., 1977, Incomplete particles of adenovirus type 2. III. Viral and cellular DNA sequences in incomplete particles, *Virology* **76**:365.

To, C. M., Eisenstark, A., and Töreci, B., 1966, Structure of mutatorphage mu 1 of *E. coli, J. Ultrastruct. Res.* **14**:441.

Todaro, G. J., 1972, "Spontaneous" release of type C viruses from clonal lines of "spontaneously" transformed BALB/3T3 cells, *Nature (London)* **240**:157.

Todaro, G. J., and Green, H., 1966*a*, High frequency of SV40 transformation of mouse cell line 3T3, *Virology* **28**:756.

Todaro, G. J., and Green, H., 1966*b*, Cell growth and the initiation of transformation by SV40, *Proc. Natl. Acad. Sci. USA* **55**:302.

Todaro, G. J., Green, H., and Swift, M. R., 1966*c*, Susceptibility of human diploid fibroblast strains to transformation by SV40 virus, *Science* **153**:1252.

Todaro, G. J., Benveniste, R. E., Callahan, R., Lieber, M. M., and Sherr, C. J., 1974, Endogenous primate and feline type C viruses, *Cold Spring Harbor Symp. Quant. Biol.* **39**:1159.

Tolun, A., and Pettersson, U., 1975, Termination sites of adenovirus type 2 DNA replication, *J. Virol.* **16**:759.

Tomizawa, J., and Ogawa, T., 1968, Replication of phage lambda DNA, *Cold Spring Harbor Symp. Quant. Biol.* **33**:533.

Tonegawa, S., Walter, G., Bernardini, A., and Dulbecco, R., 1970, Transcription of the SV40 genome in transformed cells and during lytic infection, *Cold Spring Harbor Symp. Quant. Biol.* **35**:823.

Tooze, J., ed., 1973, *The Molecular Biology of Tumor Viruses,* Cold Spring Harbor Laboratory, Cold Spring Harbor, N.Y.

Torti, F., Barksdale, C., and Abelson, J., 1970, Mu-1 bacteriophage DNA, *Virology* **41**:567.

Tournier, P., Cassingena, R., Wicker, R., Coppey, J., and Suarez, H., 1967, Etude du mécanisme de l'induction chez des cellules de hamster Syrien transformées par le virus SV40. I. Propriétés d'une lignée cellulaire clonale, *Int. J. Cancer* **2**:117.

Toussaint, A., 1969, Insertion of phage mu-1 within prophage λ: A new approach for studying the control of the late functions in bacteriophage λ, *Mol. Gen. Genet.* **106**:89.

Toussaint, A., Faelen, M., and Couturier, M., 1971, Mu-1 promoted integration of a λgal phage in the chromosome of *E. coli, Mol. Gen. Genet.* **113**:367.

Trentin, J. J., Yabe, Y., and Taylor, G., 1962, The quest for human cancer viruses, *Science* **137**:835.

Tseui, D., Fujinaga, K., and Green, M., 1972, The mechanism of viral carcinogenesis by DNA mammalian viruses: RNA transcripts containing viral and highly reiterated cellular base sequences in adenovirus-transformed cells, *Proc. Natl. Acad. Sci. USA* **69**:427.

van de Putte, P., and Gruijthuijsen, M., 1972, Chromosome-mobilization and integration of F-factors in the chromosome of Rec A strains of *E. coli* under the influence of bacteriophage mu-1, *Mol. Gen. Genet.* **118**:173.

van der Eb, A. J., 1973, Intermediates in type 5 adenovirus DNA replication, *Virology* **51**:11.

van der Eb, A. J., van Kesteren, L. W., and van Bruggen, E. F. J., 1969, Structural properties of adenovirus DNA's, *Biochim. Biophys. Acta* **182**:530.

van der Vliet, P. C., and Sussenbach, J. S., 1972, The mechanism of adenovirus-DNA-synthesis in isolated nuclei, *Eur. J. Biochem.* **30**:584.

Varmus, H. E., Weiss, R. A., Friis, R. R., Levinson, W., and Bishop, J. M., 1972, Detection of avian tumor virus-specific nucleotide sequences in avian cell DNA, *Proc. Natl. Acad. Sci. USA* **69**:20.

Varmus, H. E., Bishop, J. M., and Vogt, P. K., 1973a, Appearance of virus-specific DNA in mammalian cells following transformation by Rous sarcoma virus, *J. Mol. Biol.* **74**:613.

Varmus, H. E., Vogt, P. K., and Bishop, J. M., 1973b, Integration of deoxyribonucleic acid specific for Rous sarcoma virus after infection of permissive and nonpermissive hosts, *Proc. Natl. Acad. Sci. USA* **70**:3067.

Varmus, H. E., Guntaka, R. V., Deng, C. T., and Bishop, J. M., 1974, Synthesis, structure, and function of avian sarcoma virus-specific DNA in permissive and nonpermissive cells, *Cold Spring Harbor Symp. Quant. Biol.* **39**:987.

Varmus, H. E., Heasley, S., Linn, J., and Wheeler, K., 1976, Use of alkaline sucrose gradients in a zonal rotor to detect integrated and unintegrated avian sarcoma virus-specific DNA in cells, *J. Virol.* **18**:574.

Vigier, P., 1970, RNA oncogenic viruses: Structure, replication and oncogenicity, *Prog. Med. Virol.* **12**:240.

Vlak, J. M., Rozijn, T. H., and Sussenbach, J. S., 1975, Studies on the mechanism of replication of adenovirus DNA. IV. Discontinuous DNA chain propagation, *Virology* **63**:168.

Waggoner, B. T., González, N. S., and Taylor, A. L., 1974, Isolation of heterogeneous circular DNA from induced lysogens of bacteriophage mu-1, *Proc. Natl. Acad. Sci. USA* **71**:1255.

· Waldeck, W., Kammer, K., and Sauer, G., 1973, Preferential integration of simian virus 40 deoxyribonucleic acid into a particular size class of CV-1 cell deoxyribonucleic acid, *Virology* **54**:452.

Wall, R., and Darnell, J. E., 1971, Presence of cell and virus-specific sequences in the same molecules of nuclear RNA from virus transformed cells, *Nature (London) New Biol.* **232**:73.

Wall, R., Philipson, L., and Darnell, J. E., 1972, Processing of adenovirus specific nuclear RNA during virus replication, *Virology* **50**:27.

Wall, R., Weber, J., Gage, Z., and Darnell, J. E., 1973, Production of viral m-RNA in adenovirus-transformed cells by the post-transcriptional processing of heterogeneous nuclear RNA containing viral and cell sequences, *J. Virol.* **11**:953.

Wang, J. C., and Kaiser, A. D., 1973, Evidence that the cohesive ends of mature λ DNA are generated by the gene A product, *Nature (London) New Biol.* **241**:16.

Watkins, J. F., and Dulbecco, R., 1967, Production of SV40 virus in heterokaryons of transformed and susceptible cells, *Proc. Natl. Acad. Sci. USA* **58**:1396.

Weber, J., 1974, Absence of adenovirus-specific repressor in adenovirus tumour cells, *J. Gen. Virol.* **22**:259.

Weber, J., and Mak, S., 1972, Synthesis of viral components in hybrids of differentially permissive cells infected with adenovirus type 12, *Exp. Cell Res.* **74**:423.

Weil, J., and Signer, E. R., 1968, Recombination in bacteriophage λ. II. Site-specific recombination promoted by the integration system, *J. Mol. Biol.* **34**:273.

Weil, R., Salomon, C., May, E., and May, P., 1974, A simplifying concept in tumor virology: Virus-specific "pleiotropic effectors," *Cold Spring Harbor Symp. Quant. Biol.* **39**:381.

Weinberg, R. A., Warnaar, S. O., and Winocour, E., 1972, Isolation and characterization of SV40 RNA, *J. Virol.* **10**:193.

Weinberg, R. A., Ben-Ishai, Z., and Newbold, J. E., 1974, SV40-transcription in productively infected and transformed cells, *J. Virol.* **13**:1263.

Weingärtner, B., Winnacker, E.-L., Tolun, A., and Pettersson, U., 1976, Two complementary strand-specific termination sites for adenovirus DNA replication, *Cell* **9**:259.

Weisberg, R. A., and Gallant, J. A., 1967, Dual function of the prophage repressor, *J. Mol. Biol.* **25**:537.

Weisberg, R. A., and Gottesman, M. E., 1971, The stability of *int* and *xis* functions, in: *The Bacteriophage Lambda* (A. D. Hershey, ed.), pp. 489–500, Cold Spring Harbor Laboratory, Cold Spring Harbor, N.Y.

Weiss, M. C., 1970, Further studies on loss of T-antigen from somatic hybrids between mouse cells and SV40-transformed human cells, *Proc. Natl. Acad. Sci. USA* **66**:79.

Weiss, M. C., Ephrussi, B., and Scaletta, L. J., 1968, Loss of T-antigen from somatic hybrids between mouse cells and SV40-transformed human cells, *Proc. Natl. Acad. Sci. USA* **59**:1132.

Weiss, R. A., Friis, R. R., Katz, E., and Vogt, P. K., 1971, Induction of avian tumor viruses in normal cells by physical and chemical carcinogens, *Virology* **46**:920.

Westphal, H., 1970, SV40 DNA strand selection by *Escherichia coli* RNA polymerase, *J. Mol. Biol.* **50**:407.

Westphal, H., and Dulbecco, R., 1968, Viral DNA in polyoma- and SV40-transformed cell lines, *Proc. Natl. Acad. Sci. USA* **59**:1158.

Westphal, H., and Kiehn, E. D., 1970, The in vitro product of SV40 DNA transcription and its specific hybridization with DNA of SV40-transformed cells, *Cold Spring Harbor Symp. Quant. Biol.* **35**:819.

Wetmur, J. G., and Davidson, N., 1968, Kinetics of renaturation of DNA, *J. Mol. Biol.* **31**:349.

Wijffelman, C. A., Westmaas, G. C., and van de Putte, P., 1972, Vegetative recombination of bacteriophage mu-1 in *Escherichia coli, Mol. Gen. Genet.* **116**:40.

Wijffelman, C. A., Westmaas, G. C., and van de Putte, P., 1976, Similarity of vegetative map and prophage map of bacteriophage mu-1, *Virology* **54**:125.

Wilson, D. A., and Thomas, C. A., Jr., 1974, Palindromes in chromosomes, *J. Mol. Biol.* **84**:115.

Winnacker, E. L., 1974, Origins and termini of adenovirus type 2 DNA replication, *Cold Spring Harbor Symp. Quant. Biol.* **39**:547.

Winnacker, E. L., 1975, Adenovirus type 2 DNA replication. I. Evidence for discontinuous DNA synthesis, *J. Virol.* **15**:744.

Winocour, E., 1967, On the apparent homology between DNA from polyoma virus and normal mouse synthetic RNA, *Virology* **31**:15.

Winocour, E., 1971, The investigation of oncogenic viral genomes in transformed cells by nucleic acid hybridization, *Adv. Cancer Res.* **14**:37.

Winocour, E., Frenkel, N., Lavi, S., Osenholts, M., and Rozenblatt, S., 1974, Host substitution in SV40 and polyoma DNA, *Cold Spring Harbor Symp. Quant. Biol.* **39**:101.

Wolf, H., Werner, J., and zur Hausen, H., 1974, EBV DNA in nonlymphoid cells of nasopharyngeal carcinomas and in a malignant lymphoma obtained after inoculation of EBV into cottontop marmosets, *Cold Spring Harbor Symp. Quant. Biol.* **39**:791.

Wolfson, J., and Dressler, D., 1972, Adenovirus 2-DNA contains an inverted terminal repetition, *Proc. Natl. Acad. Sci. USA* **69**:3054.

Wu, R., and Taylor, E., 1971, Nucleotide sequence analysis of DNA. II. Complete nucleotide sequence of the cohesive ends of bacteriophage λ DNA, *J. Mol. Biol.* **57**:491.

Wyke, J. A., 1975, Temperature sensitive mutants of avian sarcoma viruses, *Biochim. Biophys. Acta* **417**:91.

Yamamoto, H., Shimojo, H., and Hamada, C. H., 1972, Less tumorigenic (*cyt*) mutants of adenovirus 12 defective in induction of cell surface change, *Virology* **50**:743.

Yoshiike, K., 1968, Studies on DNA from low-density particles of SV40. I. Heterogeneous defective virions produced by successive undiluted passages, *Virology* **34**:391.

Yoshiike, K., Furuno, A., and Uchida, S., 1974, Rescue of defective SV40 from a transformed mouse 3T3 cell line: Selection of a specific defective, *Virology* **60**:342.

Young, E. T., and Sinsheimer, R. L., 1964, Novel intra-cellular forms of lambda DNA, *J. Mol. Biol.* **10**:562.

Zaenen, I., van Larebeke, N., Teuchy, H., van Montagu, M., and Schell, J., 1974, Supercoiled circular DNA in crown-gall inducing Agrobacterium strains, *J. Mol. Biol.* **86**:109.

Zain, B. S., Dhar, R., Weissman, S. M., Lebowitz, P., and Lewis, A. M., Jr., 1973, Preferred site for initiation of RNA transcription by *Escherichia coli* RNA polymerase within the simian virus 40 DNA segment of the nondefective adenovirus–simian virus 40 hybrid viruses Ad2$^+$ND$_1$ and Ad2$^+$ND$_3$, *J. Virol.* **11**:682.

Zeldis, J. B., Bukhari, A. I., and Zipser, D., 1973, Orientation of prophage mu, *Virology* **55**:289.

Zhdanov, V. M., 1975, Integration of viral genomes, *Nature (London)* **256**:471.

Zissler, J., 1967, Integration-negative (*int*) mutants of phage λ, *Virology* **31**:189.

zur Hausen, H., 1967, Induction of specific chromosomal aberrations by adenovirus type 12 in human embryonic kidney cells, *J. Virol.* **1**:1174.

zur Hausen, H., 1968, Association of adenovirus type 12 DNA with host cell chromosomes, *J. Virol.* **2**:218.

zur Hausen, H., 1972, Epstein-Barr virus in human tumor cells, *Int. Rev. Exp. Pathol.* **11**:233.

zur Hausen, H., 1973, Interactions of adenovirus type 12 with host cell chromosomes, *Prog. Exp. Tumor Res.* **18**:240.

zur Hausen, H., 1975, Oncogenic herpes viruses, *Biochim. Biophys. Acta* **417**:25.

zur Hausen, H., and Sokol, F., 1969, Fate of adenovirus type 12 genomes in nonpermissive cells, *J. Virol.* **4**:256.

zur Hausen, H., and Schulte-Holthausen, H., 1970, Presence of EB virus nucleic acid homology in a "virus-free" line of Burkitt tumour cells, *Nature (London)* **227**:245.

zur Hausen, H., Schulte-Holthausen, H., Klein, G., Henle, W., Henle, G., Clifford, P., and Santesson, L., 1970, EBV DNA in biopsies of Burkitt tumours and anaplastic carcinomas of the nasopharynx, *Nature (London)* **228**:1056.

Cell Transformation by RNA Tumor Viruses

Hidesaburo Hanafusa

The Rockefeller University
New York, New York 10021

1. INTRODUCTION

Along with other carcinogens of physical or chemical origin, viruses are known to be associated with and to cause tumors in a variety of experimental animals. RNA tumor viruses in particular are widespread in many species of animals, and frequently cause sarcomas or leukemia. The isolation of fowl sarcoma-leukemia virus in the early 1910s by a number of investigators marked the first successful identification of such tumor viruses (Ellermann and Bang, 1909; Rous, 1911; Fujinami and Inamoto, 1914). The cell–virus systems used for the studies of basic aspects of the pathogenesis of avian tumor viruses illustrate the progress in methodology for studying animal viruses in general. Early work in animal hosts was gradually replaced by a system using the chorioallantoic membrane of eggs (Keogh, 1938), then by tissue culture cells (Manaker and Groupé, 1956). The establishment of an assay system for Rous sarcoma virus (RSV) in tissue culture cells (Temin and Rubin, 1958) eventually led to an era of quantitative studies on the mechanism of cellular alteration by viruses.

Cells altered *in vitro* by RNA tumor viruses, particularly by sarcoma viruses, are essentially the same in many properties (e.g., morphology and less restricted growth) as cells cultured from tumors induced in animals by the same virus. Thus focus formation in fibroblast cultures is considered to mimic tumor formation. The changes in cellular properties at the level of cell culture are collectively called "cell transformation." Early reports of RSV infection also used the term "cell conversion." The term "transformation" has been generally accepted since the Cold Spring Harbor meeting in 1962. On the other hand, leukemia viruses, with some exceptions, cause neither solid tumors *in vivo* nor morphological changes in infected fibroblast cells *in vitro*. These leukemia viruses do not contain the transformation gene detectable in sarcoma viruses. For this reason, the leukemia viruses are often called "nontransforming viruses" as opposed to "transforming" sarcoma viruses. However, leukemia viruses clearly cause malignancy in blood cells. Moreover, *in vitro* transformation systems using hematopoietic cells have been developed for a number of strains of this class of viruses. To avoid confusion, a distinction has been made in this chapter between acute leukemia viruses, which cause transformation in tissue culture, and lymphatic leukemia viruses, which do not cause such alteration (see Section 2.2). This distinction is not always clear, and the terms "transforming" and "nontransforming" should be used with caution to avoid oversimplification of these properties.

RNA tumor viruses possess attractive features for study of the mechanisms involved in cell transformation. These viruses cause rapid and highly efficient transformation which is reproducible under well-defined cellular conditions. The changes can be synchronously induced so that sequential events can be analyzed. Furthermore, the virus introduces a relatively small piece of genetic material, and it has become clear that the product(s) coded by one viral gene is responsible for the induction and maintenance of cell transformation. It now appears that the intensified efforts to identify the key product(s) and elucidate its functional role in the alteration of cellular regulatory mechanisms could lead to an understanding of the crucial steps in neoplastic transformation by virus. Therefore, in this chapter, emphasis will be placed on the nature of viral genes responsible for transformation and on the cellular properties altered by these agents.

2. TRANSFORMING AGENTS

Most of the RNA tumor viruses thus far tested cause neoplastic diseases in host animals. The original classification of viruses was based on the type of pathogenic changes they induced. While this nomenclature is still maintained to distinguish different groups of viruses, recent biochemical and biological studies have made it possible to classify RNA tumor viruses in a single family, based on their unique structural components and their replicative cycles (described in Volume 4, Chapter 4, of *Comprehensive Virology*). In this section, a brief history of RNA tumor viruses will be given. Further, because the rest of the chapter is primarily concerned with sarcoma virus-induced transformation, the leukemia viruses will be discussed in some detail here.

2.1. Sarcoma Viruses

In 1911, Rous isolated a filtrable agent which produced sarcomas in chickens. This virus, originally called the "agent of chicken tumor I" but later named "Rous sarcoma virus" (RSV), has been widely used as a prototype of sarcoma viruses. Many different strains of RSV are now known (Table 1), all of which are considered to be derived from the

TABLE 1
Avian and Murine RNA Tumor Viruses

	Sarcoma virus	Acute leukemia virus	Lymphatic leukemia virus
Avian	Rous sarcoma virus (RSV) Bryan high-titer strain (BH-RSV) Prague strain (PR-RSV) Schmidt-Ruppin strain (SR-RSV) Harris strain (HA-RSV) Carr-Zilber strain (CZ-RSV) Avian sarcoma virus—Blatislava 77 (B77) Fujinami sarcoma virus	Avian myeloblastosis virus (AMV) Avian erythroblastosis virus (AEV) Myelocytomatosis virus (MC29)	Avian lymphomatosis virus strain RPL12 Resistance-inducing factor (RIF) Rous-associated viruses (RAV)
Murine	Harvey sarcoma virus (HaSV) Moloney sarcoma virus (MoSV) Kirsten sarcoma virus (KiSV) Finkel-Biskis-Jinkins sarcoma virus (FBJSV) Gazdar sarcoma virus (GaSV)	Friend leukemia virus Rauscher leukemia virus Abelson leukemia virus	Murine leukemia virus Gross, Graffi, Moloney, Kirsten, Tennant, and other strains Radiation-induced virus LLV associated with acute leukemia viruses LLV isolated from AKR and other strains of mice

original RSV isolate. The different strains are generally named after
the researcher involved at some stage in the derivation of the particular
strain. A partial history of these viruses was given by Morgan (1964),
but a precise history as to how each strain evolved is not clear. The
serotype (neutralization specificity, determined by the envelope glycoprotein in the case of this class of viruses) cannot be used as a single
practical marker to identify strains, because the glycoprotein gene of
sarcoma viruses is known to be exchanged by recombination with other
viruses, notably avian leukosis viruses, which often exist as passenger
viruses in host animals. At present, the strains of RSV are distinguishable only by the morphology of the cells they transform and by the
morphology of the foci. In addition to RSV, some independent isolates
of avian sarcoma virus are known: notable examples are Fujinami sarcoma virus (Fujinami and Inamoto, 1914) and B77 sarcoma virus
(Thurzo *et al.,* 1963). Some strains of Rous sarcoma virus such as
Bryan strain are known to be defective in virus reproduction and
require helper virus (Hanafusa *et al.,* 1963), but many strains of avian
sarcoma virus are competent (nondefective) in replication.

Avian sarcoma virus can produce neoplastic lesions at the site of
injection within 48–72 h. The tumor cells can be cultured in dishes and
assume a spindle or round morphology which is indistinguishable from
that of *in vitro* transformed cells. In young chickens, the tumors
generally increase in size quickly and metastasize into lungs, liver, or
spleen (Beard, 1963). Tumor cells generally produce virus continuously.
The development of tumors is greatly influenced by the age of the
chickens. If adult chickens are inoculated with relatively small doses of
RSV, tumors may appear but will often regress because of host immunological responses.

In contrast to avian sarcoma viruses, which have been isolated
from naturally occurring solid tumors, most of the murine sarcoma
viruses were derived from tumors experimentally produced by infection
of rats or mice with murine leukemia virus (MuLV). Harvey (1964)
passaged the Moloney strain of MuLV (Moloney, 1960) in rats and
found that the plasma obtained from leukemic rats produced
pleomorphic sarcomas in mice. In an independent experiment,
Moloney (1966) observed development of multiple rhabdomyosarcomas
following inoculation of high titers of Moloney strain of MuLV into
newborn mice. The Kirsten sarcoma virus (KiSV) was also obtained
from rat cells. Kirsten and Mayer (1967) originally found that a virus
isolated from spontaneous lymphomas in mice induced thymic lymphomas in newborn rats, and after serial passage in rats both eryth-

roblastosis and the formation of lymphosarcomas occurred. A new sarcoma virus was recovered from the lymphosarcoma. Thus two strains, KiSV and Harvey sarcoma virus (HaSV), have histories of serial passage in rats, while Moloney sarcoma virus (MoSV) was passaged only in mice. This distinction appears to be reflected in their genome compositions, as described later. Ball *et al.* (1973*a*) passaged Moloney MuLV in JLS-V9 mouse cells and obtained sarcoma virus. Only two strains of murine sarcoma virus were isolated from spontaneously occurring sarcomas (Finkel *et al.*, 1966, Gazdar *et al.*, 1972), but the properties of these viruses have not been well examined. All of the murine sarcoma viruses are defective in replication and require murine leukemia virus as helper virus.

Sarcoma viruses have also been isolated from spontaneous feline fibrosarcomas (Snyder and Theilen, 1969; Gardner *et al.*, 1971) and from a fibrosarcoma in monkey (Theilen *et al.*, 1971). These sarcoma viruses are similar to murine sarcoma virus in their pathogenicity, physical structure, and biological functions.

The host range of the sarcoma viruses is not necessarily restricted to the animal species from which they have been isolated. Murine sarcoma virus can infect many species of mammals, and a pseudotype of xenotropic murine leukemia virus [MSV(x-MuLV)] has been shown to be infectious also for avian cells (Levy, 1973, 1975). Infectivity of avian sarcoma virus for mammals has been documented since 1958 (Svoboda and Hlozánek, 1970).

2.2. Leukemia Viruses

In contrast to the sarcoma viruses, a large number of leukemia viruses have been isolated from various species of animals. Leukemia viruses are essentially the same as sarcoma viruses in both structure and replication. They are distinguishable primarily by the responses they elicit in the cells or animals they infect. In fact, in one case in the avian system, a mutant which is deleted in the sarcoma gene has proved to be leukemogenic (Biggs *et al.*, 1973).

The leukemia viruses may be divided into two classes (Table 1). The first group of viruses cause acute leukemia. This group includes viruses such as avian myeloblastosis virus (AMV) (Beard, 1956), avian erythroblastosis virus (AEV) (Engelbreth-Holm and Rothe-Meyer, 1935), myelocytomatosis virus (strain MC29) (Ivanov *et al.*, 1964), and Abelson (Abelson and Rabstein, 1970), Friend (Friend, 1957), and

Rauscher (Rauscher, 1962) murine leukemia viruses. The second group of viruses cause lymphoid leukosis or lymphoma, both of which develop more slowly than the acute leukemias. The primary replication site for these viruses appears to be the bursa in the case of chickens (Peterson *et al.*, 1966). Except for the length of the latent period required for leukemia formation, the distinction between these two classes of viruses is not always clear-cut, probably because two classes of viruses are often present together in a virus preparation. However, certain characteristic features of the two groups of viruses have gradually emerged from studies on their biological composition and their interaction with stem cells or hematopoietic cells.

The first common feature of the acute leukemic viruses is their ability to cause morphological alteration in specific types of cells either *in vivo* or *in vitro*. The first studies of this type involved the conversion of bone marrow cells to myeloblasts by AMV (Beaudreau *et al.*, 1960; Baluda and Goetz, 1961). The ability of AMV to transform fibroblast cells was described by Moscovici *et al.*, (1969), but the cell population in the fibroblast cultures could have contained a specific type of susceptible cell (Graf, 1973).

Moscovici *et al.* (1975) described a quantitative assay for AMV involving transformation of yolk sac macrophages. The transformed cells have all the characteristics of myeloblasts and generally produce virus. Avian myelocytomatosis virus, strain MC29, causes another myeloid neoplasm (Ivanov *et al.*, 1964). This virus was shown to transform fibroblast cells *in vitro* (Langlois and Beard, 1967).

The analysis by Graf (1973) of target cells for MC29 transformation indicates that MC29 transforms hematopoietic cells (spleen cells) into leukemic cells which form large colonies in agar suspension cultures. This type of target cell is apparently present in a normal population of chick embryo "fibroblast" cells. Another target for MC29 seems to be the fibroblast. Morphological changes by MC29 are not as apparent as those caused by sarcoma viruses, but MC29-infected fibroblastic cells grow into small colonies in agar. The composition of the cell population changes upon passage of cultures. Thus a more defined population of host cells would be required for quantitative studies. A similar transformation of chicken fibroblast cells by avian erythroblastosis virus (strain R) was reported (Ishizaki and Shimizu, 1970).

In the murine system, Friend and Rauscher strains of MuLV cause rapid development of erythroid leukemia in mice (Rauscher, 1962; Friend and Rossi, 1968). These viruses are known to produce discrete areas of proliferating cells, called "foci," on the surface of the spleen

(Axelrad and Steeves, 1964; Pluznick and Sachs, 1964); therefore, a component of this virus is named "spleen focus-forming virus" (Dawson et al., 1968; Fieldsteel et al., 1969). Some leukemic cells induced by Friend virus can be maintained in tissue culture and produce virus (Mirand, 1966; Friend et al., 1971). Direct transformation of bone marrow cells has been shown for Friend virus (Clarke et al., 1975); similarly, Abelson strain of MuLV, which was isolated from mice pretreated with prednisolone and then infected with Moloney MuLV (Abelson and Rabstein, 1970), was shown to transform lymphatic cells (Rosenberg et al., 1975; Sklar et al., 1974) or even transform 3T3 mouse fibroblasts (Scher and Siegler, 1975). These transformed cells can form colonies in soft agar culture, and injection of a clone of transformed cells produced fibrosarcomas in newborn mice. Furthermore, Abelson MuLV-transformed lymphoblasts have been shown to be B cell in origin, and to synthesize small amounts of immunoglobulin (Rosenberg et al., 1975; Premkumar et al., 1975).

The second characteristic of the acute leukemic viruses is their defectiveness in replication. The nature of the defectiveness in these viruses is not known, but the viruses are similar to defective sarcoma virus in their dependence on helper lymphatic leukemia virus for reproduction. Proof of the defectiveness is indirect for some viruses. Transformed cells producing no infectious progeny have been isolated after infection with MC29 (Ishizaki et al., 1971), AMV (Moscovici and Zanetti, 1970) Abelson-MuLV (Scher and Siegler, 1975), and Friend MuLV (Fieldsteel et al., 1969). The defectiveness of AEV and Rauscher spleen focus-forming virus is shown by the requirement of double infection with two different viruses for productive transformation (Ishizaki and Shimizu, 1970; Bentvelzen et al., 1972).

Thus far, none of these acute leukemic viruses has been isolated in a form free of helper virus. Isolation of such virus particles may be possible considering the precedents with avian or murine defective sarcoma viruses. Molecular biological techniques which have been applied to sarcoma virus must also be tried with these viruses in order to determine whether they contain unique nucleic acid sequences responsible for acute leukemia.

In contrast to the acute leukemic viruses, a class of lymphatic leukemia or lymphoid leukosis viruses such as Moloney MuLV, Gross MuLV (Gross, 1957), and many laboratory strains of Rous-associated virus (RAV) cause a much slower response in infected animals (see Gross, 1970). The major manifestation of infection is the proliferation of lymphoid cells and their infiltration into the liver, lungs, or other

visceral organs. In tissue culture fibroblasts, no significant morphological alteration is induced by these viruses. Many investigators
have observed, however, that infection with RAV seems to prolong the
life span of chicken cells in culture. After long cultivation, one can
occassionally obtain cells that are slightly altered morphologically
(Calnek, 1964). This is most likely the result of selection of cells.
Another change frequently seen in RAV-infected cells is a temporal
cytotoxic effect (Kawai and Hanafusa, 1972a; Graf, 1972). Thus far, no
transformation of hematopoietic cells with these lymphatic leukemia
virus (LLV) has been reported.

In order to distinguish this class of viruses from the defective acute
leukemic viruses, they will be called "LLV-type virus" in this chapter.
The LLV-type virus would include all C-type viruses capable of replicating but unable to cause sarcoma or acute leukemia in animals or to
induce cell transformation in tissue culture cells. Thus, by definition,
many laboratory viruses (such as RAV) whose biological activities in
animals have not been examined can tentatively be included in this
class. "Helper virus" required for replication of most sarcoma and
acute leukemia viruses would therefore be considered LLV-type virus.
The life cycles of LLV and nondefective sarcoma virus appear to be
identical. As described later, LLV can be obtained from nondefective
sarcoma virus by deletion of the gene for sarcomagenic transformation
(see Sections 4.1 and 5.2).

3. EVENTS IN TRANSFORMATION

3.1. General Features of Sarcoma Virus Transformation

Since sarcoma viruses produce solid tumors in animals and
produce foci of transformed cells in cultured cells, these viruses present
the most clear example of cell transformation by RNA tumor virus and
have been the subject of most studies on the process of transformation.

The earliest tissue culture system for studying sarcoma virus-
induced transformation was the RSV–chicken cell system established
by Temin and Rubin (1958), based on the observation by Manaker and
Groupé (1956) that RSV-infected chicken fibroblasts acquire altered
morphology and gain a new capacity to grow in cultures under conditions that suppress the growth of normal cells.

Chicken cells transformed by RSV retain the capacity to divide. A
single cell infected with a single infectious unit of RSV will therefore

multiply and form a cluster of transformed cells called a focus, even if the originally infected cell does not produce infectious virus. However, if nondefective RSV (which produces infectious progeny in the absence of helper virus) infects cells, a focus is formed both by multiplication of the infected cells and by new transformation of neighboring cells infected with progeny virus. The size of the foci thus tends to be larger in the latter case.

The number of foci is directly proportional to the input concentration of virus, indicating that a single virus particle is sufficient to induce transformation (Temin and Rubin, 1958). This is the basis of the focus assay for sarcoma virus. Under optimum cellular conditions, most if not all transformed cells develop into foci: cultures exposed to dilutions containing less than one focus-forming unit do not transform in later passages and do not release infectious RSV.

One condition considered important for good focus formation is the cell density at the time of infection. If the cell density is too high, the number of foci is reduced. This suggests that the initial round of cell division is important for focus development, and in fact we know now that mitosis is essential for establishment of virus infection (see below).

When a culture is infected with a high multiplicity of virus, transformation can be recognized in it as early as 16 h after infection, and most infected cells are transformed within 24–48 h (Trager and Rubin, 1966; Hanafusa, 1969; Salzberg et al., 1973) (Fig. 1). By the use of high titers of sarcoma virus, therefore, the entire cell population can be transformed within a relatively short period of time. The time periods required for cell transformation and for progeny virus production are essentially the same. This can be considered to reflect the fact that both require translational products of viral genes.

Production of foci in the mammalian sarcoma virus (SV) system takes place in essentially the same manner. One difference is that most mammalian SVs are defective in replication. The only exception thus far known is a mutant of SV isolated from a stock of MoSV by Ball et al. (1973b). Furthermore, mammalian SV-infected cells often cannot form foci by cell division alone. In the initial attempts to demonstrate focus formation by MoSV, Hartley and Rowe (1966) showed that focus formation in mouse embryo fibroblasts required infection with two particles, MoSV and helper LLV. Thus, unlike RSV, simultaneous infection with helper virus seemed to be essential at an early stage of infection by MoSV. However, later demonstration of single-hit MoSV infection of other cells such as 3T3 or rat NRK cells (Aaronson et al.,

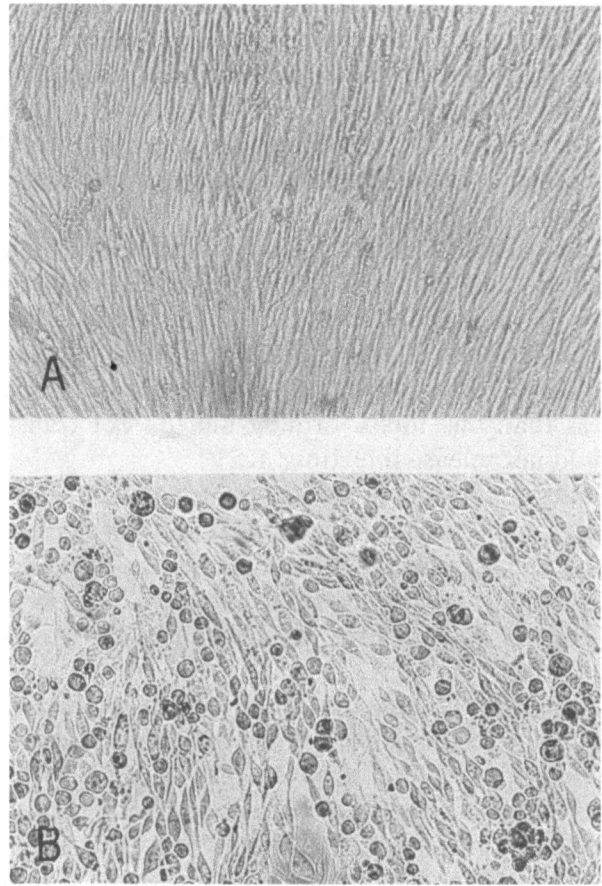

Fig. 1. Transformation of chick embryo fibroblasts by SR-RSV. A: Uninfected. B: SR-RSV-infected cells at 30 h after infection (multiplicity of infection was 5).

1970; Aaronson and Rowe, 1970) indicated that helper virus is not essential for MoSV transformation. Under the conditions employed in earlier studies, virus production from transformed cells, and the spread of virus to surrounding cells, was required for formation of easily recognizable foci.

3.2. Establishment of Infection

The life cycle of avian RNA tumor viruses has been described by Bader (Vol. 4, this series). Integration of the virus genetic material into the host chromosome appears to be required for successful transformation to occur. Briefly, the replication process is as follows: (1) the virion attaches to the cell surface in a rather nonspecific fashion (Hanafusa, 1975); (2) the virus penetrates the cell membrane by a mechanism involving a highly specific interaction between the viral glycoprotein and a cellular structure on the membrane; (3) viral genetic information is transferred by the RNA-dependent DNA polymerase from genome RNA to double-stranded DNA (Temin and Baltimore, 1972), which is probably converted to a closed circular form (Varmus *et al.,* 1975; Gianni *et al.,* 1975); (4) viral DNA is then inserted into and covalently linked with chromosomal DNA; (5) viral DNA is transcribed into RNA by a cellular enzyme; and (6) viral RNA is translated to form viral proteins, including structural proteins as well as a protein which is presumably responsible for cell transformation. The nature of this putative transforming protein and its function in cells are not known.

Blockage of any of these events will prevent cell transformation as well as virus replication. For example, penetration can be blocked by the use of genetically resistant host cells (see below). Addition of an inhibitor of DNA synthesis to infected cultures will block provirus synthesis, and actinomycin D will stop transcription of provirus DNA (Temin, 1963, 1964*a,b*; Bader, 1964, 1965; Vigier and Goldé, 1964). These processes are identical in both sarcoma virus and leukemia virus infection.

Natural cellular resistance to RNA tumor virus infection in the avian system and that in the murine system occur at different steps in the virus life cycle. In the avian system, the only known restriction occurs at the penetration step (Piraino, 1967; Steck and Rubin, 1965). This restriction is profound: virus is almost totally prevented from entering resistant cell. The virus structure that determines host range specificity is the envelope glycoprotein. This specificity provides the basis for subgroup classification of avian viruses (Vogt and Ishizaki, 1965). The cellular factor involved in virus penetration is the presumptive receptor on the cell surface. Susceptibility of a cell to a given virus is a genetically determined cellular property, with susceptibility dominant over resistance (Payne and Biggs, 1964; Rubin, 1965; Crittenden *et al.,* 1967). Perhaps resistant cells lack a specific receptor or have defective receptors. In the avian system, no restriction is known to occur at any step beyond virus penetration.

A good example of cell surface restriction can be found in infection of chicken cells with defective Bryan RSV. This virus, which is defective in the synthesis of glycoprotein, is released from infected cells without the envelope glycoprotein and is therefore noninfectious for any type of chicken cell. Addition of UV-Sendai virus, known to be a fusion agent, allows this virus to bypass the penetration barrier. Once incorporated into the cells, Bryan RSV can cause transformation (Robinson *et al.*, 1967; Weiss, 1969; T. Hanafusa *et al.*, 1970).

Two different types of restriction are seen in the mammalian virus–cell system. First, a block against virus entry, similar to that in avian cells, appears to play a role in the restriction of transspecies infection (such as feline leukemia virus infecting mouse cells) (Sarma *et al.*, 1970). The resistance of mouse cells to xenotropic virus infection is also probably at this level (Levy, 1973; Stephenson *et al.*, 1974). Second, certain virus strains are restricted at a postpenetration step controlled by the so-called *Fv-1* gene of the cell (Lilly and Pincus, 1973). The *Fv-1* product suppresses virus production in resistant cells, and suppression is dominant over susceptibility in heterozygotes (Hartley *et al.*, 1970; Pincus *et al.*, 1971a,b). The resistance conferred by this mechanism is partial: virus production is generally suppressed to one-hundredth of that in susceptible cells.

The product of the *Fv-1* gene of mouse cells seems to act at some step following penetration, but before RNA transcription (Huang *et al.*, 1973; Eckner, 1973; Krontiris *et al.*, 1973; Sveda *et al.*, 1974; Jolicoeur and Baltimore, 1976). The precise mechanism of this host range determination is not known. One report has described successful transfer of the resistance factor from resistant cells to sensitive cells by a cell-free extract in which RNA appears to be an active principle (Tennant *et al.*, 1974).

Murine sarcoma virus and Friend spleen focus-forming virus can form pseudotypes with murine LLV. The host range (N,B tropism) of these pseudotypes of these transforming viruses is determined by the helper virus (Eckner, 1973; Bassin *et al.*, 1975). This result seems to eliminate the possibility that the *Fv-1* product interacts with viral nucleic acids since the RNA of transforming viruses is not considered to be altered by the helper virus. However, in the experiment determining N,B tropism of mammalian SV, infectivity of SV pseudotypes was assayed in the presence of N,B-tropic helper virus to avoid the influence of helper virus on focus formation due to virus spread. Since helper virus can also cause interference, the deliberate addition of helper may not be ideal in this type of study. Perhaps a more definitive answer may be obtained by plating MSV-infected cells on dually susceptible host cells for infective center assay.

3.3. Formation of Provirus

The requirement for synthesis and integration of provirus DNA in RNA tumor virus infection has been established (Boettiger and Temin, 1970; Balduzzi and Morgan, 1970; Varmus *et al.*, 1973*a,b*; Guntaka *et al.*, 1975). Recent studies revealed the existence of closed circular double-stranded viral DNA as well as linear or open circular forms (Gianni and Weinberg, 1975; Varmus *et al.*, 1975). The free DNA was shown to be infectious by transfection (Smotkin *et al.*, 1975). The synthesis of free viral DNA was demonstrated in enucleated cells (Varmus *et al.*, 1974*b*). Apparently, a large number of unintegrated viral DNA copies (more than 100 copies per haploid cell genome) can be accumulated in cells shortly after infection, but the number declines with time and eventually leaves only a limited number, possibly one to three copies per haploid cell genome, in the integrated form (Khoury and Hanafusa, 1976).

The synthesis of viral DNA takes place in the absence of cellular DNA synthesis (Boettiger and Temin, 1970; Balduzzi and Morgan, 1970). According to Varmus (Bishop and Varmus, 1975), the synthesized DNA is integrated in the absence of mitosis, but it is not known whether or not cell DNA synthesis is required. The enzymatic mechanisms involved in the integration are unknown. Integration is suppressed in the presence of ethidium bromide (Varmus *et al.*, 1975), an agent known to intercalate with linear duplex DNA to prevent its circularization. In fact, the amount of closed circular DNA seems to be greatly reduced in the presence of ethidium bromide (Guntaka *et al.*, 1975).

While the formation of closed circular DNA has been established, and its infectivity has been demonstrated, some reports have presented evidence suggesting that viral RNA preserved in a RNA-DNA heteroduplex might be directly integrated (Leis *et al.*, 1975; Sveda *et al.*, 1974; Takano and Hatanaka, 1975). At the present time, one cannot exclude the possibility that a small fraction of input RNA remains with the duplex DNA.

Hill and Hillova (1972) showed that DNA isolated from rat XC cells, a cell line which was originally transformed by Prague strain of RSV but does not produce virus particles, is capable of causing transformation and virus production after transfection to normal chicken cells. The minimum size of DNA infectious by this method was estimated as about 6×10^6 daltons (double-strand molecular weight), corresponding to the size of the viral genome (Hill *et al.*, 1975). These transfection experiments demonstrate most directly the existence of viral genetic information in cellular DNA. The fact that both transformation and virus production always take place in the recipient cells argues against integration of viral genes as smaller fragments.

Molecular hybridization between labeled viral RNA or virus-complementary DNA and cellular DNA has been extensively used for the estimation of the number of viral DNA copies synthesized. The presence of endogenous viral DNA causes some inaccuracy in measurements of the number of copies added after virus infection; estimates vary from 1 to 15 (Neiman *et al.*, 1975; Wright and Neiman, 1974; Schincariol and Joklik, 1973; Shoyab *et al.*, 1974; Varmus *et al.*, 1974*a*; Khoury and Hanafusa, 1976). The fact that genetic interaction takes place frequently in doubly infected cells strongly indicates that a cell should be able to accommodate more than one provirus. However, this does not exclude possible integration of two viruses at the same site of the DNA in the diploid chromosomes. Whether or not the viruses integrate at different sites or at the same site by being linked to each other tandemly is not known. No information has been obtained about the relative chromosomal location of endogenous and exogenous provirus.

3.4. Dependence of Virus Infection on Cell Cycle

Cell density appears to be an important factor for establishment of infection with RSV (Temin and Rubin, 1958; Rubin, 1960*a*; Nakata and Bader, 1968; Yoshikura, 1970; Weiss, 1971). The capacity of cells to support RSV replication is very sensitive to X-ray or UV irradiation (Rubin and Temin, 1959; Bader, 1966). The synthesis of provirus DNA can take place in the absence of cellular DNA synthesis (Balduzzi and Morgan, 1970; Boettiger and Temin, 1970), but cell transformation or virus reproduction does not occur until cells are allowed to proceed to further steps in the cell cycle. These facts strongly imply that either cell DNA synthesis or mitosis is required for the establishment of infection of RNA tumor virus.

Temin (1967*b*) suggested that cell DNA synthesis may not be essential since successful infection can be established by infecting cells in G_2 phase. Hobom-Schnegg *et al.* (1970) observed no difference in the efficiency of infection at different stages in the cell cycle, but the latent period required for virus production was shortest when cells were exposed to virus in S phase. Goldé *et al.* (1973*a,b*) presented evidence indicating that a cell event in G_2 phase is critical for establishment of infection. Thus it appears that cell DNA synthesis is not essential for the "fixation" of virus infection. It should also be noted that, unlike DNA tumor virus infection, no new induction of cellular DNA synthesis has been observed in RNA tumor virus-infected cells.

Using stationary cultures made by serum depletion, Humphries and Temin (1972, 1974) demonstrated that viral RNA synthesis cannot be initiated until serum is added to the cultures. Viral transcription is also inhibited by colchicine, which does not affect S phase but interacts with the mitotic spindle at metaphase (Humphries and Temin, 1974; Fischinger *et al.*, 1975). Interestingly, once the initial transcription is allowed, serum depletion at a later stage does not reduce the production of virus (Humphries and Temin, 1972, 1974). These data strongly suggest that cell mitosis activates the transcription of newly integrated viral DNA by some unknown mechanism. However, Bader (1972*a*, 1973) reported that cell mitosis is not essential for either cell transformation or virus replication, and suggested the possibility that unintegrated DNA is transcribed. He attributed the difference between his results and those of others to the difference in the multiplicity of input virus, but this explanation has not been tested.

3.5. Expression of Provirus

The synthesis of virus-specific RNA can be monitored by molecular hybridization with DNA complementary to viral RNA (Leong *et al.*, 1972; Tsuchida *et al.*, 1972; Hayward and Hanafusa, 1973; Fan and Baltimore, 1973; Schincariol and Joklik, 1973). Transcription appears to be carried out by cellular RNA polymerase (Jacquet *et al.*, 1974; Rymo *et al.*, 1974; Dinowitz, 1975), and the synthesis of RNA in the nuclei and its transport to the cytoplasm have been demonstrated (Parsons *et al.*, 1973). In early studies, no complementary strand of RNA was found in infected cells (Coffin and Temin, 1972; Leong *et al.*, 1972). The presence of a small amount of negative strand viral RNA in infected cells was recently described (Bishop *et al.*, 1976), but the negative strand appears to represent only 30–40% of the genome. These results support the view that viral RNA is exclusively synthesized by transcription of provirus DNA with no involvement of a double-stranded intermediate form of RNA.

The amount of virus-specific RNA varies with the cell–virus

system. In some infected chicken cells, approximately 10,000 copies of viral RNA are detectable (Hayward and Hanafusa, 1976). This would roughly correspond to about 0.7% of the total cellular RNA and perhaps more than 10% of the total messenger RNA (Hayward and Hanafusa, 1973). Analysis by Humphries and Coffin (1976) showed similar values and also demonstrated that the rate of viral RNA synthesis is unaffected by cell cycle. There is no direct estimate of the rate of transcription, but considering the presence of only a few integrated copies of provirus DNA, and the constant utilization of this RNA pool for the formation of virions, the transcription of the provirus must be exceedingly active.

Two sizes of viral RNA (30–40 S and 20–24 S) have been observed in infected cells (Tsuchida *et al.,* 1972; Fan and Baltimore, 1973; Schincariol and Joklik, 1973; Gielkens *et al.,* 1974; Shanmugam *et al.,* 1974; Bishop *et al.,* 1976; Hayward *et al.,* 1976). No 60–70 S RNA can be found inside cells. Since a complete complementary strand is not synthesized, the plus strand must serve as a messenger as well as being incorporated into virions. Both 35 S and 24 S viral RNA have been found associated with polysomes. The functional role of the smaller-sized RNA is not known. Attempts have been made to translate various sizes of viral RNA, and successful formation of viral structural proteins has been reported using virion 35 S RNA (or heated 60–70 S RNA) as well as intracellular viral messenger RNA (Siegert *et al.,* 1972; Twardzik *et al.,* 1973; Von der helm and Duesberg, 1975; Naso *et al.,* 1975b; Salden and Bloemendal, 1976; Gielkens *et al.,* 1976).

Vogt and Eisenman (1973) first reported the *in vivo* synthesis of precursor forms of viral structural proteins. Thus at least a portion of the viral genome [the *gag* gene coding for structural proteins, Baltimore, 1975 (see Section 4.4)] serves as a monocistronic messenger whose primary translational product is processed by proteolytic cleavage into each of the structural proteins (Vogt *et al.,* 1975; Eisenman *et al.,* 1975). This has been demonstrated in both avian and mouse virus-infected cells (Naso *et al.,* 1975a; Van Zaane *et al.,* 1975). However, no definitive evidence has been obtained for the synthesis of polyproteins other than the precursor of structural proteins. This could be partly due to the fact that monospecific antibodies against structural proteins are more readily available but may also reflect the fact that these are the major viral products in infected cells.

Only *gag* products, precursors of structural proteins, have been identified as translational products of 35 S viral RNA in the cell-free system. This may be attributable to the inefficiency of translation of genes located near the 3′ end of the RNA molecule. The *gag* gene appears to be located close to the 5′ end, where translation is thought to be initiated. The smaller 20–24 S RNA may serve as a more efficient messenger for translation of genes close to the 3′ end. The amounts of translational products such as virus structural proteins in cells have been determined by radioimmunoassay. There seems to be a great variation in the rate of translation of individual viral genes. For example, the ratio of the number of molecules of P27 (one of the major structural proteins) and RNA-dependent DNA polymerase in the cell appears to be approximately 20 (Chen and Hanafusa, 1974; Panet *et al.*, 1975*a*). In transformed cells, the product(s) of the gene responsible for transformation (*src* gene for sarcoma virus, Wang *et al.*, 1976) should be formed, but this protein unique to sarcoma virus has not been demonstrated in virions.

3.6. Abortive Expression

The resistance of certain chicken cells to avian viruses (restriction at entry) and of mouse cells to murine viruses (restriction by the *Fv-1* gene product) has been discussed. Another example of restriction can be seen in mammalian cells for infection with avian viruses. Some strains of RSV can produce tumors in a variety of mammals such as hamsters and mice (see Svoboda and Hlozánek, 1970). Efficiency of transformation is very low, and virus is generally not recoverable even from transformed cells. However, the synthesis and integration of provirus were shown by molecular hybridization in a large fraction of cells in which no transformation takes place (Varmus *et al.*, 1975). This indicates that viral expression is strongly suppressed in these cells by unknown mechanisms. The presence of complete viral genetic information in RSV-transformed mammalian cells has been shown both by the recovery of infectious RSV following fusion of rat XC cells with permissive chicken cells (Svoboda, 1964; Machala *et al.*, 1970; Coffin, 1972) and by transfection with transformed XC cell DNA (Hill and Hillova, 1972). The tumor cells generally produce *gs* antigen (structural viral antigens) (Huebner *et al.*, 1964) and contain viral RNA (Coffin and Temin, 1972; Deng *et al.*, 1974), but only 23 S RNA species are found in the cytoplasm (Bishop *et al.*, 1976). These transformed cells generally lack both envelope glycoprotein and viral DNA polymerase (Coffin and Temin, 1972; Livingston *et al.*, 1972). Eisenman *et al.* (1975) described that in some transformed hamster cells the processing of the polyprotein precursor of the *gs* antigens appears to be inhibited. Fusion with chicken cells may provide the proteolytic factor responsible for processing, but the precise mechanism of rescue by cell fusion is unknown. The possibility that chicken endogenous virus genes are involved in the rescue process has been excluded (Svoboda, 1972).

3.7. Assay

Expression of the transformed state in the cell or the culture can be monitored by the use of various phenotypic properties known to be associated with cell transformation (see Sections 6.1–6.11). Transforming viruses can be assayed biologically by focus formation and colony formation.

3.7.1. Focus Formation

Focus formation has already been described briefly in Section 3.1. Here, some technical aspects which may be related to the basic problems of cell transformation will be considered. All of the early work by Temin and Rubin (1958) to establish focus assays was done with the Bryan RSV, which is defective in production of infectious progeny (Hanafusa *et al.*, 1963). Thus all conditions utilized in the above system are designed to allow the initially infected transformed cells to divide and develop into a recognizable mass of cells. Figure 2 shows one large focus formed by BH-RSV. As will be described later (see Section 6.10), in the fluid cultures with optimal nutritional supplies, the rate of growth of embryonic fibroblast cells and that of RSV-transformed cells are essentially the same. If the infected cells grow at the same rate as the surrounding normal cells, the cluster of transformed cells is very difficult to recognize. Rubin (1960*b*) showed that the presence of fetal calf serum, which enhances cell growth better than normal calf serum, made foci unrecognizable, although it did not inhibit infection by RSV. Temin (1966, 1968*b*) showed that transformed cells require less serum growth factor(s) than normal fibroblast cells. The requirement of serum is also known to be substantially different for uninfected and transformed 3T3 cells (Jainchill and Todaro, 1970). The use of agarose or highly purified agar generally inhibits the formation of clear foci (Vigier, 1970, 1973). Crude agar is known to contain sulfated polysaccharide, and the addition of dextran sulfate to agarose restores focus formation. One hypothesis is that the growth factor in serum is bound by sulfated polysaccharides present in agar so that the level of the effective serum factors becomes too low for multiplication of uninfected cells but is sufficient for sustaining growth of transformed cells. In fact, under agar the growth of normal fibroblasts is greatly inhibited. However, Vigier (1970) observed that the addition of a polyanion to the RSV-infected cultures kept in agarose allowed transformed cells to form recognizable foci, even after cultures became confluent. Thus the sulfated polysaccharide may have a direct positive effect to promote the expression of the transformed state. The suppression of the growth of normal cells under agar might be one manifestation of the sensitivity of normal cells to contact with semisolid surroundings.

While its role is not clearly understood, the presence of agar differentiates the growth of normal and transformed cells (Temin, 1965; Hanafusa, 1969). Thus the overlay with agar not only is important in

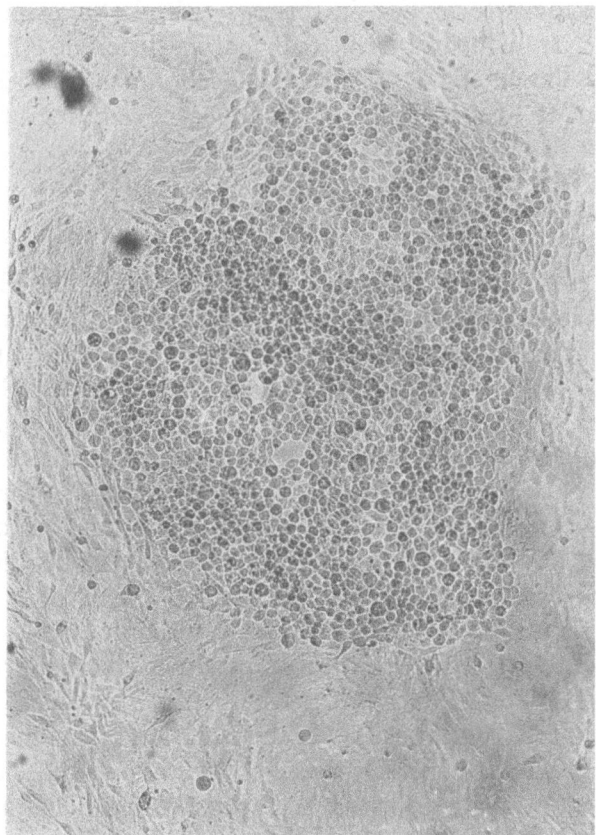

Fig. 2.   A large focus formed by BH-RSV under agar overlay.

suppressing virus spread in the virus assay but also serves as the most useful means for selective growth of transformed cells (Hanafusa, 1968). A similar explanation can be given for successful focus formation by MoSV on 3T3 cells. The large difference in the requirement of normal and transformed cells for serum permits selective growth of the transformed cell (Aaronson *et al.*, 1970). On the other hand, focus assay of MoSV in fresh mouse embryo fibroblasts depends on the virus spread in fluid medium, because mouse cell cultures are kept in fluid with high serum content, in which uninfected and infected fibroblast cells may be similar in growth rate and final cell density. Some other undefined nutritional components such as tryptose phosphate broth and beef embryo extract are known to enhance clear focus formation in chicken cells (Prince, 1962; Vigier, 1970), but their roles for enhanced morphological changes in transformed cells are not known.

The transformation by nondefective RSV is less influenced by physiological conditions (Vigier, 1970), since reinfection by the progeny is involved in focus formation in this case. While the essential features are the same, transformed cells induced by different strains of virus may differ in the extent of alteration of cellular properties that would also contribute to the variations in the appearance of foci they induce.

3.7.2. Colony Formation

Macpherson and Montagnier (1964) developed a method to assay transformation by polyoma virus, utilizing the difference in growth capacity in agar suspension. Most fibroblasts do not grow without attachment to a solid surface (anchorage dependence) (Stoker *et al.*, 1968), but transformed cells do. This can be applied to RNA tumor virus-infected cells (Rubin, 1966). This method is particularly useful for isolation of colonies derived from a single cell, and also for selection of transformed cells. This technique is superior to focus formation for isolating possible nonproducer clones from among producer clones (Bassin *et al.*, 1970; Kawai and Yamamoto, 1970). The size of colonies of transformed cells is independent of virus production. Thus isolation of colonies is nonselective with respect to producer and nonproducer cells. In contrast, the size of foci is larger when virus is produced and spreads to neighboring cells, and thus favors the isolation of producer foci.

Although agar is generally used for suspending cells, the principal factor differentiating the normal and transformed cells in this case is probably the anchorage dependence property, rather than serum dependence, since other supporting media such as methylcellulose can substitute for agar (Vogel and Pollack, 1974).

4. GENES FOR TRANSFORMATION

4.1. RNA in the Virion

The RNA extracted from RNA tumor virus by sodium dodecylsulfate and phenol contains both high molecular weight (60–70 S) and low molecular weight (4–10 S) components (Robinson et al., 1965). The molecular weight of the heavier RNA, which appears to contain all the viral genetic information, was originally estimated at approximately 10 million. The low molecular weight RNA contains several different RNA species, including transfer RNAs (Erikson and Erickson, 1971), but their functional roles are not known. The 60–70 S RNA can be dissociated into smaller 30–40 S subunits by heating or treatment with dimethylsulfoxide (Duesberg, 1968). At the same time, additional smaller RNAs (4, 5, and 7 S) can be dissociated from the 60–70 S RNA at various temperatures (Faras et al., 1973). One of the 4 S RNA species is known to be tightly associated with 35 S RNA and to function as a primer for RNA-dependent DNA synthesis (Canaani and Duesberg, 1972; Sawyer et al., 1974; Faras et al., 1974).

For some time, the structural organization of these RNAs was unknown. In particular, it was not known whether each of the 35 S RNA subunits in the 60–70 S complex contained different genetic information ("haploid" model) or whether these subunits were identical ("polyploid" model) (Vogt, 1973). Recent results obtained by several approaches seem to indicate that the 35 S RNA is the genetic unit of the virus, as originally proposed in the polyploid model. First, radioactively labeled oligonucleotides, obtained by RNase T_1 digestion of viral RNA, were separated by two-dimensional chromatography. If the viral RNA were uniformly labeled with ^{32}P, then the amount of radioactivity in each unique oligonucleotide should be proportional to the length of the oligonucleotide. Therefore, the molecular weight of viral RNA can be calculated by determining the molecular weight of each of the oligonucleotides, the amount of radioactive label in each oligonucleotide, and the radioactivity of the total RNA. The results of this type of analysis, reported by several laboratories, showed that the "complexity" of the virus genome corresponds to a molecular weight of approximately 3×10^6 (Beemon et al., 1974; Billeter et al., 1974;

Quade *et al,* 1974*a*). Second, the minimum size of infectious DNA
recovered from infected cells and measured by transfection experiments
was shown to be 6×10^6 (double stranded) (Montagnier and Vigier,
1972; Hill *et al.,* 1975; Cooper and Temin, 1974; Smotkin *et al.,* 1975).
Third, the complexity of the genome was also measured by molecular
hybridization. This is probably the least accurate method of the three
because of the many factors affecting the kinetics of hybridization.
Baluda *et al.* (1975) obtained a value of about 3×10^6 daltons, but two
earlier determinations favored a genome size of more than 9×10^6
daltons (Taylor *et al.,* 1974; Fan and Paskind, 1974).

The oligonucleotide mapping of the RNase T_1 digest of 35 S RNA
showed that specific oligonucleotides (specific spots) are located at
unique distances from the 3′ end of the RNA molecule (see Section
4.2; Joho *et al.,* 1975; Wang *et al.,* 1975, 1976). This establishes that all
of the subunits have the same order of nucleotide sequences and
eliminates the possibility of circular permutation.

The molecular weight of 60–70 S RNA has been reexamined by
several different methods, including sedimentation equilibrium (Riggin
et al., 1975), determination of particle weight (Bellamy *et al.,* 1974),
and electron microscopy (Mangel *et al.,* 1974; Delius *et al.,* 1975; Kung
et al., 1975). These results indicate a molecular weight of $6–7 \times 10^6$ for
the 60–70 S complex, which is substantially lower than the original
estimate, corresponding to probably two 30–40 S RNA subunits
per virion. The linkage between the 30–40 S RNA subunits to form 60–
70 S RNA is not known. Evidence has been presented that RNA is
incorporated into virions in the 30–40 S form and that final "maturation" of the 60–70 S complex takes place within the assembled virions
(Cheung *et al.,* 1972; Canaani *et al.,* 1973; Stoltzfus and Snyder, 1975).

As with most animal cell or viral messengers, the 35 S RNA
contains poly(A) at the 3′ terminus (Lai and Duesberg, 1972*a*; Rho and
Green, 1974; Quade *et al.,* 1974*b*; Wang and Duesberg, 1974). This
structure has been used for the purification of specific RNA molecules
or RNA fragments, based on its affinity to poly(U)-sepharose or oligo
(dT)-cellulose. The 5′ end of the viral RNA contains a sequence of
methylated bases, with an inverted methylated base at the terminus
(Furuichi *et al.,* 1975; Keith and Fraenkel-Conrat, 1975). The primer
tryptophan tRNA (Dahlberg *et al.,* 1974) binds to a site close to the 5′
end of 35 S RNA (Taylor and Illmensee, 1975). The primer tRNA also
binds strongly to the DNA polymerase molecule (Panet *et al.,* 1975*b*).
These findings suggest that the initiation of DNA synthesis may begin
at a site near the 5′ end. Since DNA synthesis proceeds from the 3′

end to the 5′ end of the viral RNA template (Smoler *et al.*, 1971), some mechanism, such as the circularization of viral RNA, would be necessary to traverse from the 5′ end to the 3′ end of the RNA to synthesize a complete DNA transcript of the viral genome.

In summary, recent studies have established that all of the genetic information of the RNA tumor virus is contained in a single 35 S RNA subunit, with a molecular weight of about 3×10^6. The mature virions contain 60–70 S RNA which is made up of two identical subunits of 35 S RNA. The nature of the linkage between these subunits is unknown, but hydrogen bonds must be the major force involved. One model drawn from electron micrographs of RNA from the cat virus, RD114, suggests that 60–70 S RNA contains two 35 S RNA subunits linked at the 5′ ends of both molecules (Kung *et al.*, 1975), but this model needs confirmation by other analytical methods.

4.2. Presence of the Transforming Gene

Since sarcoma viruses can produce tumors in animals and cell transformation characteristic of each virus strain in tissue culture cells, it has been assumed that the product of a virus gene is responsible for malignant changes. However, a question remained of whether a temporary virus–cell interaction is sufficient to cause a permanent alteration in the cell (essentially a "hit-and-run" phenomenon) or whether the virus genome must act continuously in descendant cells to maintain the transformed state (Dulbecco, 1967). In the RNA tumor virus system, the presence of the viral genome in infected cells has never been doubted since virus has been recovered even from apparently "nonproducer" cultures. Yet the first definitive proof for the direct participation of a virus gene product in maintaining cell transformation was obtained by Martin (1970), who demonstrated that cells transformed by a temperature-sensitive virus mutant underwent reversible transformation, dependent on temperature, while virus production was not temperature dependent.

Another equally important result was the observation that LLV-type viruses segregated spontaneously from a clone of nondefective sarcoma virus (Vogt, 1971; Kawai and Hanafusa, 1972*b*). In some sarcoma virus preparations, this segregation occurs at high frequencies. Although a more complete comparison must be made between these segregated LLV-type viruses and the "naturally" occurring leukosis viruses, these two classes of viruses are indistinguishable in biological properties which have thus far been tested. In fact, one of the LLV-type viruses obtained from a mutagenized sarcoma virus population was shown to have leukemogenic potential in chickens (Biggs *et al.*, 1973). Thus spontaneous changes of this type may play an important role in the natural evolution of leukemia viruses. The studies which will be described below have shown that the segregation of nontransforming viruses is due to the deletion of the gene responsible for sarcoma formation. The sarcomagenic virus and its nonsarcomagenic segregant provide an ideal system for studying the structure and function of the portion of the genome responsible for sarcoma formation.

4.3. Size of the Transforming Gene

Comparing the size of 30–40 S RNAs from sarcoma and LLV-type viruses by gel electrophoresis, Duesberg and Vogt (1970) observed a consistent difference: most sarcoma virus RNAs showed two peaks, one corresponding to the single peak obtained with LLV RNA, and another about 15% larger than the LLV RNA (Martin and Duesberg, 1972). The larger and smaller RNA species were called *a* and *b* subunits, respectively. Since the organization of the viral genome had not been established at that time, the term "subunit" included the possibility of the presence of both *a* and *b* subunits in the same RSV virion. However, when viral RNA was obtained from freshly cloned RSV which contained fewer segregants of the LLV type, the RNA migrated as an essentially homogeneous *a* subunit (Duesberg and Vogt, 1973a,b). The direct comparison of PR-RSV and its spontaneous nontransforming derivative [this will be called a "transformation-defective" (*td*) mutant] confirmed this size difference, which was estimated at 12%. The differences in size between RNAs of various nondefective sarcoma viruses and *td* mutants or LLV-type viruses are on the average about 15%. The demonstration of two peaks (*a* and *b*) in RSV RNA in the original study can be explained by the presence of *td* virus segregants in the RSV preparation.

Interestingly, the size of the RNA from two viruses defective in replication [Bryan strain of RSV (BH-RSV) and a mutant of Schmidt-Ruppin strain, SR-N8] was comparable to or even smaller than that of LLV RNA (Duesberg and Vogt, 1970; Scheele and Hanafusa, 1972; Duesberg *et al.*, 1975). These two viruses contain deletions within the gene coding for the envelope glycoprotein. Harvey and Moloney sarcoma viruses which are more profoundly defective have RNA smaller than murine LLV RNA (Maisel *et al.*, 1973; Tsuchida *et al.*, 1974).

Gel analysis of RNA also showed that recombinant sarcoma viruses obtained by recombination between RSV and LLV contained only *a*-size RNA subunits (Beemon *et al.*, 1974).

The sequences specific for sarcoma virus have also been detected by nucleic acid hybridization. Neiman *et al.* (1974) estimated that the RNAs from PR-RSV and its *td* mutant differed by about 13%.

As will be described below, the sequences missing in *td* mutants, corresponding to about 14% of the virus genome, mapped in a single location on the viral genome. If one assumes that the size of the deleted gene roughly corresponds to the size of the transforming gene, this gene would code for one or more proteins with a total molecular weight of 30,000–45,000.

4.4. Location of the Sarcoma-Specific Gene within the Virus Genome

When 35 S RNA of RSV was digested with RNase T_1 and subjected to two-dimensional electrophoresis and homochromatography, clusters of spots, representing large oligonucleotides that separated from the mass of smaller oligonucleotides, produced a specific pattern on the autoradiograms (Horst *et al.*, 1972; Lai *et al.*, 1973). The map of the spots was characteristic of the specific virus strain, but importantly some of the spots characteristic of nondefective sarcoma viruses were missing on the maps obtained with transformation-defective derivatives. This finding raised the possibility that these oligonucleotides originated from the portion of the genome responsible for cell transformation.

In order to determine the location of these specific oligonucleotides within the genome, ^{32}P-labeled RNA was randomly degraded by alkali, and the RNA fragments were subjected to oligo(dT)-cellulose to selectively trap fragments containing poly(A) (i.e., fragments containing the 3′ end of the RNA genome) (Wang *et al.*, 1975; Coffin and Billeter, 1976). The poly(A)-containing RNA fragments, ranging from very small to full genome-length species, were fractionated according to size by sucrose gradient sedimentation. Several size classes of poly(A)-containing fragments were pooled and analyzed by oligonucleotide mapping. All sarcoma virus-specific oligonucleotides were detectable in the fingerprints derived from about 10–20 S poly(A)-containing fragments (Wang *et al.*, 1975, 1976). This would indicate that the sarcoma-specific sequence starts at about 6.5% and extends to about 20% of the distance from the poly(A) terminus of sarcoma virus RNA. Sequences (termed *c*) common to all avian virus RNAs including both sarcoma and leukemia viruses were found in the region

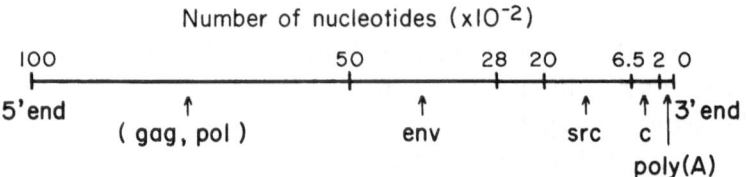

Fig. 3.   Genetic map of nondefective avian sarcoma virus.

between the poly(A) sequence and the *src* gene. Using RSV mutants which contain deletions in the envelope glycoprotein gene (*env*), this gene was mapped next to the *src* gene. From all the data thus far obtained (Wang and Duesberg, 1974; Wang *et al.*, 1975, 1976; Duesberg *et al.*, 1975), the relative location of the known viral genes can be drawn as shown in Fig. 3. The relative positions of the polymerase (*pol*) and structural protein (*gag*) genes have not been determined, although they both appear to be located toward the 5′ end of the genome. Joho *et al.* (1975) reached essentially the same conclusions by comparing the oligonucleotide maps of sarcoma virus, LLV-type virus, and recombinants between these two viruses.

4.5.   Origin of the Transforming Gene

Since the LLV-type virus contains all the genetic information necessary for virus replication, one possible theory to explain the origin of sarcoma virus is that the gene responsible for sarcoma transformation was originally derived from the host cell sequences. No direct proof has been obtained for this hypothesis, but indirect evidence from both mammalian and avian virus–cell systems has been presented.

As mentioned before, most strains of MSV were obtained from tumors that developed after consecutive passage of LLV type virus in rats or mice. Both Kirsten and Harvey strains were derived from LLV passaged in rats, and Moloney strains were recovered after passage in mice (see Section 2.1). Further, Ball *et al.* (1973*a*) confirmed more recently that sarcomagenic virus can be recovered from mice bearing tumors which were formed by a clone-purified murine LLV-type virus. Hybridization analyses of the rat-passaged KiSV and HaSV showed that these viruses contain some sequences homologous to Moloney leukemia virus and others homologous to RNA of either rat endogenous C-type virus or normal rat cells. The latter set of sequences, which were not homologous to murine viruses, are considered to represent, at least in part, sequences specific for sarcomagenesis (Scolnick *et al.*, 1973; Scolnick and Parks, 1974; Roy-Burman and Klement, 1975; Anderson and Robbins, 1976). The "sarcoma-specific" sequences of MoSV (prepared by hybridizing cDNA with LLV-type virus RNA and recovering unhybridized DNA, a technique similar to that used for avian virus cDNA$_{sarc}$; see below) do not show significant hybridization with HaSV RNA. These results suggest that the sequences responsible for cell transformation were gained from host cells, possibly by recombination, during the passage of LLV-type virus. The presence of rat-specific sequences in HaSV and KiSV and mouse-specific sequences in MoSV is consistent with the passage histories of these viruses.

Using nondefective RSV and its *td* derivative, Stehelin *et al.* (1976*a*) have succeeded in preparing cDNA which specifically hybridizes with avian sarcoma virus RNA. The principle was to first prepare a large DNA transcript from the Prague strain of RSV and then hybridize this DNA with RNA from a *td* mutant of Prague strain of RSV. The nonhybridized DNA was separated from hybridized DNA by hydroxylapatite chromatography. After "recycling" was repeated, this cDNA (cDNA$_{sarc}$) hybridized only with sarcoma virus RNA, not with LLV-type virus RNA. The cDNA$_{sarc}$ protects 10–15% of the viral RNA from RNase, consistent with the known size of the *src* gene. This selected DNA is highly homologous to RNA of various strains of RSV, and to RNA from other sarcoma viruses (B77 and Fujinami strain) which were isolated independently of Rous sarcoma virus, suggesting a very similar origin of the *src* genes of the different isolates (Stehelin *et al.*, 1976*a*).

Moreover, homology was also demonstrated between this cDNA$_{sarc}$ and various species of avian cell DNA (Stehelin *et al.*, 1976*b*). The extent of homology between the RSV *src* gene and the endogenous sequences present in normal cells from different avian species, determined by measuring the melting temperatures of the various hybrids, was found to correlate with the evolutionary relatedness of the given avian species to chicken. Even chicken cell DNA shows some mismatching with the cDNA$_{sarc}$. This may be considered to indicate the divergence of this particular cellular sequence during evolution.

While the hypothesis that the viral *src* gene is derived from this homologous cellular DNA sequence is quite attractive, no example has been shown in the avian system for the formation of transforming virus by passaging LLV-type virus. Acquisition of new host cell sequences by passaging B77 in duck cells has been reported by Shoyab *et al.* (1975), but the possibility of a selection of duck-tropic virus was not totally excluded.

5. MUTANTS OF TRANSFORMING VIRUS

Genetic alteration, particularly gene deletion and genetic recombination, takes place at rather high frequencies in RNA tumor viruses. The molecular basis for these changes, which result in a wide diversity of genetic materials, will not be discussed here. However, the realization of these changes and an understanding of possible modifications in the virus population are of great practical importance in studies of cell transformation using virus mutants as a tool. Therefore, in the following sections, various important mutants will be described with particular attention to their variability.

5.1. Mutants Defective in Replication

As surveyed in Sections 2.1 and 2.2, many strains of avian sarcoma virus, essentially all of the mammalian sarcoma viruses, and perhaps most acute leukemia viruses are replication-defective (*rd*) viruses. These viruses are detectable by their transforming or tumor-inducing capacity but are incapable of producing infectious progeny without assistance from a helper virus.

The best-characterized virus in this class is the Bryan high-titer strain of RSV (BH-RSV). This virus and another virus (SR-N8) isolated as a mutant of SR-RSV (Kawai and Hanafusa, 1973) are defective for glycoprotein synthesis because of deletions in the gene for envelope glycoprotein (*env* gene) but are capable of producing all other viral products (see review by Bader, Vol. 4, this series; Hanafusa, 1975). Therefore, cells infected with this virus become transformed but produce virions which lack the envelope glycoprotein and are thus noninfectious (Scheele and Hanafusa, 1971; Ogura and Friis, 1975; DeGiuli *et al.*, 1975). The glycoprotein synthesized by other viruses, either sarcoma virus or LLV (many are known as RAV), can complement the defect of BH-RSV virions grown in the same cells, resulting in the production of infectious pseudotypes (Rubin, 1965). The glycoprotein made by endogenous viral genes can also be utilized by the defective virus to yield an infectious pseudotype (Weiss, 1969; H. Hanafusa *et al.,* 1970). It is well known that many RSV properties are determined by the nature of the glycoprotein (see Tooze, 1973; Hanafusa, 1975). The yield of noninfectious physical particles from helper-free transformed cells is comparable to that of infectious particles produced by cells infected with competent viruses. Crittenden (1968) showed that coculture of lethally irradiated BH-RSV(RAV-1) infected cells with C/A cells, which are genetically resistant to BH-RSV(RAV-1), resulted in the synthesis of BH-RSV(RAV-1). This fact suggests that the particles released from transformed cells might occasionally be taken up by directly adjacent resistant cells.

The mechanism by which deletion mutants are generated is unknown. In fact, the *rd* mutant N8 is the only mutant of this class that has been well characterized. This type of mutant seems to be produced at a low frequency (about 1%) from the stock of SR-RSV from which SR-N8 was isolated (Kawai and Yamamoto, 1970; Kawai and Hanafusa, 1973). Recombination between these defective viruses and LLV-type virus is known, but restoration of the glycoprotein deletion through recombination has not been observed.

In a population of *env⁻* sarcoma virus, a fraction (5–10%) of the viruses lack the ability to synthesize an active RNA-dependent DNA polymerase. These types of viruses are called BH-RSVα or SR-N8α (Hanafusa and Hanafusa, 1968, 1971; Kawai and Hanafusa, 1973). The *pol⁻* characteristic of these mutants is genetically stable, and can be maintained through infection with pseudotypes of α virus, which in this case have both envelope glycoprotein and DNA polymerase provided by a helper virus.

The defectiveness of murine sarcoma virus seems to be more profound. KiSV, HaSV, and one variant of MoSV are defective in the synthesis of the envelope glycoprotein, structural proteins, and viral polymerase (Hartley and Rowe, 1966; Huebner *et al.,* 1966; Aaronson and Weaver, 1971; Aaronson *et al.,* 1972; Harvey and East, 1971). Since cells transformed by these viruses do not form virus particles, they are called NP (nonproducer) cells (Aaronson and Rowe, 1970). In some NP rat cells induced by KiSV, glycoprotein can be detected by radioimmunoassay (Bilello *et al.,* 1974). Infectious SVs can be recovered by infecting NP cells with LLV-type virus. Pseudotypes of these SVs with LLV viruses from different species can be produced by cell fusion between transformed cells and cells of another species infected with LLV. Feline leukemia virus pseudotypes of MoSV were formed by this method (Sarma *et al.,* 1970).

Cells transformed by another variant of MoSV do synthesize viral structural proteins and produce small amounts of virus particles with low levels of DNA polymerase activity. These transformed cells are referred to as $S^+ L^-$ cells (Bassin *et al.*, 1970, 1971*a*). Thus the genome of this variant must contain the *src, pol,* and *gag* genes, although they may not be fully expressed. The characteristics of all of these sarcoma virus strains are genetically stable: the pseudotype KiSV or MoSV obtained from NP- or $S^+ L^-$-transformed cells will establish the same type of infection in newly transformed cells (Aaronson, 1971; Aaronson *et al.*, 1972).

The RNA of Harvey strain of MSV was shown to be smaller than RNA of LLV-type murine leukemia virus (Maisel *et al.*, 1973, 1975). The analysis of RNA from murine sarcoma virus is more difficult than the analysis of nondefective RSV RNA because of the presence of helper LLV-type virus, generally in excess, in most murine sarcoma virus stocks. Further analysis of Harvey or Kirstein virus RNA would be of great interest since these viruses are defective in all replication functions.

Ball *et al.* (1973*b*) described the isolation of a nondefective variant of MoSV from a single colony of MoSV-infected JLS-V9 cells. The colony produced a virus capable of transforming mouse cells but no LLV detectable by the XC test. Some mouse cells transformed by the progeny virus produced only MoSV, which was XC negative, but many cells produced a mixture of defective MoSV and XC-positive LLV. The nondefective MoSV is unique in its extremely high rate of segregation of both defective MoSV and LLV. The reason for these variations is unknown. LLV-type viruses which do not produce XC plaques have been isolated (Hopkins and Jolicoeur, 1975; Rapp and Nowinski, 1976), and these plaque variants are known to convert to plaque producers. It would seem possible that Ball's MoSV could be contaminated with this type of helper virus. Lo and Ball (1974) showed that RNA of the nondefective MoSV is larger than that of defective MoSV but indistinguishable in size from that of Moloney LLV. Interpretation of the RNA analysis becomes more complicated if nondefective virus stock contains XC-negative LLV, whose RNA is probably larger than that of defective MoSV.

Many strains of acute leukemia viruses have been recognized as defective in replication (see Section 2.2). As seen with sarcoma viruses, all these leukemia viruses can be recovered as pseudotypes by infecting them with LLV, but the nature of their defectiveness and the composition of their genomes are not understood.

5.2. Mutants Defective in Transformation

The first examples of transformation-defective (*td*) mutants were isolated by treatment of biologically pure nondefective RSV with physical or chemical mutagens (Goldé, 1970; Toyoshima *et al.*, 1970; Graf *et al.*, 1971). These results raised the possibility mentioned above, that the LLV-type viruses often found in stocks of cloned RSV (Dougherty and Rasmussen, 1964; Hanafusa and Hanafusa, 1966; Duff and Vogt, 1969) may be spontaneous mutants. Vogt (1971) proved the validity of this explanation by demonstrating that about 4–17% of the population of various stocks of clone-purified RSV were LLV-type virus. The *td* derivatives retain all viral functions other than cell transformation and thus can be considered to be equivalent to naturally occurring LLV-type leukemia viruses. Biggs *et al.* (1973) showed that one of the *td* derivatives of B77 avian sarcoma virus is leukemogenic.

The actual frequency of *td* deletion formation is difficult to estimate, and the cause of this deletion is not understood. The size of the deletion is similar in most such mutants (Duesberg and Vogt, 1973*a,b*) (see Section 4.2). However, one *td* mutant obtained after hydroxylamine treatment of SR-RSV has been shown to have RNA larger than that of other *td* mutants (Bolognesi and Graf, 1971; Stone *et al.*, 1975). Recently, Kawai *et al.* (unpublished) found that some *td* derivatives have smaller deletions and can recombine with a temperature-sensitive transformation mutant to produce wild-type nondefective RSV.

5.3. Temperature-Sensitive Mutants

Many temperature-sensitive mutants of avian and mammalian sarcoma viruses have been isolated. The class "R" (replication) *ts* mutants have one or more temperature-sensitive lesions affecting virus reproduction. Since their *ts* function is not directly related to cell transformation, their characteristics will not be discussed here. The mutants classified as "*C*" (coordinated) class are temperature sensitive in a function which is required for both cell transformation and virus replication. Thus far, four C class mutants, *LA*335, *LA*336, *LA*337, and *NY*21, have been isolated (Wyke and Linial, 1973; Linial and Mason, 1973; Toyoshima and Vogt, 1969; Friis *et al.*, 1971; Metroka *et al.*, unpublished). All of these mutants contain temperature-sensitive

defects in DNA polymerase. Two of these, *LA*336 and *NY*21, contain
additional *ts* lesions, the former being temperature sensitive in a function required for maintenance of transformation (Friis *et al.*, 1971) and
the latter being *ts* for a late replication function related to the structural
proteins (Metroka *et al.*, unpublished). Both DNA polymerase and
RNase H activities of the α subunits of *LA*335 and *LA*337 have been
shown to be heat labile at elevated temperatures (Verma *et al.*, 1974;
Verma, 1975). In order to obtain successful infection with these
mutants, infected cells must be maintained at 35°C (permissive
temperature) until the provirus is synthesized. Once the infection is
established, however, both cell transformation and virus production are
normal at 41°C (nonpermissive temperature) (Linial and Mason, 1973).
The temperature dependence of early infection by these mutants and
the lack of infectivity of nonconditional polymerase-defective mutants
(BH-RSVα and SR-N8α) provide convincing evidence for the requirement of DNA polymerase for the establishment of virus infection.

The third class of *ts* mutants, called "T" (transformation) class,
are temperature sensitive only for transformation. All of these mutants
were isolated from stocks of virus (or virus-infected cells) treated with
mutagens. The first mutant isolated by Toyoshima and Vogt (1969) was
found to have a second mutation in a late replication function (Friis *et
al.*, 1971; Owada and Toyoshima, 1973; Hunter and Vogt, 1976).
However, many mutants isolated later (Martin, 1970, 1971; Biquard
and Vigier, 1970, 1972; Kawai and Hanafusa, 1971; Kawai *et al.*, 1972;
Bader and Brown, 1971; Bader, 1972*b*; Wyke, 1973; Toyoshima *et al.*,
1973) produce normal amounts of virus at the nonpermissive temperature, but do not cause transformation. Thus transformation is independent of virus replication and is clearly under the influence of a viral
gene(s). Representative mutants of this class are listed in Table 2.

Following a shift from nonpermissive to permissive temperature,
the *ts* mutant-infected cultures rapidly express the transformed
phenotype. The conversion between the nontransformed and
transformed states can be reversed at any time simply by shifting
temperature and without changing any other parameters of the cultures. Thus these cultures provide an ideal system for investigating the
biochemical events associated with cell transformation. The results of
these studies will be discussed in Sections 6.1–6.11. If a *ts* mutant is a
point mutant, it will revert to wild type with a certain frequency.
Mutants which are less leaky and do not revert easily may contain
multiple mutations in the *src* gene. In this respect, the mutants thus far
obtained require more rigorous genetic analysis of their *ts* lesions.

However, for the biochemical studies, stable and less leaky mutants are useful. Since revertant wild-type virus may have an advantage in replication (because transformed cells infected with revertants may grow better under certain conditions) and since *td* virus may be spontaneously formed, even stable mutants must be purified by cloning after a few passages.

In considering the mechanism of cell transformation, ·it is important to know how many gene products are coded by the *src* gene. If more than one product is involved, different mutants would be expected to complement each other to restore transformation at the nonpermissive temperature. Initial experiments indicated that there were more than two (Kawai *et al.*, 1972) or four complementation groups (Wyke, 1973). However, further studies by Wyke *et al.* (1975) suggested that the presumed complementation actually resulted from the formation of recombinant wild-type virus, rather than from true complementation in the original doubly infected cells. A crucial point in this study was the correlation between the frequency of wild-type virus formation and the level of cooperative transformation for a given combination of mutants. This explanation for the apparent "complementation" is also supported in a recent observation by Kawai (personal communication) that clones of *NY*68-infected cells kept at 41°C can be "complemented" by *NY*10 at low multiplicity, which allows virus progeny to reinfect surrounding cells, but not at high multiplicity, which prevents reinfection. Therefore, from the more recent data obtained in genetic studies, the assumption of multiple gene products coded by the *src* gene seems to be unnecessary.

The properties of two T class mutants, *LA*30 (Wyke and Linial, 1973) and *MI*100 (Bookout and Sigel, 1975), require comment. Like

TABLE 2

Avian Sarcoma Virus *ts* Mutants (T Class)

Virus strain	Mutant
BH-RSV	*BE*1 (Bader, 1972*b*)
BH-RSV	*RO*1–*RO*137 (Balduzzi, 1976)
PR-RSV	*LA*22–*LA*30 (Wyke, 1973)
SR-RSV	*MI*100 (Bookout and Sigel, 1975)
SR-RSV	*NY*10, *NY*19 (Kawai *et al.*, 1972)
SR-RSV	*NY*68 (Kawai and Hanafusa, 1971)
SR-RSV	*OS*122 (Toyoshima *et al.*, 1973)
SR-RSV	*PA*19 (Biquard and Vigier, 1970)
SR-RSV	*BK*1, *BK*5 (Martin, 1970, 1971)

other T class mutants, these two mutants can replicate but cannot transform cells at 41°C. If infected cultures are shifted from 37°C to 41°C, they lose the transformed state. However, if cells are kept at 41°C during the first 24 h after infection, these cells do not become transformed following a shift to the lower temperature despite the fact that the virus production proceeds unaffected. The results suggest that, in addition to the function maintaining the transformed state, the mutants are *ts* in some early function required only for transformation. However, virus production was very poor under these conditions, and it is possible that virus infection was established too slowly to permit detection of transformed cells. Since no such early function affecting only transformation has been shown in previous genetic and biochemical studies, these results are potentially important and deserve careful examination. Both mutants can recombine with LLV-type virus to form a T class mutant which causes cell transformation reversible at different temperatures, indicating that *LA*30 and *MI*100 are double mutants.

Three nonproducer cell lines containing *ts* mutants of KiSV have been isolated from cells infected with mutagenized virus stocks (Scolnick *et al.*, 1972). When superinfected with LLV-type virus, two lines produce pseudotypes of *ts*KiSV but one line produces wild-type virus. The reason for the formation of wild-type virus is not known. The yield of pseudotypes of *ts* mutants from the two lines was very poor. They may have more than one mutation so that the capacity to be rescued might have been impaired. No further physiological studies were performed on these isolates.

A cold-sensitive mutant of MoSV has been isolated from rat (NRK) cells transformed by MoSV (Somers and Kit, 1973). Cells can be transformed by this mutant at 39°C but not at 33°C, and MoSV can be rescued as an LLV pseudotype. Transformation after shift-up is not blocked by inhibitors of DNA or RNA synthesis but is blocked by protein synthesis inhibitors (Somers *et al.*, 1973).

No *ts* mutants have been isolated from acute leukemia viruses. Since *in vitro* transformation has been established for some of these viruses, the isolation of such mutants is now technically feasible. The successful isolation of such a mutant would provide a direct demonstration of the existence of the gene responsible for leukemia induction. The hybridization studies by Scolnick *et al.* (1975) indicated that Abelson leukemia viruses have no sequences homologous to the *src* gene of MoSV.

5.4. Revertants of Transformed Cells

Since established lines of mammalian cells can be infected with RNA tumor viruses, individual transformed cells can be cloned and maintained for long periods of time. During these cultivations, flat revertants with morphology similar to that of normal cells sometimes appear among subclones of transformed cells. The first such case was described by Macpherson (1965) with BHK cells transformed by SR-RSV. Since RSV could not be rescued by cocultivation with chicken cells, he suggested that reversion may have occurred through loss of the virus genome. This problem was again analyzed by Boettiger (1974), with a more sensitive rescue technique using UV-Sendai virus. RSV was rescued from all revertant subclones, although the frequency of rescue was less than one-tenth of that from transformed clones. Deng *et al.* (1974) further substantiated this finding by the detection of one copy of viral DNA in almost every cell with varying degrees of transcription into viral RNA. A good correlation was found between the rescue frequency and the amount of viral RNA. They also found that treatment with 5-bromodeoxyuridine increased the concentration of viral RNA and the frequency of rescue.

Similar morphological revertants have been reported with mouse cells infected with KiSV and MoSV (Stephenson *et al.*, 1972, 1973; Nomura *et al.*, 1972, 1973: Fischinger *et al.*, 1974). With most of the revertant clones isolated by Nomura, MoSV could not be rescued by infecting with LLV-type virus or cocultivating with susceptible host cells. These clones often retransform spontaneously, and some retransformants produce MSV particles upon rescue, suggesting that at least some of these flat revertants and retransformants have retained sarcoma specific genes (Nomura *et al.*, 1973).

Revertants obtained with KiSV in rat (NRK) cells or mouse (BALB) cells appear to retain sarcoma virus genetic information judged by hybridization (Stephenson *et al.*, 1972; Greenberger *et al.*, 1974). These revertant cells show a low frequency of spontaneous retransformation. Sarcoma-specific sequences are present in these revertants in amounts comparable to those in transformed cells, but the level of KiSV rescued by infection with LLV is very low. Since spontaneous retransformation is associated with the formation of wild-type sarcoma virus, reversion seems to be due to a point mutation in *src* gene of KiSV integrated in transformed cells.

6.   ALTERATION OF CELLS BY TRANSFORMATION

Infection with nontransforming RNA tumor viruses causes little alteration in basic cellular structure or in the content of cellular macromolecules. However, cells infected with sarcoma virus undergo remarkable changes in both cell morphology and growth regulation. While the identification of the primary product of the *src* gene remains as one of the most important tasks for tumor virologists, the changes in cellular components and the functions of these components are under intensive study. Obviously, the study of cellular changes associated with transformation includes many areas of cell biology and is not restricted to RNA tumor virus infection. An extensive survey of all of these areas would be beyond the scope of this chapter. Thus an attempt will be made to concentrate on the major topics that are directly relevant to RNA tumor viruses and the *src* gene product. Temperature-sensitive mutants of sarcoma viruses have proven to be valuable in these studies.

A comprehensive review of animal cell growth, membrane structure, and the effect of cell transformation has been presented by Tooze (1973). Two publications from the Cold Spring Harbor Laboratory, edited by Clarkson and Baserga (1974) and by Reich *et al.* (1975), are also extremely valuable.

6.1.   Carbohydrates, Lipids, and Proteins

The accumulation of hyaluronic acid and sulfated mucopolysaccharides in Rous sarcomas and RSV-transformed cells was observed many years ago (Kabat, 1939; Harris *et al.*, 1954; Erichsen *et al.*, 1961). An increased level of hyaluronic acid synthetase has been demonstrated in RSV-infected chicken cells (Ishimoto *et al.*, 1966), and the production of hyaluronic acid has been correlated with a morphological transformed state in *BE*1-infected cells (Bader, 1972*b*). The increased production of hyaluronic acid is considered to be one factor causing difficulty in demonstrating lectin agglutination of RSV-infected cells (Burger and Martin, 1972). Although some contradictory results have been reported on the content of sialic acid in virus-transformed cells (Perdue *et al.*, 1972; Hartmann *et al.*, 1972; Grimes, 1973), it was observed by Bosmann *et al.* (1974*a*) that more sialic acid can be released by neuraminidase from RSV-transformed cells. They attributed the increased levels of both hyaluronic acid and sialic acid to the increased negative charge of *NY*68-transformed cells at permissive temperatures. Bosmann *et al.* (1974*a*) also showed that the level of glycosyltransferase activity and the number of acceptors on the cell surface increase following RSV infection. The amount of glycopeptides released by protease treatment of plasma membranes of RSV-transformed cells is always larger than that from normal chicken cells (Buck *et al.,* 1970; Warren *et al.,* 1972*a*; Lai and Duesberg, 1972*b*). Since this difference is abolished if the glycopeptides are subjected to treatments which hydrolyze sialic acid, Warren *et al.* (1972*b*) concluded that sialic acid is mainly responsible for this difference. They also demonstrated that cells transformed with *BK*5 (Martin, 1971) had increased levels of sialic acid transferase activity at 35°C but not at 40°C.

Alterations in glycolipid synthesis have been observed in transformed cells (Hakomori *et al.,* 1971; Hakomori, 1975; Fishman *et al.,* 1974). Generally the synthesis of ganglioside is blocked, resulting in the accumulation of the precursor ceramide. This alteration in RSV-infected cells (Hakomori *et al.,* 1971) was not found by Warren *et al.* (1972*a*). Hakomori (1975) attributed this to the difference in techniques used and presented data (Wyke *et al.,* quoted in Hakomori, 1975) showing that *LA*24 and *LA*25 caused a block in the synthesis of hematosides.

The development of SDS-polyacrylamide gel electrophoresis for protein analysis has made it possible to detect changes in some species of glycoproteins located on the plasma membrane (Bussell and Robinson, 1973; Wickus and Robbins, 1973; Stone *et al.,* 1974; Vaheri and Rouslahti, 1974; Hogg, 1974; Isaka *et al.,* 1975; Robbins *et al.,* 1975; Yamada *et al.,* 1975; Hynes and Wyke, 1975). The molecular weights reported by the various groups differ somewhat, possibly because of the difference in the carbohydrate moiety of the glycosylated proteins. The most noticeable change is a drastic decrease in the content of a large glycoprotein with a molecular weight of about 250,000. This protein appears to be exposed on the cell surface since it can be labeled by iodination in the presence of lactoperoxidase (Hynes, 1973; Hogg, 1974) or by the galactose oxidase-tritiated borohydride method (Critchley, 1974; Gahmberg and Hakamori, 1973) and is sensitive to trypsin treatment (Hynes, 1974; Wickus *et al.,* 1974). Since the disappearance of this protein can also be seen in DNA virus-transformed cells, it is called the "large external transformation-sensitive (LETS) protein" (Hynes and Bye, 1974) or Z protein (Robbins *et al.,* 1975). The removal and reappearance of the LETS protein were shown in chicken cells infected with various T class *ts* mutants. The content of this protein in normal cells is highest during G_1 phase, and it is also higher in confluent resting cells than in sparsely growing cells (Hynes, 1974: Hynes and Bye, 1974; Gahmberg and Hakomori, 1974; Pearlstein and Waterfield, 1974). This raised the possibility that the LETS protein may play a role in regulating cell growth.

The loss of LETS protein in transformed cells could be explained by a block in synthesis, by leakage from the cells, or by removal with proteolytic enzymes. Using *ts* mutant-infected cells, Hynes and Wyke (1975) showed that cycloheximide does not inhibit the reappearance of iodinatable LETS protein after shift to the nonpermissive temperature, suggesting that the LETS protein is synthesized in transformed cells but not present at the cell surface. The intracellular protein may be in a nonglycosylated form (Hynes *et al.,* 1975). These authors (Hynes *et al.,* 1975) also failed to find evidence for the involvement of proteolytic enzymes, including plasmin, which is produced in cultures of transformed cells (Unkeless *et al.,* 1973*b*). First, the involvement of plasmin appeared unlikely because the LETS protein was lost from transformed cells in the absence of serum or plasminogen. Second, the medium from transformed cells was incapable of accelerating turnover of the LETS protein, and, finally, various protease inhibitors did not prevent the loss of this protein. Hynes *et al.* (1975) suggested that a faster turnover was the major cause of the decrease in LETS protein. However, Yamada and Weston (1975) showed that the rate of turnover of LETS protein is not higher than that of other cellular proteins. Teng and Chen (1975) showed that treatment of cells with thrombin, at a concentration sufficient to induce DNA synthesis in quiescent cells, does not cause a loss of LETS protein. Consistent with these findings, Vaheri and Ruoslahti (1975) presented evidence that the loss of LETS protein in transformed cells is due to changes in another structure, possibly a protein associated with microfilaments under the plasma membranes, which is required for the attachment of LETS protein.

The appearance of LETS protein after shift-up of *ts* mutant-infected cells is detectable within 2–4 h and complete by 24 h, but the decrease after shift-down is undetectable for 8 h and LETS protein can be found at a relatively high level even after 2 days (Robbins *et al.,* 1975; Hynes and Wyke, 1975). These facts suggest that LETS proteins are not *ts* products of the virus and may not directly control the growth of cells. Yamada *et al.* (1975) found that LETS protein causes cell agglutination and suggested a role in adhesiveness of cells. This was supported by the experiments of Zetter *et al.* (1976) using various proteases.

Changes in the amounts of other cell proteins have also been reported: increases in a 70,000–95,000 dalton protein (Stone *et al.*, 1974; Isaka *et al.*, 1975; Vaheri and Ruoslahti, 1974), decreases in a 40,000–50,000 dalton protein (Wickus and Robbins, 1973; Stone *et al.*, 1974; Isaka *et al.*, 1975), and decrease in a 200,000 dalton protein (Robbins *et al.*, 1975; Vaheri and Ruoslahti, 1974). The significance of these changes has not been clarified.

6.2. Protease

Mild treatment of normal cell cultures with protease can cause various transient effects mimicking transformation. In addition to the loss of LETS protein, protease causes increased sugar uptake (Section 6.4), increased agglutinability by lectin (Section 6.5), induction of DNA synthesis in resting cells (Section 6.8), changes in the particle distribution in freeze-fractured membranes (Section 6.6), and depolymerization of cytoskeleton molecules (Section 6.7). Thus protease may be playing one of the key roles in regulation of cell growth (Talmadge *et al.*, 1974; Hynes, 1974). Cathepsinlike protease activity was found to be elevated in RSV-infected chicken cells (Bosmann *et al.*, 1974*b*). Rubin (1970) has suggested that "overgrowth stimulating factor," released from RSV transformed cells, is a protease.

A unique protease released from transformed cells was discovered by Unkeless *et al.* (1973*a,b*). This enzyme converts serum plasminogen to its active form, plasmin. Plasmin was detected by its fibrinolytic activity. The original assay for plasminogen activator consisted of spreading cells on a plate coated with ^{125}I-labeled fibrin. If transformed cells are present in the culture, the fibrin is digested by activated plasmin and soluble ^{125}I is released. Chen and Buchanan (1975) claimed that fibrinolysis can occur in the absence of plasminogen and proposed that the enzyme released from transformed cells may activate other zymogens in serum or that more than one enzyme is released from transformed cells. Chick embryo cells infected with *NY*68 showed rapid synthesis of plasminogen activator after shift-down (complete by 8 h) and a rapid loss of activator after shift-up (in 4 h) (Rifkin *et al.*, 1975). Both the synthesis and the disappearance of intracellular plasminogen activator are blocked if mRNA synthesis is inhibited by 5-bromotubercidin or actinomycin D (Rifkin *et al.*, 1975).

While the level of plasminogen activator is low in primary embryo cells and early-passage cultures, rather high activity is found in many established cell lines including some 3T3 lines (Ossowski *et al.*, 1973*a*). Some rare morphologically transformed cells contain low levels of activator (Rifkin, personal communication; Wolf and Goldberg, 1976), suggesting that its presence is not essential for expression of transformation.

Despite the existence of these exceptional cases, the plasminogen activator production provides a very sensitive assay for cell transformation. The high sensitivity of the plaque assay developed on the basis of this phenomenon allows detection of the early appearance of a small number of transformed cells (Goldberg, 1974; Balduzzi and Murphy, 1975; Jones *et al.*, 1975).

6.3. Tumor-Specific Surface Antigen

The presence of unique tumor-specific transplantation antigens (TSTA) was demonstrated in RNA tumor virus-infected cells by essentially the same techniques utilized for polyoma virus (Sjögren and Jonsson, 1963; Jonsson and Sjögren, 1966; Koldovský *et al.*, 1966). In these studies, the establishment of transplantation immunity was observed in mice immunized against challenge inoculation with syngeneic cells transformed by avian sarcoma virus. Since RSV-transformed mammalian cells do not produce the viral envelope glycoprotein, the established resistance is not due to immunity against virus particles. The resistance was specific to avian RNA tumor viruses but was not strain specific (Bubenik and Bauer, 1967). More interestingly, mice immunized with avian myeloblastosis virus were shown to be resistant to sarcoma cell transplantation (Bauer *et al.*, 1969).

In vitro techniques for detection of antigens specific to transformed cells have been developed mainly by Bauer and his associates, and detectable antigens were named "tumor-specific cell surface antigen" (TSSA). No rigorous proof exists, but TSTA and TSSA are considered to be identical. Two early techniques for detecting TSSA were immunoferritin methods (Hämmerling *et al.*, 1968; Gelderblom *et al.*, 1972) and cellular microcytotoxicity tests (Kurth and Bauer, 1972). A more recent method uses the binding of iodinated IgG to glutaraldehyde-fixed cells (Kurth *et al.*, 1975). Immune sera generally contain antibody against both TSSA and viral envelope glycoprotein (Ve). Studies have shown that Ve is subgroup specific but TSSA is common to transformed cells induced by any strain of RSV. Chick embryo fibroblast cells infected with LLV-type virus or even AMV are not reactive with TSSA IgG. However, TSSA is present in transformed myeloblasts, confirming the results obtained earlier in the mouse system (Bauer *et al.*, 1975). TSSAs are not related to lectin receptors or embryonic antigens (Kurth and Bauer, 1973*a,b*).

Using immunoprecipitation and gel electrophoresis, TSSAs in the avian sarcoma systems were identified as glycoproteins with molecular weights of about 100,000 and 32,000 (Rohrschneider *et al.*, 1975). In tests with cells infected by a number of T class *ts* mutants of RSV, Kurth *et al.* (1975) found that some mutant-infected cells (three out of nine) expressed TSSA at the nonpermissive temperature. Two explanations were offered. First, TSSA could be the *ts* gene product of the mutants, and at 41°C could still be immunologically intact but functionally defective. Second, TSSA formation could be one of the prerequisites for cell transformation, and *ts* mutants producing TSSA at 41°C might be defective at a step following TSSA formation. There has been no direct proof that TSSA is the *src* gene product. In fact, the observation that TSSAs of sarcoma cells and myeloblast cells are cross-reactive would tend to exclude this hypothesis, unless AMV RNA is shown to contain sequences homologous to $cDNA_{sarc}$. Mouse anti-TSSA sera against RSV-transformed cells were used in cytotoxicity tests and shown to be reactive with chicken myeloblasts (Kurth and Bauer, 1972*b*, 1975), but the reaction seems to be weak and needs further characterization. If the TSSAs synthesized by the same virus in different species share antigenicity, it would strongly suggest that TSSA is a virus-coded molecule.

The expression of TSSA was also shown in RSV-infected cells treated with cyclic AMP. These cells were normal in morphology, in agglutinability, and in cell growth (Kurth and Bauer, 1973*b*). Further analysis of TSSA in relation to other properties of tumor cells would be extremely interesting since TSSA is one of the few characterized molecules which appear upon cell transformation.

The presence of sarcoma-specific antigens in murine sarcoma virus-transformed cells was demonstrated by the immunoferritin technique (Aoki *et al.*, 1973, 1974).

6.4. Transport

Changes in the plasma membrane upon transformation result in an increase in the uptake of certain components of the growth medium. The first clear example of this was the increased uptake of glucose and its derivatives, demonstrated by Hatanaka (for review, see Hatanaka,

1974) in mammalian and avian sarcoma virus-transformed cells (Hatanaka *et al.,* 1969; Hatanaka and Hanafusa, 1970). In general, sugar
uptake increases about three- to fourfold, and is associated with morphological changes in the cells. Because the assay is technically simple,
this change has been examined with most types of *ts* mutant-infected
cells (Martin *et al.,* 1971; Kawai and Hanafusa, 1971; Bader, 1972*b*;
Hynes and Wyke, 1975). In every case, the level of sugar uptake correlated with the morphological characteristics of the cells at different
temperatures. The uptake of glucose, 2-deoxyglucose, 3-*O*-methylglu-
cose, mannose, galactose, and glucosamine increases in transformed
cells, whereas that of fructose, some pentoses, sucrose, and glucose-
phosphates does not (Hatanaka, 1974).

The increase in the uptake of 3-*O*-methylglucose is considered to
indicate that changes in transport, rather than phosphorylation, are
responsible for the increase, since this sugar cannot be phosphorylated
by hexokinase (Weber, 1973; Venuta and Rubin, 1973). The increase
seems to be due to an increased maximum velocity rather than to
changes in the Michaelis constant (Weber, 1973; Venuta and Rubin,
1973; Kletzien and Perdue, 1974), but some conflicting results have
been obtained in measurements of K_m (Hatanaka, 1974). According to
Isselbacher (1972), the increase in V_{max} reflects an increase in the
number of sites involved in transport while the changes in K_m reflect
qualitative changes in the transport sites.

Kletzien and Perdue (1975) showed that an increased V_{max}
equivalent to that in transformed cells can be obtained with NY68-
infected cells kept at 41°C or with normal uninfected cells if glucose is
depleted from the culture medium or if serum is added to serum-
depleted cultures. Combining this observation with the fact that the K_m
is unchanged under all these conditions, they concluded that neither the
function nor the synthesis of the transport system is temperature sensitive. It was suggested that some other function regulating the transport
activity, perhaps involving the organization of the membranes, is
affected by the temperature-sensitive product.

Somewhat inconsistent results have been obtained on the effects of
various metabolic inhibitors on the increase in sugar transport following a shift to permissive temperature. Actinomycin D blocked the
increase in *BE*1-infected cells (Bader, 1972*b*) but not in *NY*68-infected
cells (Kletzien and Perdue, 1974). However, Kletzien and Perdue (1975)
found that cordycepin inhibited the increase with *NY*68-infected cells.
With both *NY*68 and *BK*5, inhibitors of protein synthesis (cycloheximide or puromycin) blocked both the increase in sugar transport and

morphological alteration following temperature shift (Kawai and Hanafusa, 1971; Martin *et al.*, 1971), indicating a requirement for new protein synthesis for the establishment of these transformed phenotypes.

6.5. Agglutination by Lectin

A difference in surface membranes of normal and transformed
cells was also shown in terms of agglutination of cells by plant lectin
(Aub *et al.*, 1963, 1965; Burger and Goldberg, 1967; Burger, 1969).
This difference was not unequivocally shown in early studies with
chicken cells transformed by RSV (Moore and Temin, 1971), probably
because of the accumulation of mucopolysaccharides on the surface of
the cells, as shown by Burger and Martin (1972). An increased
agglutinability was demonstrated by many investigators (Kapeller and
Doljanski, 1972; Biquard, 1973; Graf and Friis, 1973). The lectin
agglutinability is not a property specific to transformed cells but is
considered to be masked in normal cells. A brief exposure to trypsin
makes normal cells agglutinable like transformed cells (Burger, 1969).
Subsequently, trypsin-treated normal cells and transformed cells were
shown to contain equivalent numbers of sites for lectin binding (for
review, see Robbins and Nicolson, 1975). Nicolson (1971, 1973)
proposed that the difference in agglutinability may be due to a cluster-
ing of lectin-binding sites so that multiple lectin cross-bridging between
cells becomes possible. This view is based on the fluid mosaic structure
of membranes (Singer and Nicolson, 1972) and is supported by the
results of lectin binding and the "capping" phenomenon with
lymphocytes (Taylor *et al.*, 1971; Yahara and Edelman, 1972). The
capping phenomenon with fibroblasts, described by Edidin and Weiss
(1972), correlated with the development of ruffled membranes (see Section 6.7). Thus agglutination by lectin seems to reflect a greater
mobility of lectin-binding sites within the membrane of transformed
cells. The mobility of binding sites or antigens on the surface of
transformed cells may be due to (1) an increase in the intrinsic fluidity
of the lipid bilayer, (2) a modification of the structure of the binding
sites which allows these structures to be more readily cross-linked or to
diffuse more rapidly in the membranes, or, more likely, (3) alterations
of peripheral structures attached at the inner or outer membrane surface which influence the mobility of the binding sites (Robbins and
Nicolson, 1975).

6.6. Membrane Fluidity

The fluid mosaic structure of plasma membranes (Nicolson, 1971; Singer and Nicolson, 1972) explains the movement of structures such as antigens or lectin-binding sites on the membranes. The differences in the fluidity between normal and transformed cells are recognizable by the freeze-fracture technique (Branton, 1966). This technique allows the frozen membrane to be split along the central plane of the bilayer so that the fractured faces, containing particles that represent protein-containing structures, are exposed. Scott and others (Scott *et al.*, 1973; Barnett *et al.*, 1974) demonstrated that 3T3 cells transformed by either DNA or RNA tumor virus showed randomly dispersed intramembranous particles, whereas particles in normal 3T3 cell membranes were more aggregated. However, these differences in distribution were not confirmed in two other reports (Pinto da Silva and Martinez-Palomo, 1975; Torpier *et al.*, 1975). Gilula *et al.* (1975) demonstrated that this discrepancy is due to differences in procedure. Fixation with glutaraldehyde before glycerol treatment apparently blocks the aggregation of particles in normal cell membranes, which takes place in the glycerol solution. In contrast, transformed cell membranes did not show clearly aggregated particles even without fixation. Random and nonrandom distributions were seen in *NY*68-infected cells at 36°C and 41°C, respectively. The aggregate formation in normal cells was susceptible to trypsin (Gilula *et al.*, 1975).

The membrane fluidity determined by measurements of electron paramagnetic resonance of spin-labeled cells was found to be increased in SV40-transformed cells (Barnett *et al.*, 1974), but there was no significant difference between normal and *NY*68-transformed cells (Robbins *et al.*, 1975). Therefore, while the difference in the distribution of particles in the fractured faces seems to reflect a difference in the mobility of particles in normal and transformed cells, more work is required to understand the meaning of these changes.

6.7. Microtubules and Microfilaments

According to the fluid mosaic membrane model (Singer and Nicolson, 1972), integral globular proteins are inserted into the lipid bilayer and are capable of moving laterally. The integral proteins seem to be attached to peripheral proteins, such as spectrin in the case of erythrocytes. The fibrous proteins (microtubules, microfilaments, and

10 nm filaments) are considered to be responsible for cell movement, ruffling, and cytokinesis (Porter, 1966; Goldman and Follett, 1969; Olmsted and Borisy, 1973). The detailed arrangement and mutual interactions of these fiber systems in nonmuscle cells are not clear, but increasing amounts of information suggest that these proteins play important roles in both structural and functional changes in cell transformation.

The microtubules, tubular structures with a diameter of about 22 nm, are formed by assembly of tubulin subunits, each of which has a molecular weight of 55,000. Microtubules are distinguished from microfibers, which consist of actin, by their uniform tubular structure and by their sensitivity to alkaloids, colchicine, and vinblastine. The microtubules form a flexible cytoskelton in resting normal cells and are disassembled during cell mitosis. Evidence suggests that microtubules are functionally linked to surface membrane glycoproteins (Berlin and Ukena, 1972; Yahara and Edelman, 1972, 1974).

By immunofluorescent staining with antitubulin, Brinkley et al. (1975) showed that SV40-transformed cells contain fewer microtubules than normal 3T3 cells. In addition, it was shown by Edelman and Yahara (1976) that the microtubule content of NY68-infected cells was much lower at the permissive temperature. Both assembly and disassembly of microtubules by temperature shift occur quite rapidly: the changes in both directions are essentially complete within 2 h. Brinkley et al. (1975) speculated that Ca^{2+} and cyclic AMP concentrations play a role in the regulation of microtubule formation.

The microfilaments are thin fibers of about 6 nm diameter consisting of polymerized actin, a protein with a molecular weight of 42,000. The microfilament is very sensitive to the action of cytochalasin B. The organization of microfilament bundles in nonmuscle cells is still under investigation, but the presence of actin, myosin (Lazarides and Weber, 1974; Pollack et al., 1975; Willingham et al., 1974; Weber and Groeschel-Stewart, 1974), and tropomyosin (Lazarides, 1975) in the bundles suggests the possibility that the interaction of actin and myosin may be regulated by tropomyosin in a manner similar to that in skeletal muscle. The filaments can be present singly or in bundles which are sometimes called "actin cables." Wickus et al. (1975) found that total actin content in transformed chicken cells (infected with wild-type SR-RSV or with NY68) is only slightly reduced compared to that in nontransformed cells. However, profound qualitative differences can be seen; most of the microfilament bundles disappear in transformed cells (Wang and Goldberg, 1976) (Fig. 4). Edelman and Yahara (1976)

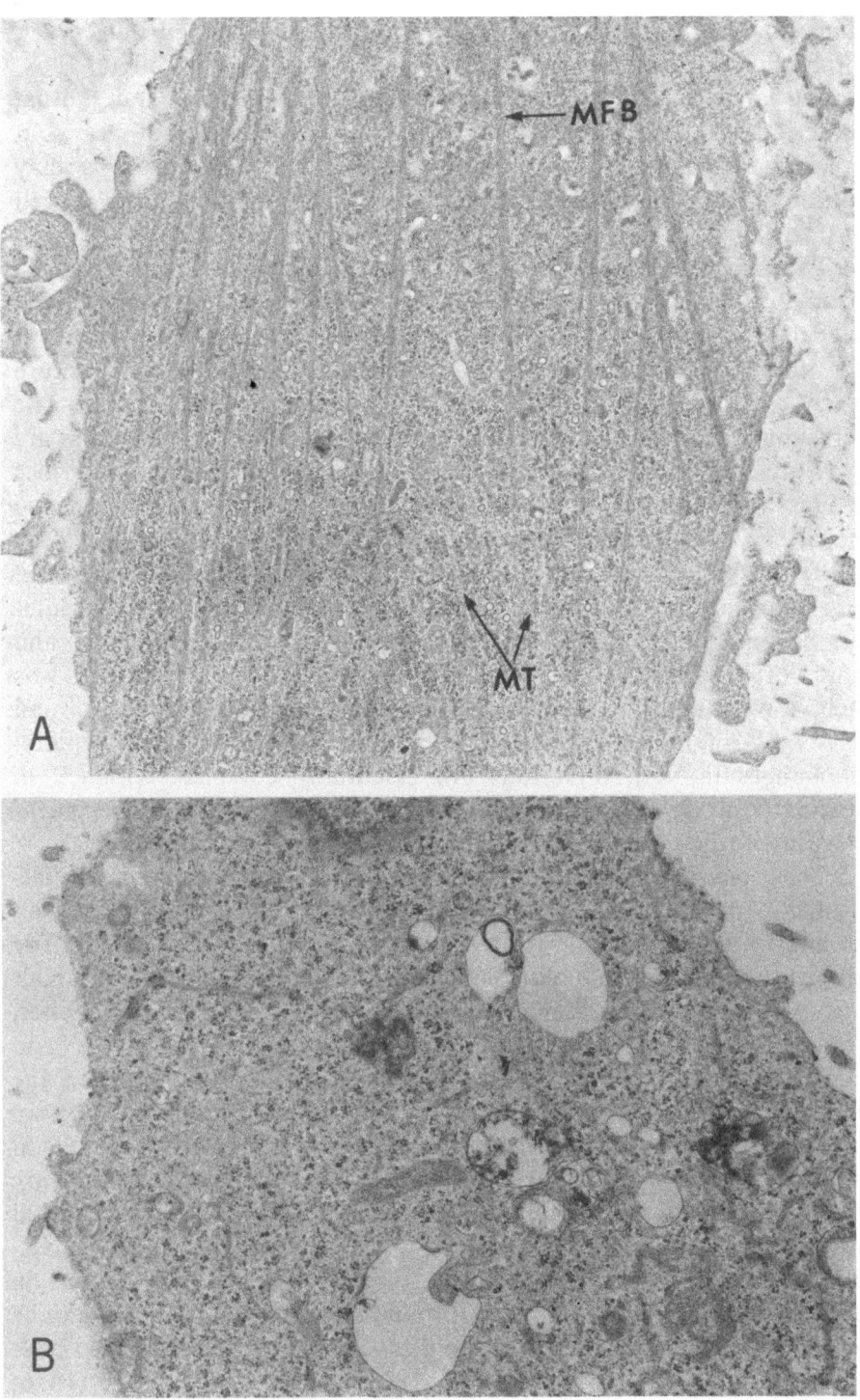

demonstrated that microfilament bundles detectable by fluorescent staining appear and disappear reversibly at permissive and restrictive temperatures in the *ts* mutant-infected cells. Pollack and Rifkin (1975) postulated that increased protease in transformed cells may be responsible for the loss of actin cables.

6.8. Cyclic AMP and Its Metabolism

Since the observations of Hsie and Puck (1971), Johnson *et al.* (1971), and Sheppard (1971) on the effect of cAMP on cell morphology and cell growth, cAMP has been considered to be a factor which might play an important role in functional changes of cells. For example, Hsie and Puck (1971) have shown that dibutyryl cAMP (dbcAMP) caused a loss of transformed morphology in Chinese hamster ovarian cells. This effect was reversed by removal of dbcAMP and was prevented by colchicine or vinblastine, both of which are known to disassemble or inhibit assembly of microtubules (Olmsted and Borisy, 1973).

However, the role of cAMP in RNA tumor virus transformation has not been decisively demonstrated. Otten *et al.* (1972) found that chicken cells transformed by BH-RSV had lowered levels of cAMP and this correlated with the transformed state in temperature-shift experiments using *BE*1. The difference in the amount of cAMP in these cells correlated with the levels of intracellular adenylate cyclase activity (Anderson *et al.*, 1973a). This observation was extended to cells transformed with *BK*5 (Anderson *et al.*, 1973b), but the changes observed were only twofold and appeared very slowly. Yoshida *et al.* (1975) essentially confirmed Anderson's results and extended them to cells transformed with other strains of avian sarcoma virus. Changes in the K_m of the adenylate cyclase upon transformation were observed with BH-RSV and B77 strains but not with SR-RSV or PR-RSV (Anderson *et al.*, 1973b; Yoshida *et al.*, 1975). The basis for this variation among strains is not clear. Hovi *et al.* (1974), on the other hand, showed no significant change in the level of cAMP in RSV-transformed and

Fig. 4. Microfilaments in uninfected and RSV-infected cells. A: A thin section through the cell–substratum region of an uninfected chick cell. Note the abundant parallel bundles of microfilaments (MFB). Microtubules (MT) can also be seen ($\times 6160$). B: A thin section of a *NY*68-infected cell at 24 h after temperature shift (41–37°C) shows the loss of organized microfilaments and microtubules ($\times 16,660$). Courtesy of Drs. E. Wang and A. R. Goldberg.

normal chicken cells. There have been no reports that exogenously supplied cAMP derivatives cause a disappearance of the transformed state
in RSV-infected cells. The finding that differentiation can be induced in
Friend virus-infected erythroleukemic cells by butyric acid might have
some relevance to the changes seen in dbcAMP-treated cells (Leder
and Leder, 1975).

Because of these ambiguities, the role of cAMP should be carefully examined further. Although the difference in cAMP content of
normal and transformed cells is not great, there may be more dramatic
differences in localized critical cell compartments, or, as a regulatory
compound, cAMP may function within a very narrow range of
concentrations.

6.9. Cell Morphology

Morphological alterations, particularly those visible in the optical
microscope, were the first changes to be recognized in early studies of
cell transformation. Although the definition of "morphological alteration" is loose, and most qualitative in nature, morphological changes
are still the basic criteria used to distinguish transformed cells from
their normal counterparts in tissue cultures. These changes involve both
the morphology of the individual cell and the behavior of the cell population. Upon transformation, fibroblast cells generally become round or
spindle shaped, with increased refractivity due to a smoother cell surface. As a population, transformed cells generally lose the parallel
orientation of normal fibroblasts and have an increased tendency to pile
up on each other or to make overlapping "criss-cross" patterns. All of
these characteristics of transformed cells are seen when cells from sarcoma are cultured *in vitro*. The extent of these changes seems to be
dependent on both the virus strains and the cells. Different strains
produce changes that are more or less characteristic of the specific
strain. Mutants causing different cell morphologies have also been
isolated (Temin, 1960; Yoshii and Vogt, 1970). Figures 1 and 2
illustrate the difference in the morphologies of SR-RSV-infected and
BH-RSV-infected chicken cells. In addition, the heterogeneity in a
population of embryonic fibroblasts contributes to variation in the morphologies of individual transformed cells. The heterogeneity is very
apparent from the fact that isolated foci or colonies maintain the morphologies of the original single cells. If viruses recovered from morphologically different foci are used to infect new cells, they generally

induce changes that are morphologically indistinguishable from each other.

When cells fully infected with T class *ts* mutants are shifted from nonpermissive to permissive temperature, increase in the refractivity and changes in shape can be recognized within 1 h; the expression of transformed morphology in the individual cell becomes more profound with time and reaches the same level as in wild-type transformed cells in 12 h (Martin, 1970; Kawai and Hanafusa, 1971; Biquard and Vigier, 1970). The transformed cells become flat within 1 h after shift-up to nonpermissive temperature, but about 12 h is needed to regain the fully stretched fibroblastic shape of the normal cell (Fig. 5).

It has been suggested that the morphology of transformed cells is a reflection of changes in cell membranes and their related structures. Rounding of cells during mitosis, or after treatment with proteolytic enzymes or colchicine, may be related to the changes in transformed cells. In fact, as stated above, transformation causes increased levels of protease activity (Unkeless *et al.*, 1973a,b; Ossowski *et al.*, 1973b), which may cause disassembly of microfilaments or microtubules maintaining the cell shape (Edelman and Yahara, 1976). Other structural or functional changes discussed above are also associated

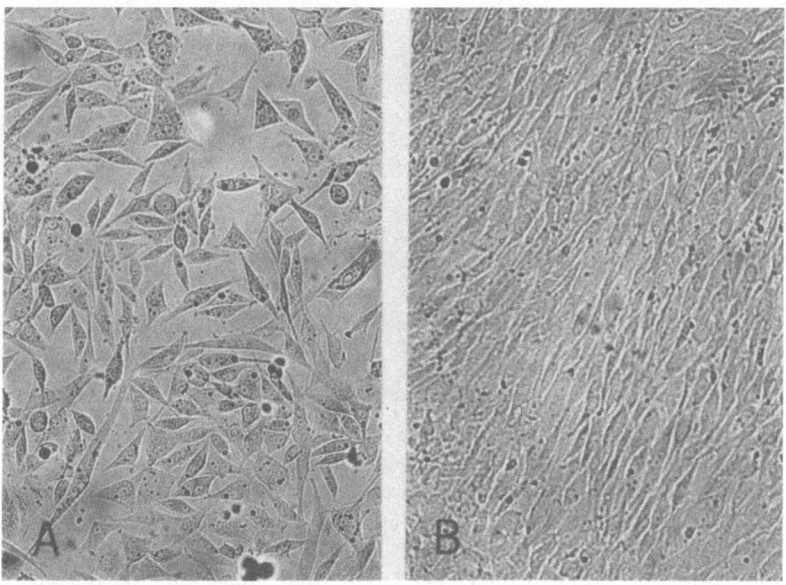

Fig. 5.   A clone of cells transformed with *NY*68 at 37°C (A) and a parallel culture 24 h after shift to 41°C (B). Courtesy of Dr. S. Kawai.

with the cell surface. Whether these changes play a primary or second-ary role in the transformation events remains to be seen.

Another example of these surface changes is the increased ruffling on transformed cells, which has been demonstrated by scanning electron microscopy (Boyde and Weiss, 1963; Stone *et al.*, 1974; Hale *et al.*, 1975; Ambros *et al.*, 1975). Figure 6 shows the surface changes which occur in *NY*68-infected cells following a temperature shift

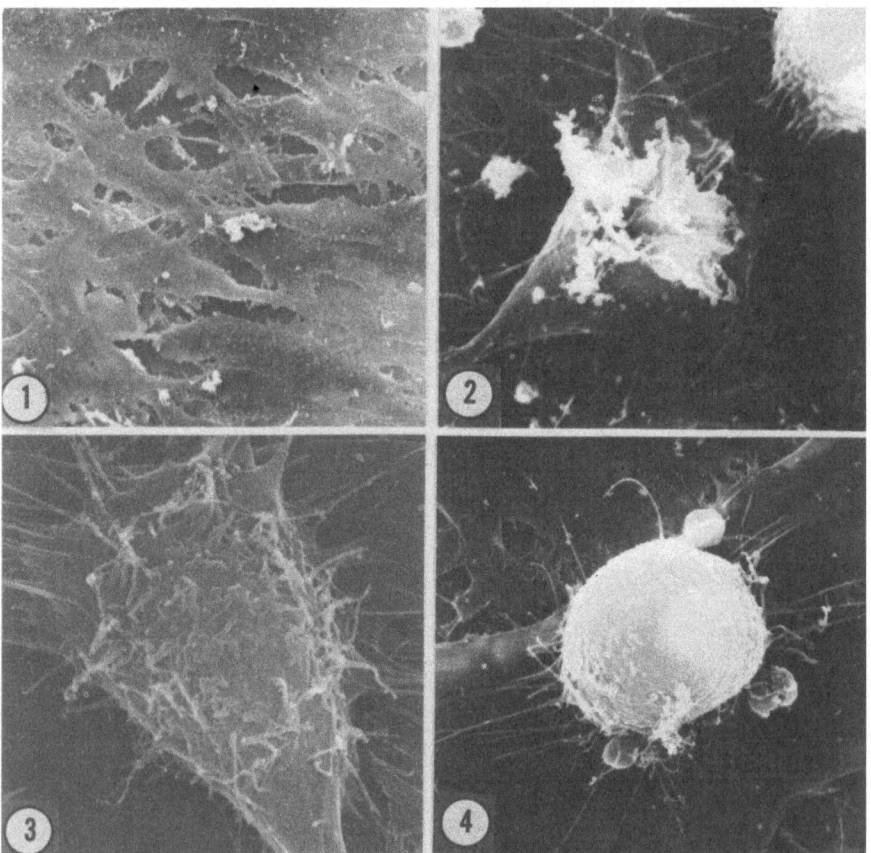

Fig. 6. Temperature-sensitive mutant infected cells viewed by scanning electron microscopy. (1) *NY*68-infected cells kept at 41°C. Note the flat, smooth cell surface. The small dots are probably budding viruses (×200). (2) *NY*68-infected cultures fixed at 1 h after the shift-down to 37°C. Flowerlike surface ruffles appear near the nuclear region (×800). (3) Twelve hours after shift-down. This cell shows numerous microvilli, blebs, and retraction fibers (×1000). (4) Twenty-four hours after shift-down. The cell displays round form with less apparent surface ruffling. Retraction fibers, blebs, and microvilli are seen at the adhesion site (×800). Courtesy of Drs. E. Wang and A. R. Goldberg.

(Wang and Goldberg, 1976). As shown by Ambros *et al.* (1975), ruffle formation is easily detectable within 1 h after shift from 41°C to 37°C. The morphology of completely rounded transformed cells is essentially indistinguishable from that of normal cells in mitotic phase. Porter *et al.* (1974) indicated that flattening of transformed cells treated with cyclic AMP results from a drastic reduction in movement of the cell surface.

6.10. Changes in Growth Characteristics

Cells in sarcomas are apparently less restricted in growth than cells in uninfected normal tissues. However, the growth conditions *in vivo* are difficult to reproduce in tissue culture. Under normal cell culture conditions, both transformed and uninfected normal cells grow exponentially until they reach a certain cell density, which is determined by the availability of growth-promoting substances in the serum and medium (Temin *et al.*, 1972; Holley, 1975). This is called "density-dependent inhibition." Originally, cell contact was considered to be a critical controlling factor. Cessation of growth upon contact with neighboring cells, which is very clear with some mammalian cells such as 3T3 cells, was called "contact inhibition." However, the addition of fresh medium to contact-inhibited cultures releases them from the growth inhibition, and no definitive proof for the existence of contact-induced control has been found (Holley, 1972). The cause of density-dependent inhibition is not clear, but Holley (1975) postulated that the supply of growth factors is reduced because of the limited surface area and the increased destruction of growth factors by the cells.

Normal cells in the resting state can be activated to proceed to S phase by addition of serum growth-promoting factors (Holley and Kiernan, 1974*a*), insulin (Temin, 1967*a*; Vaheri *et al.*, 1973), cyclic nucleotides (Sheppard and Prescott, 1972; Seifert and Rudland, 1974), or proteolytic enzymes (Burger, 1970; Sefton and Rubin, 1970; Blumberg and Robbins, 1975). The growth of normal cells is also influenced by the level of phosphates, glucose, amino acids (Holley and Kiernan, 1974*b*), and calcium ions (Balk *et al.*, 1973; Dulbecco and Elkington, 1975), and by changes in pH (Ceccarini and Eagle, 1971; Rubin and Koide, 1975).

Normal and transformed cells generally grow at the same rate if sufficient amounts of nutrients and serum are supplied. The major difference appears to be a reduced requirement by transformed cells for serum growth factors (Temin, 1965; Colby and Rubin, 1969; Jainchill and Todaro, 1970; Hanafusa, 1969; Balk *et al.*, 1973). The components in serum are complex, and both high and low molecular weight substances are involved (Temin *et al.*, 1972; Holley, 1972). Normal cells cease to grow under conditions of limiting serum, and accumulate at G_1 or G_0 phase of the cell cycle. Thus transformation is considered to represent a state in which this restriction is lost or reduced (Temin *et al.*, 1974; Pardee, 1974; Pardee *et al.*, 1974). Apparently, some mechanism established in the transformed cells allows them to respond more efficiently to serum factors so that cells can proceed to S phase. Whether this is related to changes in intracellular or extracellular processes is unknown. Recent findings that serum can be replaced by hormones for growth of various types of cells (Hayashi and Sato, 1976) and that chick embryo cells have unique receptors (Rechler *et al.*, 1976) for a family of peptides called "multiplication-stimulating activity" (MSA) (Dulak and Temin, 1973) in serum suggest that growth-promoting substances which are hormone in nature are principal factors controlling normal and transformed cell growth. This indicates that changes in the cellular receptors may be the key difference between normal and transformed cells. Thus many regulatory compounds, including low molecular weight substances and substances affecting pH, may be important in modifying the cell membranes that contain the receptors.

The different growth characteristics in the untransformed and transformed states have been demonstrated with *LA*24 RSV-infected chicken cells (Bell *et al.*, 1975). When depleted of serum, *ts* mutant-infected cells became quiescent after 2 days at 41°C, as expected for nontransformed cells. When the temperature was shifted to 35°C without addition of serum, the cells assumed transformed morphology and new synthesis of cell DNA was observed, as expected for transformed cells. One interesting question is whether this new round of cell growth requires serum factors, which might be retained on the cell surface even in the depleted medium, or whether the expression of transformation and DNA synthesis are totally independent of serum. Further analysis of this system would be of great interest.

Finally, the growth of normal cells requires attachment to a solid surface. Without this attachment, normal cells accumulate at one stage in the cell cycle, possibly at G_2 (Yaoi *et al.*, 1972), whereas transformed cells can continue to grow in agar suspension cultures (Macpherson and Montagnier, 1964; Rubin, 1966; Kawai and Yamamoto, 1970; Bassin *et al.*, 1970) or in liquid suspension cultures (Bader, 1968). The ability of normal cells to grow only on a solid substratum is called "anchorage dependence of multiplication" (Stoker *et al.*, 1968). Colony formation in agar is temperature sensitive with most, but not all, *ts* mutants (Bauer *et al.*, 1975). In general, more technical difficulties are encountered in studies of the temperature dependence of colony formation than in studies of focus formation. Foci of transformed cells will revert to normal morphology following a shift to the nonpermissive temperature, even after foci have been fully developed, whereas colonies already formed cannot be eliminated by the temperature shift. Thus either leaky expression of the *ts* mutants or temporary failure to maintain the nonpermissive temperature can cumulatively result in the limited growth of colonies.

6.11. Other Properties

In accord with Warburg's theory that increased aerobic glycolysis and reduced respiration are common features of tumor cells (Warburg, 1931), increased glycolysis has been observed with Rous sarcoma tissues (Burk *et al.*, 1941). However, studies on various types of tumors have shown that aerobic glycolysis is not necessarily an essential feature of malignancy. Careful examination of glycolysis in cultured normal and RSV-transformed chicken cells showed no significant difference when cells were growing exponentially (Steck *et al.*, 1968; Temin, 1968*a*). Therefore, increased lactic acid synthesis in transformed cells was attributed to the preferential growth of transformed cells under certain conditions. Under one such condition (in semiconfluent cultures), Singh *et al.* (1974) found that the level of activities of several enzymes involved in the glycolytic pathway increased in RSV-transformed cells. It seems likely, therefore, that increased sugar uptake in transformed cells under these conditions is influenced more profoundly by the increased glycolytic activities than by changes in the transport system.

The relationship between virus infection and cell differentiation has been studied in a variety of RNA virus-infected cell systems. Infection of iris epithelial cells with RSV results in virus reproduction and increased production of mucopolysaccharides, but pigment formation is blocked (Ephrussi and Temin, 1960; Temin, 1965). For myoblasts infected with RSV, it was originally reported that neither cell transformation nor differentiation into myotubes was prevented (Kaighn *et al.*, 1966; Lee *et al.*, 1966), but more recent studies indicate that viral expression related to cell transformation can block myogenesis at certain stages (Easton and Reich, 1972; Fiszman and Fuchs, 1975; Holtzer *et al.*, 1975). If myoblasts infected with T class *ts* mutants were maintained at 41°C, the formation of contracting postmitotic myotubes was normal, but myotube formation was blocked when the cultures were shifted to the permissive temperature. Successful transformation of neuroretina cells by RSV can occur with cells from 7-day-old embryos but not with cells from 10-day-old embryos (Pessac and Calothy, 1974).

Little is known about which cellular genes are activated following cell transformation. However, one such expression in chicken cells was shown to be embryonic globin message (Groudine and Weintraub, 1975). This messenger RNA is present in RSV-transformed cells (100–500 copies per cell) but is not present in cells infected with *td* mutants. Whether the expression of the normally quiescent globin gene plays any role in transformation should be first clarified. To this end, the transcription of this gene upon temperature shift of *ts* mutant-infected cells should be analyzed and correlated with various other expressions of the transformed state, using various physiological conditions including the presence of metabolic inhibitors.

Various compounds are known to have somewhat specific lethal effects on transformed cells. Rifkin and Reich (1971) demonstrated a selective lysis of RSV-transformed cells by a local anesthetic, dibucaine. This is extremely interesting in view of the importance of cell surface changes in the transformation process. A recent report suggests that the action of these anesthetic compounds is equivalent to the combined action of colchicine and cytochalasin B (Poste *et al.*, 1975). Rifampicin derivatives (Vaheri and Hanafusa, 1971; Bissell *et al.*, 1974) and diphtheria toxin (Iglewski *et al.*, 1975) are also known to preferentially inhibit the growth of RSV-transformed cells, but their mode of action is not clear.

An increased frequency of dimeric forms of mitochondrial DNA in leukemic cells has been reported (Clayton and Vinograd, 1969). Nass (1973) found increased amounts of dimeric or oligomeric forms in *BK*5-infected cells at permissive temperatures. Bosmann *et al.* (1974*c*) observed that mitochondrial synthesis of DNA, RNA, and protein is elevated at permissive temperatures in *NY*68-infected cells.

Finally virus-induced transformation appears to sensitize cells to a lytic response to infection by certain strains of LLV-type virus (Bassin *et al.* 1971*b*; Kawai and Hanafusa, 1972*a*; Wyke and Linial, 1973). Some clonal lines of BALB 3T3 cells (e.g., UCI-B) (Hackett and Sylvester, 1972) become transformed after infection with LLV-type virus. The foci produce LLV but not sarcoma virus. The mechanism of these LLV–cell interactions is not understood, but these responses have been used for the assay of LLV-type virus infection.

7. CONCLUDING REMARKS

We have seen a remarkable concentration of efforts to analyze the process of cell transformation by RNA tumor viruses. Molecular biological approaches combined with virus genetics have considerably broadened our knowledge about the gene responsible for transformation. We now know something about its size and its location in the virus genome. An outline of the processes involved in integration of the virus genome and the basic requirements for its expression have been given. However, we also realize how little we know about the details of these processes. We would need to understand whether or not the virus integrates at a specific site in the cellular chromosomes, and how its expression is related to the regulation of adjacent cellular genes.

The question of the origin of the transforming genes is of basic importance. Is the *src* gene of the virus derived from host cell genetic information? If so, what types of interactions are involved? It is also mysterious why most sarcoma viruses are defective in replicating functions. Research on the acute leukemic viruses is expanding rapidly with the availability of new *in vitro* systems for studying the transformation of hematopoietic cells. It should soon be possible to analyze the genome composition of these viruses and to study the expression of specific genes, as has been done with sarcoma viruses. A challenging question is the basis for the target cell specificity of these viruses. The gene specific for leukemogenesis is reproduced during multiplication of these viruses in fibroblasts, generally without causing transformation. Thus the specificity of the transforming function in these viruses must be somehow related to the differentiated properties of cells. The product of the transforming gene of leukemia viruses may interact with cellular structures unique to the target cells. In this respect, investigations into possible qualitative differences in fibroblasts transformed by sarcoma viruses and by leukemia viruses such as Abelson virus or MC29 might be interesting.

Lymphatic leukemia virus was treated essentially as a helper virus in this chapter. Most of these viruses do cause leukemia, and a consideration of the mechanism involved in this process is obviously important. Considering the difference in nucleotide sequences between sarcoma or acute leukemia viruses and lymphatic leukemia viruses, and the coding capacity of virus genomes, it seems conceivable that leukosis induction by LLV is not due to the expression of an additional viral gene, but due to virus replication itself. Perhaps modification of the lymphatic cell surface by viral structural components (such as the viral glycoprotein) may elicit proliferation of certain clones of lymphoid cells. If this is the case, viruses which have structural components common to an endogenous virus present in the host may not be oncogenic.

Isolation of the product of the *src* gene is an important task for opening up further approaches for analyzing cell transformation. While the problem is difficult, it is basically technical in nature. For example, if techniques for translation in cell-free systems could be improved, so that translational products could be obtained without contamination by cellular proteins, then the identification of the *src* product would not be difficult.

The advances in our understanding of the intracellular events involved in transformation appear very encouraging, although we may be far from the final solution. Many hypotheses can be made to explain the process of cell transformation. The various models can be divided into three general classes. First, the *src* gene product may directly interact with a structure related to cell membranes. This change could cause modifications of the membranes that in turn modify the regulation of cell growth. Second, the *src* gene product may interact directly, or indirectly (by changing effectors or deblocking mechanisms), with cell DNA to induce the transcription of specific cellular genes. The products of these induced cellular genes could then be responsible for the modification of the cell membrane. Differences in cell growth might be determined by changes taking place on the membrane—for example, by changes in the receptors for serum factors. Third, a process similar to that described in the second hypothesis may activate cellular genes coding for products related to the regulation of the cell cycle. Changes in the membranes might be only secondary events resulting from the modification of growth regulation. Currently one cannot exclude any of these possibilities. A crucial experiment would be to demonstrate specific binding or interaction of the *src* gene product with an identifiable cellular component.

In considering these mechanisms, the further analysis of *ts* mutant-infected cells may be quite useful. Many expressions of the transformed state have been studied with T class *ts* mutant-infected cells. One approach is to compare the time sequence of the various events following the temperature shift. Early changes observed following shift-down include the disassembly of microtubules and microfilaments, synthesis of plasminogen activator, changes in the particle distribution in freeze-fractured cell membranes, and ruffling. Morphological changes detectable in the optical microscope may also be included in this class. Slower changes described are the disappearance of LETS proteins and increases in the level of cAMP. However, the accuracy of detection of each of these parameters is not equivalent, and the time course involved in the structural changes (such as disappearance of LETS protein) does not necessarily correlate with that of functional changes. Nevertheless, events occurring at early stages should be related more closely to the primary events responsible for transformation.

Results obtained with metabolic inhibitors in temperature-shift experiments have not always been consistent. Some properties such as ruffling (Ambros *et al.*, 1975) or vacuole formation (Bader *et al.*, 1974) were shown to occur without protein synthesis after shift-down of temperature, indicating that the *ts* protein is reversibly inactivated. On the other hand, plasminogen activator synthesis (Rifkin *et al.*, 1975) or increase in sugar uptake (Bader, 1972*b*) seems to require new DNA-dependent RNA synthesis. In another study, the increased sugar uptake upon shift-down was shown to be blocked by cordycepin but not by actinomycin D in *NY*68-infected cells (Kletzien and Perdue, 1975). In some cases, the effects of the same inhibitors on the same properties were different in cells infected with different mutants. There are a number of possible explanations for these discrepancies. First, the stability of the temperature-sensitive protein, the reversibility of the temperature inactivation, or the life span of its messenger RNA may be variable with different mutants. Second, the intracellular concentrations of these products may depend on the physiological state of the cells before the temperature shift. Third, as shown by Bell *et al.* (1975), *ts* mutant-infected cells can undergo drastic metabolic changes caused by the temperature shift if cells have been in the resting state. Also, the effects of inhibitors on the synthesis of mRNA (particularly *src* RNA) must be studied, together with analysis of the kinetics of decay and the utilization of viral RNA for virus assembly. Finally, effects of a given inhibitor on the synthesis of viral gene products must be carefully distinguished from effects on the synthesis of the cellular components involved. If all of these factors are carefully investigated, the effects of inhibitors on the changes of each cellular property may give useful information concerning the processes involved in transformation.

ACKNOWLEDGMENTS

I thank Dr. S. Kawai for photomicrographs, Drs. E. Wang and A. R. Goldberg for electron micrographs, Drs. G. Edelman, A. R. Goldberg, D. B. Rifkin, and I. Yahara for providing information on their unpublished work, and Drs. S. A. Aaronson, P. H. Duesberg, C. Moscovici, S. Nomura, D. B. Rifkin, H. M. Temin, P. K. Vogt, and I. Yahara for kindly providing their comments on the manuscript. I am deeply indebted to Dr. W. S. Hayward, Ms. Emily Silva, and Ms. Grazina Boecke for assistance in the preparation of the manuscript. Research by the author was supported by Research Grant CA14935 from the National Cancer Institute.

8. REFERENCES

Aaronson, S. A., 1971, Isolation of a rat-tropic helper virus from M-MSV-0, *Virology* **44**:29.

Aaronson, S. A., and Rowe, W. P., 1970, Nonproducer clones of murine sarcoma virus transformed BALB/3T3 cells, *Virology* **42**:9.

Aaronson, S. A., and Weaver, C. A., 1971, Characterization of murine sarcoma virus (Kirsten) transformation of mouse and human cells, *J. Gen. Virol.* **13**:245.

Aaronson, S. A., Jainchill, J. L., and Todaro, G. J., 1970, Murine sarcoma virus transformation of BAB/3T3 cells: Lack of dependence on murine leukemia virus, *Proc. Natl. Acad. Sci. USA* **66**:1236.

Aaronson, S. A., Bassin, R. H., and Weaver, C., 1972, Comparison of murine sarcoma viruses in nonproducer and S$^+$L$^-$-transformed cells, *J. Virol.* **9**:701.

Abelson, H. T., and Rabstein, L. S., 1970, Lymphosarcoma: Virus-induced thymic-independent disease in mice, *Cancer Res.* **30**:2213.

Ambros, V. R., Chen, L. B., and Buchanan, J. M., 1975, Surface ruffles as markers for studies of cell transformation by Rous sarcoma virus, *Proc. Natl. Acad. Sci. USA* **72**:3144.

Anderson, G. R., and Robbins, K. C., 1976, Rat sequences of the Kirsten and Harvey murine sarcoma virus genomes: Nature, origin, and expression in rat tumor RNA, *J. Virol.* **17**:335.

Anderson, W. B., Johnson, G. S., and Pastan, I., 1973a, Transformation of chick embryo fibroblasts by wild type and temperature sensitive Rous sarcoma virus alters adenylate cyclase activity, *Proc. Natl. Acad. Sci. USA* **70**:1055.

Anderson, W. B., Lovelace, E., and Pastan, I., 1973b, Adenylate cyclase activity is decreased in chick embryo fibroblasts transformed by wild type and temperature sensitive Schmidt-Ruppin Rous sarcoma virus, *Biochem. Biophys. Res. Commun.* **52**:1293.

Aoki, T., Stephenson, J. R., and Aaronson, S. A., 1973, Demonstration of a cell surface antigen associated with murine sarcoma virus by immunoelectron microscopy, *Proc. Natl. Acad. Sci. USA* **70**:742.

Aoki, T., Stephenson, J. R., Aaronson, S. A., and Hsu, K. C., 1974, Surface antigens of mammalian sarcoma virus-transformed nonproducer cells, *Proc. Natl. Acad. Sci. USA* **71**:3445.

Aub, J. C., Tieslau, C., and Lankester, A., 1963, Reaction of normal and tumor cell surfaces to enzymes. I. Wheat-germ lipase and associated mucopolysaccharides, *Proc. Natl. Acad. Sci. USA* **50**:613.

Aub, J. C., Sanford, B. H., and Cote, M. N., 1965, Studies on reactivity of tumor and normal cells to a wheat germ agglutinin, *Proc. Natl. Acad. Sci. USA* **54**:396.

Axelrad, A. A., and Steeves, R. A., 1964, Assay for Friend leukemia virus: Rapid quantitative method based on enumeration of macroscopic spleen foci in mice, *Virology* **24**:513.

Bader, J. P., 1964, The role of deoxyribonucleic acid in the synthesis of Rous sarcoma virus, *Virology* **22**:462.

Bader, J. P., 1965, The DNA requirement of Rous sarcoma and Rous-associated viruses, *Virology* **26**:253.

Bader, J. P., 1966, Metabolic requirements for infection by Rous sarcoma virus. I. Transient requirement for DNA synthesis, *Virology* **29**:444.

Bader, P., 1968, A method for the propagation of large amounts of Rous sarcoma virus, *Virology* **36**:140.

Bader, J. P., 1972a, Metabolic requirements for infection by Rous sarcoma virus. IV. Virus reproduction and cellular transformation without cellular division, *Virology* **48**:494.

Bader, J. P., 1972b, Temperature-dependent transformation of cells infected with a mutant of Bryan Rous sarcoma virus, *J. Virol.* **10**:267.

Bader, J. P., 1973, Virus-induced transformation without cell division. *Science* **180**:1069.

Bader, J. P., and Brown, N. R., 1971, Induction of mutations in an RNA tumor virus by an analogue of a DNA precursor, *Nature (London) New Biol.* **234**:11.

Bader, J. P., Ray, D. A., and Brown, N. R., 1974, Accumulation of water during transformation of cells by an avian sarcoma virus, *Cell* **3**:307.

Balduzzi, P. C., 1976, Cooperative transformation studies with temperature-sensitive mutants of Rous sarcoma virus, *J. Virol.* **18**:332.

Balduzzi, P., and Morgan, H. R., 1970, Mechanism of oncogenic transformation by Rous sarcoma virus. I. Intracellular inactivation of cell-transforming ability of Rous sarcoma virus by 5-bromodeoxyuridine and light, *J. Virol.* **5**:470.

Balduzzi, P. C., and Murphy, H., 1975, Plaque assay of avian sarcoma viruses using casein, *J. Virol.* **16**:707.

Balk, S. D., Whitfield, J. F., Youdale, T., and Braun, A. C., 1973, Role of calcium, serum, plasma, and folic acid in the control of proliferation of normal and Rous sarcoma virus-infected chicken fibroblasts, *Proc. Natl. Acad. Sci. USA* **70**:675.

Ball, J. K., Harvey, D., and McCarter, J. A., 1973a, Evidence for naturally occurring murine sarcoma virus, *Nature (London)* **241**:272.

Ball, J. K., McCarter, J. A., and Sunderland, S. M., 1973b, Evidence for helper independent murine sarcoma virus. I. Segregation of replication-defective and transformation-defective viruses, *Virology* **56**:268.

Baltimore, D., 1975, Tumor viruses: 1974, *Cold Spring Harbor Symp. Quant. Biol.* **39**:1187.

Baluda, M. A., and Goetz, I. E., 1961, Morphological conversion of cell cultures by avian myeloblastosis virus, *Virology* **15**:185.

Baluda, M. A., Shoyab, M., Markham, P. D., Evans, R. M., and Drohan, W. N., 1975, Base sequence complexity of 35 S avian myeloblastosis virus RNA determined by molecular hybridization kinetics, *Cold Spring Harbor Symp. Quant. Biol.* **39**:869.

Barnett, R. E., Furch, L. T., and Scott, R. E., 1974, Differences in membrane fluidity and structure in contact-inhibited and transformed cells, *Proc. Natl. Acad. Sci. USA* **71**:1992.

Bassin, R. H., Tuttle, N., and Fischinger, P. J., 1970, Isolation of murine sarcoma virus-transformed mouse cells which are negative for leukemia virus from agar suspension cultures, *Int. J. Cancer* **6**:95.

Bassin, R. H., Phillips, L. A., Kramer, M. J., Haapala, D. K., Peebles, P. T., Nomura, S., and Fischinger, P. J., 1971a, Transformation of mouse 3T3 cells by murine sarcoma virus: Release of virus-like particles in the absence of replicating murine leukemia helper virus, *Proc. Natl. Acad. Sci. USA* **68**:1520.

Bassin, R. H., Tuttle, N., and Fischinger, P. J., 1971b, Rapid cell culture assay technique for murine leukemia viruses, *Nature (London)* **229**:564.

Bassin, R. H., Gerwin, B. I., Duran-Troise, G., Gisselbrecht, S., and Rein, A., 1975, Murine sarcoma virus pseudotypes acquire a determinant specifying N or B tropism from leukemia virus during rescue, *Nature (London)* **256**:223.

Bauer, H., Bubenik, J., Graf, T., and Allgaier, C., 1969, Induction of transplatation resistance to Rous sarcoma isograft by avian leukosis virus, *Virology* **39**:482.

Bauer, H., Kurth, R., Rosrschneider, L. R., Pauli, G., Friis, R. R., and Gelderblom, H., 1975, On the role of cell surface changes in RNA-tumor virus transformed cells, *Cold Spring Harbor Symp. Quant. Biol.* **39**:57.

Beard, J. W., 1956, Virus of avian myeoblastic leukosis, *Poultry Sci.* **35**:203.

Beard, J. W., 1963, Avian virus growths and their etiologic agents. *Adv. Cancer Res.* **7**:1.

Beaudreau, G. S., Becker, C., Bonar, R. A., Wallbank, A. M., Beard, D., and Beard, J. W., 1960, Virus of avian myeloblastosis. XIV. Neoplastic response of normal chicken bone marrow treated with the virus in tissue culture. *J. Natl. Cancer Inst.* **24**:395.

Beemon, K., Duesberg, P., and Vogt, P., 1974, Evidence for crossing-over between avian tumor viruses based on analysis of viral RNAs, *Proc. Natl. Acad. Sci. USA* **71**:4254.

Bell, J. G., Wyke, J. A., and Macpherson, I. A., 1975, Transformation by a temperature sensitive mutant of Rous sarcoma virus in the absence of serum, *J. Gen. Virol.* **27**:127.

Bellamy, A. R., Gillies, S. C., and Harvey, J. D., 1974, Molecular weight of two oncornavirus genomes: Derivation from particle molecular weights and RNA content, *J. Virol.* **14**:1388.

Bentvelzen, P., Aarssen, A. M., and Brinkhof, J., 1972, Defectivity of Rauscher murine erythroblastosis virus, *Nature (London) New Biol.* **239**:122.

Berlin, R. D., and Ukena, T. E., 1972, Effect of colchicine and vinblastine on the agglutination of polymorphonuclear leucocytes by concanavalin A, *Nature (London) New Biol.* **238**:120.

Biggs, P. M., Milne, B. S., Graf, T., and Bauer, H., 1973, Oncogenicity of non-transforming mutants of avian sarcoma viruses, *J. Gen. Virol.* **18**:399.

Bilello, J. A., Strand, M., and August, J. T., 1974, Murine sarcoma virus gene expression: Transformants which express viral envelope glycoprotein in the absence of the major internal protein and infectious particles, *Proc. Natl. Acad. Sci. USA* **71**:3234.

Billeter, M. A., Parsons, J. T., and Coffin, J. M., 1974, The nucleotide sequence complexity of avian tumor virus RNA, *Proc. Natl. Acad. Sci. USA* **71**:3560.

Biquard, J.-M., 1973, Agglutinability of Rous cells by concanavalin A: Study with a temperature-sensitive RSV mutant and inhibitors of macromolecular synthesis, *Intervirology* **1**:220.

Biquard, J-M., and Vigier, P., 1970, Isolement et etude d'un mutant conditionnel du virus de Rous a capacite transformante thermosensible, *C. R. Acad. Sci. Paris Ser. D* **271**:2430.

Biquard, J.-M., and Vigier, P., 1972, Characteristics of a conditional mutant of Rous sarcoma virus defective in ability to transform cells at high temperature, *Virology* **47**:444.

Bishop, J. M., and Varmus, H. E., 1975, The molecular biology of RNA tumor viruses, in *Cancer,* Vol. 2 (F. F. Becker, ed.), p. 3, Plenum, New York.

Bishop, J. M., Deng, C. T., Mahy, B. W. J., Quintrell, N., Stavnezer, E., and Varmus, H. E., 1976, Synthesis of viral RNA in cells infected with avian sarcoma virus, in: *Animal Virology,* ICN-UCLA Symposium on Molecular and Cellular Biology, (D. Baltimore, A. S. Huang, and C. F. Fox, eds.), p. 1, Academic Press, New York.

Bissell, M. J., Hatie, C., Tischfer, A. N., and Calvin, M., 1974, Preferential inhibition of the growth of virus transformed cells in culture by Rifazone-8_2, a new rifamycin derivative, *Proc. Natl. Acad. Sci. USA* **71**:2520.

Blumberg, P. M., and Robbins, P. W., 1975, Effect of proteases on activation of resting chick embryo fibroblasts and on cell surface proteins. *Cell* **6**:137.

Boettiger, D., 1974, Reversion and induction of Rous sarcoma virus expression in virus-transformed baby hamster kidney cells, *Virology* **62**:522.

Boettiger, D., and Temin, H. M., 1970, Light inactivation of focus formation by chicken embryo fibroblasts infected with avian sarcoma virus in the presence of 5-bromodeoxyuridine, *Nature (London)* **228**:622.

Bolognesi, D. P., and Graf, T., 1971, Size differences among the high molecular weight RNA's of avian tumor viruses, *Virology* **43**:214.

Bookout, J. B., and Sigel, M. M., 1975, Characterization of a conditional mutant of Rous sarcoma virus with alterations in early and late functions of cell transformation, *Virology* **67**:474.

Bosmann, H. B., Case, K. R., and Morgan, H. R., 1974*a*, Surface biochemical changes accompanying primary infection with Rous sarcoma virus. I. Electrokinetic properties of cells and cell surface glycoprotein: glycosyltransferase activities, *Exp. Cell Res.* **83**:15.

Bosmann, H. B., Lockwood, T., and Morgan, H. R., 1974*b*, Surface biochemical changes accompanying primary infection with Rous sarcoma virus. II. Proteolytic and glycosidase activity and sublethal autolysis, *Exp. Cell Res.* **83**:25.

Bosmann, H. B., Myers, M. W., and Morgan, H. R., 1974*c*, Synthesis of DNA-RNA, protein, and glycoprotein in mitochondria of cells transformed with Rous sarcoma viruses, *Biochem. Biophys. Res. Commun.* **56**:75.

Boyde, A., and Weiss, R. A., 1973, in: *The Molecular Biology of Tumor Viruses* (J. Tooze, ed.), p. 666, Cold Spring Harbor Laboratory, Cold Spring Harbor, N.Y.

Branton, D., 1966, Fracture faces of frozen membranes, *Proc. Natl. Acad. Sci. USA* **55**:1048.

Brinkley, B. R., Fuller, G. M., and Highfield, D. P., 1975, Cytoplasmic microtubules in normal and transformed cells in culture: Analysis by tubulin antibody immuno-fluorescence, *Proc. Natl. Acad. Sci. USA* **72**:4981.

Bubenik, J., and Bauer, H., 1967, Antigenic characteristics of the interaction between Rous sarcoma virus and mammalian cells: Complement-fixing and transplantation antigens, *Virology* **31**:489.

Buck, C. A., Glick, M. C., and Warren, L., 1970, A comparative study of glycopro-teins from the surface of control and Rous sarcoma virus transformed hamster cells, *Biochemistry* **9**:4567.

Burger, M. M., 1969, A difference in the architecture of the surface membrane of normal and virally-transformed cells, *Proc. Natl. Acad. Sci. USA* **62**:994.

Burger, M. M., 1970, Proteolytic enzymes initiating cell division and escape from contact inhibition of growth, *Nature (London)* **227**:170.

Burger, M. M., and Goldberg, A. R., 1967, Identification of a tumor specific determinant on neoplastic cell surfaces, *Proc. Natl. Acad. Sci. USA* **57**:359.

Burger, M. M., and Martin, G. S., 1972, Agglutination of cells transformed by Rous sarcoma virus by wheat germ agglutinin and concanavalin A, *Nature (London) New Biol.* **237**:9.

Burk, D., Springe, H., Spangeler, J. M., Kabat, E. A., Furth, J., and Claude, A., 1941, The metabolism of chicken tumors, *J. Natl. Cancer Inst.* **2**:201.

Bussell, R. H., and Robinson, W. S., 1973, Membrane proteins of uninfected and Rous sarcoma virus-transformed avian cells, *J. Virol.* **12**:320.

Calnek, B. W., 1964, Morphological alteration of RIF-infected chick embryo fibro-blasts, *Natl. Cancer Inst. Monogr.* **17**:425.

Canaani, E., and Duesberg, P., 1972, Role of subunits of 60 to 70 S avian tumor virus ribonucleic acid in its template activity for the viral deoxyribonucleic acid polymerase, *J. Virol.* **10**:23.

Canaani, E., Helm, K. V. D., and Duesberg, P., 1973, Evidence for 30–40 S RNA as precursor of the 60–70 S RNA of Rous sarcoma virus, *Proc. Natl. Acad. Sci. USA* **72**:401.

Ceccarini, C., and Eagle, H., 1971, pH as a determinant of cellular growth and contact inhibition, *Proc. Natl. Acad. Sci. USA* **68**:229.

Chen, J. H., and Hanafusa, H., 1973, Detection of a protein of avian leukoviruses in uninfected chick cells by radioimmunoassay, *J. Virol.* **13**:340.

Chen, L. B., and Buchanan, J. M., 1975, Plasminogen-independent fibrinolysis by proteases produced by transformed chick embryo fibroblasts, *Proc. Natl. Acad. Sci. USA* **72**:1132.

Cheung, K.-S., Smith, R. E., Stone, M. P., and Joklik, W. K., 1972, Comparison of immature (rapid harvest) and mature Rous sarcoma virus particles, *Virology* **50**:851.

Clarke, B. J., Axelrad, A. A., Shreeve, M. M., and McLeod, D. L., 1975, Erythroid colony induction without erythropoietin by Friend leukemia virus, *in vitro, Proc. Natl. Acad. Sci. USA* **72**:3556.

Clarkson, B., and Baserga, R., eds., 1974, *Control of Proliferation in Animal Cells*, Cold Spring Harbor Laboratory, Cold Spring Harbor, N.Y.

Clayton, D. A., and Vinograd, J., 1969, Complex mitochondrial DNA in leukemic and normal human myeloid cells, *Proc. Natl. Acad. Sci. USA* **62**:1077.

Coffin, J. M., 1972, Rescue of Rous sarcoma virus from Rous sarcoma virus transformed mammalian cells, *J. Virol.* **10**:153.

Coffin, J. M., and Billeter, M. A., 1976, A physical map of the Rous sarcoma virus genome, *J. Mol. Biol.* **100**:293.

Coffin, J. M., and Temin, H. M., 1972, Hybridization of Rous sarcoma virus deoxyribonucleic acid polymerase product and ribonucleic acid from chicken and rat cells infected with Rous sarcoma virus, *J. Virol.* **9**:766.

Colby, C., and Rubin, H., 1969, Growth and nucleic acid synthesis in normal cells and cells infected with Rous sarcoma virus, *J. Natl. Cancer Inst.* **43**:437.

Cooper, G. M., and Temin, H. M., 1974, Infectious Rous sarcoma virus and reticuloendotheliosis virus DNAs, *J. Virol.* **14**:1132.

Critchley, D. R., 1974, Cell surface proteins of NIL 1 hamster fibroblasts labeled by a galactose oxidase, tritiated borohydride method, *Cell* **3**:121.

Crittenden, L. B., 1968, Observation on the nature of a genetic cellular resistance to avian tumor viruses, *J. Natl. Cancer Inst.* **41**:145.

Crittenden, L. B., Stone, H. A., Reamer, R. H., and Okazaki, W., 1967, Two loci controlling genetic cellular resistance to avian leukosis-sarcoma viruses, *J. Virol.* **1**:898.

Dahlberg, J. E., Sawyer, R. C., Taylor, J. M., Faras, A. J., Levinson, W. E., Goodman, H. M., and Bishop, J. M., 1974, Transcription of DNA from the 70 S RNA of Rous sarcoma virus: Identification of a specific 4 S RNA which serves as primer, *J. Virol.* **13**:1126.

Dawson, P. J., Tacke, R. B., and Fieldsteel, A. H., 1968, Relationship between Friend virus and an associated lymphatic leukemia virus, *Br. J. Cancer* **22**:569.

DeGiuli, C., Kawai, S., Dales, S., and Hanafusa, H., 1975, Absence of surface projections on some noninfectious forms of RSV, *Virology* **66**:253.

Delius, H., Duesberg, P. H., and Mangel, W. F., 1975, Electron microscope measurements of Rous sarcoma virus RNA, *Cold Spring Harbor Symp. Quant. Biol.* **39**:835.

Deng, C. T., Boettiger, D., Macpherson, I., and Varmus, H. E., 1974, The persistence and expression of virus-specific DNA in revertants of Rous sarcoma virus-transformed BHK-21 cells, *Virology* **62**:512.

Dinowitz, M., 1975, Inhibition of Rous sarcoma virus by α-amanitin: Possible role of cell DNA-dependent RNA polymerase form II, *Virology* **66**:1.

Dougherty, R. M., and Rasmussen, T., 1964, Properties of a strain of Rous sarcoma virus that infects mammals, *Natl. Cancer Inst. Monogr.* **17**:337.

Duesberg, P. H., 1968, Physical properties of RSV RNA, *Proc. Natl. Acad. Sci. USA* **60**:1511.

Duesberg, P. H., and Vogt, P. K., 1970, Differences between the ribonucleic acids of transforming and nontransforming avian tumor viruses, *Proc. Natl. Acad. Sci. USA* **67**:1673.

Duesberg, P. H., and Vogt, P. K., 1973a, Gel electrophoresis of avian leukosis and sarcoma viral RNA in formamide: Comparison of other viral and cellular RNA species, *J. Virol.* **12**:594.

Duesberg, P. H., and Vogt, P. K., 1973b, RNA species obtained from clonal lines of avian sarcoma and avian leukosis virus, *Virology* **54**:207.

Duesberg, P. H., Kawai, S., Wang, L.-H., Vogt, P. K., Murphy, H. M., and Hanafusa, H., 1975, RNA of replication-defective strains of Rous sarcoma virus, *Proc. Natl. Acad. Sci. USA* **72**:1569.

Duff, R. G., and Vogt, P. K., 1969, Characteristics of two new avian tumor virus subgroups, *Virology* **39**:18.

Dulak, N. C., and Temin, H. M., 1973, Multiplication-stimulating activity for chicken embryo fibroblasts from rat liver cell conditioned medium: A family of small polypeptides, *J. Cell. Physiol.* **81**:161.

Dulbecco, R., 1967, The induction of cancer by viruses, *Sci. Am.* **216**:28.

Dulbecco, R., and Elkington, J., 1975, Induction of growth in resting fibroblastic cell cultures by Ca^{++}, *Proc. Natl. Acad. Sci. USA* **72**:1584.

Easton, T. G., and Reich, E., 1972, Muscle differentiation in cell culture, effect of nucleoside inhibitors and Rous sarcoma virus, *J. Biol. Chem.* **247**:6420.

Eckner, R. J., 1973, Helper-dependent properties of Friend spleen focus-forming virus: Effect of the *Fv-1* gene on the late stages in virus synthesis, *J. Virol.* **12**:523.

Edelman, G. M., and Yahara, I., 1976, Temperature-sensitive changes in surface modulating assemblies of fibroblasts transformed by mutants of Rous sarcoma virus, *Proc. Natl. Acad. Sci. USA* **73**:2047.

Edidin, M., and Weiss, A., 1972, Antigen cap formation in cultured fibroblasts: A reflection of membrane fluidity and of cell motility, *Proc. Natl. Acad. Sci. USA* **69**:2456.

Eisenman, R., Vogt, V. M., and Diggelmann, H., 1975, Synthesis of avian RNA tumor virus structural proteins, *Cold Spring Harbor Symp. Quant. Biol.* **39**:1067.

Ellermann, V., and Bang, O., 1909, Experimentelle Leukämie bei Hühnern, *Z. Hyg. Infektionskr.* **63**:231.

Engelbreth-Holm, J., and Rothe-Meyer, A., 1935, On the connection between erythroblastosis (haemocytoblastosis), myelosis and sarcoma in chicken, *Acta Pathol. Microbiol. Scand.* **12**:352.

Ephrussi, B., and Temin, H. M., 1960, Infection of chick iris epithelium with the Rous sarcoma virus *in vitro, Virology* **11**:547.

Erichsen, S., Eng, J., and Morgan, H. R., 1961, Comparative studies in Rous sarcoma with virus, tumor cells and chick embryo cells transformed *in vitro* by virus. I. Production of mucopolysaccharides, *J. Exp. Med.* **114**:435.

Erikson, E., and Erikson, R. L., 1971, Association of 4 S ribonucleic acid with oncornavirus ribonucleic acids, *J. Virol.* **8**:254.

Fan, H., and Baltimore, D., 1973, RNA metabolism of murine leukemia virus: Detection of virus-specific RNA sequences in infected and uninfected cells and identification of virus-specific messenger RNA, *J. Mol. Biol.* **80**:93.

Fan, H., and Paskind, M., 1974, Measurement of the sequence complexity of cloned Moloney murine leukemia virus 60–70 S RNA: Evidence for a haploid genome, *J. Virol.* **14**:421.

Faras, A. J., Garapin, A. C., Levinson, W. E., Bishop, J. M., and Goodman, H. M., 1973, Characterization of low molecular weight RNAs associated with the 70 S RNA of Rous sarcoma virus, *J. Virol.* **12**:334.

Faras, A. J., Dahlberg, J. E., Sawyer, R. C., Harada, F., Taylor, J. M., Levinson, W. E., and Bishop, J. M., 1974, Transcription of DNA from the 70 S RNA of Rous sarcoma virus: Structure of a 4 S RNA primer, *J. Virol.* **13**:1134.

Fieldsteel, A. H., Kurahara, C., and Dawson, P. J., 1969, Moloney leukemia virus as a helper in retrieving Friend virus from a non-infectious reticulum cell sarcoma, *Nature (London)* **223**:1274.

Finkel, M. P., Biskis, B. O., and Jinkins, P. B., 1966, Virus induction of osteosarcomas in mice, *Science* **151**:698.

Fischinger, P. J., Nomura, S., Tuttle-Fuller, N., and Dunn, K. J., 1974, Revertants of mouse cells transformed by murine sarcoma virus, *Virology* **59**:217.

Fischinger, P. J., Tuttle-Fuller, N., Hüper, G., and Bolognesi, D. P., 1975, Mitosis is required for production of murine leukemia virus and structural proteins during *de novo* infection, *J. Virol.* **16**:267.

Fishman, P. H., Brady, R. O., Bradley, R. M., Aaronson, S. A., and Todaro, G. J., 1974, Absence of a specific ganglioside galactosyltransferase in mouse cells transformed by murine sarcoma virus, *Proc. Natl. Acad. Sci. USA* **71**:298.

Fiszman, M. Y., and Fuchs, P., 1975, Temperature-sensitive expression of differentiation in transformed myoblasts, *Nature (London)* **254**:429.

Friend, C., 1957, Cell-free transmission in adult Swiss mice of a disease having the character of a leukemia, *J. Exp. Med.* **105**:307.

Friend, C., and Rossi, G. B., 1968, The phenomenon of differentiation in murine virus-induced leukemic cells, *Can. Cancer Conf.* **8**:171.

Friend, C., Scher, W., Holland, J. G., and Sato, J., 1971, Hemoglobin synthesis in murine virus-induced leukemic cells *in vitro*: Stimulation of erythroid differentiation by dimethyl sulfoxide, *Proc. Natl. Acad. Sci. USA* **68**:378.

Friis, R. R., Toyoshima, K., and Vogt, P. K., 1971, Conditional lethal mutants of avian sarcoma viruses. I. Physiology of *ts*75 and *ts*149, *Virology* **43**:375.

Fujinami, A., and Inamoto, K., 1914, Ueber Geschwülste bei japanischen Haushühnern insbesondere über einen transplantablen Tumor, *Z. Krebsforsch.* **14**:94.

Furuichi, Y., Shatkin, A. J., Stavnezer, E., and Bishop, J. M., 1975, Blocked, methylated 5′-terminal sequence in avian sarcoma RNA, *Nature (London)* **257**:618.

Gahmberg, C. G., and Hakomori, S., 1974, Organization of glycolipid and glycoprotein in surface membranes dependency on cell cycle and on transformation, *Biochem. Biophys. Res. Commun.* **59**:283.

Gardner, M. B., Arnstein, P., Johnson, E., Rongey, R. W., Charman, H. P., and Huebner, R. J., 1971, Feline sarcoma virus tumor induction in cats and dogs, *J. Am. Vet. Med. Assoc.* **158**:1046.

Gazdar, A. F., Chopra, H. C., and Sarma, P. S., 1972, Properties of a murine sarcoma virus isolated from a tumor arising in an NZW/NZB F$_1$ hybrid mouse. I. Isolation and pathology of tumors induced in rodents, *Int. J. Cancer* **9**:219.

Gelderblom, H., Bauer, H., and Graf, T., 1972, Cell-surface antigens induced by avian RNA tumor viruses: Detection by immunoferritin technique, *Virology* **47**:416.

Gianni, A. M., and Weinberg, R. A., 1975, Partially single-stranded form of free Moloney viral DNA, *Nature (London)* **255**:646.

Gianni, A. M., Smotkin, D., and Weinberg, R. A., 1975, Murine leukemia virus: Detection of unintegrated double-stranded DNA forms of the provirus, *Proc. Natl. Acad. Sci. USA* **72**:447.

Gielkens, A. L. J., Salden, M. H. L., and Bloemendal, H., 1974, Virus-specific messenger RNA on free and membrane-bound polyribosomes from cells infected with Rauscher leukemia virus, *Proc. Natl. Acad. Sci. USA* **71**:1093.

Gielkens, A. L. J., Van Zaane, D., Bloemers, H. P. J., and Bloemendal, H., 1976, Synthesis of Rauscher murine leukemia virus-specific polypeptides *in vitro, Proc. Natl. Acad. Sci. USA* **73**:356.

Gilula, N. B., Eger, R. R., and Rifkin, D. B., 1975, Plasma membrane alteration associated with malignant transformation in culture, *Proc. Natl. Acad. Sci. USA* **72**:3594.

Goldberg, A. R., 1974, Increased protease levels in transformed cells: A casein overlay assay for the detection of plasminogen activator production, *Cell* **2**:95.

Goldé, A., 1970, Radio-induced mutants of the Schmidt-Ruppin strain of Rous sarcoma virus, *Virology* **40**:1022.

Goldé, A., Villaudy, J., and Aghion, J., 1973*a*, Characteristics of infection by RSV. I. Non-requirement for the S phase of cellular DNA synthesis, *Intervirology* **1**:242.

Goldé, A., Villaudy, J., and Aghion, J., 1973*b*, Characteristics of infection by RSV. II. Dependence of cell events which can occur in the G_2 phase and after mitosis, *Intervirology* **1**:250.

Goldman, R. D., and Follett, E. A. C., 1969, The structure of the major cell processes of isolated BHK-21 fibroblasts, *Exp. Cell Res.* **57**:263.

Graf, T., 1972, A plaque assay for avian RNA tumor viruses, *Virology* **50**:567.

Graf, T., 1973, Two types of target cells for transformation with avian myelocytomatosis virus, *Virology* **54**:398.

Graf, T., and Friis, R. R., 1973, Differential expression of transformation in rat and chicken cells infected with an avian sarcoma virus *ts* mutant, *Virology* **56**:369.

Graf, T., Bauer, H., Gelderblom, H., and Bolognesi, D. P., 1971, Studies on the reproductive and cell-converting abilities of avian sarcoma viruses, *Virology* **43**:427.

Greenberger, J. S., Anderson, G. R., and Aaronson, S. A., 1974, Transformation-defective virus mutants in a class of morphologic revertants of sarcoma virus transformed nonproducer cells, *Cell* **2**:279.

Grimes, W. J., 1973, Glycosyl transferase and static acid levels of normal and transformed cells, *Biochemistry* **12**:990.

Groudine, M., and Weintraub, H., 1975, Rous sarcoma virus activates embryonic globin genes in chicken fibroblasts, *Proc. Natl. Acad. sci. USA* **72**:4464.

Gross, L., 1957, Development and serial cell-free passage of a highly potent strain of mouse leukemia virus, *Proc. Soc. Exp. Biol. Med.* **94**:767.

Gross, L., 1970, *Oncogenic Viruses*, 2nd ed., Pergamon Press, New York.

Guntaka, R. V., Mahy, B. W., Bishop, J. M., and Varmus, H. E., 1975, Ethidium bromide inhibits the appearance of closed circular viral DNA and integration of virus-specific DNA in duck cells infected by avian sarcoma virus, *Nature (London)* **253**:507.

Hackett, A. J., and Sylvester, S. S., 1972, Cell line derived from Balb/3T3 that is transformed by murine leukaemia virus: A focus assay for leukemia virus, *Nature (London) New Biol.* **239**:164.

Hakomori, S., 1975, Structures and organization of cell surface glycolipids: Dependency on cell growth and malignant transformation, *Biochim. Biophys. Acta* **417**:55.

Hakomori, S., Saito, T., and Vogt, P. K., 1971, Transformation by Rous sarcoma virus: Effects on cellular glycolipids, *Virology* **44**:609.

Hale, A. H., Winkelhake, J. L., and Weber, M. J., 1975, Cell surface changes and Rous sarcoma virus gene expression in synchronized cells, *J. Cell Biol.* **64**:398.

Hämmerling, U., Aoki, T., De Harven, E., Boyse, E. A., and Old, L. J., 1968, Use of hybrid antibody with anti-γG and anti-ferritin specificities in locating cell surface antigens by electron microscopy, *J. Exp. Med.* **128**:1461.

Hanafusa, H., 1968, Methods for the study of defective viruses, in: *Methods in Virology*, Vol. 4 (K. Maramorosch and H. Koprowski, eds.), p. 321, Academic Press, New York.

Hanafusa, H., 1969, Rapid transformation of cells by Rous sarcoma virus, *Proc. Natl. Acad. Sci. USA* **63**:318.

Hanafusa, H., 1975, Avian RNA tumor viruses, in: *Cancer,* Vol. 2 (F. F. Becker, ed.), p. 49, Plenum, New York.

Hanafusa, H., and Hanafusa, T., 1966, Determining factor in the capacity of Rous sarcoma virus to induce tumors in mammals, *Proc. Natl. Acad. Sci. USA* **55**:532.

Hanafusa, H., and Hanafusa, T., 1968, Further studies on RSV production from transformed cells, *Virology* **34**:630.

Hanafusa, H., and Hanafusa, T., 1971, Noninfectious RSV deficient in DNA polymerase, *Virology* **43**:313.

Hanafusa, H., Hanafusa, T., and Rubin, H., 1963, The defectiveness of Rous sarcoma virus, *Proc. Natl. Acad. Sci. USA* **49**:572.

Hanafusa, H., Miyamoto, T., and Hanafusa, T., 1970, A cell-associated factor essential for formation of an infectious form of Rous sarcoma virus, *Proc. Natl. Acad. Sci. USA* **66**:314.

Hanafusa, H., Aoki, T., Kawai, S., Miyamoto, T., and Wilsnack, R. E., 1973, Presence of antigen common to avian tumor viral envelope antigen in normal chick embryo cells, *Virology* **56**:22.

Hanafusa, T., Miyamoto, T., and Hanafusa, H., 1970, A type of chick embryo cell that fails to support formation of infectious RSV, *Virology* **40**:55.

Harris, R. J., Malmgreen, H., and Sylven, B., 1954, The polysaccharides of Rous sarcoma No. 1, *Br. J. Cancer* **8**:141.

Hartley, J. W., and Rowe, W. P., 1966, Production of altered cell foci in tissue culture by defective Moloney sarcoma virus particles, *Proc. Natl. Acad. Sci. USA* **55**:780.

Hartley, J. W., Rowe, W. P., and Huebner, R. J., 1970, Host-range restrictions of murine leukemia viruses in mouse embryo cell cultures, *J. Virol.* **5**:221.

Hartmann, J. F., Buck, C. A., Defendi, V., Glick, M. C., and Warren, L., 1972, The carbohydrate content of control and virus-transformed cells, *J. Cell. Physiol.* **80**:159.

Harvey, J. J., 1964, An unidentified virus which causes the rapid production of tumours in mice, *Nature (London)* **204**:1104.

Harvey, J. J., and East, J., 1971, The murine sarcoma virus (MSV), *Int. Rev. Exp. Pathol.* **10**:265.

Hatanaka, M., 1974, Transport of sugars in tumor cell membranes, *Biochim. Biophys. Acta* **355**:77.

Hatanaka, M., and Hanafusa, H., 1970, Analysis of a functional change in membrane in the process of cell transformation by Rous sarcoma virus; alteration in the characteristics of sugar transport, *Virology* **41**:647.

Hatanaka, M., Huebner, R. J., and Gilden, R. V., 1969, Alterations in characteristics of sugar uptake by mouse cells transformed by murine sarcoma viruses, *J. Natl. Cancer Inst.* **43**:1091.

Hayashi, I., and Sato, G. H., 1976, Replacement of serum by hormomes permits growth of cells in a defined medium, *Nature (London)* **259**:132.

Hayward, W. S., and Hanafusa, H., 1973, Detection of avian tumor virus RNA in uninfected chicken embryo cells, *J. Virol.* **11**:157.

Hayward, W. S., and Hanafusa, H., 1976, Independent regulation of endogenous and exogenous avian RNA tumor virus genes, *Proc. Natl. Acad. Sci. USA* **73**:2259.

Hayward, W. S., Wang, S. Y., Urm, E., and Hanafusa, H., 1976, Transcription of the avian RNA tumor virus glycoprotein gene in uninfected and infected cells, in: *Animal Virology,* ICN-UCLA Symposium on Molecular and Cellular Biology, (D. Baltimore, A. S. Huang, and C. F. Fox, eds.), p. 21, Academic Press, New York.

Hill, M., and Hillova, J., 1972, Virus recovery in chicken cells tested with Rous sar-
    coma cell DNA, *Nature (London) New Biol.* **237**:35.
Hill, M., Hillova, J., Dantchev, D., Mariagr, R., and Goubin, G., 1975, Infectious viral
    DNA in Rous sarcoma virus-transformed nonproducer are producer animal cells,
    Cold Spring Harbor Symp. Quant. Biol. **39**:1015.
Hobom-Schnegg, B., Robinson H. L., and Robinson, W. S., 1970, Replication of Rous
    sarcoma virus in synchronized cells, *J. Gen. Virol.* **7**:85.
Hogg, N. M., 1974, A comparison of membrane proteins of normal and transformed
    cells by lactoperoxidase labeling, *Proc. Natl. Acad. Sci. USA* **71**:489.
Holley, R. W., 1972, A unifying hypothesis concerning the nature of malignant growth,
    Proc. Natl. Acad. Sci. USA **69**:2840.
Holley, R. W., 1975, Control of growth of mammalian cells in cell cultures, *Nature
    (London)* **258**:487.
Holley, R. W., and Kiernan, J. A., 1974a, Control of the inhibition of DNA synthesis
    in 3T3 cells: Serum factors, *Proc. Natl. Acad. Sci. USA* **71**:2908.
Holley, R. W., and Kiernan, J. A., 1974b, Control of the initiation of DNA synthesis in
    3T3 cells: Low-molecular-weight nutrients, *Proc. Natl. Acad. Sci. USA* **71**:2942.
Holtzer, H., Biehl, J., Yeoh, G., Meganathan, R., and Kaji, A., 1975, Effect of onco-
    genic virus on muscle differentiation, *Proc. Natl. Acad. Sci. USA* **72**:4051.
Hopkins, N., and Jolicoeur, P., 1975, Variants of N-tropic leukemia virus derived from
    Balb/c mice, *J. Virol.* **16**:991.
Horst, J., Keith, J., and Fraenkel-Conrat, H., 1972, Characteristic two-dimensional
    patterns of enzymatic digests of oncorna and other viral RNAs, *Nature (London)
    New Biol.* **240**:105.
Hovi, T., Keski-Oja, J., and Vaheri, A., 1974, Growth control in chick embryo fibro-
    blasts; no evidence for a specific role for cyclic purine nucleotides, *Cell* **2**:235.
Hsie, A. W., and Puck, T. T., 1971, Morphological transformation of Chinese hamster
    cells by dibutyryl adenosine cyclic 3':5'-monophosphate and testosterone, *Proc.
    Natl. Acad. Sci. USA* **68**:358.
Huang, A. S., Besman, P., Chu, L., and Baltimore, D., 1973, Growth of pseudotypes of
    vesticular stomatitis virus with N-tropic murine leukemic virus coats in cells resistant
    to N-tropic viruses, *J. Virol.* **12**:659.
Huebner, R. J., Armstrong, D., Okuyan, M., Sarma, P. S., and Turner, H. C., 1964,
    Specific complement-fixing viral antigens in hamster and guinea pig tumors induced
    by the Schmidt-Ruppin strain of avian sarcoma, *Proc. Natl. Acad. Sci. USA* **51**:742.
Huebner, R. J., Hartley, J. W., Rowe, W. P., Lane, W. T., and Capps, W. I., 1966,
    Rescue of the defective genome of Moloney sarcoma virus from a noninfectious
    hamster tumor and the production of pseudotype sarcoma viruses with various
    murine leukemia viruses, *Proc. Natl. Acad. Sci. USA* **56**:1164.
Humphries, E. H., and Coffin, J. M., 1976, Rate of virus-specific RNA synthesis in
    synchronized chicken embryo fibroblasts infected with avian leukosis virus, *J. Virol.*
    17:393.
Humphries, E. H., and Temin, H. M., 1972, Cell cycle-dependent activation of Rous
    sarcoma virus-infected stationary chicken cells: Avian leukosis virus group-specific
    antigens and ribonucleic acid, *J. Virol.* **10**:82.
Humphries, E. H., and Temin, H. M., 1974, Requirement for cell division for initiation
    of transcription of Rous sarcoma virus RNA, *J. Virol.* **14**:531.
Hunter, E., and Vogt, P. K., 1976, Temperature sensitive mutants of avian sarcoma

viruses: Genetic recombination with wild type sarcoma virus and physiological analysis of multiple mutants, *Virology* **69**:23.

Hynes, R. O., 1973, Alteration of cell-surface proteins by viral transformation and by proteolysis, *Proc. Natl. Acad. Sci. USA* **70**:3170.

Hynes, R. O., 1974, Role of surface alterations in cell transformation: The importance of proteases and surface proteins, *Cell* **1**:147.

Hynes, R. O., and Bye, J. M., 1974, Density and cell cycle dependence of cell surface proteins in hamster fibroblasts, *Cell* **3**:113.

Hynes, R. O., and Wyke, J. A., 1975, Alterations in surface proteins in chicken cells transformed by temperature-sensitive mutants of Rous sarcoma virus, *Virology* **64**:492.

Hynes, R. O., Wyke, J. A., Bye, J. M., Humphryes, K. C., and Pearlstein, E. S., 1975, Are proteases involved in altering surface proteins during viral transformation? in *Proteases and Biological Control* (E. Reich, D. B. Rifkin, and E. Shaw, eds.), p. 931, Cold Spring Harbor Laboratory, Cold Spring Harbor, N.Y.

Iglewski, B. H., Rittenberg, M. B., and Igrewski, W. J., 1975, Preferential inhibition of growth and protein synthesis in Rous sarcoma virus transformed cell by diphtheria toxin, *Virology* **65**:272.

Isaka, T., Yoshida, M., Owada, M., and Toyoshima, K., 1975, Alterations in membrane polypeptides of chick embryo fibroblasts induced by transformation with avian sarcoma viruses, *Virology* **65**:226.

Ishimoto, N., Temin, H. M., and Strominger, J. L., 1966, Studies on carcinogenesis by avian sarcoma viruses. II. Virus-induced increase in hyaluronic acid synthetase in chicken fibroblasts, *J. Biol. Chem.* **241**:2052.

Ishizaki, R., and Shimizu, T., 1970, Heterogeneity of strain R avian erythroblastosis virus, *Cancer Res.* **30**:2827.

Ishizaki, R., Langlois, A. J., Chabot, J., and Beard, J. W., 1971, Component of strain MC29 avian leukosis virus with the property of defectiveness, *J. Virol.* **8**:821.

Isselbacher, K. J., 1972, Increased uptake of amino acids and 2-deoxy-D-glucose by virus-transformed cells in culture, *Proc. Natl. Acad. Sci. USA* **69**:585.

Ivanov, X., Mladenov, Z., Nedyalkov, S., Todorov, T. G., and Yakimov, M., 1964, Experimental investigations into avian leucoses. V. Transmission, haematology and morphology of avian myelocytomatosis, *Bull. Inst. Patho. Comp. Anim.* **10**:5.

Jacquet, M., Groner, Y., Monroy, G., and Hurwitz, 1974, The *in vitro* synthesis of avian myeloblastosis viral RNA sequences, *Proc. Natl. Acad. Sci. USA* **71**:3045.

Jainchill, J. L., and Todaro, G. J., 1970, Stimulation of cell growth *in vitro* by serum with and without growth factor; relation to contact inhibition and viral transformation, *Exp. Cell Res.* **59**:137.

Johnson, G. S., Friedman, R. M., and Pastan, I., 1971, Restoration of several morphological characteristics of normal fibroblasts in sarcoma cells treated with adenosine-3′:5′-cyclic monophosphate and its derivatives, *Proc. Natl. Acad. Sci. USA* **68**:425.

Joho, R. H., Billeter, M. A., and Weissmann, C., 1975, Mapping of biological functions on RNA of avian tumor viruses: Location of regions required for transformation and determination of host range, *Proc. Natl. Acad. Sci. USA* **72**:4772.

Jolicoeur, P., and Baltimore, D., 1976, Effect of *Fv-1* gene product on synthesis of N-tropic and B-tropic murine leukemia viral RNA, *Cell* **7**:33.

Jones, P., Benedict, W., Strickland, S., and Reich, E., 1975, Fibrin overlay methods for the detection of single transformed cells and colonies of transformed cells, *Cell* **5**:323.

Jonsson, N., and Sjögren, H. O., 1966, Specific transplantation immunity in relation to Rous sarcoma virus tumorigenesis in mice, *J. Exp. Med.* **123**:487.

Kabat, E. A., 1939, A polysaccharide in tumors due to a virus of leucosis and sarcoma of fowls, *J. Biol. Chem.* **130**:143.

Kaighn, M. E., Ebert, J. D., and Stott, P. M., 1966, The susceptibility of differentiating muscle clones to Rous sarcoma virus, *Proc. Natl. Acad. Sci. USA* **56**:133.

Kapeller, M., and Doljanski, F., 1972, Agglutination of normal and Rous sarcoma virus-transformed chick embryo cells by concanavalin A and wheat germ agglutinin, *Nature (London) New Biol.* **235**:184.

Kawai, S., and Hanafusa, H., 1971, The effects of reciprocal changes in temperature on the transformed state of cells infected with a Rous sarcoma virus mutant, *Virology* **46**:470.

Kawai, S., and Hanafusa, H., 1972*a*, Plaque assay for some strains of avian leukosis virus, *Virology* **48**:126.

Kawai, S., and Hanafusa, H., 1972*b*, Genetic recombination with avian tumor virus, *Virology* **49**:37.

Kawai, S., and Hanafusa, H., 1973, Isolation of defective mutant of avian sarcoma virus, *Proc. Natl. Acad. Sci. USA* **70**:3493.

Kawai, S., and Yamamoto, T., 1970, Isolation of different kinds of non-virus producing chick cells transformed by Schmidt-Ruppin strain (subgroup A) of Rous sarcoma virus, *Jpn. J. Exp. Med.* **40**:243.

Kawai, S., Metroka, C. E., and Hanafusa, H., 1972, Complementation of functions required for cell transformation by double infection with RSV mutants, *Virology* **49**:302.

Keith, J., and Fraenkel-Conrat, H., 1975, Identification of the 5′ end of Rous sarcoma virus RNA, *Proc. Natl. Acad. Sci. USA* **72**:3347.

Keogh, E. V., 1938, Ectodermal lesions produced by the virus of Rous sarcoma, *Br. J. Exp. Pathol.* **19**:1.

Khoury, A. T., and Hanafusa, H., 1976, Synthesis and integration of viral DNA in chick cells at different times after infection with various multiplicities of avian oncornavirus, *J. Virol.* **18**:383.

Kirsten, W. H., and Mayer, L. A., 1967, Morphologic responses to a murine erythroblastosis virus, *J. Natl. Cancer Inst.* **39**:311.

Kletzien, R. F., and Perdue, J. F., 1974, Sugar transport in chick embryo fibroblasts. II. Alterations in transport following transformation by a temperature-sensitive mutant of the Rous sarcoma virus, *J. Biol. Chem.* **249**:3375.

Kletzien, R. F., and Perdue, J. F., 1975, Regulation of sugar transport in chick embryo fibroblasts infected with a temperature-sensitive mutant of RSV, *Cell* **6**:513.

Koldovský, P., Svoboda, J., and Bubenik, J., 1966, Further studies on the immunobiology of tumor RNA$_2$ induced by RSV in C57Bl strain mice, *Folia Biol.* **12**:1.

Krontiris, T. G., Soeiro, R., and Fields, B. N., 1973, Host restriction of Friend leukemia virus: Role of the viral outer coat, *Proc. Natl. Acad. Sci. USA* **70**:2549.

Kung, H.-J., Bailey, J. M., Davidson, N., Nicolson, M. O., and McAllister, R. M., 1975, Structure, subunit composition, and molecular weight of RD-114 RNA, *J. Virol.* **16**:397.

Kurth, R., and Bauer, H., 1972a, Cell-surface antigens induced by avian RNA tumor viruses: Detection by a cytotoxicity microassay, *Virology* **47**:426.

Kurth, R., and Bauer, H., 1972b, Common tumor specific antigens on cells of different species transformed by avian RNA tumor viruses, *Virology* **49**:145.

Kurth, R., and Bauer, H., 1973a, Avian oncornavirus-induced tumor antigens of embryonic and unknown origin, *Virology* **56**:496.

Kurth, R., and Bauer, H., 1973b, Influence of dibutyryl cyclic AMP and theophylline on cell surface antigens of oncornavirus transformed cells, *Nature (London) New Biol.* **243**:243.

Kurth, R., and Bauer, H., 1975, Avian RNA tumor viruses: A model for studying tumor associated cell surface alterations, *Biochim. Biophys. Acta* **417**:1.

Kurth, R., Friis, R. R., Wyke, J. A., and Bauer, H., 1975, Expression of tumor-specific surface antigens on cells infected with temperature-sensitive mutants of avian sarcoma virus, *Virology* **64**:400.

Lai, M. M. C., and Duesberg, P. H., 1972a, Adenylic acid-rich sequence in RNAs of Rous sarcoma virus and Rauscher mouse leukaemia virus, *Nature (London)* **235**:383.

Lai, M. M., and Duesberg, P. H., 1972b, Differences between the envelope glycoproteins and glycopeptides of avian tumor virus released from transformed and from nontransformed cells, *Virology* **50**:359.

Lai, M. M.-C., Duesberg, P. H., Horst, J., and Vogt, P. K., 1973, Avian tumor virus RNA: A comparison of three sarcoma viruses and their transformation-defective derivatives by oligonucleotide fingerprinting and DNA-RNA hybridization, *Proc. Natl. Acad. Sci. USA* **70**:2266.

Langlois, A. J., and Beard, J. W., 1967, Converted-cell focus formation in culture by strain MC29 avian leukosis virus, *Proc. Soc. Exp. Biol. Med.* **126**:718.

Lazarides, E., 1975, Tropomyosin antibody: The specific localization of tropomyosin in nonmuscle cells, *J. Cell Biol.* **65**:549.

Lazarides, E., and Weber, K., 1974, Actin antibody: The specific visualization of actin filaments in non-muscle cells, *Proc. Natl. Acad. Sci. USA* **71**:2268.

Leder, A., and Leder, P., 1975, Butyric acid, a potent inducer of erythroid differentiation in cultured erythroleukemic cells, *Cell* **5**:319.

Lee, H. H., Kaighn, M. E., and Ebert, J. D., 1966, Viral antigens in differentiating muscle colonies after infection with Rous sarcoma virus *in vitro*, *Proc. Natl. Acad. Sci. USA* **56**:521.

Leis, J., Schincariol, A., Ishizaki, R., and Hurwitz, J., 1975, RNA-dependent DNA polymerase activity of RNA tumor viruses. V. Rous sarcoma virus single-stranded RNA-DNA covalent hybrids in infected chicken embryo fibroblast cells, *J. Virol.* **15**:484.

Leong, J. A., Garapin, A. C., Jackson, N., Fanshier, L., Levinson, W., and Bishop, J. M., 1972, Virus specific ribonucleic acid in cells producing Rous sarcoma virus: Detection and characterization, *J. Virol.* **9**:891.

Levy, J. A., 1973, Xenotropic viruses: Murine leukemia viruses associated with NIH Swiss, NZB, and other mouse strains, *Science* **182**:1151.

Levy, J. A., 1975, Host range of murine xenotropic virus: Replication in avian cells, *Nature (London)* **253**:140.

Lilly, F., and Pincus, T., 1973, Genetic control of murine viral leukemogenesis, *Adv. Cancer Res.* **17**:231.

Linial, M., and Mason, W. S., 1973, Characterization of two conditional early mutants of Rous sarcoma virus, *Virology* **53**:258.

Livingston, D. M., Parks, W. P., Scolnick, E. M., and Ross, J., 1972, Affinity chromatography of avian type C viral reverse transcriptase: Studies with Rous sarcoma virus transformed rat cells, *Virology* **50**:388.

Lo, A. C. H., and Ball, J. K., 1974, Evidence for helper-independent murine sarcoma virus. II. Differences between the ribonucleic acids of clone-purified leukemia virus, helper-independent and helper-dependent sarcoma viruses, *Virology* **59**:545.

Machala, O., Donner, L., and Svoboda, J., 1970, A full expression of the genome of Rous sarcoma virus in heterokaryons formed after fusion of virogenic mammalian cells and chicken fibroblasts, *J. Gen. Virol.* **8**:219.

Macpherson, I. A., 1965, Reversion in hamster cells transformed by Rous sarcoma virus, *Science* **148**:1731.

Macpherson, I., and Montagnier, L., 1964, Agar suspension culture for the selective assay of cells transformed by polyoma virus, *Virology* **23**:291.

Maisel, J., Klement, V., Lai, M. M.-C., Ostertag, W., and Duesberg, P., 1973, ribonucleic acid components of murine sarcoma and leukemia viruses, *Proc. Natl. Acad. Sci. USA* **70**:3536.

Maisel, J., Scolnick, E. M., and Duesberg, P., 1975, Base sequence differences between the RNA components of Harvey sarcoma virus, *J. Virol.* **16**:749.

Manaker, R. A., and Groupé, V., 1956, Discrete foci of altered chick embryo cells associated with Rous sarcoma virus in tissue culture, *Virology* **2**:838.

Mangel, W. F., Delius, H., and Duesberg, P. H., 1974, Structure and molecular weight of the 60–70 S RNA and the 30–40 S RNA of Rous sarcoma virus, *Proc. Natl. Acad. Sci. USA* **71**:4541.

Martin, G. S., 1970, Rous sarcoma virus: A function required for the maintenance of the transformed state, *Nature (London)* **227**:1021.

Martin, G. S., 1971, Mutants of the Schmidt-Ruppin strain of Rous sarcoma virus, in: *The Biology of Oncogenic Viruses* (L. G. Silvestri, ed.), p. 320, North-Holland, Amsterdam.

Martin, G. S., and Duesberg, P. H., 1972, The *a* subunit in the RNA of transforming avian tumor viruses. I. Occurrence in different virus strains. II. Spontaneous loss resulting in nontransforming variants, *Virology* **47**:494.

Martin, G. S., Venuta, S., Weber, M., and Rubin, H., 1971, Temperature-dependent alterations in sugar transport in cells infected by a temperature-sensitive mutant of Rous sarcoma virus, *Proc. Natl. Acad. Sci. USA* **68**:2739.

Mirand, E. A., 1966, Erythropoietic response of animals infected with various strains of Friend virus, *Natl. Cancer Inst. Monogr.* **22**:483.

Moloney, J. B., 1960, Biological studies on a lymphoid leukemia virus extracted from sarcoma S37. I. Origin and introductory investigations, *J. Natl. Cancer Inst.* **24**:933.

Moloney, J. B., 1966, A virus-induced rhabdomyosarcoma of mice, *Natl. Cancer Inst. Monogr.* **22**:139.

Moutagnier, L., and Vigier, P., 1972, Un intermédiare ADN infectieux et transformant du virus du sarcome de Rous dans les cellules de Poule transformées par ce virus, *C. R. Acad. Sci. (Paris) Ser. D* **274**:1977.

Moore, E. G., and Temin, H. M., 1971, Lack of correlation between conversion by RNA tumour viruses and increased agglutinability of cells by concanavalin A and wheat germ agglutinin, *Nature (London)* **231**:117.

Morgan, H. R., 1964, Origin of Rous sarcoma strains, *Natl. Cancer Inst. Monogr.* **17**:392.

Moscovici, C., and Zanetti, M., 1970, Studies of single foci of hematopoietic cells transformed by avian myeloblastosis virus, *Virology* **42**:61.

Moscovici, C., Moscovici, M. G., and Zanetti, M., 1969, Transformation of chick embryo fibroblast cultures with avian myeloblastosis virus, *J. Cell. Physiol.* **73**:105.

Moscovici, C., Gazzolo, L., and Moscovici, M. G., 1975, Focus assay and defectiveness of avian myeloblastosis virus, *Virology* **68**:173.

Nakata, Y., and Bader, J. P., 1968, Transformation by murine sarcoma virus: Fixation (deoxyribonucleic acid synthesis) and development, *J. Virol.* **2**:1255.

Naso, R. B., Arcement, L. J., and Arlinghaus, R. B., 1975a, Biosynthesis of Rauscher leukemia viral proteins, *Cell* **4**:31.

Naso, R. B., Arcement, L. J., Wood, T. G., Saunders, T. E., and Arlinghaus, R. B., 1975b, The cell-free translation of Rauscher leukemia virus RNA into high molecular weight polypeptides, *Biochim. Biophys. Acta* **383**:195.

Nass, M. M. K., 1973, Temperature-dependent formation of dimers and oligomers of mitochondrial DNA in cell transformed by a thermosensitive mutant of Rous sarcoma virus, *Proc. Natl. Acad. Sci. USA* **70**:3739.

Neiman, P. E., Wright, S. E., McMillin, C., and MacDonnell, D., 1974, Nucleotide sequence relationships of avian RNA tumor viruses: Measurement of the deletion in a transformation-defective mutant of Rous sarcoma virus, *J. Virol.* **13**:837.

Neiman, P. E., Purchase, H. G., and Okazaki, W., 1975, Chicken leukosis virus genome sequences in DNA from normal chick cells and virus induced bursal lymphomas, *Cell* **4**:311.

Nicholson, G. L., 1971, Difference in the topology of normal and tumor cell membranes as shown by different distributions of ferritin-conjugated concanavalin A on their surfaces, *Nature (London) New Biol.* **233**:244.

Nicholson, G. L., 1973, Temperature-dependent mobility of concanavalin A sites on tumour cell surfaces, *Nature (London) New Biol.* **243**:218.

Nomura, S., Fischinger, P. J., Mattern, C. F. T., Peebles, P. T., Bassin, R. H., and Friedman, G. P., 1972, Revertants of mouse cells transformed by murine sarcoma virus. I. Characterization of flat and transformed sublines without a rescuable murine sarcoma virus, *Virology* **50**:51.

Nomura, S., Fishchinger, P. J., Mattern, C. F. T., Gerwin, B., and Dunn, K. J., 1973, Revertants of mouse cells transformed by murine sarcoma virus. II. Flat variants induced by fluorodeoxyuridine and colcemid, *Virology* **56**:152.

Ogura, H., and Friis, R., 1975, Further evidence for the existence of a viral envelope protein defect in Bryan high-titer strain of Rous sarcoma virus, *J. Virol.* **16**:443.

Olmsted, J. B., and Borisy, G., 1973, Microtubules, *Annu. Rev. Biochem.* **42**:507.

Ossowski, L., Unkeless, J. C., Tobia, A., Quigley, J. P., Rifkin, D. B., and Reich, E., 1973a, An enzymatic function associated with transformation of fibroblasts by oncogenic viruses, *J. Exp. Med.* **137**:112.

Ossowski, L., Quigley, J. P., Kellerman, G. M., and Reich, E., 1973b, fibrinolysis associated with oncogenic transformation: Requirement of plasminogen for correlated changes in cellular morphology, colony formation in agar, and cell migration, *J. Exp. Med.* **138**:1056.

Otten, J., Bader, J., Johnson, G. S., and Pastan, I., 1972, A mutation in a Rous sarcoma virus gene that controls adenosine 3′, 5′-monophosphate levels and transformation. *J. Biol. Chem.* **247**:1632.

Owada, M., and Toyoshima, K., 1973, Analysis on the reproducing and cell-transform-
    ing capacities of a temperature sensitive mutant (*ts*334) of avian sarcoma virus B77,
    Virology **54**:170.
Panet, A., Baltimore, D., and Hanafusa, T., 1975*a*, Quantitation of avian RNA tumor
    virus reverse transcriptase by radioimmunoassay, *J. Virol.* **16**:146.
Panet, A., Haseltine, W. A., Baltimore, D., Peters, G., Harada, F., and Dahlberg, J.
    E., 1975*b*, Specific binding of tryptophan transfer RNA to avian myeloblastosis virus
    RNA-dependent DNA polymerase (reverse transcriptase), *Proc. Natl. Acad. Sci.
    USA* **72**:2535.
Pardee, A. B., 1974, A restriction point for control of normal animal cell proliferation,
    Proc. Natl. Acad. Sci. USA **71**:1286.
Pardee, A. B., Jimenez de Asúa, L., and Rozengurt, E., 1974, Functional membrane
    changes and cell growth: Significance and mechanism, in: *Control of Proliferation in
    Animal Cells* (B. Clarkson and R. Baserga, eds.), p. 547, Cold Spring Harbor Labo-
    ratory, Cold Spring Harbor, N.Y.
Parsons, J. T., Coffin, J. M., Haroz, R. K., Bromley, P. A., and Weissman, C., 1973,
    Quantitative determination and location of newly synthesized virus-specific ribonu-
    cleic acid in chicken cells infected with Rous sarcoma virus, *J. Virol.* **11**:761.
Payne, L. N., and Biggs, P. M., 1964, Differences between highly inbred lines of
    chickens in the response to Rous sarcoma virus of the chorioallantoic membrane and
    of embryonic cells in tissue culture, *Virology* **24**:610.
Pearlstein, E., and Waterfield, M. D., 1974, Metabolic studies on ^{125}I-labeled baby
    hamster kidney cell plasma membranes, *Biochim. Biophys. Acta* **362**:1.
Perdue, J. F., Kletzien, R., and Wray, V. L., 1972, The isolation and characterization
    of plasma membrane from cultured cells. IV. The carbohydrate composition of
    membranes isolated from oncogenic RNA virus-converted chick embryo fibroblasts,
    Biochim. Biophys. Acta **266**:505.
Pessac, B., and Calothy, G., 1974, Transformation of chick embryo neuroretinal cells
    by Rous sarcoma virus *in vitro*: Induction of cell proliferation, *Science* **185**:709.
Peterson, R. D. A., Purchase, H. G., Burmester, B. R., Cooper, M. D., and Good, R.
    A., 1966, Relationships among visceral lymphomatosis, bursa of Fabricius, and
    bursa-dependent lymphoid tissue of the chicken, *J. Natl. Cancer Inst.* **36**:585.
Pincus, T., Hartley, J. W., and Rowe, W. P., 1971*a*, A major genetic locus affecting
    resistance to infection with murine leukemia viruses. I. Tissue culture studies, of
    naturally occurring viruses, *J. Exp. Med.* **133**:1219.
Pincus, T., Rowe, W. P., and Lilly, F., 1971*b*, A major genetic locus affecting
    resistance to infection with murine leukemia viruses. II. Apparent identity to a major
    locus described for resistance to Friend murine leukemia virus, *J. Exp. Med.*
    133:1234.
Pinto da Silva, P., And Martinez-Palomo, A., 1975, Distribution of membrane parti-
    cles and gap junctions in normal and treated with concanavalin A, *Proc. Natl. Acad.
    Sci. USA* **72**:572.
Piraino, F., 1967, The mechanism of genetic resistance of chick embryo cells to infec-
    tion by Rous sarcoma virus—Bryan strain, *Virology* **32**:700.
Pluznick, D. H., and Sachs, L., 1964, Quantitation of a murine leukemia virus with a
    spleen colony assay, *J. Natl. Cancer Inst.* **33**:535.
Pollack, R., and Rifkin, D., 1975, Actin-containing cables within anchorage-dependent
    rat embryo cells are dissociated by plasmin and trypsin, *Cell* **6**:495.

Pollack, R., Osborn, M., and Weber, K., 1975, Pattern of organization of actin and myosin in normal and transformed culture cells, *Proc. Natl. Acad. Sci. USA* **72**:994.

Porter, K. R., 1966, Cytoplasmic microtubules and their functions, in: *Principles of Biomolecular Organization* (G. E. W. Wolstenholme and M. O'Connor, eds.), p. 308, Churchill, London.

Porter, K. R., Puck, T. T., Hsie, A. W., and Kelley, D., 1974, An electron microscope study of the effects of dibutyryl cyclic AMP on Chinese hamster ovary cells, *Cell* **2**:145.

Poste, G., Papahadjopoulos, D., and Nicolson, G. L., 1975, Local anesthetics affect transmembrane cytoskeletal control of mobility and distribution of cell surface receptors, *Proc. Natl. Acad. Sci. USA* **72**:4430.

Premkumar, E., Potter, M., Singer, P. A., and Sklar, M. D., 1975, Synthesis, surface deposition, and secretion of immunoglobulins by Abelson virus-transformed lymphosarcoma cell lines, *Cell* **6**:149.

Prince, A. M., 1962, Factors influencing the determination of cellular morphology in cells infected with Rous sarcoma virus, *Virology* **18**:524.

Quade, K., Smith, R. E., and Nichols, J. L., 1974*a*, Evidence for common nucleotide sequences in the RNA subunits comprising Rous sarcoma virus 70 S RNA, *Virology* **61**:287.

Quade, K., Smith, R. E., and Nichols, J. L., 1974*b*, Poly(riboadenylic acid) and adjacent nucleotides in Rous sarcoma virus RNA, *Virology* **62**:60.

Rapp, U. R., and Nowinski, R. C., 1976, Endogenous mouse type C viruses deficient in replication and production of XC plaques, *J. Virol.* **18**:411.

Rauscher, F. J., 1962, A virus induced disease of mice characterized by erythrocytopoiesis and lymphoid leukemia, *J. Natl. Cancer Inst.* **29**:515.

Rechler, M. M., Podskalny, J. M., and Nissley, S. P., 1976, Interaction of multiplication-stimulating activity with chick embryo fibroblasts demonstrates a growth receptor, *Nature (London)* **259**:134.

Reich, E., Rifkin, D. B., and Shaw, E., eds., 1975, *Proteases and Biological Control.* Cold Spring Harbor Laboratory, Cold Spring Harbor, N.Y.

Rho, H. M., and Green, M., 1974, The homopolyadenylate and adjacent nucleotides at the 3′-terminus of 30–40 S RNA subunits in the genome of murine sarcoma-leukemia virus, *Proc. Natl. Acad. Sci. USA* **71**:2386.

Rifkin, D. B., and Reich, E., 1971, Selective lysis of cells transformed by Rous sarcoma virus, *Virology* **45**:172.

Rifkin, D. B., Beal, L. P., and Reich, E., 1975, Macromolecular determinants of plasminogen activator synthesis, in: *Proteases and Biological Control* (E. Reich, D. B. Rifkin, and E. Shaw, eds.), p. 841, Cold Spring Harbor Laboratory, Cold Spring Harbor, N.Y.

Riggin, C. H., Bondurant, M., and Mitchell, W. M., 1975, Physical properties of Moloney murine leukemia virus high-molecular-weight RNA: A two subunit structure, *J. Virol.* **16**:1528.

Robbins, J. C., and Nicolson, G. L., 1975, Surfaces of normal and transformed cells, in: *Cancer,* Vol. 4, (F. F. Becker, ed.), p. 1, Plenum, New York.

Robbins, P. W., Wickus, G. G., Branton, P. E., Gaffney, B. J., Hirschberg, C. B., Fuchs, P., and Blumberg, P. M., 1975, The chick fibroblast cell surface after transformation by Rous sarcoma virus, *Cold Spring Harbor Symp. Quant. Biol.* **39**:1173.

Robinson, W. S., Piktanen, A., and Rubin, H., 1965, The nucleic acid of the Bryan strain of Rous sarcoma virus: Purification of the virus and isolation of the nucleic acid, *Proc. Natl. Acad. Sci. USA* **54**:137.

Robinson, W. S., Robinson, H. L., and Duesberg, P. H., 1967, tumor virus RNA's, *Proc. Natl. Acad. Sci. USA* **58**:825.

Rohrschneider, L. R., Kurth, R., and Bauer, H., 1975, Biochemical characterization of tumor-specific cell surface antigens on avian oncornavirus transformed cells, *Virology* **66**:481.

Rosenberg, N., Baltimore, D., and Scher, C. D., 1975, *In vitro* transformation of lymphoid cells by Abelson murine leukemia virus, *Proc. Natl. Acad. Sci. USA* **72**:1932.

Rous, P., 1911, A sarcoma of the fowl transmissible by an agent separable from the tumor cells, *J. Exp. Med.* **13**:397.

Roy-Burman, P., and Klement, V., 1975, Derivation of mouse sarcoma virus (Kirsten) by acquisition of genes from heterologous host, *J. Gen. Virol.* **28**:193.

Rubin, H., 1960a, An analysis of the assay of Rous sarcoma cells *in vitro* by the infective center technique, *Virology* **10**:29.

Rubin, H., 1960b, The suppression of morphological alterations in cells infected with Rous sarcoma virus, *Virology* **12**:14.

Rubin, H., 1965, Genetic control of cellular susceptibility to pseudotypes of Rous sarcoma virus, *Virology* **26**:270.

Rubin, H., 1966, The inhibition of chick embryo cell growth by medium obtained from cultures of Rous sarcoma cells, *Exp. Cell Res.* **41**:149.

Rubin, H., 1970, Overgrowth stimulating factor released from Rous sarcoma cells, *Science* **167**:1271.

Rubin, H., and Koide, T., 1975, Early cellular responses to diverse growth stimuli independent of protein and RNA synthesis, *J. Cell. Physiol.* **86**:47.

Rubin, H., and Temin, H. M., 1959, A radiological study of cell–virus interaction in the Rous sarcoma, *Virology* **7**:75.

Rymo, L., Parsons, J. T., Coffin, J. M., and Weissmann, C., 1974, *In vitro* synthesis of Rous sarcoma virus-specific RNA is catalyzed by a DNA-dependent RNA polymerase, *Proc. Natl. Acad. Sci. USA* **71**:2782.

Salden, M. H. L., and Blöemendal, H., 1976, Translation of avian myeloblastosis virus RNA in a reticulocyte cell-free system, *Biochem. Biophys. Res. Commun.* **68**:249.

Salzberg, S., Robin, M. S., and Green, M., 1973, Appearance of virus specific RNA, virus particles, and cell surface changes in cells rapidly transformed by the murine sarcoma virus, *Virology* **53**:186.

Sarma, P. S., Log, T., and Huebner, R. J., 1970, Trans-species rescue of defective genomes of murine sarcoma virus from hamster tumor cells with helper feline leukemia virus, *Proc. Natl. Acad. Sci. USA* **65**:81.

Sawyer, R. C., Harada, F., and Dahlberg, J. E., 1974, An RNA primer for Rous sarcoma virus DNA synthesis: Isolation from uninfected host cells, *J. Virol.* **13**:1302.

Scheele, C. M., and Hanafusa, H., 1971, Proteins of helper-dependent RSV, *Virology* **45**:401.

Scheele, C. M., and Hanafusa, H., 1972, Electrophoretic analysis of the RNA of avian tumor viruses, *Virology* **50**:753.

Scher, C. G., and Siegler, R., 1975, Direct transformation of 3T3 cells by Abelson murine leukaemia virus, *Nature (London)* **253**:729.

Schincariol, A. L., and Joklik, W. E., 1973, Early synthesis of virus-specific RNA and DNA in cells rapidly transformed with Rous sarcoma virus, *Virology* **56**:532.

Scolnick, E. M., and Parks, W. P., 1974, Harvey sarcoma virus: A second murine type C sarcoma virus with rat genetic information, *J. Virol.* **13**:1211.

Scolnick, E. M., Stephensen, J. R., and Aaronson, S. A., 1972, Isolation of temperature sensitive mutants of murine sarcoma virus, *J. Virol.* **10**:653.

Scolnick, E. M., Rands, E., Williams, D., and Parks, W. P., 1973, Studies on the nucleic acid sequences of Kirsten sarcoma virus: A model for formation of a mammalian RNA-containing sarcoma virus, *J. Virol.* **12**:458.

Scolnick, E. M., Howk, R. S., Anisowicz, A., Peebles, P. T., Scher, C. D., and Parks, W. P., 1975, Separation of sarcoma virus-specific and leukemia virus-specific genetic sequences of Moloney sarcoma virus, *Proc. Natl. Acad. Sci. USA* **72**:4650.

Scott, R. E., Furcht, L. E., and Kersey, J. H., 1973, Changes in membrane structure associated with cell contact, *Proc. Natl. Acad. Sci. USA* **70**:3631.

Sefton, B. M., and Rubin, H., 1970, Release from density-dependent growth inhibition by proteolytic enzymes, *Nature (London)* **227**:843.

Seifert, W. E., and Rudland, P. S., 1974, Possible involvement of cyclic GMP in growth control of cultured mouse cells, *Nature (London)* **248**:138.

Shanmugam, G., Bhaduri, S., and Green, M., 1974, The virus-specific RNA species in free and membrane bound polyribosomes of transformed cells replicating murine sarcoma-leukemia viruses, *Biochem. Biophys. Res. Commun.* **56**:697.

Sheppard, J. R., 1971, Restoration of contact-inhibited growth to transformed cells by dibutyryl adenosine 3′:5′-cyclic monophosphate, *Proc. Natl. Acad. Sci. USA* **68**:1316.

Sheppard, J. R., and Prescott, D. M., 1972, Cyclic AMP levels in synchronized mammalian cells, *Exp. Cell Res.* **75**:293.

Shoyab, M., Evans, R. M., and Baluda, M. A., 1974, Aquisition of new DNA sequences after infection of chicken cells with avian myeloblastosis virus, *J. Virol.* **13**:331.

Shoyab, M., Markham, P. D., and Baluda, M. A., 1975, Host induced alteration of avian sarcoma virus B-77 genome, *Proc. Natl. Acad. Sci. USA* **72**:1031.

Siegert, W., Konings, R. N. J., Bauer, H., and Hofschneider, P. H., 1972, Translation of avian myeloblastosis virus RNA in a cell-free lysate of *Escherichia coli, Proc. Natl. Acad. Sci. USA* **69**:888.

Singer, S. J., and Nicolson, G. L., 1972, The fluid mosaic model of the structure of cell membranes, *Science* **175**:720.

Singh, V. N., Singh, M., August, J. T., and Horecker, B. L., 1974, Alterations in glucose metabolism in chick embryo cells transformed by Rous sarcoma virus: Intracellular levels of glycolytic intermediates, *Proc. Natl. Acad. Sci. USA* **71**:4129.

Sjögren, H. O., and Jonsson, N., 1963, Resistance against isotransplantation of mouse tumors induced by Rous sarcoma virus, *Exp. Cell Res.* **32**:618.

Sklar, M. D., White, B. J., and Rowe, W. P., 1974, Initiation of oncogenic transformation of mouse lymphocytes *in vitro* by Abelson leukemia virus, *Proc. Natl. Acad. Sci. USA* **71**:4077.

Smoler, D., Molineux, I., and Baltimore, D., 1971, Direction of polymerization of the avian myeloblastosis virus DNA polymerase, *J. Biol. Chem.* **246**:7697.

Smotkin, D., Gianni, A. M., Rozenblatt, S., and Weinberg, R. A., 1975, Infectious viral DNA of murine leukemia virus, *Proc. Natl. Acad. Sci. USA* **72**:4910.

Snyder, S. P., and Theilen, G. H., 1969, Transmissible feline lymphosarcoma, *Nature (London)* **221**:1074.

Somers, K., and Kit, S., 1973, Temperature-dependent expression of transformation by a cold-sensitive mutant of murine sarcoma virus, *Proc. Natl. Acad. Sci. USA* **70**:2206.

Somers, K. D., May, J. T., and Kit, S., 1973, Control of gene expression in rat cells transformed by a cold-sensitive murine sarcoma virus (MSV) mutant, *Intervirology* **1**:176.

Steck, F. T., and Rubin, H., 1965, The mechanism of interference between an avian leukosis virus and Rous sarcoma virus. II. Early steps of infection by RSV of cells under conditions of interference, *Virology* **29**:642.

Steck, T. L., Kaufman, S., and Bader, J. P., 1968, Glycolysis in chick embryo cell cultures transformed by Rous sarcoma virus, *Cancer Res.* **28**:1611.

Stehelin, D., Guntaka, R. V., Varmus, H. E., and Bishop, J. M., 1976*a*, Purification of DNA complementary to nucleotide sequences required for neoplastic transformation of fibroblasts by avian sarcoma viruses, *J. Mol. Biol.* **101**:349.

Stehelin, D., Varmus, H. E., Bishop, J. M., and Vogt, P. K., 1976*b*, DNA related to the transforming genes of avian sarcoma viruses is present in normal avian DNA, *Nature (London)* **260**:170.

Stephenson, J. R., Scolnick, E. M., and Aaronson, S. A., 1972, Genetic stability of the sarcoma viruses in murine and avian sarcoma virus transformed nonproducer cells, *Int. J. Cancer* **9**:577.

Stephenson, J. R., Reynolds, R. K., and Aaronson, S. A., 1973, Characterization of morphologic revertants of murine and avian sarcoma virus-transformed cells, *J. Virol.* **11**:218.

Stephenson, J. R., Crow, J. D., and Aaronson, S. A., 1974, Differential activation of biologically distinguishable endogenous mouse type C RNA viruses: Interaction with host cell regulatory factors, *Virology* **61**:411.

Stoker, M., O'Neill, C., Berryman, S., and Waxman, V., 1968, Anchorage and growth regulation in normal and virus-transformed cells, *Int. J. Cancer* **3**:683.

Stoltzfus, C. M., and Snyder, P. N., 1975, Structure of B77 sarcoma virus RNA: Stabilization of RNA after packaging, *J. Virol.* **16**:1161.

Stone, K. R., Smith, R. E., and Joklik, W. K., 1974, Changes in membrane polypeptides that occur when chick embryo fibroblasts and NRK cells are transformed with avian sarcoma viruses, *Virology* **58**:86.

Stone, M. P., Smith, R. E., and Joklik, W. K., 1975, 35 S *a* and *b* RNA subunits of avian RNA tumor virus strains cloned and passaged in chick and duck cells, *Cold Spring Harbor Symp. Quant. Biol.* **39**:859.

Sveda, M. M., Fields, B. N., and Soeiro, R., 1974, Host restriction of Friend leukemia virus; fate of input virion RNA, *Cell* **2**:271.

Svoboda, J., 1964, Malignant interaction of Rous virus with mammalian cells *in vivo* and *in vitro*, *Natl. Cancer Inst. Monogr.* **17**:277.

Svoboda, J., 1972, The biology of avian tumor viruses and the role of host cell in the modification of avian tumor virus expression, in: *RNA Viruses and Host Genome in Oncogenesis* (P. Emmelot and P. Bentvelzen, eds.), p. 81, North-Holland, Amsterdam, American Elsevier, New York.

Svoboda, J., and Hlozánek, I., 1970, Role of cell association in virus infection and virus rescue, *Adv. Cancer Res.* **13**:217.

Takano, T., and Hatanaka, M., 1975, Fate of viral RNA of murine leukemia virus after infection, *Proc. Natl. Acad. Sci. USA* **72**:343.

Talmadge, K. W., Noonan, K. D., and Burger, M. M., 1974, The transformed cell surface: An analysis of the increased lectin agglutinability and the concept of growth control by surface proteases, in: *Control of Proliferation in Animal Cells* (B. Clarkson and R. Baserga, eds.), p. 313, Cold Spring Harbor Laboratory, Cold Spring Harbor, N.Y.

Taylor, J. M., and Illmensee, R., 1975, Site on the RNA of an avian sarcoma virus at which primer is bound, *J. Virol.* **16**:553.

Taylor, J. M., Varmus, H. E., Faras, A. J., Levinson, W. E., and Bishop, J. M., 1974, Evidence for non-repetitive subunits in the genome of Rous sarcoma virus, *J. Mol. Biol.* **84**:217.

Taylor, R. B., Duffus, W. P. H., Raff, M. C., and DePetris, S., 1971, Redistribution and pinocytosis of lymphocyte surface immunoglobulin molecules induced by anti-immunoglobulin antibody, *Nature (London) New Biol.* **233**:225.

Temin, H. M., 1960, The control of cellular morphology in embryonic cells infected with Rous sarcoma virus *in vitro*, *Virology* **10**:182.

Temin, H. M., 1963, The effect of actinomycin D on growth of Rous sarcoma virus *in vitro*, *Virology* **20**:577.

Temin, H. M., 1964a, The participation of DNA in Rous sarcoma virus production, *Virology* **23**:486.

Temin, H. M., 1964b, The nature of the provirus of Rous sarcoma, *Natl. Cancer Inst. Monogr.* **17**:557.

Temin, H. M., 1965, The mechanism of carcinogenesis by avian sarcoma viruses, I. Cell multiplication and differentiation, *J. Natl. Cancer Inst.* **35**:679.

Temin, H. M., 1966, Studies on carcinogenesis by avian sarcoma viruses. III. The differential effect of serum and polyanions on multiplication of uninfected and converted cells, *J. Natl. Cancer Inst.* **37**:167.

Temin, H. M., 1967a, Studies on carcinogenesis by avian sarcoma viruses. IV. Differential multiplication of uninfected and of converted cells in response to insulin, *J. Cell. Physiol.* **69**:377.

Temin, H. M., 1967b, Studies on carcinogenesis by avian sarcoma viruses. V. Requirement for new DNA synthesis and for cell division, *J. Cell. Physiol.* **69**:53.

Temin, H. M., 1968a, Studies on carcinogenesis by avian sarcoma viruses. VIII. Glycolysis and cell multiplication, *Int. J. Cancer* **3**:273.

Temin, H. M., 1968b, Carcinogenesis by avian sarcoma viruses. X. The decreased requirement for insulin-replaceable activity in serum for cell multiplication, *Int. J. Cancer* **3**:771.

Temin, H. M., and Baltimore, D., 1972, RNA-directed DNA synthesis and RNA tumor viruses, *Adv. Virus Res.* **17**:129.

Temin, H. M., and Rubin, H., 1958, Characteristics of an assay for Rous sarcoma virus and Rous sarcoma cells in tissue culture, *Virology* **6**:669.

Temin, H. M., Pierson, R. W., Jr., and Dulak, N. C., 1972, The role of serum in the control of multiplication of avian and mammalian cells in culture, in: *Growth, Nutrition, and Metabolism of Cells in Culture*, Vol. 1 (G. H. Rothblat and V. J. Cristofalo, eds.), p. 49, Academic Press, New York.

Temin, H. M., Smith, G. L., and Dulak, N. C., 1974, Control of multiplication of normal and Rous sarcoma virus-transformed chicken embryo fibroblasts by purified multiplication stimulating activity with nonsuppressible insulin-like and sulfation factor activities, in: *Control of Proliferation in Animal Cells* (B. Clarkson and R. Baserga, eds.), p. 19, Cold Spring Harbor Laboratory, Cold Spring Harbor, N.Y.

Teng, N. N. H., and Chen, L. B., 1975, The role of surface proteins in cell proliferation as studied with thrombin and other proteases, *Proc. Natl. Acad. Sci. USA* **72**:413.

Tennant, R. W., Schulter, B., Yang, W.-K., and Brown, A., 1974, Reciprocal inhibition of mouse leukemia virus infection by *Fv-1* allele cell extracts, *Proc. Natl. Acad. Sci. USA* **71**:4241.

Theilen, G. T., Gould, D., Fowler, M., and Dungworth, D. L., 1971, C-type virus in tumor tissue of a woolly monkey (*Lagothrix* spp.) with fibrosarcoma, *J. Natl. Cancer Inst.* **47**:881.

Thurzo, V., Smida, J., Smidova-Kovarova, V., and Simkovic, D., 1963, Some properties of the fowl virus tumor B77, *Acta Unio Int. Contra Cancrum* **19**:304.

Tooze, J., ed., 1973, *The Molecular Biology of Tumor Viruses,* Cold Spring Harbor Laboratory, Cold Spring Harbor, N.Y.

Torpier, G., Montagnier, L., Biquard, J. M., and Vigier, P., 1975, A structural change of the plasma membrane induced by oncogenic viruses: Quantitative studies with the freeze-fracture technique, *Proc. Natl. Acad. Sci. USA* **72**:1695.

Toyoshima, K., and Vogt, P. K., 1969, Temperature sensitive mutants of an avian sarcoma virus, *Virology* **39**:930.

Toyoshima, K., Friis, R. R., and Vogt, P. K., 1970, The reproductive and cell-transforming capacities of avian sarcoma virus B77: Inactivation with UV light, *Virology* **42**:163.

Toyoshima, K., Owada, M., and Kozai, Y., 1973, Tumor producing capacity of temperature sensitive mutants of avian sarcoma viruses in chicks, *Biken J.* **16**:103.

Trager, G. W., and Rubin, H., 1966, Mixed clones produced following infection of chick embryo cell cultures with Rous sarcoma virus, *Virology* **30**:275.

Tsuchida, N., Robin, M. S., and Green, M., 1972, Viral RNA subunits in cells transformed by RNA tumor virses, *Science* **176**:1418.

Tsuchida, N., Long, C., and Hatanaka, M., 1974, Viral RNA of murine sarcoma virus produced by a hamster mouse somatic cell hybrid, *Virology* **60**:200.

Twardzik, D., Simonds, J., Oskarsson, M., and Portugal, F., 1973, Translation of AKR-murine leukemia viral RNA in an *E. coli* cell-free system, *Biochem. Biophys. Res. commun.* **52**:1108.

Unkelless, J. C., Tobia, A., Ossowski, L., Quigley, J. P., Rifkin, D. B., and Reich, E., 1973*a,* An enzymatic function associated with transformation of fibroblasts by oncogenic viruses. I. Chick embryo fibroblast cultures transformed by avian RNA tumor viruses, *J. Exp. Med.* **137**:85.

Unkeless, J. C., Dano, K., Kellerman, G. M., and Reich, E., 1973*b,* Fibrinolysis associated with oncogenic transformation: Partial purification and characterization of the cell factor—a plasminogen activator, *J. Biol. Chem.* **249**:4295.

Vaheri, A., and Hanafusa, H., 1971, Effect of rifampicin and a derivative on cells transformed by Rous sarcoma virus, *Cancer Res.* **31**:2032.

Vaheri, A., and Ruoslahti, E., 1974, Disappearance of a major cell-type specific surface glycoprotein antigen (SF) after transformation of fibroblasts by Rous sarcoma virus, *Int. J. Cancer* **13**:579.

Vaheri, A., and Ruoslahti, E., 1975, Fibroblast surface antigen molecules and their loss from virus-transformed cells: A major alteration in cell surface, in: *Protease and Biological Control* (E. Reich, D. B. Rifkin, and E. Shaw, eds.), p. 967, Cold Spring Harbor Laboratory, Cold Spring Harbor, N.Y.

Vaheri, A., Ruoslahti, E., Hovi, T., and Nordling, S., 1973, Stimulation of density inhibited cell cultures by insulin, *J. Cell. Physiol.* **81**:355.

Van Zaane, D., Gielkens, A. L. J., Dekker-Michielsen, M. J. A., and Bloemers, H. P. J., 1975, Virus-specific precursor polypeptides in cells infected with Rauscher leukemia virus, *Virology* **67**:544.

Varmus, H. E., Bishop, J. M., and Vogt, P. K., 1973*a*, Appearance of virus-specific DNA in mammalian cells following transformation by Rous sarcoma virus, *J. Mol. Biol.* **74**:613.

Varmus, H. E., Vogt, P. K., and Bishop, J. M., 1973*b*, Integration of deoxyribonucleic acid specific for Rous sarcoma virus after infection of permissive and nonpermissive hosts, *Proc. Natl. Acad. Sci. USA* **70**:3067.

Varmus, H. E., Hearsley, S., and Bishop, J. M., 1974*a*, Use of DNA-DNA annealing to detect new virus-specific DNA sequences in chicken embryo fibroblasts after infection by avian sarcoma virus, *J. Virol.* **14**:895.

Varmus, H. E., Guntaka, R. V., Fan, W. J. W., Heasley, S., and Bishop, J. M., 1974*b*, Synthesis of viral DNA in the cytoplasm of duck embryo fibroblasts and in enucleated cells after infection by avian sarcoma virus, *Proc. Natl. Acad. Sci. USA* **71**:3874.

Varmus, H. E., Guntaka, R. V., Deng, C. T., and Bishop, J. M., 1975, Synthesis, structure and function of avian sarcoma virus-specific DNA in permissive and nonpermissive cells, *Cold Spring Harbor Symp. Quant. Biol.* **39**:987.

Venuta, S., and Rubin, H., 1973, Sugar transport in normal and Rous sarcoma virus-transformed chick-embryo fibroblasts, *Proc. Natl. Acad. Sci. USA* **70**:653.

Verma, I. M., 1975, Studies on reverse transcriptase of RNA tumor viruses. I. Localization of thermolabile DNA polymerase and RNase H activities on one polypeptide, *J. Virol.* **15**:121.

Verma, I. M., Mason, W. S., Drost, S. D., and Baltimore, D., 1974, DNA polymerase activity from two temperature-sensitive mutants of Rous sarcoma virus is thermolabile, *Nature (London)* **251**:27.

Vigier, P., 1970, Effect of agar, calf embryo extract and polyanions on production of foci of transformed cells by Rous sarcoma virus, *Virology* **40**:179.

Vigier, P., 1973, Effect of nonspecific factors on focus formation by Rous sarcoma virus. II. Polyanions, divalent cations, temperature, and tryptose phosphate broth, *Intervirology* **1**:338.

Vigier, P., and Goldé, A., 1964, Effects of actinomycin D and of mitomycin C on the development of Rous sarcoma virus, *Virology* **23**:511.

Vogel, A., and Pollack, R., 1974, Isolation and characterization of revertant cell lines, *J. Virol.* **14**:1404.

Vogt, P. K., 1971, Spontaneous segregation of nontransforming viruses from cloned sarcoma viruses, *Virology* **46**:939.

Vogt, P. K., 1973, The genome of avian RNA tumor viruses: A discussion of four models, in: *Possible Episomes in Eukaryotes* (Proceedings of the Fourth Lepetit Colloquium, L. Silvestri, ed.), p. 35, North-Holland, Amsterdam.

Vogt, P. K., and Ishizaki, R., 1965, Reciprocal patterns of genetic resistance to avian tumor viruses in two lines of chickens, *Virology* **26**:664.

Vogt, V. M., and Eisenman, R., 1973, Identification of a large polypeptide precursor of avian oncornavirus proteins, *Proc. Natl. Acad. Sci. USA* **70**:1734.

Vogt, V. M., Eisenman, R., and Diggelmann, H., 1975, Generation of avian myeloblastosis virus structural proteins by proteolytic cleavage of a precursor polypeptide, *J. Mol. Biol.* **96**:471.

Von der Helm, K., and Duesberg, P. H., 1975, Translation of Rous sarcoma virus

RNA in a cell-free system from ascites Krebs II cells, *Proc. Natl. Acad. Sci. USA* **72**:614.

Wang, E., and Goldberg, A. R., 1976, Changes in mircofilament organization and surface topography upon transformation of chick embryo fibroblasts with Rous sarcoma virus, *Proc. Natl. Acad. Sci. USA* **73**:4065.

Wang, L.-H., and Duesberg, P., 1974, Properties and location of poly(A) in Rous sarcoma virus RNA, *J. Virol.* **14**:1515.

Wang, L.-H., Duesberg, P., Beemon, K., and Vogt, P. K., 1975, Mapping RNase T$_1$-resistant oligonucleotides of avian tumor virus RNAs: Sarcoma-specific oligonucleotides are near the poly(A) end and oligonucleotides common to sarcoma and transformation-defective viruses are at the poly(A) end, *J. Virol.* **16**:1051.

Wang, L.-H., Duesberg, P. H., Kawai, S., and Hanafusa, H., 1976, The location of envelope-specific and sarcoma-specific oligonucleotides on the RNA of Schmidt-Ruppin Rous sarcoma virus, *Proc. Natl. Acad. Sci. USA* **73**:447.

Warburg, O., 1931, *The Metabolism of Tumors,* R. R. Smith, New York.

Warren, L., Critchley, D., and Macpherson, I., 1972*a*, Surface glycoproteins and glycolipids of chicken embryo cells transformed by a temperature-sensitive mutant of Rous sarcoma virus, *Nature (London)* **235**:275.

Warren, L., Fuhrer, J. P., and Buck, C. A., 1972*b*, Surface glycoproteins of normal and transformed cells: A difference determined by sialic acid and a growth-dependent sialyl transferase, *Proc. Natl. Acad. Sci. USA* **69**:1838.

Weber, K., and Groeschel-Stewart, U., 1974, Myosin antibody: The specific visualization of myosin containing filaments in nonmuscle cells, *Proc. Natl. Acad. Sci. USA* **71**:4561.

Weber, M. J., 1973, Hexose transport in normal and in Rous sarcoma virus-transformed cells, *J. Biol. Chem.* **248**:2978.

Weiss, R. A., 1969, The host range of Bryan strain Rous sarcoma virus synthesized in the absence of helper virus, *J. Gen. Virol.* **5**:511.

Weiss, R. A., 1971, Cell transformation induced by Rous sarcoma virus: Analysis of density dependence, *Virology* **46**:209.

Wickus, G. G., and Robbins, P. W., 1973, Plasma membrane proteins of normal and Rous sarcoma virus-transformed chick embryo fibroblasts, *Nature (London) New Biol.* **245**:65.

Wickus, G. G., Branton, P. E., and Robbins, P. W., 1974, Rous sarcoma virus transformation of the chick cell surface, in: *Control of Proliferation in Animal Cells* (B. Clarkson and R. Baserga, eds.), p. 541, Cold Spring Harbor Laboratory, Cold Spring Harbor, N.Y.

Wickus, G., Gruenstein, E., Robbins, P., and Rich, A., 1975, Decrease in membrane-associated actin of fibroblasts after transformation by Rous sarcoma virus, *Proc. Natl. Acad. Sci. USA* **72**:746.

Willingham, M. C., Ostlund, R. E., and Pastan, I., 1974, Myosin is a component of the cell surface of cultured cells, *Proc. Natl. Acad. Sci. USA* **71**:4144.

Wolf, B. A., and Goldberg, A. R., 1976, Rous sarcoma virus-transformed fibroblasts having low levels of plasminogen activator, *Proc. Natl. Acad. Sci. USA* **73**:3613.

Wright, S. E., and Neiman, P. E., 1974, Base-sequence relationships between avian ribonucleic acid endogenous and sarcoma viruses assayed by competitive ribonucleic acid–deoxyribonucleic acid hybridization, *Biochemistry* **13**:1549.

Wyke, J. A., 1973, Complementation of transforming functions by temperature-sensitive mutants of avian sarcoma virus, *Virology* **54**:28.

Wyke, J. A., and Linial, M., 1973, Temperature-sensitive avian sarcoma viruses: A physiological comparison of twenty mutants, *Virology* **53**:152.

Wyke, J. A., Bell, J. G., and Beamand, J. A., 1975, Genetic recombination among temperature-sensitive mutants of Rous sarcoma virus, *Cold Spring Harbor Symp. Quant. Biol.* **39**:897.

Yahara, I., and Edelman, G. M., 1972, Restriction of the mobility of lymphocyte immunoglobulin receptors by concanavalin A, *Proc. Natl. Acad. Sci. USA* **69**:608.

Yahara, I., and Edelman, G. M., 1974, Modification of lymphocyte receptor mobility by concanavalin A and colchicine, *Ann. N.Y. Acad. Sci.* **253**:455.

Yamada, K. M., and Weston, J. A., 1975, The synthesis, turnover, and artificial restoration of a major cell surface glycoprotein, *Cell* **5**:75.

Yamada, K. M., Yamada, S. S., and Pastan, I., 1975, The major cell surface glycoprotein of chick embryo fibroblasts is an agglutinin, *Proc. Natl. Acad. Sci. USA* **72**:3158.

Yaoi, Y., Onoda, T., and Takahashi, H., 1972, Inhibition of mitosis in suspension culture of chick embryo cells, *Nature (London)* **237**:285.

Yoshida, M., Owada, M., and Toyoshima, K., 1975, Strain specificity of changes in adenylate cyclase activity in cells transformed by avian sarcoma viruses, *Virology* **63**:68.

Yoshii, S., and Vogt, P. K., 1970, A mutant of Rous sarcoma virus (type O) causing fusiform cell transformation, *Proc. Soc. Exp. Biol. Med.* **135**:297.

Yoshikura, H., 1970, Dependence of murine sarcoma virus infection on the cell cycle, *J. Gen. Virol.* **6**:183.

Zetter, B. R., Chen, L. B., and Buchanan, J. M., 1976, Effects of protease treatment on growth morphology, adhesion, and cell surface proteins of secondary chick embryo fibroblasts, *Cell* **7**:407.

Index